T0349928

Undergraduate Texts in Mathematics

Editors
S. Axler
F.W. Gehring
K.A. Ribet

Springer
New York
Berlin
Heidelberg
Barcelona
Hong Kong
London
Milan
Paris
Singapore
Tokyo

Undergraduate Texts in Mathematics

(continued after index)

Murray H. Protter   Charles B. Morrey, Jr.

A First Course in Real Analysis

Second Edition

With 143 Illustrations

 Springer

Murray H. Protter
Department of Mathematics
University of California at Berkeley
Berkeley, CA 94720
USA

Charles B. Morrey, Jr. (*Deceased*)

Mathematics Subject Classifications (2000): 26-01, 26A03, 26A15, 26A24, 26A42, 26B12

Library of Congress Cataloging-in-Publication Data
Protter, Murray H.
 A first course in real analysis / Murray H. Protter, Charles B.
 Morrey. — 2nd ed.
  p. cm. — (Undergraduate texts in mathematics)
 Includes index.
  1. Mathematical analysis. I. Morrey, Charles Bradfield, 1907–1984.
 II. Title. III. Series.
 QA300.P968 1991
 515—dc20   90-46562

Printed on acid-free paper.

Typeset by Asco Trade Typesetting Ltd., Hong Kong.
Printed and bound by R.R. Donnelley & Sons, Harrisonburg, VA.
Printed in the United States of America.

9 8 7 6

ISBN 0-387-97437-7
ISBN 3-540-97437-7   SPIN 10794504

Springer-Verlag New York Berlin Heidelberg
A member of BertelsmannSpringer Science+Business Media GmbH

To Ruth

Preface to the Second Edition

Many changes have been made in this second edition of *A First Course in Real Analysis*. The most noticeable are the addition of many problems and the inclusion of answers to most of the odd-numbered exercises. There are now more than one thousand exercises throughout the book. In going over the text, I improved the book's readability by further clarifying many of the proofs, providing additional explanatory remarks, improving many of the figures, and changing some notation.

The first course in analysis that follows elementary calculus is critical for those students who wish to go on to advanced mathematics. Because in a calculus course the emphasis rests on problem solving and the development of manipulative skills, students frequently get a misleading impression that mathematics is simply manipulative; thus they often are unaware of the theoretical basis of analysis and higher mathematics.

In *A First Course in Real Analysis* we present a theoretical foundation of analysis that is suitable for students who have completed a standard course in calculus. The sixteen chapters contain enough material for a one-year course, but the material is so arranged that an instructor teaching a one-semester or a one- or two-quarter course can easily find a selection of topics that he or she thinks students should have.

The first chapter, on the real number system, serves two purposes. First, it provides an opportunity for the student to develop facility in proving theorems. Since most students entering the course have had little or no experience in devising proofs, the knowledge of how one goes about this task will prove useful in the remainder of the course.

Second, for those instructors who wish to give a comprehensive course in analysis, we provide a fairly complete treatment of the real number system, including a section on mathematical induction. We also supply additional material on absolute value and the solution of algebraic inequalities in Appendixes 1 and 2.

If an instructor who is teaching a short course in analysis does not wish to discuss the real number system, he or she may choose to begin with Chapter 2, since the seventy pages that comprise Chapters 2 through 5 cover the basic theory upon which elementary calculus is based. Here we prove most of the theorems that are "stated without proof" in the standard first-year calculus course.

Critical to an understanding of analysis (and many other mathematical topics) is the concept of a metric space. We discuss the fundamental properties of metric spaces in Chapter 6. Here we show that the notion of compactness is central and we prove several important results, including the Heine–Borel theorem, which are useful later on. The ninety-four pages that make up Chapters 6, 13, and 15 form a coherent unit on metric spaces. In Chapter 13 we give the theory of contraction mappings in a metric space and apply it to prove an existence theorem in differential equations. In Chapter 15 we study in detail the properties of functions defined on a metric space. We prove the Tietze extension theorem and the Stone–Weierstrass approximation theorem, results that are useful in many contexts.

Chapters 7 and 8 show how to extend the theory of differentiation and integration to $\mathbb{R}^N$, $N \geq 2$. These chapters are the natural continuations of Chapters 4 and 5 on differentiation and integration in $\mathbb{R}^1$.

Infinite sequences and infinite series are the topics of Chapters 9 and 10. The emphasis in Chapter 9 is on uniform convergence, a topic the student must master if he or she is to understand the underlying concepts of power series and Fourier series. There is also a (optional) discussion of multiple series that we treat in a unified manner. Chapter 10, on Fourier series, contains a proof of the useful Dini test for convergence and proofs of the customary theorems on term-by-term differentiation and integration of Fourier series. We also establish Bessel's inequality and the Riemann–Lebesgue lemma.

In Chapter 11 we cover the important subject of differentiation under the integral sign for both proper and improper integrals. This topic, useful in many applications, is rarely presented in such detail. Since this chapter is relatively independent of many of the others, it may easily be included or omitted according to the wishes of the instructor.

No course in Riemann integration would be complete without a treatment of the Riemann–Stieltjes integral. In Chapter 12 we define functions of bounded variation and show how they play a crucial role in the establishment of the Riemann–Stieltjes integral.

In Chapter 14 we prove the Implicit Function theorem, first for a single equation and then for a system. We provide two proofs for a single equation, one using only basic techniques and a second using the fixed point theorem of Chapter 13. The latter proof is then extended to systems. In addition, we give a detailed proof of the Lagrange multiplier rule for finding maxima and minima, a rule frequently stated but rarely proved. For completeness we give the details of the proof of the theorem on the change of variables in a multiple integral. Since the argument here is rather intricate, the instructor may wish to assign this section as optional reading for the better students.

Proofs of Green's and Stokes's theorems and the Divergence theorem in $\mathbb{R}^2$ and $\mathbb{R}^3$ are given in Chapter 16. In order to establish the result in full generality, we include a study of the properties of two-dimensional surfaces in $\mathbb{R}^3$. For those interested only in the basic theorems, much of this geometric analysis can be skipped.

While this edition is my full responsibility, I wish to acknowledge the enormous debt I owe to the late Charles B. Morrey, Jr., for his essential contribution to the first edition as well as for his guidance and advice through-out our collaboration.

Berkeley, California   MURRAY H. PROTTER

Preface to the First Edition

The first course in analysis which follows elementary calculus is a critical one for students who are seriously interested in mathematics. Traditional advanced calculus was precisely what its name indicates—a course with topics in calculus emphasizing problem solving rather than theory. As a result students were often given a misleading impression of what mathematics is all about; on the other hand the current approach, with its emphasis on theory, gives the student insight in the fundamentals of analysis.

In *A First Course in Real Analysis* we present a theoretical basis of analysis which is suitable for students who have just completed a course in elementary calculus. Since the sixteen chapters contain more than enough analysis for a one year course, the instructor teaching a one or two quarter or a one semester junior level course should easily find those topics which he or she thinks students should have.

The first Chapter, on the real number system, serves two purposes. Because most students entering this course have had no experience in devising proofs of theorems, it provides an opportunity to develop facility in theorem proving. Although the elementary processes of numbers are familiar to most students, greater understanding of these processes is acquired by those who work the problems in Chapter 1. As a second purpose, we provide, for those instructors who wish to give a comprehensive course in analysis, a fairly complete treatment of the real number system including a section on mathematical induction.

Although Chapter 1 is useful as an introduction to analysis, the instructor of a short course may choose to begin with the second Chapter. Chapters 2 through 5 cover the basic theory of elementary calculus. Here we prove many of the theorems which are "stated without proof" in the standard freshman calculus course.

Crucial to the development of an understanding of analysis is the concept of a metric space. We discuss the fundamental properties of metric spaces in Chapter 6. Here we show that the notion of compactness is central and we prove several important results (including the Heine–Borel theorem) which are useful later on. The power of the general theory of metric spaces is aptly illustrated in Chapter 13, where we give the theory of contraction mappings and an application to differential equations. The study of metric spaces is resumed in Chapter 15, where the properties of functions on metric spaces are established. The student will also find useful in later courses results such as the Tietze extension theorem and the Stone–Weierstrass theorem, which are proved in detail.

Chapters 7, 8, and 12 continue the theory of differentiation and integration begun in Chapters 4 and 5. In Chapters 7 and 8, the theory of differentiation and integration in $\mathbb{R}_N$ is developed. Since the primary results for $\mathbb{R}_1$ are given in Chapters 4 and 5 only modest changes were necessary to prove the corresponding theorems in $\mathbb{R}_N$. In Chapter 12 we define the Riemann–Stieltjes integral and develop its principal properties.

Infinite sequences and series are the topics of Chapters 9 and 10. Besides subjects such as uniform convergence and power series, we provide in Section 9.5 a unified treatment of absolute convergence of multiple series. Here, in a discussion of unordered sums, we show that a separate treatment of the various kinds of summation of multiple series is entirely unnecessary. Chapter 10 on Fourier series contains a proof of the Dini test for convergence and the customary theorems on term-by-term differentiation and integration of such series.

In Chapter 14 we prove the Implicit Function theorem, first for a single equation and then for a system. In addition we give a detailed proof of the Lagrange multiplier rule, which is frequently stated but rarely proved. For completeness we give the details of the proof of the theorem on the change of variables in a multiple integral. Since the argument here is rather intricate, the instructor may wish to assign this section as optional reading for the best students.

Proofs of Green's and Stokes' theorems and the divergence theorem in $\mathbb{R}_2$ and $\mathbb{R}_3$ are given in Chapter 16. The methods used here are easily extended to the corresponding results in $\mathbb{R}_N$.

This book is also useful in freshman honors courses. It has been our experience that honors courses in freshman calculus frequently falter because it is not clear whether the honors student should work hard problems while he learns the regular calculus topics or should omit the regular topics entirely and concentrate on the underlying theory. In the first alternative, the honors student is hardly better off than the regular student taking the ordinary calculus course, while in the second the honors student fails to learn the *simple* problem solving techniques which, in fact, are useful later on. We believe that this dilemma can be resolved by employing two texts—one a standard calculus text and the other a book such as this one which provides the theoretical basis of the calculus in one and several dimensions. In this way the honors student gets both theory and practice. Chapters 2 through 5 and Chapters 7 and 8 provide a thorough account of the theory of elementary calculus which, along with a standard calculus book, is suitable as text material for a first year honors program.

Murray H. Protter
Charles B. Morrey, Jr.

Berkeley, California, January 1977

Contents

Appendixes   495

Answers to Odd-Numbered Problems   515

CHAPTER 1

The Real Number System

1.1. Axioms for a Field

In an elementary calculus course the student learns the techniques of differentiation and integration and the skills needed for solving a variety of problems which use the processes of calculus. Most often, the principal theorems upon which calculus is based are stated without proof, while some of the auxiliary theorems are established in detail. To compensate for the missing proofs, most texts present arguments which show that the basic theorems are plausible. Frequently, a remark is added to the effect that rigorous proofs of these theorems can be found in advanced texts in analysis.

In this and the next four chapters we shall give a reasonably rigorous foundation to the processes of the calculus of functions of one variable. Calculus depends on the properties of the real number system. Therefore, to give a complete foundation for calculus, we would have to develop the real number system from the beginning. Since such a development is lengthy and would divert us from our aim of presenting a course in *analysis*, we shall assume the reader is familiar with the usual properties of the system of real numbers.

In this section we present a set of axioms that forms a logical basis for those processes of elementary algebra upon which calculus is based. Any collection of objects satisfying the axioms stated below is called a **field**. In particular, the system of real numbers satisfies the field axioms, and we shall indicate how the customary laws of algebra concerning addition, subtraction, multiplication, and division follow from these axioms.

A thorough treatment would require complete proofs of all the theorems. In this section and the next we establish some of the elementary laws of algebra, and we refer the reader to a course in higher algebra where complete proofs of most of the theorems may be found. Since the reader is familiar with the laws of algebra and their use, we shall assume their validity throughout the remainder of the text. In addition, we shall suppose the reader is familiar with many facts about finite sets, positive integers, and so forth.

Throughout the book, we use the word *equals* or its symbol = to stand for the words "is the same as." The reader should compare this with other uses for the symbol = such as that in plane geometry where, for example, two line segments are said to be equal if they have the same length.

Axioms of Addition and Subtraction

A-1. Closure property. *If a and b are numbers, there is one and only one number, denoted $a + b$, called their **sum**.*

A-2. Commutative law. *For any two numbers a and b, the equality*

$$b + a = a + b$$

holds.

A-3. Associative law. *For all numbers a, b, and c, the equality*

$$(a + b) + c = a + (b + c)$$

holds.

A-4. Existence of a zero. *There is one and only one number 0, called **zero**, such that $a + 0 = a$ for any number a.*

It is not necessary to assume in Axiom A-4 that there is *only one* number 0 with the given property. The uniqueness of this number is easily established. Suppose 0 and 0' are two numbers such that $a + 0 = a$ and $a + 0' = a$ for *every* number a. Then $0 + 0' = 0$ and $0' + 0 = 0'$. By Axiom A-2, we have $0 + 0' = 0' + 0$ and so $0 = 0'$. The two numbers are the same.

A-5. Existence of a negative. *If a is any number, there is one and only one number x such that $a + x = 0$. This number is called the **negative of** a and is denoted by $-a$.*

As in Axiom A-4, it is not necessary to assume in Axiom A-5 that there is *only one* such number with the given property. The argument which establishes the uniqueness of the negative is similar to the one given after Axiom A-4.

Theorem 1.1. *If a and b are any numbers, then there is one and only one number x such that $a + x = b$. This number x is given by $x = b + (-a)$.*

PROOF. We must establish two results: (i) that $b + (-a)$ satisfies the equation $a + x = b$ and (ii) that no other number satisfies this equation. To prove (i), suppose that $x = b + (-a)$. Then, using Axioms A-2 through A-4 we see that

$$a + x = a + [b + (-a)] = a + [(-a) + b] = [a + (-a)] + b = 0 + b = b.$$

Therefore (i) holds. To prove (ii), suppose that x is some number such that $a + x = b$. Adding $(-a)$ to both sides of this equation, we find that

$$(a + x) + (-a) = b + (-a).$$

Now,

$$(a + x) + (-a) = a + [x + (-a)] = a + [(-a) + x]$$
$$= [a + (-a)] + x = 0 + x = x.$$

We conclude that $x = b + (-a)$, and the uniqueness of the solution is established.   □

Notation. The number $b + (-a)$ is denoted by $b - a$.

Thus far addition has been defined *only for two numbers*. By means of the associative law we can define addition for three, four and, in fact, any finite number of elements. Since $(a + b) + c$ and $a + (b + c)$ are the same, we define $a + b + c$ as this common value. The following lemma is an easy consequence of the associative and commutative laws of addition.

Lemma 1.1. *If a, b, and c are any numbers, then*

$$a + b + c = a + c + b = b + a + c = b + c + a = c + a + b = c + b + a.$$

The formal details of writing out a proof are left to the reader.
The next lemma is useful in the proof of Theorem 1.2 below.

Lemma 1.2. *If a, b, c, and d are numbers, then*

$$(a + c) + (b + d) = (a + b) + (c + d).$$

PROOF. Using Lemma 1.1 and the axioms, we have

$$(a + c) + (b + d) = [(a + c) + b] + d$$
$$= (a + c + b) + d = (a + b + c) + d$$
$$= [(a + b) + c] + d = (a + b) + (c + d).$$ □

The next theorem establishes familiar properties of negative numbers.

Theorem 1.2

(i) *If a is a number, then $-(-a) = a$.*
(ii) *If a and b are numbers, then*

$$-(a + b) = (-a) + (-b).$$

PROOF. (i) From the definition of negative, we have

$$(-a) + [-(-a)] = 0, \qquad (-a) + a = a + (-a) = 0.$$

Axiom A-5 states that the negative of $(-a)$ is *unique*. Therefore, $a = -(-a)$. To establish (ii), we know from the definition of negative that

$$(a + b) + [-(a + b)] = 0.$$

Furthermore, using Lemma 1.2, we have

$$(a + b) + [(-a) + (-b)] = [a + (-a)] + [b + (-b)] = 0 + 0 = 0.$$

The result follows from the "only one" part of Axiom A-5. $\square$

Theorem 1.2 can be stated in the familiar form: (i) *The negative of* $(-a)$ *is* a, *and* (ii) *The negative of a sum is the sum of the negatives.*

Axioms of Multiplication and Division

M-1. Closure property. *If* a *and* b *are numbers, there is one and only one number, denoted by* ab *(or* $a \times b$ *or* $a \cdot b$*), called their* **product**.

M-2. Commutative law. *For every two numbes* a *and* b*, the equality*

$$ba = ab$$

holds.

M-3. Associative law. *For all numbers* a, b, *and* c, *the equality*

$$(ab)c = a(bc)$$

holds.

M-4. Existence of a unit. *There is one and only one number* u, *different from zero, such that* $au = a$ *for every number* a. *This number* u *is called the* **unit** *and (as is customary) is denoted by* 1.

M-5. Existence of a reciprocal. *For each number* a *different from zero there is one and only one number* x *such that* $ax = 1$. *This number* x *is called the* **reciprocal** *of* a *(or the inverse of* a*) and is denoted by* a^{-1} *(or* $1/a$*).*

Remarks. Axioms M-1 through M-4 are the parallels of Axioms A-1 through A-4 with addition replaced by multiplication. However, M-5 is not the exact analogue of A-5, since the additional condition $a \neq 0$ is required. The reason for this is given below in Theorem 1.3, where it is shown that the result of multiplication of any number by zero is zero. In familiar terms we say that division by zero is excluded.

Special Axiom

D. Distributive law. *For all numbers a, b, and c the equality*

$$a(b + c) = ab + ac$$

holds.

Remarks. In every logical system there are certain terms which are un-defined. For example, in the system of axioms for plane Euclidean geometry the terms *point* and *line* are undefined. Of course, we have an intuitive idea of the meaning of these two undefined terms, but in the framework of Euclidean geometry it is not possible to define them. In the axioms for algebra given above, the term *number* is undefined. We shall interpret number to mean *real number* (positive, negative, or zero) in the usual sense that we give to it in elementary courses. Actually, the above axioms for a field hold for many number systems, of which the collection of real numbers is only one. For example, all the axioms stated so far hold for the system consisting of all *complex numbers.* Also, there are many systems, each consisting of a finite number of elements (*finite fields*), which satisfy all the axioms we have stated until now.

Additional axioms are needed if we insist that the real number system be the *only* collection satisfying all the given axioms. The additional axioms required for this purpose are discussed in Sections 1.3 and 1.4 below.

Theorem 1.3. *If a is any number, then $a \cdot 0 = 0$.*

PROOF. Let b be any number. Then $b + 0 = b$, and therefore $a(b + 0) = ab$. From the distributive law (Axiom D), we find that

$$(ab) + (a \cdot 0) = (ab),$$

so that $a \cdot 0 = 0$ by Axiom A-4. $\square$

Theorem 1.4. *If a and b are numbers and $a \neq 0$, then there is one and only one number x such that $a \cdot x = b$. The number x is given by $x = ba^{-1}$.*

The proof of Theorem 1.4 is just like the proof of Theorem 1.1 with addition replaced by multiplication, 0 by 1, and $-a$ by a^{-1}. The details are left to the reader.

Notation. The expression "if and only if," a technical one used frequently in mathematics, requires some explanation. Suppose A and B stand for propo-sitions which may be true or false. To say that A is true *if* B is true means that the truth of B implies the truth of A. The statement A is true *only if* B is true means that the truth of A implies the truth of B. Thus the shorthand statement "A is true if and only if B is true" is equivalent to the *double implication*: the truth of A implies and is implied by the truth of B. As a further shorthand we use the symbol $\Leftrightarrow$ to represent "if and only if," and we write

$$A \Leftrightarrow B$$

for the two implications stated above.[1]

We now establish the familiar principle which underlies the solution by factoring of quadratic and other algebraic equations.

Theorem 1.5

(i) *We have* $ab = 0$ *if and only if* $a = 0$ *or* $b = 0$ *or both.*
(ii) *We have* $a \neq 0$ *and* $b \neq 0$ *if and only if* $ab \neq 0$.

PROOF. We must prove two statements in each of Parts (i) and (ii). To prove (i), observe that if $a = 0$ or $b = 0$ or both, then it follows from Theorem 1.3 that $ab = 0$. Going the other way in (i), suppose that $ab = 0$. Then there are two cases: either $a = 0$ or $a \neq 0$. If $a = 0$, the result follows. If $a \neq 0$, then we see that

$$b = 1 \cdot b = (a^{-1}a)b = a^{-1}(ab) = a^{-1} \cdot 0 = 0.$$

Hence $b = 0$ and (i) is established. To prove (ii), first suppose $a \neq 0$ and $b \neq 0$. Then $ab \neq 0$, because $a \neq 0$ and $b \neq 0$ is the negation of the statement "$a = 0$ or $b = 0$ or both." Thus (i) applies. For the second part of (ii), suppose $ab \neq 0$. Then $a \neq 0$ and $b \neq 0$ for, if one of them were zero, Theorem 1.3 would apply to give $ab = 0$. $\qquad\square$

We define abc as the common value of $(ab)c$ and $a(bc)$. The reader can prove the following lemmas, which are similar to Lemmas 1.1 and 1.2.

Lemma 1.3. *If a, b, and c are numbers, then*

$$abc = acb = bac = bca = cab = cba.$$

Lemma 1.4. *If a, b, c and d are numbers, then*

$$(ac) \cdot (bd) = (ab) \cdot (cd).$$

Theorem 1.6

(i) *If* $a \neq 0$, *then* $a^{-1} \neq 0$ *and* $[(a^{-1})^{-1}] = a$.
(ii) *If* $a \neq 0$ *and* $b \neq 0$, *then* $(a \cdot b)^{-1} = (a^{-1}) \cdot (b^{-1})$.

The proof of this theorem is like the proof of Theorem 1.2 with addition replaced by multiplication, 0 replaced by 1, and $(-a)$, $(-b)$ replaced by a^{-1},

---

[1] The term **"necessary and sufficient"** is frequently used as a synonym for "if and only if."

b^{-1}. The details are left to the reader. Note that if $a \neq 0$, then $a^{-1} \neq 0$ because $aa^{-1} = 1$ and $1 \neq 0$. Then Theorem 1.5 (ii) may be used with $b = a^{-1}$.

Using Theorem 1.3 and the distributive law, we easily prove the **laws of signs** stated as Theorem 1.7 below. We emphasize that the numbers a and b may be positive, negative, or zero.

Theorem 1.7. *If a and b are any numbers, then*

(i) $a \cdot (-b) = -(a \cdot b)$.
(ii) $(-a) \cdot b = -(a \cdot b)$.
(iii) $(-a) \cdot (-b) = a \cdot b$.

PROOF. (i) Since $b + (-b) = 0$, it follows from the distributive law that

$$a[b + (-b)] = a \cdot b + a \cdot (-b) = 0.$$

Also, the negative of $a \cdot b$ has the property that $a \cdot b + [-(a \cdot b)] = 0$. Hence we see from Axiom A-5 that $a \cdot (-b) = -(a \cdot b)$. Part (ii) follows from Part (i) by interchanging a and b. The proof of (iii) is left to the reader. $\square$

Corollary. *The equality $(-1) \cdot a = -a$ holds.*

We now show that the *laws of fractions*, as given in an elementary algebra course, follow from the axioms and theorems above.

Notation. We introduce the following symbols for $a \cdot b^{-1}$:

$$a \cdot b^{-1} = \frac{a}{b} = a/b = a \div b.$$

These symbols, representing an indicated division, are called **fractions**. The *numerator* and *denominator* of a fraction are defined as usual. A fraction with *denominator* zero has no meaning.

Theorem 1.8

(i) *For every number a, the equality $a/1 = a$ holds.*
(ii) *If $a \neq 0$, then $a/a = 1$.*

PROOF. (i) We have $a/1 = (a \cdot 1^{-1}) = (a \cdot 1^{-1}) \cdot 1 = a(1^{-1} \cdot 1) = a \cdot 1 = a$. (ii) If $a \neq 0$, then $a/a = a \cdot a^{-1} = 1$, by definition. $\square$

Theorem 1.9. *If $b \neq 0$ and $d \neq 0$, then $b \cdot d \neq 0$ and*

$$\left(\frac{a}{b}\right) \cdot \left(\frac{c}{d}\right) = \frac{a \cdot c}{b \cdot d}.$$

PROOF. That $b \cdot d \neq 0$ follows from Theorem 1.5. Using the notation for fractions, Lemma 1.4, and Theorem 1.6(ii), we find

$$\left(\frac{a}{b}\right) \cdot \left(\frac{c}{d}\right) = (a \cdot b^{-1}) \cdot (cd^{-1})$$

$$= (a \cdot c) \cdot (b^{-1}d^{-1}) = (a \cdot c)(bd)^{-1} = \frac{a \cdot c}{b \cdot d}. \qquad \square$$

The proofs of Theorem 1.10(i) through (v) are left to the reader.

Theorem 1.10

(i) *If $b \neq 0$ and $c \neq 0$, then*

$$\frac{a}{b} = \frac{a \cdot c}{b \cdot c}.$$

(ii) *If $c \neq 0$, then*

$$\frac{a}{c} + \frac{b}{c} = \frac{(a+b)}{c}.$$

(iii) *If $b \neq 0$, then $-b \neq 0$ and*

$$\frac{(-a)}{b} = \frac{a}{-b} = -\left(\frac{a}{b}\right).$$

(iv) *If $b \neq 0$, $c \neq 0$, and $d \neq 0$, then $(c/d) \neq 0$ and*

$$\frac{(a/b)}{(c/d)} = \frac{a \cdot d}{b \cdot c} = \left(\frac{a}{b}\right) \cdot \left(\frac{d}{c}\right).$$

(v) *If $b \neq 0$ and $d \neq 0$, then*

$$\frac{a}{b} + \frac{c}{d} = \frac{ad + bc}{bd}.$$

PROBLEMS

1. Show that in Axiom A-5 it is not necessary to assume there is only one number x such that $a + x = 0$.

2. Prove Lemma 1.1.

3. Prove, on the basis of Axioms A-1 through A-5, that

$$(a + c) + (b + d) = (a + d) + (b + c).$$

Give the appropriate reason for each step of the proof.

4. Prove Theorem 1.4.

5. Prove Lemma 1.3.

6. Prove Lemma 1.4.

7. If a and b are any numbers show that there is one and only one number x such that $x + a = b$.

8. Prove Theorem 1.6.

9. Show that the Distributive Law may be replaced by the statement: for all numbers a, b, and c, the equality $(b + c)a = ba + ca$ holds.

10. Complete the proof of Theorem 1.7.

11. If a, b, and c are any numbers show that $a - (b + c) = (a - b) - c$ and that $a - (b - c) = (a - b) + c$. Give reasons for each step of the proof.

12. Show that $a(b + c + d) = ab + ac + ad$, giving reasons for each step.

13. Assuming that $a + b + c + d$ means $(a + b + c) + d$, prove that $a + b + c + d = (a + b) + (c + d)$.

14. Assuming the result of Problem 9, prove that

$$(a + b) \cdot (c + d) = ac + bc + ad + bd.$$

15. Prove Theorem 1.10(i).

16. Prove Theorem 1.10(ii).

17. Prove Theorem 1.10(iii).

18. Prove Theorem 1.10(iv). [*Hint*: Use Theorem 1.10(i).]

19. Prove Theorem 1.10(v).

20. Prove that if $b \neq 0$, $d \neq 0$, and $f \neq 0$, then

$$\left(\frac{a}{b}\right) \cdot \left(\frac{c}{d}\right) \cdot \left(\frac{e}{f}\right) = \frac{(a \cdot c \cdot e)}{(b \cdot d \cdot f)}.$$

21. Prove that if $d \neq 0$, then

$$\frac{a}{d} + \frac{b}{d} + \frac{c}{d} = \frac{(a + b + c)}{d}.$$

22. Prove that if $b \neq 0$, $d \neq 0$, and $f \neq 0$, then

$$\frac{a}{b} + \frac{c}{d} + \frac{e}{f} = \frac{(a \cdot d \cdot f + b \cdot c \cdot f + b \cdot d \cdot e)}{(b \cdot d \cdot f)}.$$

1.2. Natural Numbers and Sequences

Traditionally, we build the real number system by a sequence of enlargements. We start with the positive integers and extend that system to include the positive rational numbers (quotients, or ratios, of integers). The system of rational numbers is then enlarged to include all positive real numbers; finally, we adjoin the negative numbers and zero to obtain the collection of all real numbers.

The system of axioms in Section 1.1 does not distinguish (or even mention) positive numbers. To establish the relationship between these axioms and the real number system, we begin with a discussion of **natural numbers**. As we know, these are the same as the positive integers.

Intuitively, the totality of natural numbers can be obtained by starting with the number 1 and then forming $1 + 1, (1 + 1) + 1, [(1 + 1) + 1] + 1$, and so on. We call $1 + 1$ the number 2; then $(1 + 1) + 1$ is called the number 3, and in this way the collection of natural numbers is generated. Actually, it is possible to give an abstract definition of natural number, one which yields the same collection and which is logically more satisfactory. This is done in Section 1.4, where the principle of mathematical induction is established and illustrated. In the meantime, we shall suppose the reader is familiar with all the usual properties of natural numbers.

The axioms for a field given in Section 1.1 determine addition and multiplication for any *two* numbers. On the basis of these axioms we were able to *define* the sum and product of three numbers. Before describing the process of defining sums and products for more than three elements, we recall several definitions and give some notations which will be used throughout the book.

Definitions. The set (or collection) of all real numbers is denoted by $\mathbb{R}^1$. The set of ordered pairs of real numbers is denoted by $\mathbb{R}^2$, the set of ordered triples by $\mathbb{R}^3$ and so on. A **relation** *from* $\mathbb{R}^1$ *to* $\mathbb{R}^1$ is a set of ordered pairs of real numbers; that is, a relation from $\mathbb{R}^1$ to $\mathbb{R}^1$ is a set in $\mathbb{R}^2$. The **domain** of this relation is the set in $\mathbb{R}^1$ consisting of all the first elements in the ordered pairs. The **range** of the relation is the set of all the second elements in the ordered pairs. Observe that the range is also a set in $\mathbb{R}^1$. A **function** f *from* $\mathbb{R}^1$ *into* $\mathbb{R}^1$ is a relation in which no two ordered pairs have the same first element. We use the notation $f: \mathbb{R}^1 \to \mathbb{R}^1$ for such a function. The word **mapping** is a synonym for function. If D is the domain of f and S is its range, we shall sometimes use the notation $f: D \to S$. A function is a relation (set in $\mathbb{R}^2$) such that for each element x in the domain there is precisely one element y in the range such that the pair (x, y) is one of the ordered pairs which constitute the function. Occasionally, a function will be indicated by writing $f: x \to y$. Also, for a given function f, the unique number in the range corresponding to an element x in the domain is written $f(x)$. The symbol $x \to f(x)$ is used for this relationship. We assume that the reader is familiar with functional notation.

A **sequence** is a function which has as its domain some or all of the natural numbers. If the domain consists of a finite number of positive integers, we say the sequence is **finite**. Otherwise, the sequence is called **infinite**. In general, the elements in the domain of a function do not have any particular order. In a sequence, however, there is a natural ordering of the domain induced by the usual order in terms of size which we give to the positive integers. For example, if the domain of a sequence consists of the numbers $1, 2, \ldots, n$, the elements of the range, that is, the *terms of the sequence*, are usually written in the same order as the natural numbers. If the sequence (function) is denoted by a, then the terms of the sequence are denoted by $a_1, a_2, \ldots, a_n$ or, sometimes by $a(1)$, $a(2), \ldots, a(n)$. The element a_i or $a(i)$ is called the ith *term of the sequence*. If the domain of a sequence a is the set of all natural numbers (so that the sequence is infinite), we denote the sequence by

$$a_1, a_2, \ldots a_n, \ldots \qquad \text{or } \{a_n\}.$$

The sum and product of a finite sequence of terms are defined inductively. The following statements are proved in Section 1.4.

Proposition 1.1 (Sum). *If $a_1, a_2, \ldots, a_n$ is a given finite sequence, then there is a unique finite sequence $b_1, b_2, \ldots, b_n$ with the properties:*

$$b_1 = a_1$$

and, in general,

$$b_{i+1} = b_i + a_{i+1} \qquad for \quad i = 1, 2, \ldots, n-1 \quad if \quad n > 1.$$

Proposition 1.2 (Product). *If $a_1, a_2, \ldots, a_n$ is a given sequence, then there is a unique finite sequence $c_1, c_2, \ldots, c_n$ with the properties:*

$$c_1 = a_1$$

and, in general,

$$c_{i+1} = c_i \cdot a_{i+1} \qquad for \quad i = 1, 2, \ldots, n-1 \quad if \quad n > 1.$$

The elements $b_1, b_2, \ldots, b_n$ are the successive *sums* of the terms of the sequence $a_1, a_2, \ldots, a_n$, and the elements $c_1, c_2, \ldots, c_n$ are the successive *products* of the terms of $a_1, a_2, \ldots, a_n$. In particular, we have

$$b_3 = b_2 + a_3 = (a_1 + a_2) + a_3, \qquad b_4 = b_3 + a_4 = [(a_1 + a_2) + a_3] + a_4,$$

$$c_3 = c_2 \cdot a_3 = (a_1 \cdot a_2) \cdot a_3, \qquad c_4 = c_3 \cdot a_4 = [(a_1 \cdot a_2) \cdot a_3] \cdot a_4,$$

and so on. Because the sequences $b_1, b_2, \ldots, b_n$ and $c_1, c_2, \ldots, c_n$ are uniquely determined we can make the following definitions:

Definitions. For all integers $n \geqslant 1$, the **sum** and **product** of the numbers a_1, $a_2, \ldots, a_n$ are defined by

$$a_1 + a_2 + \cdots + a_n \equiv b_n \qquad \text{and} \qquad a_1 \cdot a_2 \cdots \cdot a_n \equiv c_n.$$

We use the notation

$$\sum_{i=1}^n a_i = a_1 + a_2 + \cdots + a_n \qquad \text{and} \qquad \prod_{i=1}^n a_i = a_1 \cdot a_2 \cdots \cdot a_n.$$

The symbol $\prod$ is in general use as a compact notation for product analogous to the use of $\sum$ for sum. We read $\prod_{i=1}^n$ as "the product as i goes from 1 to n."

On the basis of these definitions and the above propositions, it is not difficult to establish the next result.

Proposition 1.3. *If* $a_1, a_2, \ldots, a_n, a_{n+1}$ *is any sequence, then*

$$\sum_{i=1}^{n+1} a_i = \left(\sum_{i=1}^{n} a_i \right) + a_{n+1} \quad \text{and} \quad \prod_{i=1}^{n+1} a_i = \left(\prod_{i=1}^{n} a_i \right) \cdot a_{n+1}.$$

The following proposition may be proved by using mathematical induction.[2]

Proposition 1.4. *If* $a_1, a_2, \ldots, a_m, a_{m+1}, \ldots, a_{m+n}$ *is any sequence, then*

$$\sum_{i=1}^{m+n} a_i = \sum_{i=1}^{m} a_i + \sum_{i=m+1}^{m+n} a_i$$

and

$$\prod_{i=1}^{m+n} a_i = \left(\prod_{i=1}^{m} a_i \right) \cdot \left(\prod_{i=m+1}^{m+n} a_i \right).$$

The *extended associative* law, the *extended commutative* law, and the *extended distributive* law with which the reader is familiar from elementary algebra, are stated in the next three propositions.

Proposition 1.5. *The sum of a finite sequence can be obtained by separating the given sequence into several shorter sequences, adding the terms in each of these, and then adding the results. A similar statement holds for products.*

Proposition 1.6. *The sum of a given finite sequence is independent of the order of its terms. The product of a finite sequence is independent of the order of its terms.*

Proposition 1.7. *If* a *is any number and* $b_1, b_2, \ldots, b_n$ *is any sequence, then*

$$a \cdot (b_1 + b_2 + \cdots + b_n) = (b_1 + b_2 + \cdots + b_n) \cdot a$$

$$= ab_1 + ab_2 + \cdots + ab_n.$$

An expression of the form

$$a - b - c + d - e$$

is a shorthand notation for the sum

$$a + (-b) + (-c) + d + (-e).$$

The justification for this shorthand is based on the definition $x - y = x + (-y)$ and on Theorem 1.2(ii), which states that the negative of a sum is the sum of the negatives. Using these facts and Theorem 1.2(i), which asserts

---

[2] The principle of mathematical induction is established in Section 1.4.

that $-(-a) = a$, we obtain the usual rule for the addition and subtraction of any finite number of terms. With the help of the sign laws for multiplication and the extended distributive law, we get the standard rules for the multiplication of signed sums. For example, it is easy to verify that

$$(a_1 - a_2)(b_1 - b_2 - b_3) = a_1(b_1 - b_2 - b_3) - a_2(b_1 - b_2 - b_3)$$
$$= a_1 b_1 - a_1 b_2 - a_1 b_3 - a_2 b_1 + a_2 b_2 + a_2 b_3.$$

The symbol x^n, for n a natural number, is defined in the customary way as $x \cdot x \cdots x$ with x appearing n times in the product. We assume the reader is familiar with the laws of exponents and the customary rules for adding, subtracting, and multiplying polynomials. These rules are a simple consequence of the axioms and propositions above.

The decimal system of writing numbers (or the system with any base) depends on a representation theorem which we now state. If n is any natural number, then there is *one and only one* representation for n of the form

$$n = d_0(10)^k + d_1(10)^{k-1} + \cdots + d_{k-1}(10) + d_k$$

in which k is a natural number or zero, each d_i is one of the numbers $0, 1, 2, \ldots, 9$, and $d_0 \neq 0$. The numbers $0, 1, 2, \ldots, 9$ are called **digits** of the decimal system. On the basis of such a representation, the rules of arithmetic follow from the corresponding rules for polynomials in x with $x = 10$.

For completeness, we define the terms *integer*, *rational number*, and *irrational number*.

Definitions. A real number is an **integer** if and only if it is either zero, a natural number, or the negative of a natural number. A real number r is said to be **rational** if and only if there are *integers* p and q, with $q \neq 0$, such that $r = p/q$. A real number which is not rational is called **irrational**.

It is clear that the sum and product of a finite sequence of integers is again an integer, and that the sum, product, or quotient of a finite sequence of rational numbers is a rational number.

The rule for multiplication of fractions is given by an extension of Theorem 1.9 which may be derived by mathematical induction.

We emphasize that the axioms for a field given in Section 1.1 imply only theorems concerned with the operations of addition, subtraction, multiplication, and division. The exact nature of the elements in the field is not described. For example, the axioms do not imply the *existence* of a number whose square is 2. In fact, if we interpret number to be "rational number" and consider no others, then all the axioms for a field are satisfied. The rational number system forms a *field*. An additional axiom is needed if we wish the field to contain irrational numbers such as $\sqrt{2}$.

Consider the system with five elements $0, 1, 2, 3, 4$ and the rules of addition and multiplication as given in the following tables:

Addition						Multiplication					
+	0	1	2	3	4	×	0	1	2	3	4
0	0	1	2	3	4	0	0	0	0	0	0
1	1	2	3	4	0	1	0	1	2	3	4
2	2	3	4	0	1	2	0	2	4	1	3
3	3	4	0	1	2	3	0	3	1	4	2
4	4	0	1	2	3	4	0	4	3	2	1

It can be shown that this system satisfies the axioms for a field. Many examples of fields with only a finite number of elements are easily constructed. Such systems do not satisfy the order axiom (Axiom I) described in the next section.

PROBLEMS

1. Suppose $T: \mathbb{R}^1 \to \mathbb{R}^1$ is a relation composed of ordered pairs (x, y). We define the *inverse relation* of T as the set of ordered pairs (x, y) where (y, x) belongs to T. Let a function a be given as a finite sequence: $a_1, a_2, \ldots, a_n$. Under what conditions will the inverse relation of a be a function?

2. Prove Propositions 1.1 and 1.2 for sequences with 5 terms.

3. Show that if $a_1, a_2, \ldots, a_5$ is a sequence of 5 terms, then $\prod_{i=1}^{5} a_i = 0$ if and only if at least one term of the sequence is zero.

4. Prove Proposition 1.3 under the assumption that the preceding propositions have been established.

5. Use Proposition 1.4 to establish the following special case of Proposition 1.5. Given the sequence $a_1, a_2, \ldots, a_m, a_{m+1}, \ldots, a_n, a_{n+1}, \ldots, a_p$, show that

$$\sum_{i=1}^{p} a_i = \sum_{i=1}^{m} a_i + \sum_{i=m+1}^{n} a_i + \sum_{i=n+1}^{p} a_i.$$

6. Prove Proposition 1.6 for the case of sequences with 4 terms.

7. (a) State which propositions are used to establish the formula:

$$(a + b)^5 = \sum_{i=0}^{5} \frac{5!}{i!(5 - i)!} a^{5-i} b^i.$$

   (b) Same as (a) for the general binomial formula

$$(a + b)^n = \sum_{i=0}^{n} \frac{n!}{i!(n - i)!} a^{n-i} b^i,$$

   assuming the formula is known for the case $n - 1$.

   (c) Develop a "trinomial" formula for $(a + b + c)^n$.

8. By using both positive and negative powers of 10, state a theorem on the decimal representation of rational numbers. [*Hint*: Assume the connection between rational numbers and repeating decimals.]

9. Consider the system with the two elements 0 and 1 and the following rules of addition and multiplication:

$$0 + 0 = 0, \quad 0 + 1 = 1,$$
$$1 + 0 = 1, \quad 1 + 1 = 0,$$
$$0 \cdot 0 = 0, \quad 1 \cdot 0 = 0,$$
$$0 \cdot 1 = 0, \quad 1 \cdot 1 = 1.$$

Show that all the axioms of Section 1.1 are valid; hence this system forms a field.

10. Show that the system consisting of all elements of the form $a + b\sqrt{5}$, where a and b are any rational numbers, satisfies all the axioms for a field if the usual rules for addition and multiplication are used.

11. Is it possible to make addition and multiplication tables so that the four elements 0, 1, 2, 3 form the elements of a field? Prove your statement. [*Hint*: In the multiplication table each row, other than the one consisting of zeros, must contain the symbols 0, 1, 2, 3 in some order.]

12. Consider all numbers of the form $a + b\sqrt{6}$ where a and b are rational. Does this collection satisfy the axioms for a field?

13. Show that the system of complex numbers $a + bi$ with a, b real numbers, $i = \sqrt{-1}$ satisfies all the axioms for a field.

1.3. Inequalities

The axioms for a field describe many number systems. If we wish to describe the real number system to the exclusion of other systems, additional axioms are needed. One of these, an axiom which distinguishes positive from negative numbers, is the Axiom of inequality.

Axiom I (Axiom of inequality). *Among all the numbers of the system, there is a set called the positive numbers which satisfies the conditions*: (i) *for any number a exactly one of the three alternatives holds*: *a is positive or a = 0 or −a is positive*; (ii) *any finite sum or product of positive numbers is positive.*

When Axiom I is added to those of Section 1, the resulting system of axioms is applicable only to those number systems which have a linear order.[3] For example, the system of complex numbers does not satisfy Axiom I but does

---

[3] That is, the numbers may be made to correspond to the points on a straight line as in analytic geometry.

satisfy all the axioms for a field. Similarly, it is easy to see that the system described in Problem 9 of Section 1.2 does not satisfy Axiom I. However, both the real number system and the rational number system satisfy all the axioms given thus far.

Definitions. A number a is **negative** whenever $-a$ is positive. If a and b are any numbers, we say that $a > b$ (*read: a is greater than b*) whenever $a - b$ is positive. We write $a < b$ (*read: a is less than b*) whenever $a - b$ is negative.

The following theorem is an immediate consequence of Axiom I and the definition above.

Theorem 1.11

   (i) $a > 0 \Leftrightarrow a$ *is positive,*
   (ii) $a < 0 \Leftrightarrow a$ *is negative,*
   (iii) $a > 0 \Leftrightarrow -a < 0,$
   (iv) $a < 0 \Leftrightarrow -a > 0,$
   (v) *if a and b are any numbers then exactly one of the three alternatives holds:*
       $a > b$ *or* $a = b$ *or* $a < b,$
   (vi) $a < b \Leftrightarrow b > a.$

The next five theorems yield the standard rules for manipulating inequalities. We shall prove the first of these theorems and leave the proofs of the others to the reader.

Theorem 1.12. *If a, b, and c are any numbers and if $a > b$, then*

$$a + c > b + c \qquad and \qquad a - c > b - c.$$

PROOF. Since $a > b$, it follows from the definition and Theorem 1.11(i) that $a - b > 0$. Hence $(a + c) - (b + c) = a - b > 0$ and $(a - c) - (b - c) = a - b > 0$. Therefore $(a + c) - (b + c)$ is positive, as is $(a - c) - (b - c)$. From the definition of "greater than" we conclude that $a + c > b + c$ and $a - c > b - c$. □

Theorem 1.13

   (i) *If $a > 0$ and $b < 0$, then $a \cdot b < 0$.*
   (ii) *If $a < 0$ and $b < 0$, then $a \cdot b > 0$.*
   (iii) $1 > 0.$

Theorem 1.14

   (i) *a and b are both positive or both negative if and only if $a \cdot b > 0$; a and b have opposite signs if and only if $a \cdot b < 0$.*
   (ii) *If $a \neq 0$ and $b \neq 0$, then $a \cdot b$ and a/b have the same sign.*

Theorem 1.15

(i) *If $a > b$ and $c > 0$, then*

$$ac > bc \qquad and \qquad \frac{a}{c} > \frac{b}{c}.$$

(ii) *If $a > b$ and $c < 0$, then*

$$ac < bc \qquad and \qquad \frac{a}{c} < \frac{b}{c}.$$

(iii) *If a and b have the same sign and $a > b$, then*

$$\frac{1}{a} < \frac{1}{b}.$$

Theorem 1.16. *If $a > b$ and $b > c$, then $a > c$.*

Theorems 1.12 and 1.15 yield the following statements on inequalities, ones which are familiar to most readers.

(a) *The direction of an inequality is unchanged if the same number is added to or subtracted from both sides.*
(b) *The direction of an inequality is unchanged if both sides are multiplied or divided by a positive number.*
(c) *The direction of an inequality is reversed if both sides are multiplied or divided by a negative number.*

It is convenient to adopt a geometric point of view and to associate a horizontal axis with the totality of real numbers. We select any convenient point for the origin and denote points to the right of the origin as positive numbers and points to the left as negative numbers (Figure 1.1). For every real number there will correspond a point on the line and, conversely, every point will represent a real number. Then the inequality $a < b$ may be read: *a is to the left of b*. This geometric way of looking at inequalities is frequently of help in solving problems. It is helpful to introduce the notion of an *interval of numbers* or *points*. If a and b are numbers (as shown in Figure 1.2), then the **open interval from a to b** is the collection of all numbers which are both larger than a and smaller than b. That is, an open interval consists of all numbers *between a and b*. A number x is between a and b if both inequalities $a < x$ and $x < b$ are true. A compact way of writing these two inequalities is

$$a < x < b.$$

The **closed interval from a to b** consists of all the points between a and b and

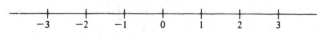

Figure 1.1

Figure 1.2. Open interval.

in addition the numbers (or points) a and b (Figure 1.3). Suppose a number x is either equal to a or larger than a, but we don't know which. We write this conveniently as $x \geqslant a$, which is read: x *is greater than or equal to a.* Similarly $x \leqslant b$ is read: x *is less than or equal to b,* and means that x may be either smaller than b or may be b itself. A compact way of designating all points x in the closed interval from a to b is

$$a \leqslant x \leqslant b.$$

An interval which contains the endpoint b but not a is said to be **half-open on the left.** Such an interval consists of all points x which satisfy the double inequality

$$a < x \leqslant b.$$

Similarly, an interval containing a but not b is called **half-open on the right,** and we write

$$a \leqslant x < b.$$

Parentheses and brackets are used as symbols for intervals in the following way:

(a, b) for the open interval $a < x < b$,
$[a, b]$ for the closed interval $a \leqslant x \leqslant b$,
$(a, b]$ for the interval half-open on the left $a < x \leqslant b$,
$[a, b)$ for the interval half-open on the right $a \leqslant x < b$.

We can extend the idea of interval to include the unbounded cases. For example, consider the set of *all* numbers larger than 7. This set may be thought of as an interval beginning at 7 and extending to infinity to the right (see Figure 1.4). Of course, infinity is not a number, but we use the symbol $(7, \infty)$ to represent all numbers larger than 7. We also use the double inequality

$$7 < x < \infty$$

to represent this set. In a similar way, the symbol $(-\infty, 12)$ stands for all numbers less than 12. We also use the inequalities $-\infty < x < 12$ to represent this set.

Figure 1.3. Closed interval.

Figure 1.4. A half infinite interval.

The equation $3x + 7 = 19$ has a unique solution, $x = 4$. The quadratic equation $x^2 - x - 2 = 0$ has two solutions, $x = -1, 2$. The trigonometric equation[4] $\sin x = 1/2$ has an infinite number of solutions, $x = \pi/6, 5\pi/6, 13\pi/6, 17\pi/6, \ldots$. The inequality $x^2 \leqslant 2$ has as its solution all numbers in the closed interval $[-\sqrt{2}, \sqrt{2}]$. The **solution** of an equation or an inequality in one unknown, say x, is the collection of all numbers which make the equation or inequality a true statement. Sometimes this set of numbers is called the **solution set**. For example, the inequality

$$3x - 7 < 8$$

has as its solution set all numbers less than 5. To demonstrate this we argue in the following way. If x is a number which satisfies the above inequality we can, by Theorem 1.16, add 7 to both sides and obtain a true statement. That is, the inequality

$$3x - 7 + 7 < 8 + 7 \qquad \text{or} \qquad 3x < 15$$

holds. Now, dividing both sides by 3 (Theorem 1.19), we obtain $x < 5$, and therefore *if* x is a solution, *then* it is less than 5. Strictly speaking, however, we have not *proved* that every number which is less than 5 is a solution. In an actual proof, we begin by supposing that x is any number less than 5; that is, $x < 5$. We multiply both sides of this inequality by 3 (Theorem 1.19), and then subtract 7 from both sides (Theorem 1.16) to get

$$3x - 7 < 8,$$

the original inequality. Since the hypothesis that x is less than 5 implies the original inequality, we have proved the result. The important thing to notice is that the proof consisted of *reversing* the steps of the original argument which led to the solution $x < 5$ in the first place. So long as each step taken is reversible, the above procedure is completely satisfactory for obtaining solutions of inequalities.

By means of the symbol $\Leftrightarrow$, we can give a solution to this example in compact form. We write

$$3x - 7 < 8 \quad \Leftrightarrow \quad 3x < 15 \qquad \text{(adding 7 to both sides)}$$

and

$$3x < 15 \quad \Leftrightarrow \quad x < 5 \qquad \text{(dividing both sides by 3).}$$

The solution set is the interval $(-\infty, 5)$.

---

[4] We assume the reader is familiar with the elementary properties of trigonometric functions. They are used mainly for illustrative purposes.

We present a second illustration of the same technique.

EXAMPLE 1. Find the solution set of the inequality:

$$-7 - 3x < 5x + 29.$$

Solution. We have

$$-7 - 3x < 5x + 29 \quad \Leftrightarrow \quad -36 < 8x \qquad \text{(adding } 3x - 29 \text{ to both sides)}$$

$$\Leftrightarrow \quad 8x > -36 \qquad \text{(Theorem 1.11(vi))}$$

$$\Leftrightarrow \quad x > -\frac{9}{2}.$$

The solution set is the interval $(-9/2, \infty)$. $\qquad\qquad\qquad\qquad\qquad$ $\square$

Notation. It is convenient to introduce some terminology and symbols concerning sets. In general, a **set** is a collection of objects. The objects may have any character (numbers, points, lines, etc.) so long as we know which objects are in a given set and which are not. If S is a set and P is an object in it, we write $P \in S$ and say that P *is an element of* S or that P *belongs to* S. If S_1 and S_2 are two sets, their **union**, denoted by $S_1 \cup S_2$, consists of all objects each of which is in at least one of the two sets. The **intersection** of S_1 and S_2, denoted by $S_1 \cap S_2$, consists of all objects each of which is in both sets. Schematically, if S_1 is the horizontally shaded set of points (Figure 1.5), and S_2 is the vertically shaded set, then $S_1 \cup S_2$ consists of the entire shaded area, and $S_1 \cap S_2$ consists of the doubly shaded area. Similarly, we may form the union and intersection of any number of sets. When we write $S_1 \cup S_2 \cup \cdots \cup S_n$ for the union S of the n sets $S_1, S_2, \ldots, S_n$, then S consists of all elements each of which is in at least one of the n sets. We also use the notation $S = \bigcup_{i=1}^{n} S_i$ as a shorthand for the union of n sets. The intersection of n sets $S_1, S_2, \ldots, S_n$ is written $S_1 \cap S_2 \cap \cdots \cap S_n$ or, briefly, $\bigcap_{i=1}^{n} S_i$. It may happen that two sets S_1 and S_2 have no elements in common. In such a case their intersection is *empty*, and we use the term **empty set** for the set which is devoid of members.

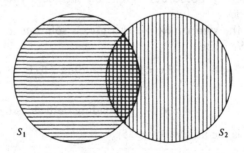

Figure 1.5. Exhibiting $S_1 \cup S_2$ and $S_1 \cap S_2$.

Most often we will deal with sets each of which is specified by some property or properties of its elements. For example, we may speak of the set of all even integers or the set of all rational numbers between 0 and 1. We employ the special symbol

$$\{x: x = 2n \text{ and } n \text{ is an integer}\}$$

to represent the set of all even integers. In this notation the letter x stands for a generic element of the set, and the properties which determine membership in the set are listed to the right of the colon. The symbol

$$\{x: x \in (0, 1) \text{ and } x \text{ is rational}\}$$

represents the set of rational numbers in the open interval $(0, 1)$. If a set has only a few elements, it may be specified by listing its members between braces. Thus the symbol $\{-2, 0, 1\}$ denotes the set whose elements are the numbers $-2, 0$, and 1. A set may be specified by any number of properties, and we use a variety of notations to indicate these properties. If a set of objects P has properties A, B, and C, we may denote this set by

$$\{P: P \text{ has properties } A, B, \text{ and } C\}.$$

Other examples are: the open interval $(0, 2)$ is denoted by

$$(0, 2) = \{x: 0 < x < 2\}$$

and the half-open interval $[2, 14)$ is denoted by

$$[2, 14) = \{t: 2 \leqslant t < 14\}.$$

To illustrate the use of the symbols for set union and set intersection, we observe that

$$[1, 3] = [1, 2\tfrac{1}{2}] \cup [2, 3]$$

and

$$(0, 1) = (0, +\infty) \cap (-\infty, 1).$$

The words *and* and *or* have precise meanings when used in connection with sets and their properties. The set consisting of elements which have property A *or* property B is the *union* of the set having property A and the set having property B. Symbolically, we write

$$\{x: x \text{ has property } A \text{ or property } B\}$$

$$= \{x: x \text{ has property } A\} \cup \{x: x \text{ has property } B\}.$$

The set consisting of elements which have both property A *and* property B is the *intersection* of the set having property A with the set having property B. In set notation, we write

$$\{x: x \text{ has property } A \text{ and property } B\}$$

$$= \{x: x \text{ has property } A\} \cap \{x: x \text{ has property } B\}.$$

If A and B are two sets and if every element of A is also an element of B, we say that A *is a* **subset** *of* B, and we write $A \subset B$. The two statements $A \subset B$ and $B \subset A$ imply that $A = B$.

We give two examples illustrating the way set notation is used.

EXAMPLE 2. Solve for x:

$$\frac{3}{x} < 5 \qquad (x \neq 0).$$

Solution. Since we don't know in advance whether x is positive or negative, we cannot multiply by x unless we impose additional conditions. We therefore separate the problem into two cases: (i) x is positive, and (ii) x is negative. The desired solution set can be written as the union of the sets S_1 and S_2 defined by

$$S_1 = \left\{ x: \frac{3}{x} < 5 \text{ and } x > 0 \right\},$$

$$S_2 = \left\{ x: \frac{3}{x} < 5 \text{ and } x < 0 \right\}.$$

Now

$$x \in S_1 \quad \Leftrightarrow \quad 3 < 5x \quad \text{and} \quad x > 0$$

$$\Leftrightarrow \quad x > \tfrac{3}{5} \quad \text{and} \quad x > 0$$

$$\Leftrightarrow \quad x > \tfrac{3}{5}.$$

Similarly,

$$x \in S_2 \quad \Leftrightarrow \quad 3 > 5x \quad \text{and} \quad x < 0$$

$$\Leftrightarrow \quad x < \tfrac{3}{5} \quad \text{and} \quad x < 0$$

$$\Leftrightarrow \quad x < 0.$$

Thus the solution set is (see Figure 1.6)

$$S_1 \cup S_2 = (\tfrac{3}{5}, \infty) \cup (-\infty, 0).$$

EXAMPLE 3. Solve for x:

$$\frac{2x - 3}{x + 2} < \frac{1}{3} \qquad (x \neq -2).$$

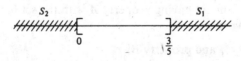

Figure 1.6

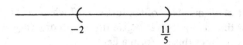

Figure 1.7. Solution set $(-2, 11/5)$.

Solution. As in Example 2, the solution set is the union $S_1 \cup S_2$, where

$$S_1 = \left\{ x: \frac{2x-3}{x+2} < \frac{1}{3} \text{ and } x + 2 > 0 \right\},$$

$$S_2 = \left\{ x: \frac{2x-3}{x+2} < \frac{1}{3} \text{ and } x + 2 < 0 \right\}.$$

For numbers in S_1, we may multiply the inequality by $x + 2$ and, since $x + 2$ is positive, the direction of the inequality is preserved. Hence

$$x \in S_1 \quad \Leftrightarrow \quad 3(2x - 3) < x + 2 \quad \text{and} \quad x + 2 > 0$$
$$\Leftrightarrow \quad 5x < 11 \quad \text{and} \quad x + 2 > 0$$
$$\Leftrightarrow \quad x < \tfrac{11}{5} \quad \text{and} \quad x > -2$$
$$\Leftrightarrow \quad x \in (-2, \tfrac{11}{5}).$$

For numbers in S_2, multiplication of the inequality by the negative quantity $x + 2$ reverses the direction. Therefore,

$$x \in S_2 \quad \Leftrightarrow \quad 3(2x - 3) > x + 2 \quad \text{and} \quad x + 2 < 0$$
$$\Leftrightarrow \quad 5x > 11 \quad \text{and} \quad x + 2 < 0$$
$$\Leftrightarrow \quad x > \tfrac{11}{5} \quad \text{and} \quad x < -2.$$

Since there are no numbers x satisfying *both* conditions $x > 11/5$ and $x < -2$, the set S_2 is empty. The solution set (Figure 1.7) consists of $S_1 = (-2, 11/5)$. $\qquad \Box$

We shall assume the reader is familiar with the notion of absolute value and elementary manipulations with equations and inequalities involving the absolute value of numbers. For those readers who wish to review this material, the basic definitions and theorems are provided in Appendix 1.1. A general method for solving polynomial inequalities and inequalities involving quotients of polynomials is presented in Appendix 1.2.

PROBLEMS

1. Consider the field consisting of all numbers of the form $a + b\sqrt{7}$ where a and b are rational. Does this field satisfy Axiom I? Justify your answer.

2. Consider the set of all numbers of the form $ai\sqrt{7}$ where a is a real number and $i = \sqrt{-1}$. Show that it is possible to give this set an ordering in such a way that it satisfies Axiom I. Does this set form a field?

3. Show that the set of all complex numbers $a + bi$, with a and b rational, satisfies all the axioms for a field.

In Problems 4 through 7 find in each case the solution set as an interval, and plot.

4. $2x - 2 < 27 + 4x$

5. $5(x - 1) > 12 - (17 - 3x)$

6. $(2x + 1)/8 < (3x - 4)/3$

7. $(x + 10)/6 + 1 - (x/4) > ((4 - 5x)/6) - 1$

In Problems 8 through 10, find the solution set of each pair of simultaneous inequalities. Verify in each case that the solution set is the intersection of the solution sets of the separate inequalities.

8. $2x - 3 < 3x - 2$ and $4x - 1 < 2x + 3$

9. $3x + 5 > x + 1$ and $4x - 3 < x + 6$

10. $4 - 2x < 1 + 5x$ and $3x + 2 < x - 7$

In Problems 11 through 14, express each given combination of intervals as an interval. Plot a graph in each case.

11. $[-1, \infty] \cap (-\infty - 2)$ 12. $(-\infty, 2) \cap (-\infty, 4)$

13. $(-1, 1) \cup (0, 5)$ 14. $(-\infty, 2) \cup (-\infty, 4)$

In Problems 15 through 19 find the solution set of the given inequality.

15. $3/x < 2/5$ 16. $(x - 2)/x < 3$

17. $(x + 2)/(x - 1) < 4$ 18. $x/(2 - x) < 2$

19. $(x + 2)/(x - 3) < -2$ 20. Prove Theorem 1.13.

21. Prove Theorem 1.14.   22. Prove Theorem 1.15.

23. Prove Theorem 1.16.

24. Using the theorems of this section prove that if $a > b$ and $c > d$, then $a + c > b + d$.

25. Using the theorems of this section show:

   (i) If $a > b > 0$ and $c > d > 0$, then $a \cdot c > b \cdot d$.
   (ii) If $a < b < 0$ and $c < d < 0$, then $a \cdot c > b \cdot d$.

1.4. Mathematical Induction

The principle of mathematical induction with which most readers are familiar can be derived as a consequence of the axioms in Section 1.1. Since the definition of natural number is at the basis of the principle of mathematical induction, we shall develop both concepts together.

Definition. A set S of numbers is said to be **inductive** if and only if

(a) $1 \in S$ and
(b) $(x + 1) \in S$ whenever $x \in S$.

Examples of inductive sets are easily found. The set of all real numbers is inductive, as is the set of all rationals. The set of all integers, positive, zero, and negative is inductive. The collection of real numbers between 0 and 10 is not inductive since it satisfies (a) but not (b). No finite set of real numbers can be inductive since (b) will be violated at some stage.

Definition. A real number n is said to be a **natural number** if and only if it belongs to *every* inductive set of real numbers. The set of *all* natural numbers will be denoted by the symbol $\mathbb{N}$.

We observe that $\mathbb{N}$ contains the number 1 since, by the definition of inductive set, 1 must always be a member of every inductive set. Intuitively we know that the set of natural numbers $\mathbb{N}$ is identical with the set of positive integers.

Theorem 1.17. *The set $\mathbb{N}$ of all natural numbers is an inductive set.*

PROOF. We must show that $\mathbb{N}$ has Properties (a) and (b) in the definition of inductive set. As we remarked above, (a) holds. Now suppose that k is an element of $\mathbb{N}$. Then k belongs to *every* inductive set S. For each inductive set, if k is an element, so is $k + 1$. Thus $k + 1$ belongs to every inductive set. Therefore $k + 1$ belongs to $\mathbb{N}$. Hence $\mathbb{N}$ has Property (b), and is inductive. $\square$

The principle of mathematical induction is contained in the next theorem which asserts that any inductive set of natural numbers must consist of the entire collection $\mathbb{N}$.

Theorem 1.18 (Principle of mathematical induction). *If S is an inductive set of natural numbers, then $S = \mathbb{N}$.*

PROOF. Since S is an inductive set, we know from the definition of natural number that $\mathbb{N}$ is contained in S. On the other hand, since S consists of natural numbers, it follows that S is contained in $\mathbb{N}$. Therefore $S = \mathbb{N}$. $\square$

We now illustrate how the principle of mathematical induction is applied in practice. The reader may not recognize Theorem 1.18 as the statement of the familiar principle of mathematical induction. So we shall prove a formula using Theorem 1.18.

EXAMPLE 1. Show that

$$1 + 2 \ldots + n = \frac{n(n + 1)}{2} \tag{1.1}$$

for every natural number n.

Solution. Let S be the set of natural numbers n for which the Formula (1.1) holds. We shall show that S is an inductive set.

(a) Clearly $1 \in S$ since Formula (1.1) holds for $n = 1$.

(b) Suppose $k \in S$. Then Formula (1.1) holds with $n = k$. Adding $(k + 1)$ to both sides we see that

$$1 + \cdots + k + (k + 1) = \frac{k(k + 1)}{2} + (k + 1) = \frac{(k + 1)(k + 2)}{2},$$

which is Formula (1.1) for $n = k + 1$. Thus $(k + 1)$ is in S whenever k is.

Combining (a) and (b), we conclude that S is an inductive set of natural numbers and so consists of all natural numbers. Therefore Formula (1.1) holds for all natural numbers.   □

EXAMPLE 2. Show that $n! > 2^n$ for each natural number $n > 3$.

Solution. Let S consist of 1, 2, 3 and all $n > 3$ for which $n! > 2^n$. We shall show that S is inductive.

(a) Clearly $1 \in S$ by the definition of S.

(b) Suppose $k \in S$. Then $k = 1, 2,$ or 3, or $k > 3$ and $k! > 2^k$. If $k = 1$, $k + 1 = 2 \in S$. If $k = 2, k + 1 = 3 \in S$. If $k = 3$, then $k + 1 = 4 \in S$ since $4! = 24 > 2^4 = 16$. If $k > 3$ and $k! > 2^k$, then $(k + 1) > 3$ obviously, and

$$(k + 1)! = (k + 1) \cdot k! > (k + 1) \cdot 2^k > 4 \cdot 2^k = 2 \cdot 2^{k+1} > 2^{k+1}$$

so that $(k + 1) \in S$. Thus $(k + 1)$ is in S whenever k is.

Therefore S is inductive and so consists of all natural numbers. If n is any natural number > 3, it follows that $n! > 2^n$.   □

Note that in this example the inequality $n! > 2^n$ is false for $n = 1, 2, 3$. This made necessary the unusual definition of our set S.

Remark. An alternative (simplier) procedure for proving statements such as those in Example 2 uses a modified form of Theorem 1.18. A set S of real numbers has the **modified inductive property** if (1) S has a smallest number and (2) $(x + 1) \in S$ whenever $x \in S$. Theorem 1.18 becomes: if S is a set of natural numbers with the modified inductive property, then S contains all the natural numbers greater than the smallest natural number in S. We see that in Example 2, we may choose for S the set of all natural numbers greater than 3. The result is the same.

Now mathematical induction can be used to establish Proposition 1.1 of Section 1.2. We state this and subsequent propositions of Section 1.2 as theorems.

Theorem 1.19. *For each finite sequence of numbers $a_1, a_2, \ldots, a_n$, there is associated a unique sequence $b_1, b_2, \ldots, b_n$ such that*

$$b_1 = a_1 \tag{1.2}$$

and

$$b_{i+1} = b_i + a_{i+1} \quad for \quad i = 1, 2, \ldots, n-1 \quad if \quad n > 1. \tag{1.3}$$

PROOF. Let S be the set of all natural numbers n for which the theorem is true. We shall show that S is inductive and hence $S = \mathbb{N}$. Clearly 1 is in S since $b_1 = a_1$, by hypothesis. Let $a_1, a_2, \ldots, a_k, a_{k+1}$ be given and suppose that $k \in S$. Then there is a *unique* sequence $b_1, b_2, \ldots, b_k$ such that

$$b_1 = a_1$$

and

$$b_{i+1} = b_i + a_{i+1} \quad for \quad i = 1, 2, \ldots, k-1 \quad if \quad k > 1.$$

If $k = 1$, then $b_1 = a_1$. We define $b_{k+1} = b_k + a_{k+1}$. Then the sequence b_1, $b_2, \ldots, b_k, b_{k+1}$ satisfies Equations (1.2) and (1.3) of the theorem, and we must show that this sequence is unique. Suppose $b_1', b_2', \ldots, b_k', b_{k+1}'$ is a second sequence with the same properties. We have $b_i' = b_i$ for $i = 1, 2, \ldots, k$ because the assumption $k \in S$ implies these b_i are unique. Now,

$$b_{k+1}' = b_k' + a_{k+1} = b_k + a_{k+1} = b_{k+1},$$

and so $(k + 1) \in S$. The set S is inductive and therefore $S = \mathbb{N}$. The theorem is true for every natural number n. $\qquad \square$

Theorem 1.20. *For each finite sequence $a_1, a_2, \ldots, a_n$ there is a unique sequence $c_1, c_2, \ldots, c_n$ such that*

$$c_1 = a_1$$

and

$$c_{i+1} = c_i \cdot a_{i+1} \quad for \quad i = 1, 2, \ldots, n-1 \quad if \quad n > 1.$$

The proof of Theorem 1.20 is similar to the proof of Theorem 1.19 and is left to the reader.

The next theorem, originally stated as Proposition 1.4 of Section 1.2, is a consequence of Theorems 1.19 and 1.20. Its proof, which employs the Principle of mathematical induction, is left to the reader.

Theorem 1.21. *If $a_1, a_2, \ldots, a_m, a_{m+1}, \ldots, a_{m+n}$ is any finite sequence, then*

$$\sum_{i=1}^{m+n} a_i = \sum_{i=1}^{m} a_i + \sum_{i=m+1}^{m+n} a_i \quad.$$

and

$$\prod_{i=1}^{m+n} a_i = \left(\prod_{i=1}^{m} a_i\right)\cdot\left(\prod_{i=m+1}^{m+n} a_i\right).$$

Since the set of natural numbers $\mathbb{N}$ is identical with the set of positive integers, the fact that a natural number is positive follows from the Axiom of inequality and the definition of $\mathbb{N}$.

Theorem 1.22 (The well-ordering principle). *Any nonempty set T of natural numbers contains a smallest element.*

PROOF. Let k be a member of T. We define a set S of natural numbers by the relation

$$S = \{p: p \in T \text{ and } p \leqslant k\}.$$

The set S contains a portion (not necessarily all) of the set consisting of the natural numbers $\{1, 2, 3, \ldots, k - 1, k\}$. Thus since S is finite it has a smallest element, which we denote by s. We now show that s is the smallest element of T. First, since $s \in S$ and $S \subset T$, then $s \in T$. Suppose t is any element of T different from s. If $t > k$, then the inequality $k \geqslant s$ implies that $t > s$. On the other hand, if $t \leqslant k$, then $t \in S$. Since s is the smallest element of S and $s \neq t$, we have $s < t$. Thus s is smaller than any other element of T, and the proof is complete.   □

PROBLEMS

In each of Problems 1 through 6 use the Principle of mathematical induction to establish the given formula.

1. $\sum_{i=1}^{n} i^2 = n(n + 1)(2n + 1)/6$

2. $\sum_{i=1}^{n} i^3 = n^2(n + 1)^2/4$

3. $\sum_{i=1}^{n} (2i - 1) = n^2$

4. $\sum_{i=1}^{n} i(i + 1) = n(n + 1)(n + 2)/3$

5. $\sum_{i=1}^{n} i(i + 2) = n(n + 1)(2n + 7)/6$

6. $\sum_{i=1}^{n} (1/i(i + 1)) = n/(n + 1)$

7. Suppose p, q, and r are natural numbers such that $p + q < p + r$. Show that $q < r$.

8. Suppose that p, q, and r are natural numbers such that $p \cdot q < p \cdot r$. Show that $q < r$.

9. (a) Show that the set of positive rational numbers is inductive.
   (b) Is the set $a + b\sqrt{5}$, a and b natural numbers, an inductive set?
   (c) Is the set of all complex numbers an inductive set?

In each of Problems 10 through 15 use the Principle of mathematical induction to establish the given assertion.

10. $\sum_{i=1}^{n} [a + (i - 1)d] = n[2a + (n - 1)d]/2$, where a and d are any real numbers.

11. $\sum_{i=1}^{n} ar^{i-1} = a(r^n - 1)/(r - 1)$, where a and r are real numbers, and $r \neq 1$.

12. $\sum_{i=1}^{n} i(i + 1)(i + 2) = n(n + 1)(n + 2)(n + 3)/4$.

13. $(1 + a)^n \geqslant 1 + na$ for $a \geqslant 0$ and n a natural number.

14. $\sum_{i=1}^{n} 3^{2i-1} = 3(9^n - 1)/8$.

15. Prove by induction that

$$(x_1 + x_2 + \cdots + x_k)^2 = \sum_{i=1}^{k} x_i^2 + 2(x_1 x_2 + x_1 x_3 + \cdots + x_1 x_k + x_2 x_3$$

$$+ x_2 x_4 + \cdots x_2 x_k + x_3 x_4 + x_3 x_5 + \cdots + x_{k-1} x_k).$$

16. Prove Theorem 1.20.

17. Prove Theorem 1.21.

18. Use mathematical induction to prove Proposition 1.5 of Section 1.2.

19. Use mathematical induction to prove Proposition 1.6 of Section 1.2.

*20.[5] We denote by $\mathbb{N} \times \mathbb{N}$ the set of all ordered pairs of natural numbers (m, n). State and prove a Principle of mathematical induction for sets contained in $\mathbb{N} \times \mathbb{N}$.

[5] An asterisk is used to indicate difficult problems.

CHAPTER 2

Continuity and Limits

2.1. Continuity

Most of the functions we study in elementary calculus are described by simple formulas. These functions almost always possess derivatives and, in fact, a portion of any first course in calculus is devoted to the development of routine methods for computing derivatives. However, not all functions possess derivatives everywhere. For example, the functions $(1 + x^2)/x$, cot x, and $\sin(1/x)$ do not possess derivatives at $x = 0$ no matter how they are defined at $x = 0$.

As we progress in the study of analysis, it is important to enlarge substantially the class of functions under examination. Functions which possess derivatives everywhere form a rather restricted class; extending this class to functions which are differentiable except at a few isolated points does not enlarge it greatly. We wish to investigate significantly larger classes of functions, and to do so we introduce the notion of a continuous functions.

Definitions. Suppose that f is a function from a domain D in $\mathbb{R}^1$ to $\mathbb{R}^1$. The function f is **continuous at** a if and only if (i) the point a is in an open interval I contained in D, and (ii) for each positive number ε there is a positive number δ such that

$$|f(x) - f(a)| < \varepsilon \quad \text{whenever} \quad |x - a| < \delta.$$

If f is continuous at each point of a set S, we say that f is **continuous on** S. A function f is called **continuous** if it is continuous at every point of its domain.

The geometric significance of continuity at a point a is indicated in Figure 2.1. We recall that the inequality $|f(x) - f(a)| < \varepsilon$ is equivalent to the double inequality

$$-\varepsilon < f(x) - f(a) < \varepsilon$$

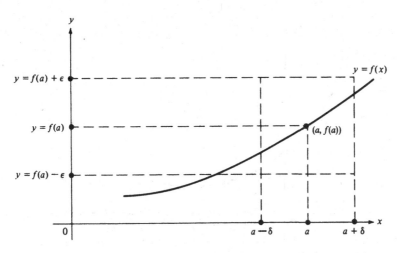

Figure 2.1. The graph of f is in the rectangle for $a - \delta < x < a + \delta$.

or

$$f(a) - \varepsilon < f(x) < f(a) + \varepsilon.$$

Similarly, the inequality $|x - a| < \delta$ is equivalent to the two inequalities

$$a - \delta < x < a + \delta.$$

We construct the four lines $x = a - \delta$, $x = a + \delta$, $y = f(a) - \varepsilon$, and $y = f(a) + \varepsilon$, as shown in Figure 2.1. The rectangle determined by these four lines has its center at the point with coordinates $(a, f(a))$. The geometric interpretation of continuity at a point may be given in terms of this rectangle. A function f is continuous at a if for each $\varepsilon > 0$ there is a number $\delta > 0$ such that the graph of f remains within the rectangle for all x in the interval $(a - \delta, a + \delta)$.

It is usually very difficult to verify continuity directly from the definition. Such verification requires that for *every* positive number ε, we exhibit a number δ and show that the graph of f lies in the appropriate rectangle. However, if the function f is given by a sufficiently simple expression, it is sometimes possible to obtain an explicit value for the quantity δ corresponding to a given number ε. We describe the method by means of two examples.

EXAMPLE 1. Given the function

$$f: x \to \frac{1}{x + 1}, \qquad x \neq -1,$$

and $a = 1$, $\varepsilon = 0.1$, find a number δ such that $|f(x) - f(1)| < 0.1$ for $|x - 1| < \delta$.

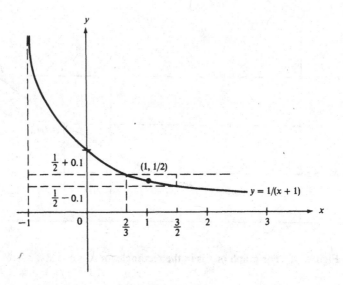

Figure 2.2

Solution. We sketch the graph of f and observe that f is decreasing for $x > -1$ (see Figure 2.2). The equations $f(x) - f(1) = 0.1$, $f(x) - f(1) = -0.1$ can be solved for x. We find

$$\frac{1}{x+1} - \frac{1}{2} = 0.1 \quad \Leftrightarrow \quad x = \frac{2}{3}$$

and

$$\frac{1}{x+1} - \frac{1}{2} = -0.1 \quad \Leftrightarrow \quad x = \frac{3}{2}.$$

Since f is decreasing in the interval $\frac{2}{3} < x < \frac{3}{2}$, it is clear that the graph of f lies in the rectangle formed by the lines $x = \frac{2}{3}$, $x = \frac{3}{2}$, $y = \frac{1}{2} - 0.1$, and $y = \frac{1}{2} + 0.1$. Since the distance from $x = 1$ to $x = \frac{2}{3}$ is smaller than the distance from $x = 1$ to $x = \frac{3}{2}$, we select $\delta = 1 - \frac{2}{3} = \frac{1}{3}$. We make the important general observation that *when a value of δ is obtained for a given quantity ε, then any smaller (positive) value for δ may also be used for the same number ε.* □

Remarks. For purposes of illustration, we assume in this chapter that $\sqrt[n]{x}$ is defined for all $x \geqslant 0$ if n is even and that $\sqrt[n]{x}$ is defined for all x if n is odd. We also suppose that for positive numbers x_1, x_2, the inequality $\sqrt{x_2} > \sqrt{x_1}$ holds whenever $x_2 > x_1$. Facts of this type concerning functions of the form $\sqrt[n]{x}$, with n a natural number, do not follow from the results given thus far. They will be proved in the next section.

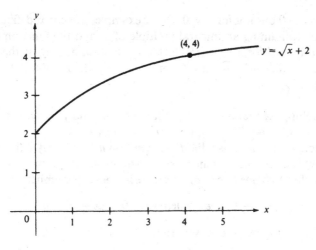

Figure 2.3

EXAMPLE 2. Consider the function

$$f: x \to \begin{cases} \dfrac{x-4}{\sqrt{x}-2} & x \geq 0, x \neq 4, \\ 4 & x = 4. \end{cases}$$

If $\varepsilon = 0.01$, find a δ such that $|f(x) - f(4)| < 0.01$ for all x such that $|x - 4| < \delta$.

Solution. If $x \neq 4$, then

$$f(x) = \frac{(\sqrt{x} - 2)(\sqrt{x} + 2)}{\sqrt{x} - 2} = \sqrt{x} + 2.$$

The graph of f is shown in Figure 2.3, and we observe that f is an increasing function. We solve the equations $f(x) - f(4) = 0.01$ and $f(x) - f(4) = -0.01$ and obtain

$$\sqrt{x} + 2 - 4 = 0.01 \quad \Leftrightarrow \quad \sqrt{x} = 2.01 \quad \Leftrightarrow \quad x = 4.0401$$
$$\sqrt{x} + 2 - 4 = -0.01 \quad \Leftrightarrow \quad \sqrt{x} = 1.99 \quad \Leftrightarrow \quad x = 3.9601.$$

Since f is increasing, it follows that $|f(x) - f(4)| < 0.01$ for $3.9601 < x < 4.0401$. Selecting $\delta = 0.0399$, we find that $|f(x) - f(4)| < \varepsilon$ for $|x - 4| < \delta$. $\qquad \square$

We shall frequently be concerned with functions which have a domain in $\mathbb{R}^1$ which consists of an interval except for a single point. This exceptional point may be an interior point or an endpoint of the interval. For example, the function $(x^2 + 1)/x$ is defined for all x except $x = 0$. The function $\log x$ is defined for $x > 0$ but not for $x = 0$. Other examples are $\cot x$, defined on any interval not containing an integral multiple of π, and the function $(x^3 + 8)/(x + 2)$ not defined for $x = -2$; in fact, any function defined as the quotient of two polynomials has excluded from its domain those values at which the denominator is zero.

Definition. Suppose that a and L are real numbers and f is a function from a domain D in $\mathbb{R}^1$ to $\mathbb{R}^1$. The number a may or may not be in the domain of f. The function f **tends to L as a limit as x tends to** a if and only if (i) there is an open interval I containing a which, except possibly for the point a, is contained in D, and (ii) for each positive number ε there is a positive number δ such that

$$|f(x) - L| < \varepsilon \quad \text{whenever} \quad 0 < |x - a| < \delta.$$

If f tends to L as x tends to a, we write

$$f(x) \to L \quad \text{as} \quad x \to a$$

and denote the number L by

$$\lim_{x \to a} f(x)$$

or by $\lim_{x \to a} f(x)$.

Remarks

(i) We see that a function f is continuous at a if and only if a is in the domain of f and $f(x) \to f(a)$ as $x \to a$.

(ii) The condition $0 < |x - a| < \delta$ (excluding the possibility $x = a$) is used rather than the condition $|x - a| < \delta$ as in the definition of continuity since f may not be defined at a itself.

PROBLEMS

In Problems 1 through 8 the functions are continuous at the value a given. In each case find a value δ corresponding to the given value of ε so that the definition of continuity is satisfied. Draw a graph.

1. $f(x) = 2x + 5, a = 1, \varepsilon = 0.01$

2. $f(x) = 1 - 3x, a = 2, \varepsilon = 0.01$

3. $f(x) = \sqrt{x}, a = 2, \varepsilon = 0.01$

4. $f(x) = \sqrt[3]{x}, a = 1, \varepsilon = 0.1$

5. $f(x) = 1 + x^2, a = 2, \varepsilon = 0.01$

6. $f(x) = x^3 - 4, a = 1, \varepsilon = 0.5$

7. $f(x) = x^3 + 3x, a = -1, \varepsilon = 0.5$

8. $f(x) = \sqrt{2x + 1}, a = 4, \varepsilon = 0.1$

In Problems 9 through 17 the functions are defined in an interval about the given value of a but not at a. Determine a value δ so that for the given values of L and ε, the statement "$|f(x) - L| < \varepsilon$ whenever $0 < |x - a| < \delta$" is valid. Sketch the graph of the given function.

9. $f(x) = (x^2 - 9)/(x + 3)$, $a = -3$, $L = -6$, $\varepsilon = 0.005$

10. $f(x) = (\sqrt{2x} - 2)/(x - 2)$, $a = 2$, $L = \frac{1}{2}$, $\varepsilon = 0.01$

11. $f(x) = (x^2 - 4)/(x - 2)$, $a = 2$, $L = 4$, $\varepsilon = 0.01$

12. $f(x) = (x - 9)/(\sqrt{x} - 3)$, $a = 9$, $L = 6$, $\varepsilon = 0.1$

13. $f(x) = (x^3 - 8)/(x - 2)$, $a = 2$, $L = 12$, $\varepsilon = 0.5$

14. $f(x) = (x^3 + 1)/(x + 1)$, $a = -1$, $L = 3$, $\varepsilon = 0.1$

*15. $f(x) = (x - 1)/(\sqrt[3]{x} - 1)$, $a = 1$, $L = 3$, $\varepsilon = 0.1$

16.[1] $f(x) = x \sin(1/x)$, $a = 0$, $L = 0$, $\varepsilon = 0.01$

*17. $f(x) = (\sin x)/x$, $a = 0$, $L = 1$, $\varepsilon = 0.1$

18. Show that $\lim_{x \to 0}(\sin(1/x))$ does not exist.

*19.[1] Show that $\lim_{x \to 0} x \log|x| = 0$.

20. The function $f(x) = x \cot x$ is not defined at $x = 0$. Can the domain of f be enlarged to include $x = 0$ in such a way that the function is continuous on the enlarged domain?

21. For all $x \in R^1$ define

$$f(x) = \begin{cases} 1 & \text{if } x \text{ is a rational number,} \\ 0 & \text{if } x \text{ is an irrational number.} \end{cases}$$

Show that f is not continuous at every value of x.

22. Suppose that f is defined in an interval about the number a and $\lim_{h \to 0}[f(a + h) - f(a - h)] = 0$. Show that f may not be continuous at a. Is it always true that $\lim_{h \to 0} f(a + h)$ exists?

2.2. Limits

The basic theorems of calculus depend for their proofs on certain standard theorems on limits. These theorems are usually stated without proof in a first course in calculus. In this section we fill the gap by providing proofs of the customary theorems on limits. These theorems are the basis for the formulas for the derivative of the sum, product and quotient of functions and for the Chain Rule.

---

[1] For purposes of illustration we assume the reader is familiar with the elementary properties of the exponential, logarithmic, trigonometric, and inverse trigonometric functions. The exponential and logarithmic functions are discussed in Section 5.3.

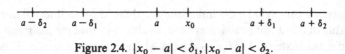

Figure 2.4. $|x_0 - a| < \delta_1, |x_0 - a| < \delta_2$.

Theorem 2.1 (Uniqueness of limits). *Suppose f is a function from $\mathbb{R}^1$ to $\mathbb{R}^1$. If $f(x) \to L$ as $x \to a$ and $f(x) \to M$ as $x \to a$, then $L = M$.*

PROOF. We shall assume that $L \neq M$ and reach a contradiction. Define $\varepsilon = \frac{1}{2}|L - M|$. According to the definition of limit, we know that for this positive number ε, there is a number $\delta_1 > 0$ such that (x is in the domain of f and)

$$|f(x) - L| < \varepsilon \quad \text{for all } x \text{ whenever} \quad 0 < |x - a| < \delta_1.$$

Similarly since $f(x) \to M$, there is a number $\delta_2 > 0$ such that (x is in the domain of f and)

$$|f(x) - M| < \varepsilon \quad \text{for all } x \text{ whenever} \quad 0 < |x - a| < \delta_2.$$

Let x_0 be a number whose distance from a is less than both δ_1 and δ_2. (See Figure 2.4.) We write

$$L - M = (L - f(x_0)) + (f(x_0) - M).$$

Therefore,

$$|L - M| \leqslant |L - f(x_0)| + |f(x_0) - M|$$

and, from the inequalities above,

$$|L - M| < \varepsilon + \varepsilon = 2\varepsilon.$$

On the other hand, we defined $\varepsilon = \frac{1}{2}|L - M|$. We have a contradiction, and so $L = M$. $\qquad\square$

Theorem 2.2 (Limit of a constant). *If c is a number and $f(x) = c$ for all x on $\mathbb{R}^1$, then for every real number a*

$$\lim_{x \to a} f(x) = c.$$

PROOF. In the definition of limit, we may choose $\delta = 1$ for every positive ε. Then

$$|f(x) - c| = |c - c| = 0 < \varepsilon \quad \text{for} \quad |x - a| < 1. \qquad\square$$

Theorem 2.3 (Obvious limit). *If $f(x) = x$ for all x on $\mathbb{R}^1$ and a is any real number, then*

$$\lim_{x \to a} f(x) = a.$$

PROOF. In the definition of limit, we may choose $\delta = \varepsilon$. Then, since $f(x) - f(a) = x - a$, we clearly have

$$|f(x) - f(a)| = |x - a| < \varepsilon \quad \text{whenever} \quad 0 < |x - a| < \varepsilon. \qquad\square$$

Theorem 2.4 (Limit of equal functions). *Suppose f and g are functions whose domains contain the set $S = \{x: 0 < |x - a| < r\}$ for some positive number r. If $f(x) = g(x)$ for all x in S and $\lim_{x \to a} g(x) = L$, then*

$$\lim_{x \to a} f(x) = L.$$

PROOF. We show that the definition of limit is satisfied for f. Let $\varepsilon > 0$ be given. Then there is a number $\delta > 0$ such that $|g(x) - L| < \varepsilon$ whenever $0 < |x - a| < \delta$. We decrease δ, if necessary, so that $\delta < r$. Then $|g(x) - L| = |f(x) - L|$ for $0 < |x - a| < \delta$ and so the definition of limit is satisfied for the function f. $\square$

Remarks. In Theorem 2.4 f and g may differ outside the set $0 < |x - a| < r$. In addition, one or both of the functions may not be defined at $x = a$.

Theorem 2.5 (Limit of a sum). *Suppose that*

$$\lim_{x \to a} f_1(x) = L_1 \quad and \quad \lim_{x \to a} f_2(x) = L_2.$$

Define $g(x) = f_1(x) + f_2(x)$. Then

$$\lim_{x \to a} g(x) = L_1 + L_2.$$

PROOF. Let $\varepsilon > 0$ be given. Then, using the quantity $\varepsilon/2$, there are positive numbers δ_1 and δ_2 such that

$$|f_1(x) - L_1| < \frac{\varepsilon}{2} \quad \text{for all } x \text{ satisfying} \quad 0 < |x - a| < \delta_1$$

and

$$|f_2(x) - L_2| < \frac{\varepsilon}{2} \quad \text{for all } x \text{ satisfying} \quad 0 < |x - a| < \delta_2.$$

Define δ as the smaller of δ_1 and δ_2. Then

$$|g(x) - (L_1 + L_2)| = |f_1(x) - L_1 + f_2(x) - L_2|$$
$$\leqslant |f_1(x) - L_1| + |f_2(x) - L_2|,$$

and for $0 < |x - a| < \delta$, it follows that

$$|g(x) - (L_1 + L_2)| < \frac{\varepsilon}{2} + \frac{\varepsilon}{2} = \varepsilon.$$

The result is established. $\square$

Corollary. *Suppose that $\lim_{x \to a} f_i(x) = L_i$, $i = 1, 2, \ldots, n$. Define $g(x) = \sum_{i=1}^{n} f_i(x)$. Then*

$$\lim_{x \to a} g(x) = \sum_{i=1}^{n} L_i.$$

The corollary may be established either by induction or directly by the method used in the proof of Theorem 2.5.

Theorem 2.6 (Limit of a product). *Suppose that*

$$\lim_{x \to a} f_1(x) = L_1 \quad and \quad \lim_{x \to a} f_2(x) = L_2.$$

Define $g(x) = f_1(x) \cdot f_2(x)$. *Then*

$$\lim_{x \to a} g(x) = L_1 L_2.$$

PROOF. Suppose that $\varepsilon > 0$ is given. We wish to show that there is a $\delta > 0$ such that

$$|g(x) - L_1 L_2| < \varepsilon \quad \text{whenever} \quad 0 < |x - a| < \delta.$$

If ε_1 and ε_2 are positive numbers (their exact selection will be made later), there are positive numbers δ_1 and δ_2 such that

$$|f_1(x) - L_1| < \varepsilon_1 \quad \text{whenever} \quad 0 < |x - a| < \delta_1 \tag{2.1}$$

and

$$|f_2(x) - L_2| < \varepsilon_2 \quad \text{whenever} \quad 0 < |x - a| < \delta_2. \tag{2.2}$$

We first show that $|f_1(x)| < |L_1| + \varepsilon_1$ for $0 < |x - a| < \delta_1$. To see this, we write

$$f_1(x) = f_1(x) - L_1 + L_1$$

and hence

$$|f_1(x)| \leqslant |f_1(x) - L_1| + |L_1| < \varepsilon_1 + |L_1|.$$

Define $M = |L_1| + \varepsilon_1$. To establish the result of the theorem, we use the identity

$$g(x) - L_1 L_2 = L_2(f_1(x) - L_1) + f_1(x)(f_2(x) - L_2).$$

Now we employ the triangle inequality for absolute values as well as Inequalities (2.1) and (2.2) to get

$$|g(x) - L_1 L_2| \leqslant |L_2| \cdot |f_1(x) - L_1| + |f_1(x)| \cdot |f_2(x) - L_2|$$

$$\leqslant |L_2| \cdot \varepsilon_1 + M \cdot \varepsilon_2.$$

Select $\varepsilon_1 = \varepsilon/2L_2$ and $\varepsilon_2 = \varepsilon/2M$. The quantities δ_1 and δ_2 are those which correspond to the values of ε_1 and ε_2, respectively. Then with δ as the smaller of δ_1 and δ_2, it follows that

$$|g(x) - L_1 L_2| < \varepsilon \quad \text{whenever} \quad 0 < |x - a| < \delta.^2 \qquad \square$$

Corollary. *Suppose that* $f_i(x) \to L_i$ *as* $x \to a$ *for* $i = 1, 2, \ldots, n$, *and suppose that* $g(x) = f_1(x) \cdots f_n(x)$. *Then* $\lim_{x \to a} g(x) = L_1 L_2 \cdots L_n$.

---

[2] This proof assumed that $L_2 \neq 0$. A slight modification of the proof establishes the result if $L_2 = 0$.

Theorem 2.7 (Limit of a composite function). *Suppose that f and g are functions on $\mathbb{R}^1$ to $\mathbb{R}^1$. Define the composite function $h(x) = f[g(x)]$. If f is continuous at L, and if $g(x) \to L$ as $x \to a$, then*

$$\lim_{x \to a} h(x) = f(L).$$

PROOF. Since f is continuous at L, we know that for every $\varepsilon > 0$ there is a $\delta_1 > 0$ such that

$$|f(t) - f(L)| < \varepsilon \quad \text{whenever} \quad |t - L| < \delta_1.$$

From the fact that $g(x) \to L$ as $x \to a$, it follows that for every $\varepsilon' > 0$, there is a $\delta > 0$ such that

$$|g(x) - L| < \varepsilon' \quad \text{whenever} \quad 0 < |x - a| < \delta.$$

In particular, we may select $\varepsilon' = \delta_1$. Then $f[g(x)]$ is defined and

$$|f[g(x)] - f(L)| < \varepsilon \quad \text{whenever} \quad 0 < |x - a| < \delta. \qquad \square$$

Remark. The result of Theorem 2.7 is false if we weaken the hypothesis on the function f and require only that f tend to a limit as $t \to L$ rather than have f continuous at L. An example exhibiting this fact is discussed in Problem 11 at the end of the section.

Corollary 1. *If f is continuous at L and g is continuous at a with $g(a) = L$, then the composite function $h(x) = f[g(x)]$ is continuous at a.*

Speaking loosely, we say that "a continuous function of a continuous function is continuous."

Corollary 2. *The function $F: x \to 1/x$, defined for $x \neq 0$ is continuous for every value of $x \neq 0$. The function $G: x \to \sqrt[n]{x}$ defined for $x > 0$ is continuous for every $x > 0$.*[3]

These statements may be proved directly by the methods of Section 2.1 and proofs are left to the reader.

Theorem 2.8 (Limit of a quotient). *Suppose that*

$$\lim_{x \to a} f(x) = L \quad \text{and} \quad \lim_{x \to a} g(x) = M.$$

Define $h(x) = f(x)/g(x)$. If $M \neq 0$, then

$$\lim_{x \to a} h(x) = L/M.$$

---

[3] This can be proved only after it is shown that $\sqrt[n]{x}$ is defined for each $x > 0$, i.e., if $x > 0$ and n is a natural number, there is a unique $y > 0$ such that $y^n = x$. We have assumed this fact in this chapter *for illustration only.*

The proof of Theorem 2.8 is a direct consequence of Theorem 2.6, 2.7, and Corollary 2. We leave the details to the reader.

Theorem 2.9 (Limit of inequalities). *Suppose that*

$$\lim_{x \to a} f(x) = L \quad and \quad \lim_{x \to a} g(x) = M.$$

If $f(x) \leqslant g(x)$ for all x in some interval about a (possibly excluding a itself), then

$$L \leqslant M.$$

PROOF. We assume that $L > M$ and reach a contradiction. Let us define $\varepsilon = (L - M)/2$; then from the definition of limit there are positive numbers δ_1 and δ_2 such that

$$|f(x) - L| < \varepsilon \quad \text{for all } x \text{ satisfying} \quad 0 < |x - a| < \delta_1$$

and

$$|g(x) - M| < \varepsilon \quad \text{for all } x \text{ satisfying} \quad 0 < |x - a| < \delta_2.$$

We choose a positive number δ which is smaller than δ_1 and δ_2 and furthermore so small that $f(x) \leqslant g(x)$ for $0 < |x - a| < \delta$. In this interval, we have

$$M - \varepsilon < g(x) < M + \varepsilon \quad \text{and} \quad L - \varepsilon < f(x) < L + \varepsilon.$$

Since $M + \varepsilon = L - \varepsilon$, it follows that

$$g(x) < M + \varepsilon = L - \varepsilon < f(x),$$

and so $f(x) > g(x)$, a contradiction.   □

Remarks. In Theorem 2.9 it is not necessary that f and g be defined at $x = a$. If f and g are continuous at a, we *conclude* from the theorem that $f(a) \leqslant g(a)$. If, in the hypothesis, we assume that $f(x) < g(x)$ in an interval about a, (excluding a itself) it is not possible to conclude that $L < M$. For example, letting $f(x) = x^3$ and $g(x) = x^2$, we observe that $f(x) < g(x)$ for $-1 < x < 1$, $x \neq 0$. However, $f(0) = g(0) = 0$.

Theorem 2.10 (Sandwiching theorem). *Suppose that f, g, and h are functions defined on the interval $0 < |x - a| < k$ for some positive number k. If $f(x) \leqslant g(x) \leqslant h(x)$ on this interval, and if*

$$\lim_{x \to a} f(x) = L, \qquad \lim_{x \to a} h(x) = L,$$

then $\lim_{x \to a} g(x) = L$.

PROOF. Given any $\varepsilon > 0$, there are positive numbers δ_1 and δ_2 (which we take smaller than k) so that

$$|f(x) - L| < \varepsilon \quad \text{whenever} \quad 0 < |x - a| < \delta_1$$

and

$$|h(x) - L| < \varepsilon \quad \text{whenever} \quad 0 < |x - a| < \delta_2.$$

In other words, $L - \varepsilon < f(x) < L + \varepsilon$ and $L - \varepsilon < h(x) < L + \varepsilon$ for all x such that $0 < |x - a| < \delta$ where δ is the smaller of δ_1 and δ_2. Therefore

$$L - \varepsilon < f(x) \leqslant g(x) \leqslant h(x) < L + \varepsilon$$

in this interval. We conclude that $|g(x) - L| < \varepsilon$ for $0 < |x - a| < \delta$, and the result is established.   □

PROBLEMS

1. Show how Theorems 2.2, 2.3, 2.5, and 2.6 may be combined to prove that for every number a, the linear function $cx + d$ has the property that
$$\lim_{x \to a} (cx + d) = ca + d.$$

2. Same as Problem 1 for the quadratic function: $cx^2 + dx + e$. Show that
$$\lim_{x \to a} (cx^2 + dx + e) = ca^2 + da + e.$$

3. Same as Problem 1 for a general polynomial of degree n:
$$a_0 x^n + a_1 x^{n-1} + \cdots + a_{n-1} x + a_n \equiv P_n(x).$$
   Prove that
$$\lim_{x \to a} P_n(x) = P_n(a).$$
   [*Hint*: Use induction.]

4. Let $P_n(x)$ and $Q_m(x)$ be polynomials of degree n and m, respectively. Use the result of Problem 3 and Theorem 2.8 to show that
$$\lim_{x \to a} \frac{P_n(x)}{Q_m(x)} = \frac{P_n(a)}{Q_m(a)}$$
   whenever a is not a root of $Q_m(x) = 0$.

*5. Suppose that $x_1, x_2, \ldots, x_n, \ldots$ is a sequence of points tending to a. Suppose f and g are functions defined in an interval about a with $f(x_n) = g(x_n)$. If $\lim_{x \to a} f(x) = L$, is it true that $\lim_{x \to a} g(x) = L$? Justify your answer.

6. Use the Sandwiching theorem to show that $\lim_{x \to 0} x^n = 0$ for every positive integer $n \geqslant 3$ by selecting $f(x) \equiv 0$, $g(x) = x^n$, $h(x) = x^2$, $-1 < x < 1$.

7. Suppose f and h are continuous at a and $f(x) \leqslant g(x) \leqslant h(x)$ for $|x - a| < k$. If $f(a) = h(a)$ show that g is continuous at a.

8. Suppose that $\lim_{x \to a} f_i(x) = L_i$, $\lim_{x \to a} g_i(x) = M_i$, $i = 1, 2, \ldots, n$. Let $a_i, b_i, i = 1, 2, \ldots, n$ be any numbers. Under what conditions is it true that
$$\lim_{x \to a} \frac{a_1 f_1(x) + \cdots + a_n f_n(x)}{b_1 g_1(x) + \cdots + b_n g_n(x)} = \frac{a_1 L_1 + \cdots + a_n L_n}{b_1 M_1 + \cdots + b_n M_n}?$$

9. Establish directly the Corollary to Theorem 2.5.

10. Prove the Corollary to Theorem 2.6.

*11. Let $f(x) = x \sin(1/x)$ for $-\pi \leqslant x \leqslant \pi$, $x \neq 0$. Sketch the graph of f. (Note that f is not defined for $x = 0$.)

   (a) Prove that $\lim_{x \to 0} f(x) = 0$.

   (b) Define $g(x) \equiv f(x)$. By observing that $g(x) = 0$ for $x_n = \pm 1/n\pi$, $n = 1, 2, \ldots$, conclude that $f[g(x)]$ is not defined for $x = x_n$. Note that $x_n \to 0$ as $n \to \infty$.

   (c) Using the result of Part (b) show that

$$\lim_{x \to 0} h(x)$$

   does not exist where $h(x) = f[g(x)]$. Compare this result with Theorem 2.7.

   (d) If we define

$$F(x) = \left\{ \begin{array}{ll} x \sin \dfrac{1}{x} & -\pi \leqslant x \leqslant \pi, x \neq 0 \\[2ex] 0 & x = 0 \end{array} \right\}$$

   use Theorem 2.7 to show that $\lim_{x \to 0} H(x) = 0$ where $H(x) = F[g(x)]$.

12. Let f and g be continuous functions from $\mathbb{R}^1$ to $\mathbb{R}^1$. Define

$$F(x) = \max[f(x), g(x)]$$

   for each $x \in \mathbb{R}^1$. Show that F is continuous.

13. Prove Theorem 2.8.

14. Prove that if $a \neq 0$ and $F: x \to 1/x$, then F is continuous at a.

15. Prove that if $a > 0$ and $G: x \to \sqrt[n]{x}$, then G is continuous at a.

2.3. One-Sided Limits

The function $f: x \to \sqrt{x}$ is continuous for all $x > 0$ and, since $\sqrt{0} = 0$, it is clear that $f(x) \to f(0)$ as x tends to 0 through positive values. Since f is not defined for negative values of x, the definition of continuity given in Section 2.1 is not fulfilled at $x = 0$. We now wish to extend the definition of continuous function so that a function such as $\sqrt{x}$ will have the natural property of continuity at the endpoint of its domain. For this purpose we need the concept of a one-sided limit.

Definition. Suppose that f is a function from a domain D in $\mathbb{R}^1$ to $\mathbb{R}^1$. The function f **tends to L as x tends to a from the right** if and only if (i) there is an open interval I in D which has a as its left endpoint and (ii) for each $\varepsilon > 0$ there is a $\delta > 0$ such that

$$|f(x) - L| < \varepsilon \quad \text{whenever} \quad 0 < x - a < \delta.$$

If f tends to L as x tends to a from the right, we write

$$f(x) \to L \quad \text{as} \quad x \to a^+$$

and we denote the number L by

$$\lim_{x \to a^+} f(x)$$

or by $\lim_{x \to a^+} f(x)$. A similar definition is employed for limits from the left. In this case the condition $0 < x - a < \delta$ is replaced by

$$0 < a - x < \delta$$

and the symbols $f(x) \to L$ as $x \to a^-$,

$$\lim_{x \to a^-} f(x) = L,$$

and $\lim_{x \to a^-} f(x) = L$ are used.

Remarks. The condition $0 < |x - a| < \delta$ is equivalent to the two conditions $0 < x - a < \delta$ and $0 < a - x < \delta$. All the theorems on limits which were stated and proved in Section 2.2 used the inequality $0 < |x - a| < \delta$. As a result, all the theorems on limits imply corresponding theorems for limits from the left and limits from the right.

The definition of one-sided limits leads to the definition of one-sided continuity.

Definitions. A function f is **continuous on the right at** a if and only if a is in the domain of f and $f(x) \to f(a)$ as $x \to a^+$. The function f is **continuous on the left at** a if and only if a is in the domain of f and $f(x) \to f(a)$ as $x \to a^-$.

If the domain of a function f is a finite interval, say $a \leqslant x \leqslant b$, then limits and continuity at the endpoints are of the one-sided variety. For example, the function $f: x \to x^2 - 3x + 5$ defined on the interval $2 \leqslant x \leqslant 4$ is continuous on the right at $x = 2$ and continuous on the left at $x = 4$. The following general definition of continuity for functions from a set in $\mathbb{R}^1$ to $\mathbb{R}^1$ declares that such a function is continuous on the closed interval $2 \leqslant x \leqslant 4$.

Definitions. Let f be a function from a domain D in $\mathbb{R}^1$ to $\mathbb{R}^1$. The function f is **continuous at** a **with respect to** D if and only if (i) a is in D, and (ii) for each $\varepsilon > 0$ there is a $\delta > 0$ such that

$$|f(x) - f(a)| < \varepsilon \quad \text{whenever} \quad x \in D \quad \text{and} \quad |x - a| < \delta. \quad (2.3)$$

A function f is **continuous on** D if it is continuous with respect to D at every point of D.

Remarks. (i) The phrase "with respect to D" is usually omitted in the definition of continuity since the context will always make the situation clear.

(ii) If a is a point of an open interval I contained in D, then the above definition of continuity coincides with that given in Section 2.1. (iii) If a is the left endpoint of an interval I in D and if the points immediately to the left of a are *not* in D, then the above definition coincides with continuity on the right. An analogous statement holds for continuity on the left. Therefore, if the domain of a function f is the closed interval $a \leqslant x \leqslant b$ and f is continuous on D, then f will be continuous on the right at a and continuous on the left at b (with respect to D).

Definition. A point a is an **isolated point** of a set D in $\mathbb{R}^1$ if and only if there is an open interval I such that the set $I \cap D$ consists of the single point $\{a\}$.

A function f from a domain D in $\mathbb{R}^1$ to $\mathbb{R}^1$ is continuous at every isolated point of D. To see this let a be an isolated point of D and suppose I is such that $\{a\} = I \cap D$. We choose $\delta > 0$ so small that the condition $|x - a| < \delta$ implies that x is in I. Then condition (ii) in the definition of continuity is always satisfied since (ii) holds when $x = a$, the only point in question.

The theorems of Section 2.2 carry over almost without change to the case of one-sided limits. As an illustration we state a one-sided version of Theorem 2.7, the limit of a composite function.

Theorem 2.11. *Suppose that f and g are functions on $\mathbb{R}^1$ to $\mathbb{R}^1$. If f is continuous at L and if $g(x) \to L$ as $x \to a^+$, then*

$$\lim_{x \to a^+} f[g(x)] = f(L).$$

A similar statement holds if $g(x) \to L$ as $x \to a^-$.

The proof of Theorem 2.11 follows exactly the line of proof of Theorem 2.7. Although we always require $x > a$ when $g(x) \to L$, it is not true that $g(x)$ remains always larger than or always smaller than L. Hence it is necessary to assume that f is continuous at L and not merely continuous on one side.

With the aid of the general definition of continuity it is possible to show that the function $g: x \to \sqrt[n]{x}$, where n is a positive integer, is a continuous function on its domain.

Theorem 2.12. *For n an even positive integer, the function $g: x \to \sqrt[n]{x}$ is continuous for x on $[0, \infty)$. For n an odd positive integer, g is continuous for x on $(-\infty, \infty)$.*

PROOF. For $x \geqslant 0$ and for any $\varepsilon > 0$, it follows that

$$|\sqrt[n]{x} - 0| < \varepsilon \quad \text{whenever} \quad |x - 0| < \varepsilon^n.$$

Therefore g is continuous on the right at 0 for every n. If n is odd, then $\sqrt[n]{-x} = -\sqrt[n]{x}$ and so g is also continuous on the left at 0.   □

It is sometimes easier to determine one-sided limits than two-sided limits. The next theorem and corollary show that one-sided limits can be used as a tool for finding ordinary limits.

Theorem 2.13 (On one- and two-sided limits). *Suppose that f is a function on $\mathbb{R}^1$ to $\mathbb{R}^1$. Then*

$$\lim_{x \to a} f(x) = L \quad \text{if and only if} \quad \lim_{x \to a^+} f(x) = L \quad \text{and} \quad \lim_{x \to a^-} f(x) = L.$$

PROOF.

(a) Suppose $f(x) \to L$ as $x \to a$. Then for every $\varepsilon > 0$ there is a $\delta > 0$ such that $f(x)$ is defined for $0 < |x - a| < \delta$ and

$$|f(x) - L| < \varepsilon \quad \text{whenever} \quad 0 < |x - a| < \delta. \tag{2.4}$$

For this value of δ we have

$$|f(x) - L| < \varepsilon \quad \text{whenever} \quad 0 < x - a < \delta.$$

This last statement implies that $f(x) \to L$ as $x \to a^+$. Similarly, since the condition $0 < |x - a| < \delta$ is implied by the inequality $0 < x - a < \delta$, it follows from Inequality (2.4) that $|f(x) - L| < \varepsilon$ whenever $0 < a - x < \delta$. Hence $f(x) \to L$ as $x \to a^-$.

(b) Now assume both one-sided limits exist. Given any $\varepsilon > 0$, there are numbers $\delta_1 > 0$ and $\delta_2 > 0$ such that $f(x)$ is defined for $0 < x - a < \delta_1$, and $0 < a - x < \delta_2$ and moreover the inequalities

$$|f(x) - L| < \varepsilon \quad \text{whenever} \quad 0 < x - a < \delta_1,$$

$$|f(x) - L| < \varepsilon \quad \text{whenever} \quad 0 < a - x < \delta_2,$$

hold. If δ is the smaller of δ_1 and δ_2 then

$$|f(x) - L| < \varepsilon \quad \text{whenever} \quad 0 < |x - a| < \delta.$$

Corollary. *Suppose f is a function on $\mathbb{R}^1$ to $\mathbb{R}^1$. Then f is continuous at a if and only if it is continuous on the left at a and continuous on the right at a.*

EXAMPLE. Given the function

$$f(x) = \begin{cases} x, & -1 \leqslant x \leqslant 2, \\ \sqrt{\dfrac{3(x^4 - 16)}{2(x^3 - 8)}}, & 2 < x \leqslant 5, \end{cases}$$

determine whether or not f is continuous at $x = 2$. Give a reason for each step of the proof.

Solution. We use the Corollary to Theorem 2.13 and show that f is continuous on the right at 2. Define

$$g_1(x) = x \quad \text{for all} \quad x.$$

Then for $-1 \leqslant x \leqslant 2$, we have $f(x) \equiv g_1(x)$. Since $g_1(x) \to 2$ as $x \to 2^-$ (obvious limit, Theorem 2.3), it follows that $f(x) \to 2$ as $x \to 2^-$. Here we have used Theorem 2.4 on the limit of equal functions, applied to one-sided limits (see Problem 14 at the end of this section.) In the interval $2 < x \leqslant 5$, f is given by

$$f(x) = \sqrt{\frac{3(x-2)(x+2)(x^2+4)}{2(x-2)(x^2+2x+4)}}.$$

Define

$$g_2(x) = \sqrt{\frac{3(x+2)(x^2+4)}{2(x^2+2x+4)}} \quad \text{for} \quad 2 < x \leqslant 5.$$

Then by Theorem 2.4 again, it follows that $\lim_{x \to 2^+} g_2(x) = \lim_{x \to 2^+} f(x)$. To find the limit of $g_2(x)$ as $x \to 2^+$ we make straightforward applications of the theorems on limits of Section 2.2, using on each occasion the one-sided version. By applying the theorems on limit of a constant, obvious limit, limit of a sum, and limit of a product, we see that

$$\lim_{x \to 2^+} 3(x+2)(x^2+4) = 3 \cdot 4 \cdot (4+4) = 96.$$

Similarly,

$$\lim_{x \to 2^+} 2(x^2+2x+4) = 2(4+4+4) = 24.$$

(Actually, these limits are found by simple substitution.) Now, using limit of a quotient and the limit of composite functions, we obtain

$$\lim_{x \to 2^+} g_2(x) = \sqrt{\frac{96}{24}} = \sqrt{4} = 2.$$

Hence, $\lim_{x \to 2^+} f(x) = 2$ and so $\lim_{x \to 2} f(x) = 2$. The function f is continuous at 2 since $f(2) = 2$ and $f(x) \to f(2)$ as $x \to 2$. $\qquad \square$

PROBLEMS

In Problems 1 through 12, in each case determine whether or not the function f is continuous at the given value of a. If it is not continuous, decide whether or not the function is continuous on the left or on the right. State reasons for each step in the argument as in the Example.

1. $f(x) = \begin{cases} x - 4, & -1 < x \leqslant 2, \\ x^2 - 6, & 2 < x < 5, \end{cases} \quad a = 2$

2. $f(x) = \begin{cases} \dfrac{x^2 - 1}{x^4 - 1}, & 1 < x < 2, \\ x^2 + 3x - 2, & 2 \leqslant x < 5, \end{cases} \quad a = 2$

3. $f: x \to x/|x|, \quad a = 0$

4. $f: x \to |x - 1|, \quad a = 1$

5. $f: x \to x(1 + (1/x^2))^{1/2}, \quad a = 0$

6. $f(x) = ((x^2 - 27)/(x^2 + 2x + 1))^{1/3}, \quad a = 3$

7. $f(x) = \begin{cases} \left(\dfrac{2x + 5}{3x - 2}\right)^{1/2}, & 1 < x < 2, \\ 2x - 1, & 2 \leqslant x < 4, \end{cases} \quad a = 2$

8. $f(x) = \begin{cases} \left(\dfrac{x^2 - 4}{x^2 - 3x + 2}\right)^{1/2}, & 1 < x < 2, \\ x^2/2, & 2 \leqslant x < 5, \end{cases} \quad a = 2$

9. $f(x) = \begin{cases} \dfrac{x^3 - 8}{x^2 - 4}, & x \neq 2, \\ 4, & x = 2, \end{cases} \quad a = 2$

10. $f(x) = \begin{cases} \left(\dfrac{x^2 - 9}{x^2 + 1}\right)^{1/2}, & 3 \leqslant x < 4, \\ x^2 - 9, & 2 < x < 3, \end{cases} \quad a = 3$

11. $f(x) = \begin{cases} \sin\left(\dfrac{1}{x}\right), & x \neq 0, \\ 0, & x = 0, \end{cases} \quad a = 0$

12. $f(x) = \begin{cases} x \cos\left(\dfrac{1}{x}\right), & x \neq 0, \\ 0, & x = 0, \end{cases} \quad a = 0$

13. Given $f(x) = \tan x, \ -\pi/2 < x < \pi/2, \ g(x) = (\pi/2) - x$. What conclusion can we draw about
$$\lim_{x \to a^+} f[g(x)]$$
when $a = \pi/4$? Same problem when $a = \pi/2$, when $a = 0$.

14. State and prove the analog of Theorem 2.4 on the limit of equal functions for limits from the left.

15. State and prove the analog of Theorem 2.5 on the limit of a sum for limits from the right.

16. State and prove the analog of Theorem 2.6 on the limit of a product for limits from the right.

17. State and prove the analog of Theorem 2.9 on the limit of inequalities for limits from the left.

18. State and prove the analog of Theorem 2.10 (the Sandwiching theorem) for limits from the right.

19. In Theorem 2.11 give an example which shows the result is false if the hypothesis on f is changed so that f is continuous on the left at L.

20. Given $f(x) = 1/x$ for $x = 1, 2, 3, \ldots, n, \ldots$. For what values of x is f continuous?

2.4. Limits at Infinity; Infinite Limits

Consider the function

$$f: x \to \frac{x}{x+1}, \qquad x \neq -1,$$

whose graph is shown in Figure 2.5. Intuitively, it is clear that $f(x)$ tends to 1 as x tends to infinity; this statement holds when x goes to infinity through increasing positive values (i.e., x tends to infinity to the right) or when x goes to infinity through decreasing negative values (x tends to infinity to the left). More precisely, the symbol

$$x \to \infty$$

means that x increases without bound through positive values and x decreases without bound through negative values. If x tends to infinity only through increasing positive values, we write

$$x \to +\infty,$$

while if x tends to infinity only through decreasing negative values, we write

$$x \to -\infty.$$

The above conventions are used to define a *limit at infinity*.

Definitions. We say that $f(x) \to L$ **as** $x \to \infty$ if and only if for each $\varepsilon > 0$ there is a number $A > 0$ such that $|f(x) - L| < \varepsilon$ for all x satisfying $|x| > A$.

We say that $f(x) \to L$ **as** $x \to +\infty$ if and only if for each $\varepsilon > 0$ there is an $A > 0$ such that $|f(x) - L| < \varepsilon$ for all x satisfying $x > A$.

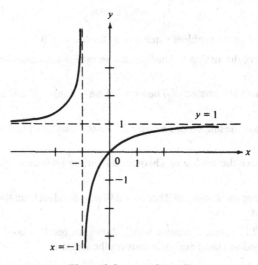

Figure 2.5. $y = x/(x+1)$.

We say that $f(x) \to L$ as $x \to -\infty$ if and only if for each $\varepsilon > 0$, there is an $A > 0$ such that $|f(x) - L| < \varepsilon$ for all x satisfying $x < -A$.

In each of the above definitions we of course suppose that x is in the domain of f.

Many of the theorems on limits in Section 2.2 have direct analogs for limits when x tends to ∞, $+\infty$, and $-\infty$. The statements and proofs merely require that $x \to a$ be replaced by $x \to \infty$, $x \to +\infty$, or $x \to -\infty$. Two exceptions to this direct analogy are Theorem 2.3 (Obvious limit) and Theorem 2.7 (Limit of a composite function). For limits at infinity, a theorem on obvious limits takes the following form.

Theorem 2.14 (Obvious limit). *If $f(x) = 1/x$ for all $x \neq 0$, then*

$$\lim_{x \to \infty} f(x) = 0, \qquad \lim_{x \to +\infty} f(x) = 0, \qquad \lim_{x \to -\infty} f(x) = 0.$$

The next theorem replaces Theorem 2.7 on composite functions.

Theorem 2.15 (Limit of a composite function). *Suppose that f and g are functions on $\mathbb{R}^1$ to $\mathbb{R}^1$. If f is continuous at L and $g(x) \to L$ as $x \to +\infty$, then*

$$\lim_{x \to +\infty} f[g(x)] = f(L).$$

Remarks. (i) Theorem 2.15 may also be stated with $x \to -\infty$ or $x \to \infty$. Similarly, every theorem in Section 2.2 has three analogs according as $x \to \infty$, $x \to +\infty$, or $x \to -\infty$. (ii) The proofs of Theorems 2.14 and 2.15 follow the pattern of the proofs of Theorems 2.3 and 2.7, respectively.

Referring to Figure 2.5 and the function $f: x \to x/(x + 1)$, $x \neq -1$, we see that f increases without bound as x tends to -1 from the left. Also, f decreases without bound as x tends to -1 from the right. We say that f has an *infinite limit* as x tends to -1; a more precise statement is given in the next definition.

Definitions. A function f **becomes infinite** as $x \to a$ if and only if for each number $A > 0$ there is a number $\delta > 0$ such that $|f(x)| > A$ for all x satisfying $0 < |x - a| < \delta$. We write $f(x) \to \infty$ as $x \to a$ for this limit. We also use the symbols

$$\lim_{x \to a} f(x) = \infty$$

and $\lim_{x \to a} f(x) = \infty$ although we must remember that *infinity is not a number and the usual rules of algebra do not apply.*

If f is defined only on one side of a number a rather than in a deleted interval containing a, we may define a one-sided infinite limit. A function f **becomes infinite** as $x \to a^+$ if and only if for each number $A > 0$ there is a number $\delta > 0$ such that $|f(x)| > A$ for all x satisfying $0 < x - a < \delta$. An *infinite limit from the left* is defined similarly. Various other possibilities may occur. For example, the function $f: x \to 1/(x - 1)^2$ has the property that $f \to +\infty$ as $x \to 1$. In such a case f has a *positive infinite limit as $x \to 1$*. Similarly, a function may have a *negative infinite limit*, and it may have both positive and negative one-sided infinite limits. Symbols for these limits are

$$\lim_{x \to a} f(x) = +\infty, \qquad \lim_{x \to a} f(x) = -\infty, \qquad \lim_{x \to a^+} f(x) = +\infty,$$

$$\lim_{x \to a^+} f(x) = -\infty, \qquad \lim_{x \to a^-} f(x) = +\infty, \qquad \lim_{x \to a^-} f(x) = -\infty.$$

The theorems on limits in Section 2.2 have analogs for infinite limits which in most cases are the same as for finite limits. However, there are differences and some of these are exhibited in the following theorems.

Theorem 2.16 (Uniqueness of limits). *If a function f on $\mathbb{R}^1$ to $\mathbb{R}^1$, becomes infinite as $x \to a$, then f does not tend to a finite limit as $x \to a$.*

PROOF. Suppose that $f \to \infty$ and $f \to L$, a finite number, as $x \to a$. We shall reach a contradiction. Since $f(x) \to L$, the definition of limit with $\varepsilon = 1$ asserts that there is a $\delta > 0$ such that $f(x)$ is defined for $0 < |x - a| < \delta$ and

$$|f(x) - L| < 1 \quad \text{for} \quad 0 < |x - a| < \delta.$$

Therefore $-1 + L < f(x) < 1 + L$, which implies that $|f(x)| < 1 + |L|$ for $0 < |x - a| < \delta$. Now, since $f(x) \to \infty$ as $x \to a$, it follows that for any number $A > 1 + |L|$ there is a δ_1 such that $f(x)$ is defined for $0 < |x - a| < \delta_1$ and

$$|f(x)| > A \quad \text{for} \quad 0 < |x - a| < \delta_1.$$

Thus for any value $\bar{x}(\neq a)$ closer to a than both δ and δ_1 we have the impossible situation: $1 + |L| < A < |f(\bar{x})| < 1 + |L|$. $\qquad\qquad\qquad\qquad\qquad\square$

    Remark. The statement and proof of Theorem 2.16 with $+\infty$ or $-\infty$ instead of ∞ and with one-sided limits instead of the two-sided limit require only simple modifications.

    The analog of Theorems 2.7 and 2.15 on the limit of a composite function takes a somewhat different form for infinite limits.

Theorem 2.17 (Limit of a composite function). *Suppose that*

$$\lim_{x \to \infty} f(x) = L \quad and \quad \lim_{x \to a} g(x) = \infty.$$

Then

$$\lim_{x \to a} f[g(x)] = L.$$

The proof is left to the reader.

    Remarks. There are many variants to Theorem 2.17 in which ∞ may be replaced by $+\infty$ or $-\infty$. Also, one-sided limits may be used so that $x \to a^+$ or a^-. Additional theorems may be stated with L replaced by ∞, $+\infty$, or $-\infty$.

In all these theorems care must be used to ensure that the composite function $f[g]$ is actually defined on an appropriate domain. The following variant of Theorem 2.17 illustrates this point.

Theorem 2.17(a). *Suppose that*

$$\lim_{x \to a} f(x) = \infty \quad and \quad \lim_{x \to -\infty} g(x) = a.$$

Suppose that for some positive number M, we have

$$g(x) \neq a \quad for \quad x < -M.$$

Then

$$\lim_{x \to -\infty} f[g(x)] = \infty.$$

Remarks. (i) If $g(x) = a$, then $f[g(x)]$ may not be defined since $f(a)$ may not exist. Hence we require that as x "nears" $-\infty$, the range of g shall eventually exclude the number a. (ii) An example of functions f and g for which Theorem 2.17(a) does not apply because the hypotheses on f and g are not met is given by

$$f(x) = \frac{1}{x}, \qquad g(x) = \frac{1}{x} \sin x, \qquad x \neq 0.$$

Then

$$f[g(x)] = \frac{x}{\sin x}, \qquad \lim_{x \to 0} f(x) = \infty, \qquad \lim_{x \to -\infty} g(x) = 0.$$

The function $f[g]$ is not defined for $x = -n\pi, n = 1, 2, \ldots$. Consequently, we cannot say that $|f[g(x)]|$ is large for *all* x less than some given number $-M$.

EXAMPLE 1. Given the function

$$f: x \to \frac{1}{x} \cos \frac{1}{x},$$

decide whether or not f tends to a limit as x tends to 0.

Solution. For the values $x_n = 1/2n\pi, n = 1, 2, \ldots$, we have $f(x_n) = 2n\pi$, while for $x_n' = 1/((\pi/2) + 2n\pi)$, we have $f(x_n') = 0$. Hence there are certain values, x_n, tending to zero as $n \to \infty$ at which f grows without bound, while on other values, x_n', also tending to zero, at which the function f always has the value zero. Therefore (see Theorem 2.20 below) f has no limit as $x \to 0$. See Figure 2.6. $\square$

EXAMPLE 2. Given the function

$$f: x \to \frac{\sqrt{x^2 + 2x + 4}}{2x + 3},$$

evaluate $\lim_{x \to +\infty} f(x)$ and $\lim_{x \to -\infty} f(x)$. Give a reason for each step.

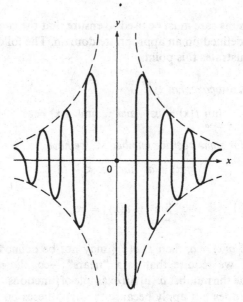

Figure 2.6. $y = (1/x) \cos (1/x)$.

Solution. For $x > 0$, division of numerator and denominator by x yields

$$f(x) = \frac{\sqrt{x^2 + 2x + 4}}{2x + 3} = \frac{\sqrt{1 + 2\left(\dfrac{1}{x}\right) + 4\left(\dfrac{1}{x^2}\right)}}{2 + 3\left(\dfrac{1}{x}\right)}.$$

We now employ the theorems on limit of a constant, obvious limit, and the limits of a sum and product to obtain

$$\lim_{x \to +\infty} \left[1 + 2 \cdot \frac{1}{x} + 4\left(\frac{1}{x^2}\right) \right] = \lim_{x \to +\infty} 1 + \lim_{x \to +\infty} 2 \cdot \lim_{x \to +\infty} \frac{1}{x}$$

$$+ \lim_{x \to +\infty} 4 \cdot \lim_{x \to +\infty} \left(\frac{1}{x^2}\right)$$

$$= 1 + 2 \cdot 0 + 4 \cdot 0$$

$$= 1;$$

also

$$\lim_{x \to +\infty} \left(2 + 3\left(\frac{1}{x}\right) \right) = 2.$$

Now, using the theorem on the limit of composite functions, we find

$$\lim_{x \to +\infty} \sqrt{1 + 2\left(\frac{1}{x}\right) + 4\left(\frac{1}{x^2}\right)} = \sqrt{1} = 1.$$

Therefore, $\lim_{x \to +\infty} f(x) = \frac{1}{2}$ by the theorem on the limit of a quotient. For $x < 0$, care must be exercised in rewriting the algebraic expression for f. Division by x in the numerator and denominator yields

$$f(x) = \frac{\sqrt{x^2 + 2x + 4}}{2x + 3} = \frac{-\sqrt{\dfrac{1}{x^2}}\,\sqrt{x^2 + 2x + 4}}{\dfrac{1}{x}\quad\; 2x + 3}$$

$$= -\frac{\sqrt{1 + 2\left(\dfrac{1}{x}\right) + 4\left(\dfrac{1}{x^2}\right)}}{2 + 3\left(\dfrac{1}{x}\right)}.$$

Now, proceeding as in the case for positive values of x, we obtain

$$\lim_{x \to -\infty} f(x) = -\frac{1}{2}. \qquad \square$$

The theorems on the limit of sums, products, and quotients of functions require special treatment when one or more of the functions has an infinite limit or when the limit of a function appearing in a denominator is zero. For example, the following theorem on the limit of a quotient shows the special hypotheses which are needed.

Theorem 2.18. *Suppose that*

$$\lim_{x \to a} f(x) = L, \qquad L \neq 0$$

and

$$\lim_{x \to a} g(x) = 0.$$

If there is a number δ such that $g(x) \neq 0$ for $0 < |x - a| < \delta$, then

$$\lim_{x \to a} \frac{f(x)}{g(x)} = \infty.$$

The proof is left to the reader.

Remarks. (i) The hypothesis $g(x) \neq 0$ in a neighborhood of a is needed in order that f/g be defined for $0 < |x - a| < \delta$. (ii) Variations of Theorem 2.18 are many. The number L may be replaced by ∞ without changing the conclusion. One-sided limits may be considered. If for example, $f \to +\infty$ and $g \to 0^+$, then we can conclude that $f/g \to +\infty$.

In the case of the limit of the sum of two functions, the appropriate theorem states that if

$$\lim_{x \to a} f(x) = L, \qquad \lim_{x \to a} g(x) = \infty,$$

then

$$\lim_{x \to a} [f(x) + g(x)] = \infty.$$

If both f and g tend to infinity as x tends to a, no conclusion can be drawn without a more detailed examination of the functions f and g. We must bear in mind that ∞ cannot be treated as a number. However, we do have the "rules"

$$+\infty + (+\infty) = +\infty \quad \text{and} \quad -\infty + (-\infty) = -\infty.$$

In the case of the limit of a product, if $f(x) \to L$, $L \neq 0$ and $g(x) \to \infty$, as $x \to a$, then $f(x)g(x) \to \infty$ as $x \to a$. However, if $L = 0$, no conclusion can be drawn without a closer investigation of the functions f and g.

PROBLEMS

In each of Problems 1 through 10 evaluate the limit or conclude that the function tends to ∞, $+\infty$ or $-\infty$.

1. $\lim_{x \to \infty} (x^2 - 2x + 3)/(x^3 + 4)$

2. $\lim_{x \to \infty} (2x^2 + 3x + 4)/(x^2 - 2x + 3)$

3. $\lim_{x \to \infty} (x^4 - 2x^2 + 6)/(x^2 + 7)$

4. $\lim_{x \to +\infty} (x - \sqrt{x^2 - a^2})$

5. $\lim_{x \to -\infty} (x - \sqrt{x^2 - a^2})$

6. $\lim_{x \to 1^+} (x - 1)/\sqrt{x^2 - 1}$

7. $\lim_{x \to 1^-} \sqrt{1 - x^2}/(1 - x)$

8. $\lim_{x \to +\infty} (x^2 + 1)/x^{3/2}$

9. $\lim_{x \to 2^-} \sqrt{4 - x^2}/\sqrt{6 - 5x + x^2}$

10. $\lim_{x \to +\infty} (\sqrt{x^2 + 2x} - x)$

11. Prove Theorem 2.14.

12. Prove Theorem 2.15.

13. State and prove an analog to Theorem 2.15 with $x \to -\infty$ instead of $x \to +\infty$.

14. State and prove an analog to Theorem 2.16 for one-sided limits from the right (i.e., $x \to a^+$ instead of $x \to a$).

15. Prove Theorem 2.17.

16. Prove Theorem 2.17(a).

17. Suppose $f(x) \to +\infty$ and $g(x) \to -\infty$ as $x \to +\infty$. Find examples of functions f and g with these properties and such that
    (a) $\lim_{x \to +\infty} [f(x) + g(x)] = +\infty$ (b) $\lim_{x \to +\infty} [f(x) + g(x)] = -\infty$
    (c) $\lim_{x \to +\infty} [f(x) + g(x)] = A$, A an arbitrary real number

18. Find the values of p, if any, for which the following limits exist.
    (a) $\lim_{x \to 0^+} x^p \sin(1/x)$ (b) $\lim_{x \to +\infty} x^p \sin(1/x)$
    (c) $\lim_{x \to -\infty} |x|^p \sin(1/x)$

19. State and prove an analog of Theorem 2.9 for the case when $x \to +\infty$ (instead of $x \to a$).

20. State and prove an analog of Theorem 2.10 (Sandwiching theorem) for the case when $x \to -\infty$ (instead of $x \to a$).

21. Prove Theorem 2.18.

2.5. Limits of Sequences

An infinite sequence may or may not tend to a limit. When an infinite sequence does tend to a limit, the rules of operation are similar to those given in Section 2.2 for limits of functions.

Definition. Given the infinite sequence of numbers $x_1, x_2, \ldots, x_n, \ldots$, we say that $\{x_n\}$ **tends to L as n tends to infinity** if and only if for each $\varepsilon > 0$ there is a natural number N such that $|x_n - L| < \varepsilon$ for all $n > N$. We also write $x_n \to L$ as $n \to \infty$ and[4] we say that the sequence $\{x_n\}$ has L as a limit. The notation

$$\lim_{n \to \infty} x_n = L$$

or $\lim_{n \to \infty} x_n = L$ will be used. If the sequence $\{x_n\}$ increases without bound as n tends to infinity, the symbol $x_n \to +\infty$ as $n \to \infty$ is used; similarly if it decreases without bound, we write $x_n \to -\infty$ as $n \to \infty$.

For many problems it is important to study the behavior of a function f on a specific sequence of numbers in the domain of f. For this purpose, almost all the theorems on limits in Section 2.2 may be used if they are interpreted appropriately. For example, the theorem on the limit of the sum of functions may be stated as follows:

Theorem 2.19 (Limit of a sum). *Suppose that $x_n \to a$ and $y_n \to b$ as $n \to \infty$. Then $x_n + y_n \to a + b$ as $n \to \infty$.*

The theorems on the limit of a constant, limit of equal functions, limit of a product, limit of a quotient, limit of inequalities, and Sandwiching theorem have corresponding statements for sequences.

A variation of the theorem on composite functions, Theorem 2.7, leads to the next result which we state without proof.

Theorem 2.20. *Suppose that f is continuous at a and that $x_n \to a$ as $n \to \infty$. Then there is an integer N such that $f(x_n)$ is defined for all integers $n > N$; furthermore, $f(x_n) \to f(a)$ as $n \to \infty$.*

EXAMPLE. Given the function

$$f : x \to \begin{cases} \dfrac{1}{x} \sin \dfrac{1}{x}, & x \neq 0, \\ 0, & x = 0, \end{cases}$$

---

[4] Strictly speaking, we should write $n \to +\infty$ instead of $n \to \infty$. However, for the natural numbers it is a long-established custom to write ∞ instead of $+\infty$; we shall continue to observe this custom.

and the sequences

$$x_n = 1/n\pi, \qquad\qquad n = 1, 2, \ldots,$$
$$y_n = 1/(2n + \tfrac{1}{2})\pi, \quad n = 1, 2, \ldots,$$
$$z_n = 1/(2n + \tfrac{3}{2})\pi, \quad n = 1, 2, \ldots,$$

find the limits of $f(x_n)$, $f(y_n)$, and $f(z_n)$ as $n \to \infty$.

Solution. Since $\sin n\pi = 0$, $n = 1, 2, \ldots$, we have $f(x_n) \to 0$ as $n \to \infty$. Also,

$$f(y_n) = (2n + \tfrac{1}{2})\pi \sin(2n + \tfrac{1}{2})\pi = (2n + \tfrac{1}{2})\pi \to +\infty \quad \text{as} \quad n \to \infty,$$

and

$$f(z_n) = (2n + \tfrac{3}{2})\pi \sin(2n + \tfrac{3}{2})\pi = -(2n + \tfrac{3}{2})\pi \to -\infty \quad \text{as} \quad n \to \infty.$$

Observing that x_n, y_n, and z_n all tend to 0 as $n \to \infty$, we note that $f(x)$ cannot tend to a limit as $x \to 0$. ☐

The axioms for a field given in Section 1.1 do not describe the real number system completely. Even when an order is introduced such as with Axiom I (Section 1.3) the real numbers are not uniquely described. For example, the collection of rational numbers satisfies both the axioms for a field and the order axiom. However, there is another difficulty. If x_n is a sequence of numbers with the property that $x_n > n$ for every natural number n, then we would expect that $x_n \to +\infty$ as $n \to \infty$. It is actually the case that there are ordered fields (i.e., fields which satisfy Axiom I) which do not have the property just stated. Such fields are called *non-Archimedean.*[5]

The next axiom, when added to those of Chapter 1, serves to describe the real number system uniquely.[6]

Axiom C (Axiom of continuity). *Suppose that an infinite sequence $x_1, x_2, \ldots,$ $x_n, \ldots$ is such that $x_{n+1} \geqslant x_n$ for all n, and there is a number M such that $x_n \leqslant M$ for all n. Then there is a number $L \leqslant M$ such that $x_n \to L$ as $n \to \infty$ and $x_n \leqslant L$ for all n.*

The situation is shown in Figure 2.7. If we plot the numbers x_n on a horizontal line, the corresponding points move steadily to the right as n

---

[5] See v. d. Waerden, *Modern Algebra*, Ungar Publishing Co., New York, page 211, for an example of a non-Archimedean ordered field. See also Problems 8, 9, and 10 at the end of this section for such an example.

[6] In the following sense: the general overriding assumption is that there is a particular number system, called the *real numbers* and denoted by $\mathbb{R}^1$. This system satisfies all the above axioms, including Axiom C. If X is any number system satisfying all the same axioms, it can be shown that there is a one-to-one correspondence between the elements of X and those of $\mathbb{R}^1$. Furthermore, the zero and unit elements correspond and the correspondence preserves sums and products. Finally, if $x_1, x_2, \ldots, x_n, \ldots$ and $r_1, r_2, \ldots, r_n, \ldots$ are corresponding elements of X and $\mathbb{R}^1$, if both are nondecreasing bounded sequences, and if $x_n \to L, r_n \to L'$, then L and L' correspond.

Figure 2.7. x_n tends to L as $n \to \infty$.

increases. It is geometrically evident that they must cluster at some value such as L which cannot exceed the number M.

Remark. If a sequence $\{y_n\}$ has the properties: $y_{n+1} \leqslant y_n$ for all n and $y_n \geqslant M$ for all n, we can conclude that there is a number $L \geqslant M$ such that $y_n \to L$ as $n \to \infty$ and $y_n \geqslant L$ for all n. To see this define $x_n = -y_n$, observe that $x_n \leqslant x_{n+1}$, $x_n \leqslant -M$ for all n, and apply Axiom C to the sequence $\{x_n\}$.

By means of Axiom C we can establish the following simple result.

Theorem 2.21

(a) *There is no real number M which exceeds all positive integers.*
(b) *If $x_n \geqslant n$ for every positive integer n, then $x_n \to +\infty$ as $n \to \infty$.*
(c) $\lim_{n \to \infty} (1/n) = 0$.

PROOF

(a) Let $x_n = n$ and suppose there is a number M such that $M \geqslant x_n$ for every n. By Axiom C there is a number L such that $x_n \to L$ as $n \to \infty$ and $x_n \leqslant L$ for all n. Choose $\varepsilon = 1$ in the definition of limit. Then for a sufficiently large integer N, it follows that $L - 1 < x_n < L + 1$ for all $n > N$. But $x_{N+1} = N + 1$ and so $L - 1 < N + 1 < L + 1$. Hence $N + 2 = x_{N+2} > L$ contrary to the statement $x_n \leqslant L$ for all n.

(b) Let M be any positive number. From Part (a) there is an integer N such that $N > M$; hence $x_n > M$ for all $n \geqslant N$.

(c) Let $\varepsilon > 0$ be given. There is an integer N such that $N > 1/\varepsilon$. Then if $n > N$, clearly $1/n < \varepsilon$. The result follows. □

PROBLEMS

1. Given the function $f: x \to x \cos x$.
   (a) Find a sequence of numbers $\{x_n\}$ such that $x_n \to +\infty$ and $f(x_n) \to 0$.
   (b) Find a sequence of numbers $\{y_n\}$ such that $y_n \to +\infty$ and $f(y_n) \to +\infty$.
   (c) Find a sequence of numbers $\{z_n\}$ such that $z_n \to +\infty$ and $f(z_n) \to -\infty$.

2. Prove Theorem 2.19.

3. If a is any number greater than 1 and n is a positive integer greater than 1, show that $a^n > 1 + n(a - 1)$.

4. If $a > 1$, show that $a^n \to +\infty$ as $n \to \infty$.

5. Prove Theorem 2.20.

6. If $-1 < a < 1$, show that $a^n \to 0$ as $n \to \infty$.

7. Suppose that $-1 < a < 1$. Define

$$s_n = b(1 + a + a^2 + \cdots + a^{n-1}).$$

Show that $\lim_{n \to \infty} s_n = b/(1 - a)$.

8. Consider all *rational expressions* of the form

$$\frac{a_0 x^m + a_1 x^{m-1} + \cdots + a_{m-1}x + a_m}{b_0 x^n + b_1 x^{n-1} + \cdots + b_{n-1}x + b_n}$$

in which m and n are nonnegative integers. We choose $b_0 = 1$ always, and assume that such fractions are reduced to their lowest terms. Show that by selecting 0/1 as the "zero" and 1/1 as the "unit," and by using the ordinary rules of algebra for addition, subtraction, multiplication, and division, the totality of such rational expressions satisfy the axioms for a field.

9. We say a rational expression of the type described in Problem 8 is *positive* whenever $a_0 > 0$. Show that with this definition, the rational expressions in Problem 8 satisfy Axiom I of Chapter 1.

10. Using the definition of order given in Problem 9, show that there is a rational expression which is larger than every positive integer $n/1$. Hence conclude that the ordered field of rational expressions is non-Archimedean.

In Problems 11 through 15 evaluate each of the limits or conclude that the given expression tends to ∞, $+\infty$, or $-\infty$.

11. $\lim_{n \to \infty} (n^2 + 2n - 1)/(n^2 - 3n + 2)$

12. $\lim_{n \to \infty} (n^3 + 4n - 1)/(2n^2 + n + 5)$

13. $\lim_{n \to \infty} (n + (1/n))/(2n^2 - 3n)$

14. $\lim_{n \to \infty} \sqrt{n^3 + 2n - 1}/\sqrt[3]{n^2 + 4n - 2}$

15. $\lim_{n \to \infty} [3 + \sin(n)]n$

16. Suppose that $x_n \leqslant y_n \leqslant z_n$ for each n and suppose that $x_n \to L$ and $z_n \to L$ as $n \to \infty$. Then $y_n \to L$ as $n \to \infty$.

CHAPTER 3

Basic Properties of Functions on $\mathbb{R}^1$

3.1. The Intermediate-Value Theorem

The proofs of many theorems of calculus require a knowledge of the basic properties of continuous functions. In this section we establish the Intermediate-value theorem, an essential tool used in the proofs of the Fundamental theorem of calculus and the Mean-value theorem.

Theorem 3.1 (Nested intervals theorem). *Suppose that*

$$I_n = \{x: a_n \leqslant x \leqslant b_n\}, \qquad n = 1, 2, \ldots,$$

is a sequence of closed intervals such that $I_{n+1} \subset I_n$ for each n. If $\lim_{n \to \infty} (b_n - a_n) = 0$, then there is one and only one number x_0 which is in every I_n.

PROOF. By hypothesis, we have $a_n \leqslant a_{n+1}$ and $b_{n+1} \leqslant b_n$. Since $a_n < b_n$ for every n, the sequence $\{a_n\}$ is nondecreasing and bounded from above by b_1 (see Figure 3.1). Similarly, the sequence b_n is nonincreasing and bounded from below by a_1. Using Axiom C, we conclude that there are numbers x_0 and x_0' such that $a_n \to x_0, a_n \leqslant x_0$, and $b_n \to x_0', b_n \geqslant x_0'$, as $n \to \infty$. Using the fact that $b_n - a_n \to 0$ as $n \to \infty$, we find that $x_0 = x_0'$ and

$$a_n \leqslant x_0 \leqslant b_n$$

for every n. Thus x_0 is in every I_n.

To show that x_0 is the *only* number in every I_n, suppose there is another number x_1 also in every I_n. Define $\varepsilon = |x_1 - x_0|$. Now, since $a_n \to x_0$ and $b_n \to x_0$, there is an integer N_1 such that $a_{N_1} > x_0 - \varepsilon$; also there is an integer N_2 such that $b_{N_2} < x_0 + \varepsilon$. These inequalities imply that I_n cannot contain x_1 for any n beyond N_1 and N_2. Thus x_1 is not in every I_n. $\qquad\square$

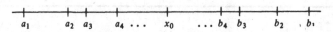

Figure 3.1. Nested intervals.

Remarks. The hypothesis that each I_n is *closed* is essential. The sequence of half-open intervals

$$I_n = \left\{ x \colon 0 < x \leqslant \frac{1}{n} \right\}$$

has the property that $I_{n+1} \subset I_n$ for every n. Since no I_n contains 0, any number x_0 in all the I_n must be positive. But then by choosing $N > 1/x_0$, we see that x_0 cannot belong to $I_N = \{ x \colon 0 < x \leqslant 1/N \}$. Although the intervals I_n are nested, they have no point in common.

Before proving the main theorem, we prove the following result which we shall use often in this chapter:

Theorem 3.2. *Suppose f is continuous on $[a, b]$, $x_n \in [a, b]$ for each n, and $x_n \to x_0$. Then $x_0 \in [a, b]$ and $f(x_n) \to f(x_0)$.*

PROOF. Since $x_n \geqslant a$ for each n, it follows from the theorem on the limit of inequalities for sequences that $x_0 \geqslant a$. Similarly, $x_0 \leqslant b$ so $x_0 \in [a, b]$. If $a < x_0 < b$, then $f(x_n) \to f(x_0)$ on account of the composite function theorem (Theorem 2.7). The same conclusion holds if $x_0 = a$ or b. A detailed proof of the last sentence is left to the reader. $\square$

We now establish the main theorem of this section.

Theorem 3.3 (Intermediate-value theorem). *Suppose f is continuous on $[a, b]$, $c \in \mathbb{R}^1$, $f(a) < c$, and $f(b) > c$. Then there is at least one number x_0 on $[a, b]$ such that $f(x_0) = c$.*

Remark. There may be more than one as indicated in Figure 3.2.

PROOF. Define $a_1 = a$ and $b_1 = b$. Then observe that $f((a_1 + b_1)/2)$ is either equal to c, greater than c or less than c. If it equals c, choose $x_0 = (a_1 + b_1)/2$ and the result is proved. If $f((a_1 + b_1)/2) > c$, then define $a_2 = a_1$ and $b_2 = (a_1 + b_1)/2$. If $f((a_1 + b_1)/2) < c$, then define $a_2 = ((a_1 + b_1)/2)$ and $b_2 = b_1$.

In each of the two last cases we have $f(a_2) < c$ and $f(b_2) > c$. Again compute $f((a_2 + b_2)/2)$. If this value equals c, the result is proved. If $f((a_2 + b_2)/2) > c$ set $a_3 = a_2$ and $b_3 = (a_2 + b_2)/2$. If $f((a_2 + b_2)/2) < c$, set $a_3 = (a_2 + b_2)/2$ and $b_3 = b_2$.

Continuing in this way, we either find a solution in a finite number of steps or we find a sequence $\{ [a_n, b_n] \}$ of closed intervals each of which is one of the

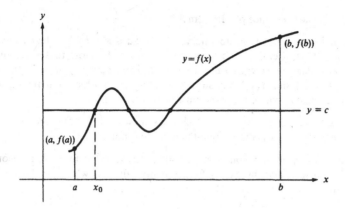

Figure 3.2. Illustrating the Intermediate-value theorem.

two halves of the preceding one, and for which we have

$$b_n - a_n = (b_1 - a_1)/2^{n-1},$$

$$f(a_n) < c, \qquad f(b_n) > c \qquad \text{for each} \qquad n. \tag{3.1}$$

From the Nested intervals theorem it follows that there is a unique point x_0 in all these intervals and

$$\lim_{n \to \infty} a_n = x_0 \qquad \text{and} \qquad \lim_{n \to \infty} b_n = x_0.$$

From Theorem 3.2, we conclude that

$$f(a_n) \to f(x_0) \qquad \text{and} \qquad f(b_n) \to f(x_0).$$

From Inequalities (3.1) and the limit of inequalities it follows that $f(x_0) \leqslant c$ and $f(x_0) \geqslant c$ so that $f(x_0) = c$. □

PROBLEMS

*1. Given the function

$$f : x \to \begin{cases} 1 & \text{if } x \text{ is rational,} \\ 0 & \text{if } x \text{ is irrational.} \end{cases}$$

(a) Show that f is not continuous at any x_0.
(b) If g is a function with domain all of $\mathbb{R}^1$, if $g(x) = 1$ if x is rational, and if g is continuous for all x, show that $g(x) \equiv 1$ for $x \in \mathbb{R}^1$.

2. Let $f : x \to a_m x^m + a_{m-1} x^{m-1} + \cdots + a_1 x + a_0$ be a polynomial of odd degree.
(a) Use the Intermediate-value theorem to show that the equation $f(x) = 0$ has at least one root.
(b) Show that the range of f is $\mathbb{R}^1$.

3. Let $f : x \to a_m x^m + a_{m-1} x^{m-1} + \cdots + a_1 x + a_0$ be a polynomial of even degree. If $a_m a_0 < 0$ show that the equation $f(x) = 0$ has at least two real roots.

4. Prove the last sentence of Theorem 3.2.

*5. A function f defined on an interval $I = \{x: a \leqslant x \leqslant b\}$ is called *increasing* $\Leftrightarrow$ $f(x_1) > f(x_2)$ whenever $x_1 > x_2$ where $x_1, x_2 \in I$. Suppose that f has the *intermediate-value property*: that is, for each number c between $f(a)$ and $f(b)$ there is an $x_0 \in I$ such that $f(x_0) = c$. Show that a function f which is increasing and has the intermediate-value property must be continuous.

6. Give an example of a function which is continuous on $[0, 1]$ except at $x = \frac{1}{2}$ and for which the Intermediate value-theorem does not hold.

7. Give an example of a nonconstant continuous function for which Theorem 3.3 holds and such that there are infinitely many values x_0 where $f(x_0) = c$.

3.2. Least Upper Bound; Greatest Lower Bound

In this section we prove an important principle about real numbers which is often taken as an axiom in place of Axiom C. We shall show, in fact, that we can prove Axiom C using this principle (and the other axioms of real numbers).

Definitions. A set S of real numbers has the **upper bound** M if and only if $x \leqslant M$ for every number x in S; we also say that the set S is *bounded above* by M. The set S has the **lower bound** m if and only if $x \geqslant m$ for every number x in S; we also say that S is *bounded below* by m. A set S is **bounded** if and only if S has an upper and a lower bound. Suppose that f is a function on $\mathbb{R}^1$ whose domain D contains the set S. We denote by $f|_S$ the **restriction of f to the set S**; that is, $f|_S$ has domain S and $f|_S(x) = f(x)$ for all x in S. A function f is bounded above, bounded below, or bounded, if the set R consisting of the range of f satisfies the corresponding condition.

The following results follow immediately from the definitions.

Theorem 3.4. *A set S in $\mathbb{R}^1$ is bounded $\Leftrightarrow$ there is a number K such that $|x| \leqslant K$ for each x in S. A function f is bounded above by M on a set $S \Leftrightarrow f(x) \leqslant M$ for all x in S; f is bounded below by m on a set $S \Leftrightarrow f(x) \geqslant m$ for all x in S. Finally, f is bounded on $S \Leftrightarrow$ there is a number K such that $|f(x)| \leqslant K$ for all x in S.*

For example, if we define

$$f(x) = \begin{cases} \dfrac{1}{2x}, & 0 < x \leqslant 1, \\ 1, & x = 0, \end{cases}$$

then f is not bounded on $S = \{x: 0 \leqslant x \leqslant 1\}$. See Figure 3.3

We now prove the fundamental principle (see Figure 3.4).

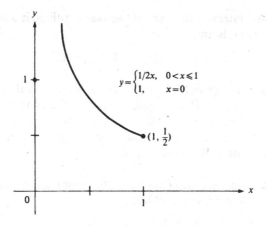

Figure 3.3

Theorem 3.5. *If a nonempty set* $S \subset \mathbb{R}^1$ *has an upper bound* M, *then it has a least upper bound* U; *that is, there is a unique number* U *which is an upper bound for* S *but is such that no number* $U' < U$ *is an upper bound for* S. *If a nonempty set has a lower bound* m, *it has a greatest lower bound* L.

PROOF. The second statement follows by applying the first to the set $S' = \{x: -x \in S\}$. We prove the first statement using the Nested intervals theorem. If $M \in S$, then we may take $U = M$ since in this case every x in S is less than or equal to U, and if $U' < U$, then U' is not an upper bound since $U \in S$ and $U > U'$. If M is not in S let $b_1 = M$ and choose a_1 as any point of S. Now either $(a_1 + b_1)/2$ is greater than every x in S or there is some x in S greater than or equal to $(a_1 + b_1)/2$. In the first case if we define $a_2 = a_1$ and $b_2 = (a_1 + b_1)/2$, then $a_2 \in S$ and b_2 is greater than every x in S. In the second case, choose for a_2 one of the numbers in S which is greater than or equal to $(a_1 + b_1)/2$ and set $b_2 = b_1$; then we again have $a_2 \in S$ and b_2 greater than every x in S. Continuing in this way, we define an infinite sequence $\{[a_n, b_n]\}$ of closed intervals such that

$$[a_{n+1}, b_{n+1}] \subset [a_n, b_n] \quad \text{and} \quad (b_{n+1} - a_{n+1}) \leqslant \tfrac{1}{2}(b_n - a_n) \quad (3.2)$$

for each n, and so for each n

$$b_n - a_n \leqslant (b_1 - a_1)/2^{n-1}, \quad a_n \in S, \quad b_n > \text{every number in } S. \quad (3.3)$$

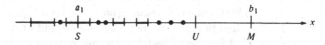

Figure 3.4. U is least upper bound for S.

From the Nested intervals theorem, it follows that there is a unique number U in all these intervals and

$$\lim_{n \to \infty} b_n = U, \qquad \lim_{n \to \infty} a_n = U.$$

Let x be any number in S. Then $x < b_n$ for each n, so that $x \leqslant U$ by the limit of inequalities. Thus U is an upper bound for S. But now, let $U' < U$ and let $\varepsilon = U - U'$. Then, since $a_n \to U$, it follows that there is an N such that $U' = U - \varepsilon < a_n \leqslant U$ for all $n > N$. But, since all the $a_n \in S$, it is clear that U' is not an upper bound. Therefore U is unique. $\square$

Definitions. The number U in Theorem 3.5, the **least upper bound**, is also called the **supremum** of S and is denoted by

$$\text{l.u.b. } S \qquad \text{or} \qquad \sup S.$$

The number L of Theorem 3.5, the **greatest lower bound**, is also called the **infimum** of S and is denoted by

$$\text{g.l.b. } S \qquad \text{or} \qquad \inf S.$$

If f is a function on $\mathbb{R}^1$ whose domain $D(f)$ contains S and whose range is denoted by $R(f)$, we define the *least upper bound* or *supremum* of f on S by

$$\text{l.u.b. } f \equiv \sup_S f = \sup R(f|_S).$$

A similar definition holds for the greatest lower bound.

Corollary. *If x_0 is the largest number in S, that is, if $x_0 \in S$ and x_0 is larger than every other number in S, then $x_0 = \sup S$. If S is not empty and $U = \sup S$, with U not in S, then there is a sequence $\{x_n\}$ such that each x_n is in S and $x_n \to U$. Also, if $\varepsilon > 0$ is given, there is an x in S such that $x > U - \varepsilon$. Corresponding results hold for $\inf S$.*

These results follow from the definitions and from the proof of Theorem 3.5.

With the help of this corollary it is possible to show that the Axiom of Continuity is a consequence of Theorem 3.5. That is, if Theorem 3.5 is taken as an axiom, then Axiom C becomes a theorem. See Problem 14 at the end of this section.

Definitions. Let f have an interval I of $\mathbb{R}^1$ as its domain and a set in $\mathbb{R}^1$ as its range. We say that f is **increasing on** I if and only if $f(x_2) > f(x_1)$ whenever $x_2 > x_1$. The function f is **nondecreasing on** I if and only if $f(x_2) \geqslant f(x_1)$ whenever $x_2 > x_1$. The function f is **decreasing on** I if and only if $f(x_2) < f(x_1)$ whenever $x_2 > x_1$. The function f is **nonincreasing on** I if and only if $f(x_2) \leqslant f(x_1)$ whenever $x_2 > x_1$. A function which has any one of these four properties is called **monotone on** I.

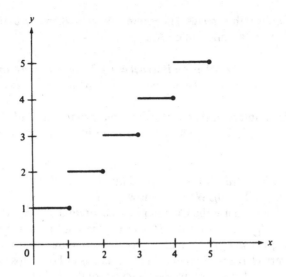

Figure 3.5. A step function.

Monotone functions are not necessarily continuous, as the following *step function* exhibits:

$$f: x \to n, \qquad n - 1 \leqslant x < n, \qquad n = 1, 2, \dots.$$

See Figure 3.5. Also, monotone functions may not be bounded. The function $f: x \to 1/(1 - x)$ is monotone on the interval $I = \{x: 0 \leqslant x < 1\}$, but is not bounded there.

The next two theorems on monotone functions follow from the definitions of this section. The proofs are left to the reader.

Theorem 3.6. *Suppose f is nondecreasing on an interval $I = \{x: a < x < b\}$ and $f(x) \leqslant M$ on I. Then there is a number $C \leqslant M$ such that*

$$\lim_{x \to b^-} f(x) = C.$$

With the help of Theorem 3.6, we can establish the next result by considering a variety of cases.

Theorem 3.7. *Suppose that f is a bounded monotone function on an interval I. Then f has a limit on the right and a limit on the left at each interior point* of I. Also, f has a one-sided limit at each endpoint of I. If I extends to infinity, then f tends to a limit as x tends to infinity in the appropriate direction.*

If a function f defined on an interval I is continuous on I, then an extension of the Intermediate-value theorem shows that the range of f is an interval. In

---

* If a and b are the endpoints of an interval, all the points between a and b are called *interior points*.

order to establish this result (Theorem 3.9 below), we use the following characterization of an interval on $\mathbb{R}^1$.

Theorem 3.8. *A set S in $\mathbb{R}^1$ is an interval $\Leftrightarrow$ (i) S contains more than one point and (ii) for every $x_1, x_2 \in S$ the number x is in S whenever $x \in (x_1, x_2)$.*

PROOF. If S is an interval, then clearly the two properties hold. Therefore, we suppose (i) and (ii) hold and show that S is an interval. We consider several cases.

Case 1: S is bounded. Define $a = \inf S$ and $b = \sup S$. Let x be an element of the open interval (a, b). We shall show $x \in S$.

Since $a = \inf S$, we use the Corollary to Theorem 3.5 to assert that there is an $x_1 \in S$ with $x_1 < x$. Also, since $b = \sup S$, there is an x_2 in S with $x_2 > x$. Hence $x \in (x_1, x_2)$ which by (ii) of the hypothesis shows that $x \in S$. We conclude that every element of (a, b) is in S. Thus S is either a closed interval, an open interval, or a half-open interval; its endpoints are a, b.

Case 2: S is unbounded in one direction. Assume S is unbounded above and bounded below, the case of S bounded above and unbounded below being similar. Let $a = \inf S$. Then $S \subset [a, \infty)$. Let x be any number such that $x > a$. We shall show that $x \in S$. As in Case 1, there is a number $x_1 \in S$ such that $x_1 < x$. Since S has no upper bound there is an $x_2 \in S$ such that $x_2 > x$. Using (ii) of the hypothesis, we conclude that $x \in S$. Therefore $S = [a, \infty)$ or $S = (a, \infty)$.

The proof when S is unbounded both above and below follows similar lines and is left to the reader.   □

We now establish a stronger form of the Intermediate-value theorem.

Theorem 3.9. *Suppose that the domain of f is an interval I and f is continuous on I. Assume that f is not constant on I. Then the range of f, denoted by J, is an interval.*

PROOF. We shall show that J has Properties (i) and (ii) of Theorem 3.8 and therefore is an interval. Since f is not constant, its range must have more than one point, and Property (i) is established. Now let $y_1, y_2 \in J$. Then there are numbers $x_1, x_2 \in I$ such that $f(x_1) = y_1$ and $f(x_2) = y_2$. We may assume that $x_1 < x_2$; the function f is continuous on $[x_1, x_2]$ and so we may apply the Intermediate-value theorem. If c is any number between y_1 and y_2, there is an $x_0 \in [x_1, x_2]$ such that $f(x_0) = c$. Thus $c \in J$, and we have established Property (ii). The set J is an interval.   □

Remarks. If f is continuous on $I = \{x: a \leqslant x \leqslant b\}$, it is not necessarily the case that J is determined by $f(a)$ and $f(b)$. Figure 3.6 shows that J may exceed the interval $[f(a), f(b)]$. In fact I may be bounded and J unbounded as is

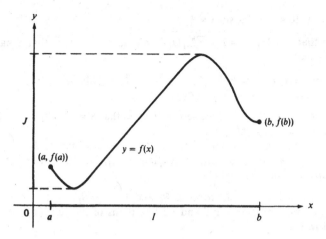

Figure 3.6

illustrated by the function $f: x \to 1/x$ with $I = \{x: 0 < x < 1\}$. The range is $J = \{y: 1 < y < \infty\}$. The function $f: x \to 1/(1 + x^2)$ with $I = \{x: 0 < x < \infty\}$ and $J = \{y: 0 < y < 1\}$ is an example of a continuous function with an unbounded domain and a bounded range.

Consider the restriction to $[0, \infty)$ of the function $f: x \to x^n$ where n is a positive integer. By Theorem 3.10, the range of f is an interval which must be $[0, \infty)$ since $f(0) = 0$, f is increasing, and $f(x) \to +\infty$ as $x \to +\infty$. Hence, for each $x \geq 0$, there is at least one number $y \geq 0$ such that $y^n = x$. If $y' > y$, then $(y')^n > y^n = x$. Also if $0 \leq y' < y$, then $(y')^n < x$. Thus the solution y is unique. We denote it by $x^{1/n}$. *Every positive number has an n-th root for every positive integer n.* The function $x \to x^{1/n}$, $x \geq 0$, can be shown to be continuous on $[0, \infty)$ by the methods of Chapter 2 (see also the Inverse function theorem, Theorem 4.19 below).

PROBLEMS

In Problems 1 through 8 find l.u.b. S and g.l.b. S. State whether or not these numbers are in S.

1. $S = \{x: 0 < x \leq 3\}$

2. $S = \{x: x^2 - 3 < 0\}$

3. $S = \{x: x^2 - 2x - 3 < 0\}$

4. $S = \{y: y = x/(x + 1), x \geq 0\}$

5. $S = \{s_n: s_n = \sum_{i=1}^{n} (1/2^i), n = 1, 2, \ldots\}$

6. $S = \{s_n: s_n = 1 + \sum_{i=1}^{n} ((-1)^i/i!), n = 1, 2, \ldots\}$

7. $S = \{x: 0 < x < 5, \cos x = 0\}$

8. $S = \{x: -10 < x < 10, \sin x = \frac{1}{2}\}$

9. Given that $S = \{s_n: s_n = 1 + \sum_{i=1}^{n} (1/i!), n = 1, 2, \ldots\}$, show that S has 3 as an upper bound.

10. Suppose that $B_1 = \text{l.u.b. } S_1$, $B_2 = \text{l.u.b. } S_2$, $b_1 = \text{g.l.b. } S_1$, and $b_2 = \text{g.l.b. } S_2$. If $S_1 \subset S_2$ show that $B_1 \leqslant B_2$ and $b_2 \leqslant b_1$.

11. Suppose that $S_1, S_2, \ldots, S_n$ are sets in $\mathbb{R}^1$; and that $S = S_1 \cup S_2 \cup \cdots \cup S_n$. Define $B_i = \sup S_i$ and $b_i = \inf S_i$, $i = 1, 2, \ldots, n$.
    (a) Show that $\sup S = \max(B_1, B_2, \ldots, B_n)$ and $\inf S = \min(b_1, b_2, \ldots, b_n)$.
    (b) If S is the union of an infinite collection of $\{S_i\}$, find the relation between $\inf S$, $\sup S$ and the $\{b_i\}$ and $\{B_i\}$.

12. Suppose that $S_1, S_2, \ldots, S_n$ are sets in $\mathbb{R}^1$ and that $S = S_1 \cap S_2 \cap \cdots \cap S_n$. If $S \neq \varnothing$, find a formula relating $\sup S$ and $\inf S$ in terms of the $\{b_i\}$, $\{B_i\}$ as defined in Problem 11.

13. Prove Theorem 3.4

14. Use the Corollary to Theorem 3.5 to show that Axiom C is a consequence of Theorem 3.5. [*Hint:* Let B be the least upper bound of the *set* of numbers $\{x_n\}$ which, as a *sequence*, are increasing. Then show that $x_n \to B$.]

15. Prove Theorem 3.6.

16. Prove Theorem 3.7.

17. Complete the proof of Theorem 3.8 by establishing the result for unbounded intervals.

18. Let S_1, S_2 be sets in $\mathbb{R}^1$. Define $S = \{x: x = x_1 + x_2, x_1 \in S_1, x_2 \in S_2\}$. Find l.u.b. S, g.l.b. S in terms of l.u.b. S_i, g.l.b. S_i, $i = 1, 2$. In particular, if $-S_1 = \{x: -x \in S_1\}$, show that l.u.b. $(-S_1) = \text{g.l.b. } S_1$.

19. Suppose that S is the union of a finite number of closed finite intervals $I_1, I_2, \ldots, I_n$, no two of which have more than one point in common. (They may be disjoint.) Suppose that f has domain S and is monotone on each I_i, $i = 1, 2, \ldots, n$. Prove the result of Theorem 3.7 for any such function f with domain S. Give two examples to show that the result is false if S is the union of an infinite collection of such closed intervals; one in case S is unbounded and another in case S is bounded.

3.3. The Bolzano–Weierstrass Theorem

Suppose that

$$x_1, x_2, \ldots, x_n, \ldots \tag{3.4}$$

is a sequence of numbers. Then the sequences $x_1, x_3, x_5, \ldots$ and $x_2, x_5, x_8, x_{11}, \ldots$ are examples of subsequences of (3.4). More generally, suppose that $k_1, k_2, k_3, \ldots k_n, \ldots$ is an *increasing* sequence of positive integers. Then we say that

$$x_{k_1}, x_{k_2}, \ldots, x_{k_n}, \ldots$$

is a **subsequence** of (3.4). The choice $k_1 = 1$, $k_2 = 3$, $k_3 = 5$, $k_4 = 7$, ... gives
the first example of a subsequence of (3.4) above, while $k_1 = 2$, $k_2 = 5$, $k_3 = 8$, ... gives the second subsequence of (3.4). To avoid double subscripts, which
are cumbersome, we will frequently write $y_1 = x_{k_1}$, $y_2 = x_{k_2}$, ..., $y_n = x_{k_n}$, ...,
in which case

$$y_1, y_2, \ldots, y_n, \ldots$$

is a subsequence of (3.4).

Remark. We easily prove by induction that if k_1, k_2, ..., k_n, ... is an
increasing sequence of positive integers, then $k_n \geqslant n$ for all n.

The sequence

$$x_1 = 0,\ x_2 = \frac{1}{2},\ x_3 = -\frac{2}{3}, \ldots, x_n = (-1)^n \left(1 - \frac{1}{n}\right), \ldots \tag{3.5}$$

has the subsequences

$$0, -\tfrac{2}{3}, -\tfrac{4}{5}, -\tfrac{6}{7}, \ldots$$

$$\tfrac{1}{2}, \tfrac{3}{4}, \tfrac{5}{6}, \tfrac{7}{8}, \ldots$$

which are obtained from (3.5) by taking the odd-numbered terms and the
even-numbered terms, respectively. Each of these subsequences is convergent
but the original sequence (3.5) is not. The notion of a convergent subsequence
of a given sequence occurs frequently in problems in analysis. The Bolzano–
Weierstrass theorem is basic in that it establishes the existence of such convergent subsequences under very simple hypotheses on the given sequence.
This theorem is a special case of a general result concerning sequences in
metric spaces which we study in Chapter 6.

Theorem 3.10 (Bolzano–Weierstrass theorem). *Any bounded infinite sequence
of real numbers contains a convergent subsequence.*

PROOF. We shall use the Nested intervals theorem (Theorem 3.1). Let $\{x_n\}$ be
a given bounded sequence. Then it is contained in some closed interval
$I = \{x : a \leqslant x \leqslant b\}$. Divide I into two equal subintervals by the midpoint
$(a + b)/2$. Then either the left subinterval contains an infinite number of the
$\{x_n\}$ or the right subinterval does (or both). Denote by $I_1 = \{x : a_1 \leqslant x \leqslant b_1\}$
the closed subinterval of I which contains infinitely many $\{x_n\}$. (If both
subintervals do, choose either one.) Next, divide I_1 into two equal parts by its
midpoint. Either the right subinterval or the left subinterval of I_1 contains
infinitely many $\{x_n\}$. Denote by I_2 the closed subinterval which does. Continue
this process, obtaining the sequence

$$I_n = \{x : a_n \leqslant x \leqslant b_n\}, \qquad n = 1, 2, \ldots,$$

with the property that each I_n contains x_p for infinitely many values of p. Since
$b_n - a_n = (b - a)/2^n \to 0$ as $n \to \infty$, we may apply the Nested intervals theorem
to obtain a unique number x_0 contained in every I_n.

We now construct a subsequence of $\{x_p\}$ converging to x_0. Choose x_{k_1} to be any member of $\{x_n\}$ in I_1 and denote x_{k_1} by y_1. Next choose x_{k_2} to be any member of $\{x_p\}$ such that x_{k_2} is in I_2 and such that $k_2 > k_1$. We can do this because I_2 has infinitely many of the $\{x_p\}$. Set $x_{k_2} = y_2$. Next, choose x_{k_3} as any member of $\{x_p\}$ in I_3 and such that $k_3 > k_2$. We can do this because I_3 also has infinitely many of the $\{x_p\}$. Set $x_{k_3} = y_3$. We continue, and by induction obtain the subsequence $y_1, y_2, \ldots, y_n, \ldots$. By the method of selection we have

$$a_n \leqslant y_n \leqslant b_n, \qquad n = 1, 2, \ldots.$$

Since $a_n \to x_0$, $b_n \to x_0$, as $n \to \infty$, we can apply the Sandwiching theorem (Theorem 2.10) to conclude that $y_n \to x_0$ as $n \to \infty$. $\square$

The proof of the following theorem is left to the reader.

Theorem 3.10′. *Suppose that $\{x_n\}$ is a convergent sequence; that is, $x_n \to x_0$ as $n \to \infty$. If $\{y_n\}$ is an infinite subsequence of $\{x_n\}$, then $y_n \to x_0$ as $n \to \infty$.*

PROBLEMS

In Problems 1 through 7 decide whether or not the given sequence converges to a limit. If it does not, find, in each case, at least one convergent subsequence. We suppose $n = 1, 2, 3, \ldots$.

1. $x_n = (-1)^n (1 - (1/n))$

2. $x_n = 1 + ((-1)^n/n)$

3. $x_n = (-1)^n (2 - 2^{-n})$

4. $x_n = \sum_{i=1}^{n} ((-1)^i/2^i)$

5. $x_n = \sum_{j=1}^{n} (1/j!)$

6. $x_n = \sin(n\pi/2) + \cos n\pi$

7. $x_n = (\sin(n\pi/3))(1 - (1/n))$

8. Prove Theorem 3.10′.

9. The sequence

$$1\tfrac{1}{2}, 2\tfrac{1}{2}, 3\tfrac{1}{2}, 1\tfrac{1}{3}, 2\tfrac{1}{3}, 3\tfrac{1}{3}, 1\tfrac{1}{4}, 2\tfrac{1}{4}, 3\tfrac{1}{4}, \ldots$$

has subsequences which converge to the numbers 1, 2, and 3.
   (a) Write a sequence which has subsequences which converge to N different numbers where N is any positive integer.
   (*b) Write a sequence which has subsequences which converge to infinitely many different numbers.

3.4. The Boundedness and Extreme-Value Theorems

In this section we establish additional basic properties of continuous functions from $\mathbb{R}^1$ to $\mathbb{R}^1$. The Boundedness and Extreme-value theorems proved below are essential in the proofs of the basic theorems in differential calculus. However, the usefulness of these results, especially in the more general setting to be established in Chapter 6, extends to many branches of analysis. The Boundedness theorem shows that a function which is continuous on a closed interval must have a bounded range. The Extreme-value theorem adds additional precise information about such functions. It states that the supremum and the infimum of the values in the range are also in the range.

Theorem 3.11 (Boundedness theorem). *Suppose that the domain of f is the closed interval $I = \{x: a \leqslant x \leqslant b\}$, and f is continuous on I. Then the range of f is bounded.*

PROOF. We shall assume the range is unbounded and reach a contradiction. Suppose for each positive integer n, there is an $x_n \in I$ such that $|f(x_n)| > n$. The sequence $\{x_n\}$ is bounded, and by the Bolzano–Weierstrass theorem, there is a convergent subsequence $y_1, y_2, \ldots, y_n, \ldots$, where $y_n = x_{k_n}$, which converges to an element $x_0 \in I$. Since f is continuous on I, we have $f(y_n) \to f(x_0)$ as $n \to \infty$. Choosing $\varepsilon = 1$, we know there is an N_1 such that for $n > N_1$, we have

$$|f(y_n) - f(x_0)| < 1 \quad \text{whenever} \quad n > N_1.$$

For these n it follows that

$$|f(y_n)| < |f(x_0)| + 1 \quad \text{for all} \quad n > N_1.$$

On the other hand,

$$|f(y_n)| = |f(x_{k_n})| > k_n \geqslant n \quad \text{for all} \quad n,$$

according to the Remark in Section 3.3. Combining these results, we obtain

$$n < |f(x_0)| + 1 \quad \text{for all} \quad n > N_1.$$

Clearly we may choose n larger than $|f(x_0)| + 1$ which is a contradiction. □

Remark. In Theorem 3.11, it is essential that the domain of f is closed. The function $f: x \to 1/(1 - x)$ is continuous on the half-open interval $I = \{x: 0 \leqslant x < 1\}$, but is not bounded there.

Theorem 3.12 (Extreme-value theorem). *Suppose that f has for its domain the closed interval $I = \{x: a \leqslant x \leqslant b\}$, and that f is continuous on I. Then there are numbers x_0 and x_1 on I such that $f(x_0) \leqslant f(x) \leqslant f(x_1)$ for all $x \in I$. That is, f takes on its maximum and minimum values on I.*

PROOF. Theorem 3.11 states that the range of f is a bounded set. Define

$$M = \sup f(x) \quad \text{for} \quad x \in I, \qquad m = \inf f(x) \quad \text{for} \quad x \in I.$$

We wish to show that there are numbers x_0, $x_1 \in I$ such that $f(x_0) = m$, $f(x_1) = M$. We prove the existence of x_1, the proof for x_0 being similar. Suppose that M is not in the range of f; we shall reach a contradiction. The function F with domain I defined by

$$F: x \to \frac{1}{M - f(x)}$$

is continuous on I and therefore (by Theorem 3.11) has a bounded range. Define $\overline{M} = \sup F(x)$ for $x \in I$. Then $\overline{M} > 0$ and

$$\frac{1}{M - f(x)} \leqslant \overline{M} \quad \text{or} \quad f(x) \leqslant M - \frac{1}{\overline{M}} \quad \text{for} \quad x \in I.$$

This last inequality contradicts the statement that $M = \sup f(x)$ for $x \in I$, and hence M must be in the range of f. There is an $x_1 \in I$ such that $f(x_1) = M$.

$\square$

Remark. A proof of Theorem 3.12 can be based on the Bolzano–Weierstrass theorem. In such a proof we select a sequence $\{y_n\}$ in the range of f which tends to M; this can be done according to the Corollary to Theorem 3.5. The reader may complete the argument (see Problem 2 at the end of Section 3.5).

The conclusion of Theorem 3.12 is false if the interval I is not closed. The function $f: x \to x^2$ is continuous on the *half-open* interval $I = \{x: 0 \leqslant x < 1\}$, but does not achieve a maximum value there. Note that f is also continuous on the *closed* interval $I_1 = \{x: 0 \leqslant x \leqslant 1\}$ and its maximum on this interval occurs at $x = 1$; Theorem 3.12 applies in this situation.

Theorem 3.12 asserts the existence of *at least* one value x_1 and one value x_0 where the maximum and minimum are achieved. It may well happen that these maxima and minima are taken on at many points of I.

3.5. Uniform Continuity

In the definition of continuity of a function f at a point x_0, it is necessary to obtain a number δ for each positive number ε prescribed in advance (see Section 2.1). The number δ which naturally depends on ε also depends on the particular value x_0. Under certain circumstances it may happen that the same value δ may be chosen for all points x in the domain. Then we say that f is uniformly continuous.

Definition. A function f with domain S is said to be **uniformly continuous** on S if and only if for every $\varepsilon > 0$ there is a $\delta > 0$ such that

$$|f(x_1) - f(x_2)| < \varepsilon \quad \text{whenever} \quad |x_1 - x_2| < \delta$$

and x_1, x_2 are in S. The important condition of uniform continuity states that the same value of δ holds for *all* x_1, x_2 in S.

Among the properties of continuous functions used in proving the basic theorems of integral calculus, that of uniform continuity plays a central role. The principal result of this section (Theorem 3.13 below) shows that under rather simple hypotheses continuous functions are uniformly continuous. A more general theorem of this type, one with many applications in analysis, is established in Chapter 6.

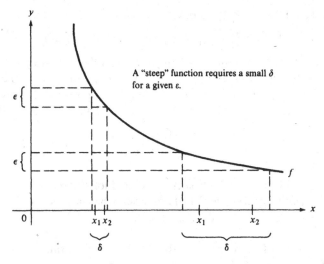

A "steep" function requires a small δ for a given ε.

Figure 3.7

A function may be continuous on a set S without being uniformly continuous. As Figure 3.7 shows, once an ε is given, the value of δ required in the definition of ordinary continuity varies according to the location of x_1 and x_2; the "steeper" the function, the smaller is the value of δ required.

As an example of a continuous function which is not uniformly continuous, consider

$$f: x \to 1/x$$

defined on the set $S = \{x: 0 < x \leqslant 1\}$. It is clear that f is continuous for each x in S. However, with ε any positive number, say 1, we shall show there is no number δ such that

$$|f(x_1) - f(x_2)| < 1 \quad \text{whenever} \quad |x_1 - x_2| < \delta$$

for *all* x_1, x_2 in S. To see this we choose $x_1 = 1/n$ and $x_2 = 1/(n+1)$ for a positive integer n. Then $|f(x_1) - f(x_2)| = |n - (n+1)| = 1$; also, we have $|x_1 - x_2| = 1/n(n+1)$. If n is very large, then x_1 and x_2 are close together. Therefore, if a δ is given, simply choose n so large that x_1 and x_2 are closer together than δ. The condition of uniform continuity is violated since $f(x_1)$ and $f(x_2)$ may not be "close."

An example of a uniformly continuous function on a set S is given by

$$f: x \to x^2 \tag{3.6}$$

on the domain $S = \{x: 0 \leqslant x \leqslant 1\}$. To see this suppose $\varepsilon > 0$ is given. We must find a $\delta > 0$ such that

$$|x_1^2 - x_2^2| < \varepsilon \quad \text{whenever} \quad |x_1 - x_2| < \delta$$

for all x_1, x_2 in S. To accomplish this, simply choose $\delta = \varepsilon/2$. Then, because

$0 \leqslant x_1 \leqslant 1$ and $0 \leqslant x_2 \leqslant 1$, we have

$$|x_1^2 - x_2^2| = |x_1 + x_2| \cdot |x_1 - x_2| < 2 \cdot \tfrac{1}{2}\varepsilon = \varepsilon.$$

This inequality holds for all x_1, x_2 on $[0, 1]$ such that $|x_1 - x_2| < \delta$.

The same function (3.6) above defined on the domain $S_1 = \{x : 0 \leqslant x < \infty\}$ is not uniformly continuous there. To see this suppose $\varepsilon > 0$ is given. Then for any $\delta > 0$, choose x_1, x_2 so that $x_1 - x_2 = \delta/2$ and $x_1 + x_2 = 4\varepsilon/\delta$. Then we have

$$|x_1^2 - x_2^2| = |x_1 + x_2| \cdot |x_1 - x_2| = \frac{4\varepsilon}{\delta} \cdot \frac{1}{2}\delta = 2\varepsilon.$$

The condition of uniform continuity is violated for this x_1, x_2.

An important criterion which determines when a function is uniformly continuous is established in the next result.

Theorem 3.13 (Uniform continuity theorem). *If f is continuous on the closed interval $I = \{x : a \leqslant x \leqslant b\}$, then f is uniformly continuous on I.*

PROOF. We shall suppose that f is not uniformly continuous on I and reach a contradiction. If f is not uniformly continuous there is an $\varepsilon_0 > 0$ for which there is *no* $\delta > 0$ with the property that $|f(x_1) - f(x_2)| < \varepsilon_0$ for all pairs x_1, $x_2 \in I$ and such that $|x_1 - x_2| < \delta$. Then for each positive integer n, there is a pair x'_n, x''_n on I such that

$$|x'_n - x''_n| < \frac{1}{n} \quad \text{and} \quad |f(x'_n) - f(x''_n)| \geqslant \varepsilon_0. \tag{3.7}$$

From the Bolzano–Weierstrass theorem, it follows that there is a subsequence of $\{x'_n\}$, which we denote $\{x'_{k_n}\}$, convergent to some number x_0 in I. Then, since $|x'_{k_n} - x''_{k_n}| < 1/n$, we see that $x''_{k_n} \to x_0$ as $n \to \infty$. Using the fact that f is continuous on I, we have

$$f(x'_{k_n}) \to f(x_0), \qquad f(x''_{k_n}) \to f(x_0).$$

That is, there are positive integers N_1 and N_2 such that

$$|f(x'_{k_n}) - f(x_0)| < \tfrac{1}{2}\varepsilon_0 \quad \text{for all} \quad n > N_1$$

and

$$|f(x''_{k_n}) - f(x_0)| < \tfrac{1}{2}\varepsilon_0 \quad \text{for all} \quad n > N_2.$$

Hence for all n larger than both N_1 and N_2, we find

$$|f(x'_{k_n}) - f(x''_{k_n})| \leqslant |f(x'_{k_n}) - f(x_0)| + |f(x_0) - f(x''_{k_n})| < \tfrac{1}{2}\varepsilon_0 + \tfrac{1}{2}\varepsilon_0 = \varepsilon_0.$$

This last inequality contradicts (3.7), and the result is established.   □

Remarks. As the example of the function $f : x \to 1/x$ shows, the requirement that I is a closed interval is essential. Also, the illustration of the function $f : x \to x^2$ on the set $S_1 = \{x : 0 \leqslant x < \infty\}$ shows that Theorem 3.13 does not apply if the interval is unbounded. It may happen that continuous functions are uniformly continuous on unbounded sets. But these must be decided on a case-by-case basis. See Problems 5 and 6 at the end of the section.

PROBLEMS

1. Suppose that f is a continuous, increasing function on a closed interval $I = \{x: a \leqslant x \leqslant b\}$. Show that the range of f is the interval $[f(a), f(b)]$.

2. Write a proof of the Extreme-value theorem which is based on the Bolzano–Weierstrass theorem (see the Remark after Theorem 3.12).

3. Suppose that f is continuous on the set $S = \{x: 0 \leqslant x < \infty\}$ and that f is bounded. Give an example to show that the Extreme-value theorem does not hold.

4. Suppose that f is uniformly continuous on the closed intervals I_1 and I_2. Show that f is uniformly continuous on $S = I_1 \cup I_2$.

5. Show that the function $f: x \to 1/x$ is uniformly continuous on $S = \{x: 1 \leqslant x < \infty\}$.

6. Show that the function $f: x \to \sin x$ is uniformly continuous on $S = \{x: -\infty < x < \infty\}$.
   [*Hint*: $\sin A - \sin B = 2 \sin((A - B)/2) \cos((A + B)/2).$]

7. Suppose that f is continuous on $I = \{x: a < x < b\}$. If $\lim_{x \to a^+} f(x)$ and $\lim_{x \to b^-} f(x)$ exist, show that by defining $f_0(a) = \lim_{x \to a^+} f(x)$, $f_0(b) = \lim_{x \to b^-} f(x)$, and $f_0(x) = f(x)$ for $x \in I$, then the function f_0 defined on $I_1 = \{x: a \leqslant x \leqslant b\}$ is uniformly continuous.

8. Consider the function $f: x \to \sin(1/x)$ defined on $I = \{x: 0 < x \leqslant 1\}$. Decide whether or not f is uniformly continuous on I.

9. Show directly from the definition that the function $f: x \to \sqrt{x}$ is uniformly continuous on $I_1 = \{x: 0 \leqslant x \leqslant 1\}$.

10. Suppose that f is uniformly continuous on the half-open interval $I = \{x: 0 < x \leqslant 1\}$. Is it true that f is bounded on I?

11. Given the general polynomial

$$f(x) = a_n x^n + a_{n-1} x^{n-1} + \cdots + a_1 x + a_0.$$

Show that f is uniformly continuous on $0 \leqslant x \leqslant 1$.

12. Show that the function

$$f(x) = \begin{cases} x \sin(1/x), & 0 < x \leqslant 1, \\ 0, & x = 0, \end{cases}$$

is uniformly continuous on $I = \{x: 0 \leqslant x \leqslant 1\}$.

3.6. The Cauchy Criterion

We recall the definition of a convergent sequence $x_1, x_2, \ldots, x_n, \ldots$. A sequence **converges to a limit** L if and only if for every $\varepsilon > 0$ there is a positive integer

N such that

$$|x_n - L| < \varepsilon \quad \text{whenever} \quad n > N. \tag{3.8}$$

Suppose we are given a sequence and wish to examine the possibility of convergence. Usually the number L is not given so that Condition (3.8) above cannot be verified directly. For this reason it is important to have a technique for deciding convergence which doesn't employ the limit L of the sequence. Such a criterion, presented below, was given first by Cauchy.

Definition. An infinite sequence $\{x_n\}$ is called a **Cauchy sequence** if and only if for each $\varepsilon > 0$, there is a positive integer N such that

$$|x_n - x_m| < \varepsilon \quad \text{for all} \quad m > N \quad \text{and all} \quad n > N.$$

Theorem 3.14 (Cauchy criterion for convergence). *A necessary and sufficient condition for convergence of a sequence $\{x_n\}$ is that it be a Cauchy sequence.*

PROOF. We first show that if $\{x_n\}$ is convergent, then it is a Cauchy sequence. Suppose L is the limit of $\{x_n\}$, and let $\varepsilon > 0$ be given. Then from the definition of convergence there is an N such that

$$|x_n - L| < \tfrac{1}{2}\varepsilon \quad \text{for all} \quad n > N.$$

Let x_m be any element of $\{x_n\}$ with $m > N$. We have

$$|x_n - x_m| = |x_n - L + L - x_m|$$
$$\leqslant |x_n - L| + |L - x_m| < \tfrac{1}{2}\varepsilon + \tfrac{1}{2}\varepsilon = \varepsilon.$$

Thus $\{x_n\}$ is a Cauchy sequence.

Now assume $\{x_n\}$ is a Cauchy sequence; we wish to show it is convergent. We first prove that $\{x_n\}$ is a *bounded* sequence. From the definition of Cauchy sequence with $\varepsilon = 1$, there is an integer N_0 such that

$$|x_n - x_m| < 1 \quad \text{for all} \quad n > N_0 \quad \text{and} \quad m > N_0.$$

Choosing $m = N_0 + 1$, we find $|x_n - x_{N_0+1}| < 1$ if $n > N_0$. Also,

$$|x_n| = |x_n - x_{N_0+1} + x_{N_0+1}|$$
$$\leqslant |x_n - x_{N_0+1}| + |x_{N_0+1}| < 1 + |x_{N_0+1}|.$$

Keep N_0 fixed and observe that all $|x_n|$ beyond x_{N_0} are bounded by $1 + |x_{N_0+1}|$, a fixed number. Now examine the *finite* sequence of numbers

$$|x_1|, |x_2|, \ldots, |x_{N_0}|, |x_{N_0+1}| + 1$$

and denote by M the largest of these. Therefore $|x_n| \leqslant M$ for *all* positive integers n, and so $\{x_n\}$ is a bounded sequence.

We now apply the Bolzano–Weierstrass theorem and conclude that there is a subsequence $\{x_{k_n}\}$ of $\{x_n\}$ which converges to some limit L. We shall show that the sequence $\{x_n\}$ itself converges to L. Let $\varepsilon > 0$ be given. Since $\{x_n\}$ is a Cauchy sequence there is an N_1 such that

$$|x_n - x_m| < \tfrac{1}{2}\varepsilon \quad \text{for all} \quad n > N_1 \quad \text{and} \quad m > N_1.$$

Also, since $\{x_{k_n}\}$ converges to L, there is an N_2 such that

$$|x_{k_n} - L| < \tfrac{1}{2}\varepsilon \quad \text{for all} \quad n > N_2.$$

Let N be the larger of N_1 and N_2, and consider any integer n larger than N. We recall from the definition of subsequence of a sequence that $k_n \geq n$ for every n. Therefore

$$|x_n - L| = |x_n - x_{k_n} + x_{k_n} - L| \leq |x_n - x_{k_n}| + |x_{k_n} - L| < \tfrac{1}{2}\varepsilon + \tfrac{1}{2}\varepsilon = \varepsilon.$$

Since this inequality holds for all $n > N$, the sequence $\{x_n\}$ converges to L. □

As an example, we show that the sequence $x_n = (\cos n\pi)/n$, $n = 1, 2, \ldots$ is a Cauchy sequence. Let $\varepsilon > 0$ be given. Choose N to be any integer larger than $2/\varepsilon$. Then we have

$$|x_n - x_m| = \left| \frac{\cos n\pi}{n} - \frac{\cos m\pi}{m} \right| \leq \frac{m|\cos n\pi| + n|\cos m\pi|}{mn} = \frac{m + n}{mn}.$$

If $m > n$, we may write

$$|x_n - x_m| \leq \frac{m + n}{mn} < \frac{2m}{mn} = \frac{2}{n}.$$

However, because $n > N$, we have $n > 2/\varepsilon$, and so $|x_n - x_m| < \varepsilon$, and the sequence is a Cauchy sequence.

3.7. The Heine–Borel and Lebesgue Theorems

In this section we establish two theorems which are useful in the further development of differentiation and integration. The Heine–Borel theorem shows that "covering" a set with a collection of special sets is particularly helpful in verifying uniform continuity. A useful generalization of this result is proved in Chapter 6.

Consider, for example, the collection of intervals

$$I_n = \left\{ x: \frac{1}{n} < x < 1 \right\}, \qquad n = 2, 3, 4, \ldots. \tag{3.9}$$

Any collection of intervals such as (3.9) is called a **family of intervals**. The symbol $\mathscr{F}$ is usually used to denote the totality of intervals in the family. Each interval is called a **member** (or **element**) of the family. Care must be exercised to distinguish the points of a particular interval from the interval itself. In the above example, the interval $J = \{x: \tfrac{1}{3} < x < \tfrac{2}{3}\}$ is *not* a member of $\mathscr{F}$. However, every point of J is in every interval I_n for $n > 2$. The points in $0 < x < 1$

are not themselves members of $\mathscr{F}$ although each point of $0 < x < 1$ is in at least one I_n.

Definitions. A family $\mathscr{F}$ of intervals is said to **cover the set** S in $\mathbb{R}^1$ if and only if each point of S is in at least one of the intervals of $\mathscr{F}$. A family $\mathscr{F}_1$ is a **subfamily** *of* $\mathscr{F}$ if and only if each member of $\mathscr{F}_1$ is a member of $\mathscr{F}$.

For example, the intervals

$$I'_n = \left\{ x: \frac{1}{2n} < x < 1 \right\}, \qquad n = 1, 2, \dots \tag{3.10}$$

form a subfamily of Family (3.9). Family (3.9) covers the set $S = \{x: 0 < x < 1\}$ since every number x such that $0 < x < 1$ is in I_n for all n larger than $1/x$. Thus every x is in at least one I_n. Family (3.10) also covers S. More generally, we may consider a collection of sets $A_1, A_2 \dots A_n, \dots$ which we call a **family of sets**. We use the same symbol $\mathscr{F}$ to denote such a family. A family $\mathscr{F}$ **covers a set** S if and only if every point of S is a point in at least one member of $\mathscr{F}$.

The family $\mathscr{F}$:

$$J_n = \{x: n < x < n + 2\}, \qquad n = 0, \pm1, \pm2, \dots$$

covers all of $\mathbb{R}^1$. It is simple to verify that if any interval is removed from $\mathscr{F}$, the resulting family fails to cover $\mathbb{R}^1$. For example, if J_1 is removed, then no member of the remaining intervals of $\mathscr{F}$ contains the number 2.

We shall be interested in families of open intervals which contain infinitely many members and which cover a set S. We shall examine those sets S which have the property that they are covered by a **finite subfamily**, i.e., a subfamily with only a finite number of members. For example, the family $\mathscr{F}$ of intervals $K_n = \{x: 1/(n + 2) < x < 1/n\}, n = 1, 2, \dots$ covers the set $S = \{x: 0 < x < \frac{1}{2}\}$, as is easily verified (see Problem 13 at the end of the section). It may also be verified that no finite subfamily of $\mathscr{F} = \{K_n\}$ covers S. However, if we consider the set $S_1 = \{x: \varepsilon < x \leqslant \frac{1}{2}\}$ for any $\varepsilon > 0$, it is clear that S_1 is covered by the finite subfamily $\{K_1, K_2, \dots, K_n\}$ for any n such that $n + 2 > 1/\varepsilon$.

We may define families of intervals indirectly. For example, suppose $f: x \to 1/x$ with domain $D = \{x: 0 < x \leqslant 1\}$ is given. We obtain a family $\mathscr{F}$ of intervals by considering all solution sets of the inequality

$$|f(x) - f(a)| < \frac{1}{3} \quad \Leftrightarrow \quad -\frac{1}{3} < \frac{1}{x} - \frac{1}{a} < \frac{1}{3} \tag{3.11}$$

for every $a \in D$. For each value a, the interval

$$L_a = \left\{ x: \frac{3a}{3 + a} < x < \frac{3a}{3 - a} \right\} \tag{3.12}$$

is the solution of Inequality (3.11). This family $\mathscr{F} = \{L_a\}$ covers the set $D = \{x: 0 < x \leqslant 1\}$, but no finite subfamily covers D. (See Problem 16 at the end of this section.)

Theorem 3.15 (Heine–Borel theorem). *Suppose that a family $\mathscr{F}$ of open intervals covers the closed interval $I = \{x: a \leqslant x \leqslant b\}$. Then a finite subfamily of $\mathscr{F}$ covers I.*

PROOF. We shall suppose an infinite number of members of $\mathscr{F}$ are required to cover I and reach a contradiction. Divide I into two equal parts at the midpoint. Then an infinite number of members of $\mathscr{F}$ are required to cover either the left subinterval or the right subinterval of I. Denote by $I_1 = \{x: a_1 \leqslant x \leqslant b_1\}$ the particular subinterval needing this infinity of members of $\mathscr{F}$. We proceed by dividing I_1 into two equal parts, and denote by $I_2 = \{x: a_2 \leqslant x \leqslant b_2\}$ that half of I_1 which requires an infinite number of members of $\mathscr{F}$. Repeating the argument, we obtain a sequence of closed intervals $I_n = \{x: a_n \leqslant x \leqslant b_n\}$, $n = 1, 2, \ldots$ each of which requires an infinite number of intervals of $\mathscr{F}$ in order to be covered. Since $b_n - a_n = (b - a)/2^n \to 0$ as $n \to \infty$, the Nested intervals theorem (Theorem 3.1) states that there is a unique number $x_0 \in I$ which is in every I_n. However, since $\mathscr{F}$ covers I, there is a member of $\mathscr{F}$, say $J = \{x: \alpha < x < \beta\}$, such that $x_0 \in J$. By choosing N sufficiently large, we can find an I_N contained in J. But this contradicts the fact that infinitely many intervals of $\mathscr{F}$ are required to cover I_N—we found that the one interval J covers I_N. $\qquad\square$

Remarks. In the Heine–Borel theorem, the hypothesis that $I = \{x: a \leqslant x \leqslant b\}$ is a *closed* interval is crucial. The open interval $J_1 = \{x: 0 < x < 1\}$ is covered by the family $\mathscr{F}$ of intervals (3.9), but no finite subfamily of $\mathscr{F}$ covers J_1. Also, Theorem 3.15 is not valid if the interval I is unbounded. For example the interval $K = \{x: 0 \leqslant x < \infty\}$ is covered by the family $\mathscr{F}_1$ given by $K_n = \{x: n - (1/2) < x < n + (3/2)\}$, $n = 0, 1, 2, \ldots$ but no finite subfamily of $\mathscr{F}_1$ covers K.

The next theorem is a useful alternate form of the Heine–Borel theorem.

Theorem 3.16 (Lemma of Lebesgue). *Suppose that a family $\mathscr{F}$ of open intervals covers the closed interval $I = \{x: a \leqslant x \leqslant b\}$. Then there is a positive number ρ such that every interval of the form $J_c = \{x: c - \rho < x < c + \rho\}$, for each c in I, is contained in some single member of $\mathscr{F}$. That is, every interval of length 2ρ with midpoint in I is contained in one member of $\mathscr{F}$.*

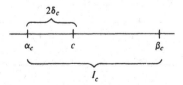

Figure 3.8

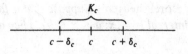

Figure 3.9

PROOF. Each c in I is in some open interval $I_c = \{x: \alpha_c < x < \beta_c\}$ of the family $\mathscr{F}$. Let $2\delta_c$ denote the distance of c to the nearest endpoint of I_c (see Figure 3.8). Then the interval $\{x: c - 2\delta_c < x < c + 2\delta_c\}$ is contained in I_c. Denote by K_c the interval $\{x: c - \delta_c < x < c + \delta_c\}$ (see Figure 3.9). Let $\mathscr{G}$ be the family of all intervals K_c for every c in I. Note that every K_c is contained in some I_c, a member of $\mathscr{F}$. From the Heine–Borel theorem, there is a finite subfamily $\mathscr{G}_1$ of $\mathscr{G}$ which covers I. Denote the members of $\mathscr{G}_1$ by

$$K_{c_1}, K_{c_2}, \ldots, K_{c_n},$$

and let δ_{c_p} be the smallest of $\delta_{c_1}, \delta_{c_2}, \ldots, \delta_{c_n}$. Choose for the ρ of our theorem this value δ_{c_p}. To see that this works let c be any point in I. Then c is in K_{c_i} for some i. We wish to show that $J_c = \{x: c - \rho < x < c + \rho\}$ is contained in some interval of $\mathscr{F}$. We know that J_c is contained in

$$I' = \{x: c_i - \delta_{c_i} - \rho < x < c_i + \delta_{c_i} + \rho\}$$

(see Figure 3.10). The interval I', in turn, is contained in

$$I'' = \{x: c_i - 2\delta_{c_i} < x < c_i + 2\delta_{c_i}\},$$

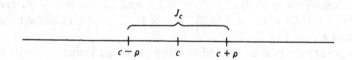

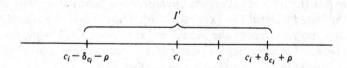

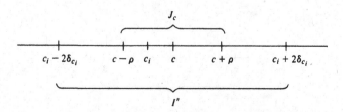

Figure 3.10

since $\rho \leqslant \delta_{c_i}$, $i = 1, 2, \ldots, n$. However, for each c_i, the interval I'' is contained in one member of $\mathscr{F}$, and the result is established.   □

The Lemma of Lebesgue may be used to give a direct proof of Theorem 3.13 on uniform continuity. See Problem 19 at the end of this section and the hint given there.

PROBLEMS

In each of Problems 1 through 6 decide whether or not the infinite sequence is a Cauchy sequence. If it is not a Cauchy sequence, find at least one subsequence which is a Cauchy sequence. In each case, $n = 1, 2, \ldots$.

1. $x_n = \sum_{j=1}^{n} (-1)^j (1/2j)$

2. $x_n = \sum_{j=1}^{n} (1/j!)$

3. $x_n = \sum_{i=1}^{n} ((-1)^j/j!)$

4. $x_n = 1 + (-1)^n + (1/n)$

5. $x_n = \sin(n\pi/3) + (1/n)$

6. $x_n = (1 + (-1)^n)n + (1/n)$

7. Suppose that $s_n = \sum_{j=1}^{n} u_j$ and $S_n = \sum_{j=1}^{n} |u_j|$, $n = 1, 2, \ldots$. If $S_n \to S$ as $n \to \infty$, show that s_n tends to a limit as $n \to \infty$.

8. Show that Theorem 3.1, the Nested intervals theorem, may be proved as a direct consequence of the Cauchy criterion for convergence (Theorem 3.14). [*Hint:* Suppose $I_n = \{x: a_n \leqslant x \leqslant b_n\}$ is a nested sequence. Then show that $\{a_n\}$ and $\{b_n\}$ are Cauchy sequences. Hence they each tend to a limit. Since $b_n - a_n \to 0$, the limits must be the same. Finally, the Sandwiching theorem shows that the limit is in every I_n.]

9. Suppose that f has a domain which contains $I = \{x: a < x < b\}$, and suppose that $f(x) \to L$ as $x \to b^-$. Prove the following: for each $\varepsilon > 0$ there is a $\delta > 0$ such that

$$|f(x) - f(y)| < \varepsilon \quad \text{for all} \quad x, y \quad \text{with} \quad b - \delta < x < b, b - \delta < y < b. \quad (3.13)$$

10. Prove the converse of the result in Problem 9. That is, suppose that for every $\varepsilon > 0$ there is a $\delta > 0$ such that Condition (3.13) is satisfied. Prove that there is a number L such that $f(x) \to L$ as $x \to b^-$.

11. Show that the result of Problem 9 holds if I is replaced by $J = \{a < x < \infty\}$ and limits are considered as $x \to +\infty$. Show that the result in Problem 10 does not hold.

12. Show that if f is uniformly continuous on $I = \{x: a < x < b\}$, then $f(x)$ tends to a limit as $x \to b^-$ and as $x \to a^+$. Hence show that if $f_0(a)$ and $f_0(b)$ are these limits and $f_0(x) = f(x)$ on I, then the extended function f_0 is uniformly continuous on $J = \{x: a \leqslant x \leqslant b\}$. Use the result of Problem 10.

13. (a) Show that the family $\mathscr{F}$ of all intervals of the form $I_n = \{x: 1/(n+2) < x < 1/n\}$ covers the interval $J = \{x: 0 < x < 1/2\}$.
    (b) Show that no finite subfamily of $\mathscr{F}$ covers J.

14. Let $\mathscr{G}$ be the family of intervals obtained by adjoining the interval $\{x: -1/6 < x < 1/6\}$ to the family $\mathscr{F}$ of Problem 13. Exhibit a finite subfamily of G which covers $K = \{x: 0 \leqslant x \leqslant 1/2\}$.

15. Let $\mathscr{F}_1$ be the family of all intevals $I_n = \{x: 1/2^n < x < 2\}$, $n = 1, 2, \ldots$. Show that $\mathscr{F}_1$ covers the interval $J = \{x: 0 < x < 1\}$. Does any finite subfamily of $\mathscr{F}_1$ cover J? Prove your answer.

16. For $f(x) = 1/x$ defined on $D = \{x: 0 < x \leqslant 1\}$, consider the inequality $|f(x) - f(a)| < 1/3$ for $a \in D$. Show that the solution set of this inequality, denoted by L_a, is given by Equation (3.12). Show that the family $\mathscr{F} = \{L_a\}$ covers D. Does any finite subfamily of $\mathscr{F}$ cover D? Prove your answer.

17. Consider $f: x \to 1/x$ defined on $E = \{x: 1 \leqslant x < \infty\}$. Let I_a be the interval of values of x for which $|f(x) - f(a)| < 1/3$. Find I_a for each $a \in E$. Show that the family $\mathscr{F} = \{I_a\}$, $a \in E$ covers E. Is there a finite subfamily of $\mathscr{F}$ which covers E? Prove your answer.

18. Prove the Boundedness theorem (Theorem 3.11) using the Heine–Borel theorem (Theorem 3.15).

19. Prove the Uniform continuity theorem (Theorem 3.13) using the Lemma of Lebesgue (Theorem 3.16). [*Hint*: Let $\varepsilon > 0$ be given. Since f is continuous at each point x_0 of the interval I, there is a δ_{x_0} such that $|f(x) - f(x_0)| < \varepsilon/2$ whenever $|x - x_0| < \delta_{x_0}$ and $x_0, x \in I$. The intervals $I_{x_0} = \{x: x_0 - \delta_{x_0} < x < x_0 + \delta_{x_0}\}$ form a family $\mathscr{F}$ to which we apply the Lemma of Lebesgue. Denote by ρ the number so obtained. With this number as the δ of uniform continuity, the inequality $|f(x_1) - f(x_2)| \leqslant |f(x_1) - f(x_0)| + |f(x_0) - f(x_2)| < \varepsilon$ can be established.]

20. All the rational numbers in $I = \{x: 0 \leqslant x \leqslant 1\}$ can be arranged in a single sequence as follows:
$$0, \tfrac{1}{1}, \tfrac{1}{2}, \tfrac{1}{3}, \tfrac{2}{3}, \tfrac{1}{4}, \tfrac{3}{4}, \tfrac{1}{5}, \tfrac{2}{5}, \tfrac{3}{5}, \tfrac{4}{5}, \tfrac{1}{6}, \tfrac{5}{6}, \tfrac{1}{7}, \ldots$$

(Note that we consider all possible fractions with denominator $= 1$, then all with denominator $= 2$, then all with denominator $= 3$, and so forth.) Let $\varepsilon > 0$ be given, and let I_n be the interval of length $\varepsilon/2^n$ which has its center at the nth rational number of the above sequence. The family $\mathscr{F} = \{I_n\}$, $n = 1, 2, \ldots$ covers all the rational numbers in I. Show that the sum of the lengths of *all* the I_n is ε. Prove that if ε is small, then $\mathscr{F}$ does not cover the entire interval I.

21. Show that $\cos(\sqrt{n}\pi)$, $n = 1, 2, \ldots$ is dense in $(-1, 1)$.

CHAPTER 4

Elementary Theory of Differentiation

4.1. The Derivative in $\mathbb{R}^1$

In a first course in calculus we learn a number of theorems on differentiation, but usually the proofs are not provided. In this section we review these elementary theorems and provide the missing proofs for the basic results.

Definition. Let f be a function on $\mathbb{R}^1$ to $\mathbb{R}^1$. The **derivative** f' is defined by the formula

$$f'(x) = \lim_{h \to 0} \frac{f(x + h) - f(x)}{h}. \tag{4.1}$$

That is, f' is the function whose domain consists of all x in $\mathbb{R}^1$ for which the limit on the right side of Equation (4.1) exists. Of course the range of f' is in $\mathbb{R}^1$.

The following seven elementary theorems on the derivative, given without proof, are simple consequences of the theorems on limits in Chapter 2. The reader should write out detailed proofs as suggested in the problems at the end of this section.

Theorem 4.1. *If f is a constant function, then $f'(x) = 0$ for all x.*

In Theorems 4.2 through 4.7 we assume that all functions involved are on $\mathbb{R}^1$ to $\mathbb{R}^1$.

Theorem 4.2. *Suppose that f is defined on an open interval I, c is a real number, and that g is defined by the equation $g(x) = cf(x)$. If $f'(x)$ exists, then $g'(x)$ does, and $g'(x) = cf'(x)$.*

Theorem 4.3. *Suppose that f and g are defined on an open interval I and that F is defined by the equation* $F(x) = f(x) + g(x)$. *If* $f'(x)$ *and* $g'(x)$ *exist, then* $F'(x)$ *does, and* $F'(x) = f'(x) + g'(x)$.

Theorem 4.4. *If* $f'(a)$ *exists for some number a, then f is continuous at a.*

Theorem 4.5. *Suppose that u and v are defined on an open interval I and that f is defined by the equation* $f(x) = u(x) \cdot v(x)$. *If* $u'(x)$ *and* $v'(x)$ *exist, then* $f'(x)$ *does, and* $f'(x) = u(x)v'(x) + u'(x)v(x)$.

Theorem 4.6. *Suppose that u and v are defined on an open interval I, that* $v(x) \neq 0$ *for all* $x \in I$, *and that f is defined by the equation* $f(x) = u(x)/v(x)$. *If* $u'(x)$ *and* $v'(x)$ *exist, then* $f'(x)$ *does, and*

$$f'(x) = \frac{v(x)u'(x) - u(x)v'(x)}{[v(x)]^2}.$$

Theorem 4.7. *Given* $f: x \to x^n$ *and n is an integer. Then*

$$f'(x) = nx^{n-1}.$$

(We assume that $x \neq 0$ *for* $n < 0$.)

We now establish the Chain rule (Theorem 4.9 below) which together with Theorems 4.1 through 4.7 forms the basis for calculating the derivatives of most elementary functions. The Chain rule is a consequence of the following lemma which states that every differentiable function is approximated by a linear function whose slope is the derivative. (See Figure 4.1.)

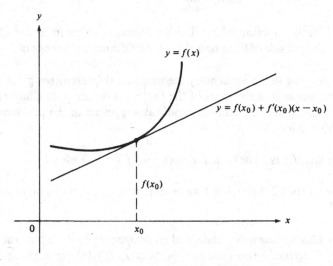

Figure 4.1. A linear approximation to f at $(x_0, f(x_0))$.

Theorem 4.8 (Fundamental lemma of differentiation). *Suppose that f has a derivative at x_0. Then there is a function η defined in an interval about 0 such that*

$$f(x_0 + h) - f(x_0) = [f'(x_0) + \eta(h)] \cdot h. \tag{4.2}$$

Also, η is continuous at 0 with $\eta(0) = 0$.

PROOF. We define η by the formula

$$\eta(h) = \begin{cases} \dfrac{1}{h}[f(x_0 + h) - f(x_0)] - f'(x_0), & h \neq 0, \\[2mm] 0, & h = 0. \end{cases}$$

Since f has a derivative at x_0, we see that $\eta(h) \to 0$ as $h \to 0$. Hence η is continuous at 0. Formula (4.2) is a restatement of the definition of η. $\qquad\square$

Theorem 4.9 (Chain rule). *Suppose that g and u are functions on $\mathbb{R}^1$ and $f(x) = g[u(x)]$. Suppose that u has a derivative at x_0 and that g has a derivative at $u(x_0)$. Then $f'(x_0)$ exists and*

$$f'(x_0) = g'[u(x_0)] \cdot u'(x_0).$$

PROOF. We use the notation $\Delta f = f(x_0 + h) - f(x_0)$, $\Delta u = u(x_0 + h) - u(x_0)$, and we find

$$\Delta f = g[u(x_0 + h)] - g[u(x_0)] = g(u + \Delta u) - g(u). \tag{4.3}$$

We apply Theorem 4.8 to the right side of Equation (4.3), getting

$$\Delta f = [g'(u) + \eta(\Delta u)]\Delta u.$$

Dividing by h and letting h tend to zero, we obtain (since $\Delta u \to 0$ as $h \to 0$)

$$\lim_{h \to 0} \frac{\Delta f}{h} = f'(x_0) = \lim_{h \to 0} [g'(u) + \eta(\Delta u)] \cdot \lim_{h \to 0} \frac{\Delta u}{h} = g'[u(x_0)] \cdot u'(x_0). \quad \square$$

The Extreme-value theorem (Theorem 3.12) shows that a function which is continuous on a closed interval I takes on its maximum and minimum values there. If the maximum value occurs at an interior point[1] of I, and if the function possesses a derivative at that point, the following result gives a method for locating maximum (and minimum) values of such functions.

Theorem 4.10. *Suppose that f is continuous on an interval I, and that f takes on its maximum value at x_0, an interior point of I. If $f'(x_0)$ exists, then*

$$f'(x_0) = 0.$$

[1] All points of an interval except the endpoints are called **interior points**.

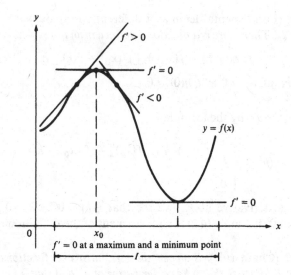

Figure 4.2

The proof follows immediately from the definition of derivative and is left to the reader. A similar result holds for functions which take on their minimum value at an interior point. See Figure 4.2.

The next two theorem, Rolle's theorem and the Mean-value theorem, form the groundwork for further developments in the theory of differentiation.

Theorem 4.11 (Rolle's theorem). *Suppose that f is continuous on the closed interval $I = \{x: a \leqslant x \leqslant b\}$, and that f has a derivative at each point of $I_1 = \{x: a < x < b\}$. If $f(a) = f(b) = 0$, then there is a number $x_0 \in I_1$ such that $f'(x_0) = 0$.*

PROOF. Unless $f(x) \equiv 0$, in which case the result is trivial, f must be positive or negative somewhere. Suppose it is positive. Then according to the Extreme-value theorem, f achieves its maximum at some interior point, say x_0. Now we apply Theorem 4.10 to conclude that $f'(x_0) = 0$. If f is negative on I_1, we apply the same theorems to the minimum point. See Figure 4.3.   □

Theorem 4.12 (Mean-value theorem). *Suppose that f is continuous on the closed interval $I = \{x: a \leqslant x \leqslant b\}$ and that f has a derivative at each point of $I_1 = \{x: a < x < b\}$. Then there exists a number $\xi \in I_1$ such that*

$$f'(\xi) = \frac{f(b) - f(a)}{b - a}.$$

PROOF. We construct the function

$$F(x) = f(x) - \frac{f(b) - f(a)}{b - a}(x - a) - f(a).$$

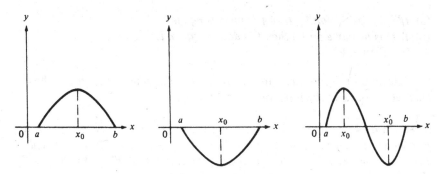

Figure 4.3. Illustrating Rolle's theorem.

We have

$$F'(x) = f'(x) - \frac{f(b) - f(a)}{b - a}.$$

By substitution, we see that $F(a) = F(b) = 0$. Thus F satisfies the hypotheses of Rolle's theorem and there is a number $\xi \in I_1$ such that $F'(\xi) = 0$. Hence

$$0 = F'(\xi) = f'(\xi) - \frac{f(b) - f(a)}{b - a},$$

and the result is established. See Figure 4.4. $\qquad\square$

The next theorem, a direct consequence of the Mean-value theorem, is useful in the construction of graphs of functions. The proof is left to the reader; see Problem 13 at the end of this section.

Theorem 4.13. *Suppose that f is continuous on the closed interval I and has a derivative at each point of I_1, the interior of I.*

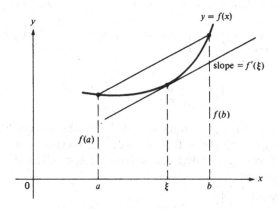

Figure 4.4. Illustrating the Mean-value theorem.

(a) *If f' is positive on I_1, then f is increasing on I.*
(b) *If f' is negative on I_1, then f is decreasing on I.*
   See Figure 4.2.

   L'Hôpital's rule, so useful in the evaluation of indeterminate forms, has its theoretical basis in the following three theorems.

Theorem 4.14 (Generalized Mean-value theorem). *Suppose that f and F are continuous functions defined on $I = \{x: a \leqslant x \leqslant b\}$. Suppose that f' and F' exist on $I_1 = \{x: a < x < b\}$, the interior of I, and that $F'(x) \neq 0$ for $x \in I_1$. Then*

(i) $F(b) - F(a) \neq 0$, *and*
(ii) *there is a number $\xi \in I_1$ such that*

$$\frac{f(b) - f(a)}{F(b) - F(a)} = \frac{f'(\xi)}{F'(\xi)}. \tag{4.4}$$

PROOF. According to the Mean-value theorem applied to F, there is a number $\eta \in I_1$ such that $F(b) - F(a) = F'(\eta)(b - a)$. Since $F'(\eta) \neq 0$, we conclude that $F(b) - F(a) \neq 0$ and (i) is proved. To prove (ii), define the function ϕ on I by the formula

$$\phi(x) = f(x) - f(a) - \frac{f(b) - f(a)}{F(b) - F(a)} [F(x) - F(a)].$$

A simple calculation shows that $\phi(a) = \phi(b) = 0$ and so Rolle's theorem can be applied to ϕ. The number $\xi \in I_1$ such that $\phi'(\xi) = 0$ yields

$$0 = \phi'(\xi) = f'(\xi) - \frac{f(b) - f(a)}{F(b) - F(a)} F'(\xi),$$

which is Equation (4.4). $\qquad\qquad\qquad\qquad\qquad\qquad\qquad\qquad\qquad\qquad\qquad\square$

Theorem 4.15 (L'Hôpital's rule for 0/0). *Suppose that f and F are continuous functions on $I = \{x: a < x < b\}$ and that f' and F' exist on I with $F' \neq 0$ on I. If*

$$\lim_{x \to a^+} f(x) = \lim_{x \to a^+} F(x) = 0 \quad and \quad \lim_{x \to a^+} \frac{f'(x)}{F'(x)} = L,$$

then

$$\lim_{x \to a^+} \frac{f(x)}{F(x)} = L.$$

PROOF. We extend the definitions of f and F to the half-open interval $I_1 = \{a \leqslant x < b\}$ by setting $f(a) = 0$, $F(a) = 0$. Since $F'(x) \neq 0$ on I and $F(a) = 0$, we see from Theorem 4.12 that $F(x) \neq 0$ on I. Let $\varepsilon > 0$ be given. Then there is a $\delta > 0$ such that

$$\left| \frac{f'(x)}{F'(x)} - L \right| < \varepsilon \quad \text{for all } x \text{ such that} \quad a < x < a + \delta.$$

Now we write

$$\left|\frac{f(x)}{F(x)} - L\right| = \left|\frac{f(x) - f(a)}{F(x) - F(a)} - L\right|,$$

and apply the Generalized Mean-value theorem to obtain

$$\left|\frac{f(x)}{F(x)} - L\right| = \left|\frac{f'(\xi)}{F'(\xi)} - L\right| < \varepsilon,$$

where ξ is such that $a < \xi < x < a + \delta$. Since ε is arbitrary the result follows.
□

Remark. The size of the interval I is of no importance in Theorem 4.15. Therefore the hypotheses need only hold in an arbitrarily small interval in which a is one endpoint.

Corollary 1. *L'Hôpital's rule holds for limits from the left as well as limits from the right. If the two-sided limits are assumed to exist in the hypotheses of Theorem 4.15, then we may conclude the existence of the two-sided limit of f/F.*

Corollary 2. *A theorem similar to Theorem 4.15 holds for limits as $x \to +\infty$, $-\infty$, or ∞.*

PROOF OF COROLLARY 2. Assume that

$$\lim_{x \to +\infty} f(x) = \lim_{x \to +\infty} F(x) = 0, \qquad \lim_{x \to +\infty} \frac{f'(x)}{F'(x)} = L, \qquad F'(x) \neq 0 \qquad \text{on } I.$$

We let $z = 1/x$ and define the functions g and G by the formulas

$$g(z) = f\left(\frac{1}{z}\right), \qquad G(z) = F\left(\frac{1}{z}\right).$$

Then

$$g'(z) = -\frac{1}{z^2} f'\left(\frac{1}{z}\right), \qquad G'(z) = -\frac{1}{z^2} F'\left(\frac{1}{z}\right),$$

so that $g'(z)/G'(z) \to L$, $g(z) \to 0$, and $G(z) \to 0$ as $z \to 0^+$. Hence we apply Theorem 4.15 to g/G and the result is established. The argument when $x \to -\infty$ or ∞ is similar.
□

Theorem 4.16 (L'Hôpital's rule for ∞/∞). *Suppose that f' and F' exist on $I = \{x : a < x < b\}$. If*

$$\lim_{x \to a^+} f(x) = \lim_{x \to a^+} F(x) = \infty, \qquad \text{and} \qquad \lim_{x \to a^+} \frac{f'(x)}{F'(x)} = L,$$

and if $F'(x) \neq 0$ on I, then

$$\lim_{x \to a^+} \frac{f(x)}{F(x)} = L.$$

PROOF. The hypotheses imply that f and F are continuous on I and that $F(x)$ does not vanish in some interval $I_h = \{z: a < x < a + h\}$. From the definition of limit we know that for every $\varepsilon > 0$ there is a $\delta > 0$ such that

$$\left| \frac{f'(\xi)}{F'(\xi)} - L \right| < \frac{1}{2}\varepsilon \quad \text{for all } \xi \text{ in } I_\delta = \{x: a < x < a + \delta\}. \tag{4.5}$$

Choose δ smaller than h and consider x and c in I_δ with $x < c$. Then from the Generalized Mean-value theorem, it follows that

$$\frac{f(x) - f(c)}{F(x) - F(c)} = \frac{f'(\xi)}{F'(\xi)}. \tag{4.6}$$

Substituting Equation (4.6) into Inequality (4.5), we get

$$\left| \frac{f(x) - f(c)}{F(x) - F(c)} - L \right| < \frac{1}{2}\varepsilon. \tag{4.7}$$

Since we are interested in small values of ε, we may take ε less than 1 and obtain

$$\left| \frac{f(x) - f(c)}{F(x) - F(c)} \right| < |L| + \frac{1}{2}\varepsilon < |L| + \frac{1}{2}. \tag{4.8}$$

An algebraic manipulation allows us to write the identity

$$\frac{f(x)}{F(x)} - \frac{f(x) - f(c)}{F(x) - F(c)} = \frac{f(c)}{F(x)} - \frac{F(c)}{F(x)} \left[\frac{f(x) - f(c)}{F(x) - F(c)} \right].$$

Taking absolute values and using Inequality (4.8), we find

$$\left| \frac{f(x)}{F(x)} - \frac{f(x) - f(c)}{F(x) - F(c)} \right| \leqslant \left| \frac{f(c)}{F(x)} \right| + \left| \frac{F(c)}{F(x)} \right| \left(|L| + \frac{1}{2} \right). \tag{4.9}$$

We keep c fixed and let x tend to a^+. Then $F(x)$ tends to ∞ and the right side of Inequality (4.9) tends to zero. Therefore, there is a $\delta_1 > 0$ (which we take less than δ) so that if $a < x < \delta_1$ we have

$$\left| \frac{f(c)}{F(x)} \right| < \frac{1}{4}\varepsilon \quad \text{and} \quad \left| \frac{F(c)}{F(x)} \right| < \frac{\varepsilon}{4(|L| + \frac{1}{2})}. \tag{4.10}$$

Finally, we use Inequalities (4.7), (4.9), and (4.10) to get

$$\left| \frac{f(x)}{F(x)} - L \right| \leqslant \left| \frac{f(x)}{F(x)} - \frac{f(x) - f(c)}{F(x) - F(c)} \right| + \left| \frac{f(x) - f(c)}{F(x) - F(c)} - L \right|$$

$$< \frac{1}{2}\varepsilon + \frac{1}{2}\varepsilon = \varepsilon. \qquad \square$$

Remarks. As in Theorem 4.15, the size of the interval I in Theorem 4.16 is of no importance. Therefore, we need only assume that $F'(x)$ does not vanish in some sufficiently small neighborhood to the right of a. Indeterminate forms such as $0 \cdot \infty$, 1^∞, and $\infty - \infty$ may frequently be reduced by algebraic manipulation and by use of the logarithmic function to $0/0$ or ∞/∞. Then we may apply Theorem 4.15 or 4.16. Sometimes $f'(x)/F'(x)$ as well as $f(x)/F(x)$ is indeterminate. In such cases, we can apply Theorem 4.15 or 4.16 to f'/F' and see whether or not f''/F'' tends to a limit. This process may be applied any number of times.

EXAMPLE. Evaluate

$$\lim_{x \to 0} \frac{1 + x - e^x}{2x^2}.$$

Solution. Set $f: x \to 1 + x - e^x$ and $F: x \to 2x^2$. Then $\lim_{x \to 0}(f(x)/F(x))$ yields the indeterminate form $0/0$. We try to apply Theorem 4.15 and find that $\lim_{x \to 0}(f'(x)/F'(x))$ also gives $0/0$. A second attempt at application of the theorem shows that

$$\lim_{x \to 0} \frac{f''(x)}{F''(x)} = \lim_{x \to 0} \frac{-e^x}{4} = -\frac{1}{4}.$$

Hence $\lim_{x \to 0}(f'(x)/F'(x)) = -1/4$ so that $\lim_{x \to 0}(f(x)/F(x)) = -1/4$.

PROBLEMS

1. Use Theorem 2.5 on the limit of a sum to write out a proof of Theorem 4.3 on the derivative of the sum of two functions.

2. Use the identity

$$u(x + h)v(x + h) - u(x)v(x) = u(x + h)v(x + h)$$
$$- u(x + h)v(x) + u(x + h)v(x) - u(x)v(x)$$

and Theorems 2.5 and 2.6 to prove the formula for the derivative of a product (Theorem 4.5).

3. The **Leibniz rule** for the nth derivative of a product is given by

$$\frac{d^n}{dx^n}(f(x)g(x)) = f^{(n)}(x)g(x) + \frac{n}{1!}f^{(n-1)}(x)g'(x)$$

$$+ \frac{n(n-1)}{2!}f^{(n-2)}(x)g^{(2)}(x) + \cdots + f(x)g^{(n)}(x),$$

where the coefficients are the same as the coefficients in the binomial expansion.

$$(A + B)^n = A^n + \frac{n}{1!}A^{n-1}B + \frac{n(n-1)}{2!}A^{n-2}B^2$$

$$+ \frac{n(n-1)(n-2)}{3!}A^{n-3}B^3 + \cdots + B^n.$$

Use mathematical induction to establish the Leibniz rule.

4. Use Theorem 2.5 and 2.8 on the limit of a sum and the limit of a quotient to prove the formula for the derivative of a quotient. Start with an identity analogous to the one in Problem 2. For Theorem 4.6 use the identity

$$\frac{u(x + h)}{v(x + h)} - \frac{u(x)}{v(x)} = \frac{u(x + h)v(x) - u(x)v(x) - (u(x)v(x + h) - u(x)v(x))}{v(x + h)v(x)}.$$

5. State hypotheses which assure the validity of the formula (the extended Chain rule)

$$F'(x_0) = f'\{u[v(x_0)]\}u'[v(x_0)]v'(x_0)$$

where $F(x) = f\{u[v(x)]\}$. Prove the result.

6. Show that if f has a derivative at a point, then it is continuous at that point (Theorem 4.4).

7. We define the **one-sided derivative from the right** and **the left** by the formulas (respectively)

$$D^+ f(x_0) = \lim_{h \to 0^+} \frac{f(x_0 + h) - f(x_0)}{h},$$

$$D^- f(x_0) = \lim_{h \to 0^-} \frac{f(x_0 + h) - f(x_0)}{h}.$$

Show that a function f has a derivative at x_0 if and only if $D^+ f(x_0)$ and $D^- f(x_0)$ exist and are equal.

8. Prove Theorem 4.7 for n a positive integer. Then use Theorem 4.6 to establish the result for n a negative integer.

9. (a) Suppose $f: x \to x^3$ and $x_0 = 2$ in the Fundamental Lemma of differentiation. Show that $\eta(h) = 6h + h^2$.
   (b) Find the explicit value of $\eta(h)$ if $f: x \to x^{-3}$, $x_0 = 1$.

10. (Partial **Converse of the Chain rule**.) Suppose that f, g, and u are related so that $f(x) = g[u(x)]$, u is continuous at x_0, $f'(x_0)$ exists, $g'[u(x_0)]$ exists and is not zero. Then prove that $u'(x_0)$ is defined and $f'(x_0) = g'[u(x_0)]u'(x_0)$. [*Hint*: Proceed as in the proof of the Chain rule (Theorem 4.9) to obtain the formula

$$\frac{\Delta u}{h} = \frac{\Delta f/h}{[g'(u) + \eta(\Delta u)]}.$$

Since (by hypothesis) u is continuous at x_0, it follows that $\lim_{h \to 0} \Delta u = 0$. Hence $\eta(\Delta u) \to 0$ as $h \to 0$. Complete the proof.]

11. Prove Theorem 4.10.

12. Suppose that f is continuous on an open interval I containing x_0, suppose that f' is defined on I except possibly at x_0, and suppose that $f'(x) \to L$ as $x \to x_0$. Prove that $f'(x_0) = L$.

13. Prove Theorem 4.13. [*Hint*: Use the Mean-value theorem.]

14. Discuss the differentiability at $x = 0$ of the function

$$f: x \to \begin{cases} x \sin \dfrac{1}{x}, & x \neq 0, \\ 0, & x = 0. \end{cases}$$

15. Discuss the differentiability at $x = 0$ of the function

$$f: x \rightarrow \begin{cases} x^n \sin \dfrac{1}{x}, & x \neq 0, \\ 0, & x = 0, \end{cases}$$

where n is an integer larger than 1. For what values of k, does the kth derivative exist at $x = 0$? (See Problem 12.)

16. If f is differentiable at x_0, prove that

$$\lim_{h \to 0} \frac{f(x_0 + \alpha h) - f(x_0 - \beta h)}{h} = (\alpha + \beta) f'(x_0).$$

17. Prove the identity in between Equations (4.8) and (4.9) on page 90.

18. Suppose $\lim_{x \to a^+} f(x) = 0$, $\lim_{x \to a^+} F(x) = \infty$. State a theorem analogous to Theorem 4.16 for $\lim_{x \to a^+} f(x) F(x)$.

19. Prove Corollary 1 to Theorem 4.15.

20. Suppose that $f(x) \to \infty$, $F(x) \to \infty$, and $f'(x)/F'(x) \to +\infty$ as $x \to +\infty$. Prove that $f(x)/F(x) \to +\infty$ as $x \to +\infty$.

21. Evaluate the following limit:

$$\lim_{x \to 0} \frac{\tan x - x}{x^3}.$$

22. Evaluate the following limit:

$$\lim_{x \to +\infty} \frac{x^3}{e^x}.$$

23. Evaluate

$$\lim_{x \to +\infty} \left(1 + \frac{1}{x}\right)^x.$$

[*Hint*: Take logarithms.]

24. If the second derivative f'' exists at a value x_0, show that

$$\lim_{h \to 0} \frac{f(x_0 + h) - 2f(x_0) + f(x_0 - h)}{h^2} = f''(x_0).$$

25. Suppose that f is defined by

$$f: x \rightarrow \begin{cases} e^{-1/x^2} & \text{if } x > 0, \\ 0 & \text{if } x \leq 0. \end{cases}$$

(a) Show that $f^{(n)}(0)$ exists for every positive integer n (see Problem 7) and has the value 0.

(b) Show that the function

$$g: x \rightarrow \begin{cases} e^{-1/x^2} & \text{if } x \neq 0, \\ 0 & \text{if } x = 0, \end{cases}$$

has the property that $g^{(n)}(0) = 0$ for all n. (See Problem 10.)

4.2. Inverse Functions in $\mathbb{R}^1$

We have seen in Section 1.2 that a **relation from** $\mathbb{R}^1$ **to** $\mathbb{R}^1$ is a set of ordered pairs of real numbers; that is, a relation is a set in $\mathbb{R}^2$. A function from $\mathbb{R}^1$ to $\mathbb{R}^1$ is a particular case of a relation, one in which no two ordered pairs have the same first element.

Definition. Suppose that S is a relation from $\mathbb{R}^1$ to $\mathbb{R}^1$. We define the **inverse relation** of S as the set of pairs (x, y) such that (y, x) is in S.

If S is the solution set in $\mathbb{R}^2$ of an equation such as $f(x, y) = 0$, the inverse of S is the solution set in $\mathbb{R}^2$ of the equation $f(y, x) = 0$. Suppose that f is a function, and we consider the ordered pairs in $\mathbb{R}^2$ which form the **graph** of the equation $y = f(x)$. Then the inverse of f is the graph of the equation $x = f(y)$. The simplest examples of functions show that the inverse of a function is a relation, and not necessarily a function. For example, the function $f: x \to x^2$, with graph $y = x^2$ has for its inverse the relation whose graph is given by $y^2 = x$. This inverse relation is not a function.

Theorem 4.17 (Inverse function theorem). *Suppose that f is a continuous, increasing function which has an interval I for domain and has range J. (See Figure 4.5.) Then,*

(a) *J is an interval.*
(b) *The inverse relation g of f is a function with domain J, and g is continuous and increasing on J.*
(c) *We have*
$$g[f(x)] = x \quad \text{for } x \in I \text{ and } f[g(x)] = x \text{ for } x \in J. \tag{4.11}$$

A similar result holds if f is decreasing on I.

PROOF. From the Intermediate-value theorem (Theorem 3.3) we know at once that J is an interval. We next establish the formulas in Part (c). Let x_0 be any

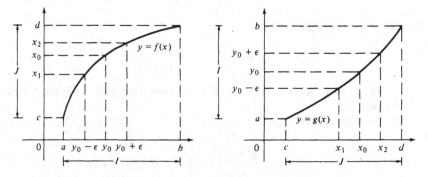

Figure 4.5. f and g are inverse functions.

point of J. Then there is a number y_0 in I such that $x_0 = f(y_0)$. Since f is increasing, y_0 is unique. Hence g is a function with domain J, and the formulas in Part (c) hold. To show that g is increasing, suppose that $x_1 < x_2$ with x_1, x_2 in J. Then it follows that $g(x_1) < g(x_2)$, for otherwise we would have $f[g(x_1)] \geqslant f[g(x_2)]$ or $x_1 \geqslant x_2$.

We now show that g is continuous. Let x_0 be an interior point of J and suppose $y_0 = g(x_0)$ or, equivalently, $x_0 = f(y_0)$. Let x_1' and x_2' be points of J such that $x_1' < x_0 < x_2'$. Then the points $y_1' = g(x_1')$ and $y_2' = g(x_2')$ of I are such that $y_1' < y_0 < y_2'$. Hence y_0 is an interior point of I. Now let $\varepsilon > 0$ be given and chosen so small that $y_0 - \varepsilon$ and $y_0 + \varepsilon$ are points of I. We define $x_1 = f(y_0 - \varepsilon)$ and $x_2 = f(y_0 + \varepsilon)$. See Figure 4.5. Since g is increasing,

$$y_0 - \varepsilon = g(x_1) \leqslant g(x) \leqslant g(x_2) = y_0 + \varepsilon \quad \text{for all } x \text{ such that} \quad x_1 \leqslant x \leqslant x_2.$$

Because $y_0 = g(x_0)$ the above inequalities may be written

$$g(x_0) - \varepsilon \leqslant g(x) \leqslant g(x_0) + \varepsilon \quad \text{for} \quad x_1 \leqslant x \leqslant x_2. \tag{4.12}$$

Choosing δ as the minimum of the distances $x_2 - x_0$ and $x_0 - x_1$, we obtain

$$|g(x) - g(x_0)| < \varepsilon \quad \text{whenever} \quad |x - x_0| < \delta.$$

That is, g is continuous at x_0, an arbitrary interior point of J. A slight modification of the above argument shows that if the range J contains its endpoints, then g is continuous on the right at c and continuous on the left at d. $\qquad\square$

The proof of Theorem 4.17 for functions which are decreasing is completely analogous. In this case the inverse function g is also decreasing. If the function f in the Inverse function theorem has a derivative, then the following result on the differentiability of the inverse function g holds.

Theorem 4.18 (Inverse differentiation theorem). *Suppose that f satisfies the hypotheses of Theorem 4.17. Assume that x_0 is a point of J such that $f'[g(x_0)]$ is defined and is different from zero. Then $g'(x_0)$ exists and*

$$g'(x_0) = \frac{1}{f'[g(x_0)]}. \tag{4.13}$$

PROOF. From Theorem 4.17 we have Formula (4.11)

$$f[g(x)] = x.$$

Using the Chain rule as given in Problem 10 of Section 4.1, we conclude that $g'(x_0)$ exists and

$$f'[g(x_0)]g'(x_0) = 1.$$

Since $f' \neq 0$ by hypothesis, the result follows. $\qquad\square$

Remarks. It is most often the case that a function f is not always increasing or always decreasing. In such situations the inverse relation of f may be

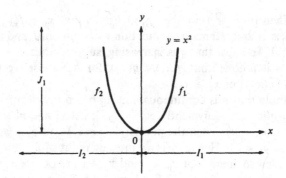

Figure 4.6

analyzed as follows: First find the intervals $I_1, I_2, \ldots$ on each of which f is always increasing or always decreasing. Denote by f_i the function f restricted to the interval $I_i, i = 1, 2, \ldots$. Then the inverse of each f_i is a function which we denote g_i. These inverses may be analyzed separately and differentiation applied to the formulas $f_i[g_i(x)] = x, i = 1, 2, \ldots$.

For example, suppose $f: x \to x^2$ is defined on $I = \{x: -\infty < x < \infty\}$. Then f is increasing on the interval $I_1 = \{x: 0 \leqslant x < \infty\}$ and decreasing on $I_2 = \{x: -\infty < x \leqslant 0\}$. The restriction of f to I_1, denoted f_1, has the inverse $g_1: x \to \sqrt{x}$ with domain $J_1 = \{x: 0 \leqslant x < \infty\}$. The restriction of f to I_2, denoted f_2, has the inverse $g_2: x \to -\sqrt{x}$, also with domain J_1. Equations (4.11) become in these two cases:

$$(\sqrt{x})^2 = x \quad \text{for } x \in J_1, \qquad \sqrt{x^2} = x \quad \text{for } x \in I_1,$$
$$(-\sqrt{x})^2 = x \quad \text{for } x \in J_1, \qquad -\sqrt{x^2} = x \quad \text{for } x \in I_2.$$

The inverse relation of f is $g_1 \cup g_2$. See Figure 4.6.

PROBLEMS

In each of Problems 1 through 12 a function f is given. Find the intervals $I_1, I_2, \ldots$ on which f is either increasing or decreasing, and find the corresponding intervals $J_1, J_2, \ldots$ on which the inverses are defined. Plot a graph of f and the inverse functions $g_1, g_2, \ldots$ corresponding to $J_1, J_2, \ldots$. Find expressions for each g_i when possible.

1. $f: x \to x^2 + 2x + 2$

2. $f: x \to (x^2/2) + 3x - 4$

3. $f: x \to 4x - x^2$

4. $f: x \to 2 - x - (x^2/2)$

5. $f: x \to 2x/(x + 2)$

6. $f: x \to (1 + x)/(1 - x)$

7. $f: x \to (4x)/(x^2 + 1)$

8. $f: x \to (x - 1)^3$

9. $f: x \to x^3 + 3x$

10. $f: x \to (2x^3/3) + x^2 - 4x + 1$

11. $f: x \to x^3 + 3x^2 - 9x + 4$ 12. $f: x \to (x^4/4) + (x^3/3) - x^2 + 2/3$

13. Suppose that f and g are increasing on an interval I and that $f(x) > g(x)$ for all $x \in I$. Denote the inverses of f and g by F and G and their domains by J_1 and J_2, respectively. Prove that $F(x) < G(x)$ for each $x \in J_1 \cap J_2$.

In each of Problems 14 through 18 a function f is given which is increasing or decreasing on I. Hence there is an inverse function g. Compute f' and g' and verify Formula (4.13) of the Inverse differentiation theorem.

14. $f: x \to 4x/(x^2 + 1), I = \{x: \frac{1}{2} < x < \infty\}$

15. $f: x \to (x - 1)^3, I = \{x: 1 < x < \infty\}$

16. $f: x \to x^3 + 3x, I = \{x: -\infty < x < \infty\}$

17. $f: x \to \sin x, I = \{x: \pi/2 < x < \pi\}$

18. $f: x \to e^{3x}, I = \{x: -\infty < x < \infty\}$

19. In the Inverse function theorem show that if the range J contains its endpoints, then g is continuous on the right at c and continuous on the left at d.

20. Given the function

$$f: x \to \begin{cases} x \sin(1/x), & 0 < x \leqslant 1, \\ 0, & x = 0, \end{cases}$$

describe the intervals $I_1, I_2, \ldots$, and the corresponding intervals $J_1, J_2, \ldots$ for which the Inverse function theorem holds.

CHAPTER 5

Elementary Theory of Integration

5.1. The Darboux Integral for Functions on $\mathbb{R}^1$

The reader is undoubtedly familiar with the idea of integral and with methods of performing integrations. In this section we define integral precisely and prove the basic theorems which justify the processes of integration employed in a first course in calculus.

Let f be a bounded function whose domain is a closed interval $I = \{x: a \leqslant x \leqslant b\}$. We subdivide I by introducing points $t_1, t_2, \ldots, t_{n-1}$ which are interior to I. Setting $a = t_0$ and $b = t_n$ and ordering the points so that $t_0 < t_1 < t_2 < \cdots < t_n$, we denote by $I_1, I_2, \ldots, I_n$ the intervals $I_i = \{x: t_{i-1} \leqslant x \leqslant t_i\}$. Two successive intervals have exactly one point in common. (See Figure 5.1.) We call such a decomposition of I into subintervals a **subdivision** of I, and we use the symbol Δ to indicate such a subdivision.

Since f is bounded on I, it has a least upper bound (l.u.b.), denoted by M. The greatest lower bound (g.l.b.) of f on I is denoted by m. Similarly, M_i and m_i denote the l.u.b. and g.l.b., respectively, of f on I_i. The length of the interval I_i is $t_i - t_{i-1}$; it is denoted by $l(I_i)$.

Definitions. The **upper Darboux sum** of f with respect to the subdivision Δ, denoted $S^+(f, \Delta)$, is defined by

$$S^+(f, \Delta) = \sum_{i=1}^{n} M_i l(I_i). \tag{5.1}$$

Similarly the **lower Darboux sum**, denoted $S_-(f, \Delta)$, is defined by

$$S_-(f, \Delta) = \sum_{i=1}^{n} m_i l(I_i). \tag{5.2}$$

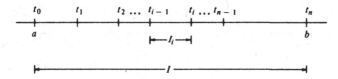

Figure 5.1. A subdivision of I.

Figure 5.2 shows the significance of a typical term in S^+ and one in S_- in case $f(x) > 0$ for x on I.

Suppose Δ is a subdivision $a = t_0 < t_1 < \cdots < t_{n-1} < t_n = b$, with the corresponding intervals denoted $I_1, I_2, \ldots, I_n$. We obtain a new subdivision by introducing additional subdivision points between the various $\{t_i\}$. This new subdivision will have subintervals $I'_1, I'_2, \ldots, I'_m$ each of which is a part (or all) of one of the subintervals of Δ. We denote this new subdivision Δ' and call it a **refinement of** Δ.

Suppose that Δ_1 with subintervals $I_1, I_2, \ldots, I_n$ and that Δ_2 with subintervals $J_1, J_2, \ldots, J_m$ are two subdivisions of an interval I. We get a new subdivision of I by taking *all* the endpoints of the subintervals of both Δ_1 and Δ_2, arranging them in order of increasing size, and then labelling each subinterval having as its endpoints two successive subdivision points. Such a subdivision is called the **common refinement of** Δ_1 **and** Δ_2, and each subinterval of this new subdivision is of the form $I_i \cap J_k$, $i = 1, 2, \ldots, n$, $k = 1, 2, \ldots, m$. Each $I_i \cap J_k$ is either empty or entirely contained in a unique subinterval (I_i) of Δ_1 and a unique subinterval (J_k) of Δ_2. An illustration of such a common refinement is exhibited in Figure 5.3.

Theorem 5.1. *Suppose that f is a bounded function with domain $I = \{x : a \leqslant x \leqslant b\}$. Let Δ be a subdivision of I and suppose that M and m are the least upper*

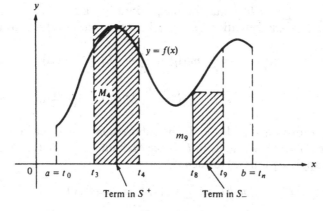

Figure 5.2

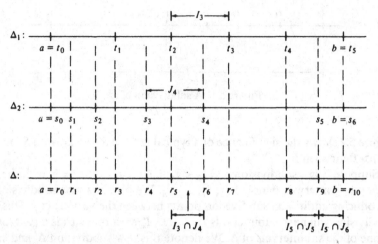

Figure 5.3. A common refinement.

bound and greatest lower bound of f on I, respectively. Then

(a) $$m(b - a) \leqslant S_-(f, \Delta) \leqslant S^+(f, \Delta) \leqslant M(b - a).$$

(b) If Δ' is a refinement of Δ, then

$$S_-(f, \Delta) \leqslant S_-(f, \Delta') \leqslant S^+(f, \Delta') \leqslant S^+(f, \Delta).$$

(c) If Δ_1 and Δ_2 are any two subdivisions of I, then

$$S_-(f, \Delta_1) \leqslant S^+(f, \Delta_2).$$

PROOF

   (a) From the definition of least upper bound and greatest lower bound, we
have

$$m \leqslant m_i \leqslant M_i \leqslant M, \qquad i = 1, 2, \ldots, n.$$

Also, $b - a = \sum_{i=1}^{n} l(I_i)$. Thus the inequalities in Part (a) are an immediate
consequence of the definition of S^+ and S_- as given in Equations (5.1) and
(5.2):

$$m(b - a) = \sum_{i=1}^{n} ml(I_i) \leqslant \sum_{i=1}^{n} m_i l(I_i) = S_-(f, \Delta)$$

$$\leqslant S^+(f, \Delta) = \sum_{i=1}^{n} M_i l(I_i) \leqslant M(b - a)$$

   (b) To prove Part (b), let Δ_i be the subdivision of I_i which consists of all
those intervals of Δ' which lie in I_i. Since each interval of Δ' is in a unique Δ_i
we have, applying Part (a) to each Δ_i,

$$S_-(f, \Delta) = \sum_{i=1}^{n} m_i l(I_i) \leqslant \sum_{i=1}^{n} S_-(f, \Delta_i) = S_-(f, \Delta')$$

$$\leqslant S^+(f, \Delta') = \sum_{i=1}^{n} S^+(f, \Delta_i) \leqslant \sum_{i=1}^{n} M_i l(I_i) = S^+(f, \Delta).$$

(c) To prove Part (c), let Δ be the common refinement of Δ_1 and Δ_2. Then, using the result in Part (b), we find

$$S_-(f, \Delta_1) \leqslant S_-(f, \Delta) \leqslant S^+(f, \Delta) \leqslant S^+(f, \Delta_2). \qquad \square$$

Definitions. If f is a function on $\mathbb{R}^1$ which is defined and bounded on $I = \{x : a \leqslant x \leqslant b\}$, we define its **upper** and **lower Darboux integrals** by

$$\overline{\int_a^b} f(x)\, dx = \inf S^+(f, \Delta) \quad \text{for all subdivisions } \Delta \text{ of } I$$

$$\underline{\int_a^b} f(x)\, dx = \sup S_-(f, \Delta) \quad \text{for all subdivisions } \Delta \text{ of } I.$$

If

$$\overline{\int_a^b} f(x)\, dx = \underline{\int_a^b} f(x)\, dx,$$

we say that f is **Darboux integrable**, or just **integrable, on** I and we designate the common value by

$$\int_a^b f(x)\, dx.$$

The following results are all direct consequences of the above definitions.

Theorem 5.2. *If $m \leqslant f(x) \leqslant M$ for all $x \in I = \{x : a \leqslant x \leqslant b\}$, then*

$$m(b - a) \leqslant \underline{\int_a^b} f(x)\, dx \leqslant \overline{\int_a^b} f(x)\, dx \leqslant M(b - a).$$

PROOF. Let Δ_1 and Δ_2 be subdivisions of I. Then from Parts (a) and (c) of Theorem 5.1, it follows that

$$m(b - a) \leqslant S_-(f, \Delta_1) \leqslant S^+(f, \Delta_2) \leqslant M(b - a).$$

We keep Δ_1 fixed and let Δ_2 vary over all possible subdivisions. Thus $S^+(f, \Delta_2)$ is always larger than or equal to $S_-(f, \Delta_1)$, and so its greatest lower bound is always larger than or equal to $S_-(f, \Delta_1)$. We conclude that

$$m(b - a) \leqslant S_-(f, \Delta_1) \leqslant \overline{\int_a^b} f(x)\, dx \leqslant M(b - a).$$

Now letting Δ_1 vary over all possible subdivisions, we see that the least upper bound of $S_-(f, \Delta_1)$ cannot exceed $\overline{\int_a^b} f(x)\, dx$. The conclusion of the theorem follows.

Corollary. *If $m \leqslant f(x) \leqslant M$ for all $x \in I = \{x : a \leqslant x \leqslant b\}$, and f is Darboux integrable, then*

$$m(b - a) \leqslant \int_a^b f(x)\, dx \leqslant M(b - a).$$

The next theorem establishes simple properties of upper and lower Darboux integrals. We first prove the following elementary facts about the supremum and infimum of functions.

Lemma 5.1. *Suppose that f is a bounded function on an interval $I = \{x: a \leqslant x \leqslant b\}$. Let k be a real number and define $g(x) = kf(x)$.*

(i) *Suppose* $k > 0$. *Then* $\inf_{x \in I} g(x) = k \inf_{x \in I} f(x)$ *and* $\sup_{x \in I} g(x) = k \sup_{x \in I} f(x)$.

(ii) *Suppose* $k < 0$. *Then* $\inf_{x \in I} g(x) = k \sup_{x \in I} f(x)$ *and* $\sup_{x \in I} g(x) = k \inf_{x \in I} f(x)$.

PROOF. We prove the last statement in Part (ii), the other proofs being similar. Let $m = \inf f(x)$ for $x \in I$. Then $f(x) \geqslant m$ for all $x \in I$. Since $k < 0$, we have $g(x) = kf(x) \leqslant km$ for all $x \in I$. Hence

$$\sup g(x) \leqslant km. \tag{5.3}$$

To show $\sup g(x) = km$, let $\varepsilon > 0$ be given. Then there is an $x_0 \in I$ such that $f(x_0) < m - (\varepsilon/k)$. Consequently, since $k < 0$, we find

$$g(x_0) = kf(x_0) > km - \varepsilon.$$

Since ε is arbitrary, we obtain

$$\sup g(x) \geqslant km. \tag{5.4}$$

The result follows from Inequalities (5.3) and (5.4).   □

Theorem 5.3. *Assuming that all functions below are bounded, we have the following formulas:*

(a) *If* $g(x) = kf(x)$ *for all* $x \in I = \{x: a \leqslant x \leqslant b\}$ *and* k *is a positive number, then*

   (i) $\underline{\int_a^b} g(x)\, dx = k \underline{\int_a^b} f(x)\, dx$ *and* $\overline{\int_a^b} g(x)\, dx = k \overline{\int_a^b} f(x)\, dx$.

   If $k < 0$, *then*

   (ii) $\underline{\int_a^b} g(x)\, dx = k \overline{\int_a^b} f(x)\, dx$ *and* $\overline{\int_a^b} g(x)\, dx = k \underline{\int_a^b} f(x)\, dx$.

(b) *If* $h(x) = f_1(x) + f_2(x)$ *for all* $x \in I$, *then*

   (i) $\underline{\int_a^b} h(x)\, dx \geqslant \underline{\int_a^l} f_1(x)\, dx + \underline{\int_a^b} f_2(x)\, dx$.

   (ii) $\overline{\int_a^b} h(x)\, dx \leqslant \overline{\int_a^b} f_1(x)\, dx + \overline{\int_a^b} f_2(x)\, dx$.

(c) *If* $f_1(x) \leqslant f_2(x)$ *for all* $x \in I$, *then*

   (i) $\underline{\int_a^b} f_1(x)\, dx \leqslant \underline{\int_a^b} f_2(x)\, dx$ *and*

   (ii) $\overline{\int_a^b} f_1(x)\, dx \leqslant \overline{\int_a^b} f_2(x)\, dx$.

(d) *If $a < b < c$ and f is bounded on $I' = \{x: a \leqslant x \leqslant c\}$, then*

(i) $\underline{\int}_a^c f(x)\, dx = \underline{\int}_a^b f(x)\, dx + \underline{\int}_b^c f(x)\, dx$ *and*

(ii) $\overline{\int}_a^c f(x)\, dx = \overline{\int}_a^b f(x)\, dx + \overline{\int}_b^c f(x)\, dx.$

We shall prove (i) of Part (a) and (ii) of Part (d), the proofs of the remaining parts being similar.

PROOF OF (i) OF PART (a). Let Δ be any subdivision of I with subintervals I_1, $I_2, \ldots, I_n$. Denote by m_i and M_i the inf f and sup f on I_i, respectively. Similarly, $\tilde{m}_i$ and $\tilde{M}_i$ are the corresponding values for g on I_i. Since $k > 0$, Lemma 5.1 shows that $\tilde{m}_i = km_i$ and $\tilde{M}_i = kM_i$. Hence

$$\sup S_-(g, \Delta) = \sup kS_-(f, \Delta) = k \sup S_-(f, \Delta).$$

Therefore

$$\underline{\int}_a^b g\, dx = k \underline{\int}_a^b f\, dx.$$

PROOF OF (ii) OF PART (d). Let $\varepsilon > 0$ be given. Since $\overline{\int}_a^b f(x)\, dx$ is defined as a greatest lower bound over all subdivisions, there is a subdivision Δ_1 of $I = \{x: a \leqslant x \leqslant b\}$ such that

$$S^+(f, \Delta_1) < \overline{\int}_a^b f(x)\, dx + \frac{1}{2}\varepsilon. \tag{5.5}$$

Similarly, there is a subdivision Δ_2 of $I'' = \{x: b \leqslant x \leqslant c\}$ such that

$$S^+(f, \Delta_2) < \overline{\int}_b^c f(x)\, dx + \frac{1}{2}\varepsilon. \tag{5.6}$$

Let Δ be the subdivision of $I' = \{x: a \leqslant x \leqslant c\}$ consisting of all subintervals of Δ_1 and Δ_2. Then because of (5.5) and (5.6)

$$\overline{\int}_a^c f(x)\, dx \leqslant S^+(f, \Delta)$$

$$= S^+(f, \Delta_1) + S^+(f, \Delta_2)$$

$$< \overline{\int}_a^b f(x)\, dx + \overline{\int}_b^c f(x)\, dx + \frac{1}{2}\varepsilon + \frac{1}{2}\varepsilon.$$

Since these inequalities are true for every $\varepsilon > 0$, we conclude that

$$\overline{\int}_a^c f(x)\, dx \leqslant \overline{\int}_a^b f(x)\, dx + \overline{\int}_b^c f(x)\, dx. \tag{5.7}$$

We now show that the reverse inequality also holds. Let $\varepsilon > 0$ be given. Then, by definition, there is a subdivision Δ of I' such that

$$S^+(f, \Delta) < \overline{\int}_a^c f(x)\, dx + \varepsilon.$$

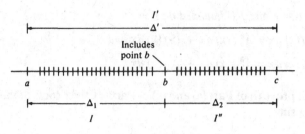

Figure 5.4

We make a refinement of Δ by introducing into Δ one more subdivision point—namely b. Calling this refinement Δ', we know that $S^+(f, \Delta') \leqslant S^+(f, \Delta)$. Next denote by Δ_1 the subintervals of Δ' contained in I, and by Δ_2 the subintervals of Δ' contained in I''. (See Figure 5.4.) Then

$$\overline{\int_a^b} f(x)\, dx + \overline{\int_b^c} f(x)\, dx \leqslant S^+(f, \Delta_1) + S^+(f, \Delta_2)$$
$$= S^+(f, \Delta')$$
$$\leqslant S^+(f, \Delta)$$
$$< \overline{\int_a^c} f(x)\, dx + \varepsilon.$$

Since these inequalities are true for every $\varepsilon > 0$, it follows that

$$\overline{\int_a^b} f(x)\, dx + \overline{\int_b^c} f(x)\, dx \leqslant \overline{\int_a^c} f(x)\, dx.$$

Combining this with Inequality (5.7) we get the result for upper integrals. □

Corollary. *If the functions considered in Theorem 5.3 are Darboux integrable, the following formulas hold:*

(a) *If $g(x) = kf(x)$ and k is any constant, then*

$$\int_a^b g(x)\, dx = k \int_a^b f(x)\, dx.$$

(b) *If $h(x) = f_1(x) + f_2(x)$, then*

$$\int_a^b h(x)\, dx = \int_a^b f_1(x)\, dx + \int_a^b f_2(x)\, dx.$$

(c) *If $f_1(x) \leqslant f_2(x)$ for all $x \in I$, then*

$$\int_a^b f_1(x)\, dx \leqslant \int_a^b f_2(x)\, dx.$$

(d) *Suppose f is Darboux integrable on $I_1 = \{x : a \leqslant x \leqslant b\}$ and on $I_2 =$*

$\{x: b \leqslant x \leqslant c\}$. Then it is *Darboux integrable on* $I = \{x: a \leqslant x \leqslant c\}$, *and*

$$\int_a^c f(x) \, dx = \int_a^b f(x) \, dx + \int_b^c f(x) \, dx. \qquad (5.8)$$

It is useful to make the definitions

$$\int_a^a f(x) \, dx = 0 \quad \text{and} \quad \int_b^a f(x) \, dx = -\int_a^b f(x) \, dx.$$

Then Formula (5.8) above is valid whether or not b is between a and c so long as all the integrals exist.

Theorem 5.4. *If f is integrable on an interval I, then it is integrable on every subinterval I' contained in I.*

The proof of this result is left to the reader.

It is important to be able to decide when a particular function is integrable. The next theorem gives a necessary and sufficient condition for integrability.

Theorem 5.5. *Suppose that f is bounded on an interval $I = \{x: a \leqslant x \leqslant b\}$. Then f is integrable $\Leftrightarrow$ for every $\varepsilon > 0$ there is a subdivision Δ of I such that*

$$S^+(f, \Delta) - S_-(f, \Delta) < \varepsilon. \qquad (5.9)$$

PROOF. Suppose that Condition (5.9) holds. Then from the definitions of upper and lower Darboux integrals, we have

$$\overline{\int_a^b} f(x) \, dx - \underline{\int_a^b} f(x) \, dx \leqslant S^+(f, \Delta) - S_-(f, \Delta) < \varepsilon.$$

Since this inequality holds for every $\varepsilon > 0$ and the left side of the inequality is independent of ε, it follows that

$$\overline{\int_a^b} f(x) \, dx - \underline{\int_a^b} f(x) \, dx = 0.$$

Hence f is integrable.

Now assume f is integrable; we wish to establish Inequality (5.9). For any $\varepsilon > 0$ there are subdivisions Δ_1 and Δ_2 such that

$$S^+(f, \Delta_1) < \overline{\int_a^b} f(x) \, dx + \frac{1}{2}\varepsilon \quad \text{and} \quad S_-(f, \Delta_2) > \underline{\int_a^b} f(x) \, dx - \frac{\varepsilon}{2}. \qquad (5.10)$$

We choose Δ as the common refinement of Δ_1 and Δ_2. Then (from Theorem 5.1)

$$S^+(f, \Delta) - S_-(f, \Delta) \leqslant S^+(f, \Delta_1) - S_-(f, \Delta_2).$$

Substituting Inequalities (5.10) into the above expression and using the fact that $\overline{\int_a^b} f(x) \, dx = \underline{\int_a^b} f(x) \, dx$, we obtain $S^+(f, \Delta) - S_-(f, \Delta) < \varepsilon$, as required. $\square$

Corollary 1. *If f is continuous on the closed interval $I = \{x: a \leqslant x \leqslant b\}$, then f is integrable.*

Corollary 2. *If f is monotone on the closed interval $I = \{x: a \leqslant x \leqslant b\}$, then f is integrable.*

PROOF OF COROLLARY 1 (SKETCH). The function f is uniformly continuous and hence for any $\varepsilon > 0$ there is a δ such that

$$|f(x) - f(y)| < \frac{\varepsilon}{b - a} \quad \text{whenever} \quad |x - y| < \delta.$$

Choose a subdivision Δ of I so that no subinterval of Δ has length larger than δ. This allows us to establish Formula (5.9) and the result follows from Theorem 5.5.   □

The proof of Corollary 2 is left to the reader. See Problem 9 at the end of the section and the hint given there.

Remark. In Corollaries 1 and 2 it is essential that the interval I is closed. For example, the function $f: x \to 1/x$ is both continuous and monotone on the half-open interval $I = \{x: 0 < x \leqslant 1\}$. However, f is not integrable there; it is not bounded there. Also, as the reader knows,

$$\int_{\varepsilon}^{1} \frac{1}{x}\, dx = \log x]_{\varepsilon}^{1} = -\log \varepsilon.$$

The integral tends to infinity as $\varepsilon \to 0$. It is easy to verify directly that $S^{+}(f, \Delta) = +\infty$ for every subdivision of I.

Theorem 5.6 (Mean-value theorem for integrals). *Suppose that f is continuous on $I = \{x: a \leqslant x \leqslant b\}$. Then there is a number ξ in I such that*

$$\int_{a}^{b} f(x)\, dx = f(\xi)(b - a).$$

PROOF. According to the Corollary to Theorem 5.2, we have

$$m(b - a) \leqslant \int_{a}^{b} f(x)\, dx \leqslant M(b - a), \tag{5.11}$$

where m and M are the minimum and maximum of f on I, respectively. From the Extreme-value theorem (Theorem 3.12) there are numbers x_0 and $x_1 \in I$ such that $f(x_0) = m$ and $f(x_1) = M$. From Inequalities (5.11) it follows that

$$\int_{a}^{b} f(x)\, dx = A(b - a)$$

where A is a number such that $f(x_0) \leqslant A \leqslant f(x_1)$. Then the Intermediate-value theorem (Theorem 3.3) shows that there is a number $\xi \in I$ such that $f(\xi) = A$. $\qquad\qquad\qquad\qquad\qquad\qquad\qquad\qquad\qquad\qquad\qquad\qquad\square$

We now present two forms of the Fundamental theorem of calculus, a result which shows that differentiation and integration are inverse processes.

Theorem 5.7 (Fundamental theorem of calculus—first form). *Suppose that* f *is continuous on* $I = \{x: a \leqslant x \leqslant b\}$, *and that* F *is defined by*

$$F(x) = \int_a^x f(t)\, dt.$$

Then F *is continuous on* I *and* $F'(x) = f(x)$ *for each* $x \in I_1 = \{x: a < x < b\}$.

PROOF. Since f is integrable on every subinterval of I, we have

$$\int_a^{x+h} f(t)\, dt = \int_a^x f(t)\, dt + \int_x^{x+h} f(t)\, dt$$

or

$$F(x + h) = F(x) + \int_x^{x+h} f(t)\, dt.$$

We apply the Mean-value theorem for integrals to the last term on the right, getting

$$\frac{F(x + h) - F(x)}{h} = f(\xi),$$

where ξ is some number between x and $x + h$. Now, let $\varepsilon > 0$ be given. There is a $\delta > 0$ such that $|f(y) - f(x)| < \varepsilon$ for all y on I_1 (taking $\xi = y$) with $|y - x| < \delta$. Thus, if $0 < |h| < \delta$, we find that

$$\left| \frac{F(x + h) - F(x)}{h} - f(x) \right| = |f(\xi) - f(x)| < \varepsilon$$

since $|\xi - x| \leqslant |h| < \delta$. Since this is true for each $\varepsilon > 0$, the result follows. $\quad\square$

Theorem 5.8 (Fundamental theorem of calculus—second form). *Suppose that* f *and* F *are continuous on* $I = \{x: a \leqslant x \leqslant b\}$ *and that* $F'(x) = f(x)$ *for each* $x \in I_1 = \{x: a < x < b\}$. *Then*

$$\int_a^b f(x)\, dx = F(b) - F(a).$$

The proof is left to the reader. See Problem 20 at the end of the section and the hint given there.

Suppose that f is a nonnegative integrable function defined on $I = \{x: a \leqslant x \leqslant b\}$. We are familiar with the fact that

$$\int_a^b f(x)\, dx \qquad\qquad\qquad\qquad (5.12)$$

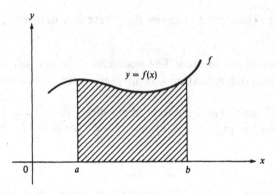

Figure 5.5

gives the area under the curve, as shown in Figure 5.5. In elementary courses, the definition of area is usually discussed intuitively. The area of rectangles, polygons, circles, etc. are determined by formulas and a precise definition of the area of more general regions is usually omitted. One way to proceed would use Integral (5.12) as a *definition* of the area of the region shown in Figure 5.5. Then more general regions can be decomposed into subregions each of the type shown in the figure, and the area of the entire region is taken as the sum of the areas of its constituent parts.

A more satisfactory approach starts from the basic principle that a square of one unit length on each side has an area of one square unit. This fact together with a limiting principle allows us to develop a precise definition of area—one which agrees fully with our intuition. We provide the details in Section 5.4.

PROBLEMS

1. Compute $S^+(f, \Delta)$ and $S_-(f, \Delta)$ for the function $f: x \to x^2$ defined on $I = \{x: 0 \leqslant x \leqslant 1\}$ where Δ is the subdivision of I into 5 subintervals of equal size.

2. (a) Given the function $f: x \to x^3$ defined on $I = \{x: 0 \leqslant x \leqslant 1\}$. Suppose Δ is a subdivision and Δ' is a refinement of Δ which adds one more point. Show that

$$S^+(f, \Delta') < S^+(f, \Delta) \quad \text{and} \quad S_-(f, \Delta') > S_-(f, \Delta).$$

   (b) Give an example of a function f defined on I such that

$$S^+(f, \Delta') = S^+(f, \Delta) \quad \text{and} \quad S_-(f, \Delta') = S_-(f, \Delta)$$

   for the two subdivisions in Part (a).

   (c) If f is a strictly increasing continuous function on I show that $S^+(f, \Delta') < S^+(f, \Delta)$ where Δ' is any refinement of Δ.

3. If $g(x) = kf(x)$ for all $x \in I = \{x: a \leqslant x \leqslant b\}$, show that (Theorem 5.3(a)):

$$\underline{\int_a^b} g(x)\, dx = k \overline{\int_a^b} f(x)\, dx \quad \text{if} \quad k < 0.$$

4. (a) If $h(x) = f_1(x) + f_2(x)$ for $x \in I = \{x: a \leqslant x \leqslant b\}$, show that (Theorem 5.3(b))

$$\int_{\underline{a}}^{b} h(x)\, dx \geqslant \int_{\underline{a}}^{b} f_1(x)\, dx + \int_{\underline{a}}^{b} f_2(x)\, dx,$$

$$\int_a^{\overline{b}} h(x)\, dx \leqslant \int_a^{\overline{b}} f_1(x)\, dx + \int_a^{\overline{b}} f_2(x)\, dx.$$

(b) If f_1 and f_2 are Darboux integrable, show that

$$\int_a^b h(x)\, dx = \int_a^b f_1(x)\, dx + \int_a^b f_2(x)\, dx.$$

5. If $f_1(x) \leqslant f_2(x)$ for $x \in I = \{x: a \leqslant x \leqslant b\}$, show that (Theorem 5.3(c))

$$\int_{\underline{a}}^{b} f_1(x)\, dx \leqslant \int_{\underline{a}}^{b} f_2(x)\, dx \quad \text{and} \quad \int_a^{\overline{b}} f_1(x)\, dx \leqslant \int_a^{\overline{b}} f_2(x)\, dx.$$

If f_1 and f_2 are Darboux integrable conclude that

$$\int_a^b f_1(x)\, dx \leqslant \int_a^b f_2(x)\, dx.$$

6. Show that

$$\int_a^c f(x)\, dx = \int_a^b f(x)\, dx + \int_b^c f(x)\, dx$$

whether or not b is between a and c so long as all three integrals exist.

7. Suppose that f is integrable on $I = \{x: a \leqslant x \leqslant b\}$. Let $I' = \{x: \alpha \leqslant x \leqslant \beta\}$ be a subinterval of I. Prove that f is integrable on I' (Theorem 5.4). [*Hint*: With $a < \alpha < \beta < b$, extend Theorem 5.3(d) to three intervals. Then subtract the lower integrals from the upper ones.]

8. Complete the proof of Corollary 1 to Theorem 5.5.

9. If f is increasing on the interval $I = \{x: a \leqslant x \leqslant b\}$, show that f is integrable. (Corollary 2 to Theorem 5.5.) [*Hint*: Use the formula

$$S^+(f, \Delta) - S_-(f, \Delta) = \sum_{i=1}^n [f(x_i) - f(x_{i-1})]l(I_i).]$$

Replace each $l(I_i)$ by the length of the *longest* subinterval getting an inequality, and then observe that the terms "telescope."

10. A function f defined on an interval I is called a **step-function** if and only if I can be subdivided into a finite number of subintervals $I_1, I_2, \ldots, I_n$ such that $f(x) = c_i$ for all x interior to I_i where the c_i, $i = 1, 2, \ldots, n$, are constants. Prove that every step-function is integrable (whatever values $f(x)$ has at the endpoints of the I_i) and find a formula for the value of the integral.

11. Given the function f defined on $I = \{x: 0 \leqslant x \leqslant 1\}$ by the formula

$$f(x) = \begin{cases} 1 & \text{if } x \text{ is rational,} \\ 0 & \text{if } x \text{ is irrational.} \end{cases}$$

Prove that $\int_{\underline{0}}^{1} f(x)\, dx = 0$ and $\int_0^{\overline{1}} f(x)\, dx = 1$.

12. Suppose that f is a bounded function on $I = \{x: a \leqslant x \leqslant b\}$. Let $M = \sup f(x)$ and $m = \inf f(x)$ for $x \in I$. Also, define $M^* = \sup |f(x)|$ and $m^* = \inf |f(x)|$ for $x \in I$.
    (a) Show that $M^* - m^* \leqslant M - m$.
    (b) If f and g are nonnegative bounded functions on I and $N = \sup g(x)$, $n = \inf g(x)$ for $x \in I$, show that

$$\sup f(x)g(x) - \inf f(x)g(x) \leqslant MN - mn.$$

13. (a) Suppose that f is bounded and integrable on $I = \{x: a \leqslant x \leqslant b\}$. Prove that $|f|$ is integrable on I. [*Hint*: See Problem 12(a).]
    (b) Show that $|\int_a^b f(x)\, dx| \leqslant \int_a^b |f(x)|\, dx$.

14. Suppose that f and g are nonnegative bounded, and integrable on $I = \{x: a \leqslant x \leqslant b\}$. Prove that fg is integrable on I. [*Hint*: See Problem 12(b).]

15. Suppose that f is defined on $I = \{x: 0 \leqslant x \leqslant 1\}$ by the formula

$$f(x) = \begin{cases} \dfrac{1}{2^n} & \text{if } x = \dfrac{j}{2^n} \text{ where } j \text{ is an odd integer} \\ & \text{and } 0 < j < 2^n, n = 1, 2, \ldots, \\ 0 & \text{otherwise.} \end{cases}$$

Determine whether or not f is integrable and prove your result.

16. Suppose that f and g are positive and continuous on $I = \{x: a \leqslant x \leqslant b\}$. Prove that there is a number $\xi \in I$ such that

$$\int_a^b f(x)g(x)\, dx = f(\xi) \int_a^b g(x)\, dx.$$

[*Hint*: Use the Intermediate-value theorem and see Problem 5.]

17. Suppose that f is continuous on $I = \{x: a \leqslant x \leqslant b\}$ except at an interior point c. If f is also bounded on I, prove that f is integrable on I. Show that the value of f at c does not affect the value of $\int_a^b f(x)\, dx$. Conclude that the value of f at any *finite* number of points cannot affect the value of the integral of f.

18. Suppose that f is continuous, nonnegative and not identically zero on $I = \{x: a \leqslant x \leqslant b\}$. Prove that $\int_a^b f(x)\, dx > 0$. Is the result true if f is not continuous but only integrable on I?

19. Suppose that f is continuous on $I = \{x: a \leqslant x \leqslant b\}$ and that $\int_a^b f(x)g(x)\, dx = 0$ for every function g continuous on I. Prove that $f(x) \equiv 0$ on I.

20. Prove Theorem 5.8. [*Hint*: Write

$$F(b) - F(a) = \sum_{i=1}^{n} \{F(x_i) - F(x_{i-1})\}$$

for some subdivision of I. Apply the Mean-value theorem to each term in the sum.]

21. Suppose that f is continuous on $I = \{x: a \leqslant x \leqslant b\}$ and that $\int_a^b f(x)\, dx = 0$. If f is nonnegative on I show that $f \equiv 0$.

22. Given f defined on $I = \{x: a \leqslant x \leqslant b\}$. Suppose $\int_a^b f^2(x)\, dx$ exists. Does it follow that $\int_a^b f(x)\, dx$ exists?

5.2. The Riemann Integral

In addition to the development of the integral by the method of Darboux, there is a technique due to Riemann, which starts with a direct approximation of the integral by a sum. The main result of this section shows that the two definitions of integral are equivalent.

Definitions. Suppose that Δ is a subdivision of $I = \{x: a \leqslant x \leqslant b\}$ with subintervals $I_1, I_2, \ldots, I_n$. We call the **mesh** of the subdivision Δ the length of the largest subinterval among $l(I_1), l(I_2), \ldots, l(I_n)$. The mesh is denoted by $\|\Delta\|$. Suppose that f is defined on I and that Δ is a subdivision. In each subinterval of Δ we choose a point $x_i \in I_i$. The quantity

$$\sum_{i=1}^{n} f(x_i)l(I_i)$$

is called a **Riemann sum.**

For nonnegative functions it is intuitively clear that a Riemann sum with very small mesh gives a good approximation to the area under the curve. (See Figure 5.6.)

Definitions. Suppose that f is defined on $I = \{x: a \leqslant x \leqslant b\}$. Then f is **Riemann integrable on** I if and only if there is a number A with the following property: for every $\varepsilon > 0$ there is a $\delta > 0$ such that for every subdivision Δ with mesh less than δ, the inequality

$$\left| \sum_{i=1}^{n} f(x_i)l(I_i) - A \right| < \varepsilon \tag{5.13}$$

holds for every possible choice of x_i in I_i. The quantity A is called the **Riemann integral of** f.

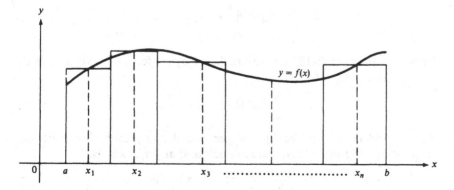

Figure 5.6. A Riemann sum.

Theorem 5.9. *The Riemann integral of a function is unique.*

The proof of Theorem 5.9 follows the outline of the proof of Theorem 2.1 on the uniqueness of limits, and is left to the reader. (See Problem 6 at the end of the section and the hint given there.)

Theorem 5.10. *If f is Riemann integrable on $I = \{x: a \leqslant x \leqslant b\}$, then f is bounded on I.*

PROOF. In the definition of Riemann integral, choose $\varepsilon = 1$ and let Δ be a subdivision such that Inequality (5.13) holds. We have

$$\left| \sum_{i=1}^{n} f(x_i) l(I_i) - A \right| < 1 \quad \text{and} \quad \left| \sum_{i=1}^{n} f(x_i') l(I_i) - A \right| < 1$$

where $x_1, x_2, \ldots, x_n$ and $x_1', x_2', \ldots, x_n'$ are any points of I such that x_i and x_i' are in I_i. Therefore

$$\left| \sum_{i=1}^{n} f(x_i) l(I_i) - \sum_{i=1}^{n} f(x_i') l(I_i) \right| < 2.$$

Now select $x_i = x_i'$ for $i = 2, 3, \ldots, n$. Then the above inequality becomes

$$|f(x_1) - f(x_1')| \, l(I_1) < 2 \quad \Leftrightarrow \quad |f(x_1) - f(x_1')| < \frac{2}{l(I_1)}.$$

Using the general inequality $|\alpha| - |\beta| < |\alpha - \beta|$, we obtain

$$|f(x_1)| < \frac{2}{l(I_1)} + |f(x_1')|.$$

Fix x_1' and observe that the above inequality is valid for every $x_1 \in I_1$. Hence f is bounded on I_1. Now the same argument can be made for $I_2, I_3, \ldots, I_n$. Therefore f is bounded on I. $\qquad\square$

Theorem 5.11. *If f is Riemann integrable on $I = \{x: a \leqslant x \leqslant b\}$, then f is Darboux integrable on I. Letting A denote the Riemann integral, we have*

$$A = \int_{a}^{b} f(x) \, dx.$$

PROOF. In Formula (5.13), we may replace ε by $\varepsilon/4$ for a subdivision Δ with sufficiently small mesh:

$$\left| \sum_{i=1}^{n} f(x_i) l(I_i) - A \right| < \frac{\varepsilon}{4}. \tag{5.14}$$

Let M_i and m_i denote the least upper bound and greatest lower bound, respectively, of f in I_i. Then there are points x_i' and x_i'' such that

$$f(x_i') > M_i - \frac{\varepsilon}{4(b-a)} \quad \text{and} \quad f(x_i'') < m_i + \frac{\varepsilon}{4(b-a)}.$$

These inequalities result just from the definition of l.u.b. and g.l.b. Then we have

$$S^+(f, \Delta) = \sum_{i=1}^{n} M_i l(I_i) < \sum_{i=1}^{n} \left[f(x_i') + \frac{\varepsilon}{4(b-a)} \right] l(I_i)$$

$$= \sum_{i=1}^{n} f(x_i') l(I_i) + \tfrac{1}{4}\varepsilon.$$

Hence, using Inequality (5.14), we find

$$S^+(f, \Delta) < A + \frac{\varepsilon}{4} + \frac{\varepsilon}{4} = A + \frac{\varepsilon}{2}. \tag{5.15}$$

Similarly,

$$S_-(f, \Delta) = \sum_{i=1}^{n} m_i l(I_i) > \sum_{i=1}^{n} \left[f(x_i'') - \frac{\varepsilon}{4(b-a)} \right] l(I_i)$$

$$= \sum_{i=1}^{n} f(x_i'') l(I_i) - \tfrac{1}{4}\varepsilon,$$

and

$$S_-(f, \Delta) > A - \frac{\varepsilon}{2}. \tag{5.16}$$

Subtraction of Inequality (5.16) from Inequality (5.15) yields

$$S^+(f, \Delta) - S_-(f, \Delta) < \varepsilon,$$

and so f is Darboux integrable. Furthermore, these inequalities show that

$$A = \int_a^b f(x)\, dx. \qquad \qquad \square$$

In order to show that every Darboux integrable function is Riemann integrable, we need the following technical result.

Lemma 5.2. *Suppose f is defined and bounded on $[a, b]$. Then for each $\varepsilon > 0$ there is a $\delta > 0$ such that*

$$S^+(f, \Delta) < \int_a^{\bar{b}} f(x)\, dx + \varepsilon \quad and \quad S_-(f, \Delta) > \int_{\underline{a}}^b f(x)\, dx - \varepsilon \tag{5.17}$$

for every subdivision Δ of mesh $< \delta$.

PROOF. We shall show that for each $\varepsilon > 0$, there is a $\delta_1 > 0$ such that the first of·Inequalities (5.17) holds for all subdivisions of mesh less than δ_1; a similar proof would show that there is a $\delta_2 > 0$ such that the second of inequalities (5.17) holds. Then δ may be chosen as the smaller of δ_1 and δ_2.

To prove the first statement above, let $\varepsilon > 0$ be given. There is a subdivision $\Delta_0 = \{I_1, \ldots, I_k\}$, where $k \geqslant 2$ and $I_i = [t_{i-1}, t_i]$ and $a = t_0 < t_1 < \cdots < t_k =$

b, such that

$$S^+(f, \Delta_0) < \int_a^{\overline{b}} f(x)\, dx + \frac{\varepsilon}{2}. \tag{5.18}$$

Now, choose any number η less than the minimum of $t_i - t_{i-1}$ and $\varepsilon/2(k-1)(M-m)$; then let Δ be a subdivision of mesh $\leqslant \eta$. There are at most $k-1$ intervals of Δ which contain any points t_i in their interiors. Let these intervals be denoted by $J_1, \ldots, J_p$ $(p \leqslant k-1)$, and let the remaining intervals of Δ be $K_1, \ldots, K_q$. Let Δ' be the common refinement of Δ_0 and Δ; then Δ' consists of the intervals $K_1, \ldots, K_q$ and the $2p$ intervals J_j' and J_j'', $j = 1, \ldots, p$ into which each J_j is divided by the point t_i in it. We find

$$S^+(f, \Delta) = \sum_{j=1}^p M_j l(J_j) + \sum_{k=1}^q M_k^* l(K_k)$$

$$S^+(f, \Delta') = \sum_{j=1}^p [M_j' l(J_j') + M_j'' l(J_j'')] + \sum_{k=1}^q M_k^* l(K_k) \leqslant S^+(f, \Delta_0) \tag{5.19}$$

where M_j, M_j', M_j'', and M_k^* are the least upper bounds of f in the corresponding subintervals. Thus, since for each j,

$$l(J_j) = l(J_j') + l(J_j''), \qquad m \leqslant M_j, M_j', M_j'' \leqslant M,$$

we see that

$$S^+(f, \Delta) - S^+(f, \Delta') \leqslant (M-m) \sum_{j=1}^p l(J_j) \leqslant (k-1)(M-m) \cdot \eta < \frac{\varepsilon}{2}. \tag{5.20}$$

Now, from Expressions (5.18), (5.19), and (5.20) it follows that

$$S^+(f, \Delta) < S^+(f, \Delta_0) + \frac{\varepsilon}{2} < \int_a^{\overline{b}} f(x)\, dx + \varepsilon. \qquad \square$$

Theorem 5.12. *A function f is Riemann integrable on $[a, b]$ if and only if it is Darboux integrable on $[a, b]$.*

PROOF. That Riemann integrability implies Darboux integrability is the content of Theorem 5.11. To show that a Darboux integrable function f is Riemann integrable, let $\varepsilon > 0$ be given and choose a δ so that the result of Lemma 5.2 above becomes applicable for a subdivision Δ with mesh less than δ. We have

$$\int_a^b f(x)\, dx - \varepsilon < S_-(f, \Delta) \leqslant \sum_{i=1}^n f(x_i) l(I_i) \leqslant S^+(f, \Delta) < \int_a^b f(x)\, dx + \varepsilon.$$

Since ε is arbitrary, we find

$$\int_a^b f(x)\, dx = A$$

where A is the Riemann integral of f. $\qquad \square$

In light of Theorem 5.12 we shall now drop the terms *Darboux* and *Riemann*, and just refer to functions as *integrable*.

One of the most useful methods for performing integrations when the integrand is not in standard form is the method known as *substitution*. For example, we may sometimes be able to show that a complicated integral has the form

$$\int_a^b f[u(x)]u'(x)\, dx.$$

Then, if we set $u = u(x)$, $du = u'(x)\, dx$, the above integral becomes

$$\int_{u(a)}^{u(b)} f(u)\, du,$$

and this integral may be one we recognize as a standard integration. In an elementary course in calculus these substitutions are usually made without concern for their validity. The theoretical foundation for such processes is based on the next result.

Theorem 5.13. *Suppose that f is continuous on an open interval I. Let u and u' be continuous on an open interval J, and assume that the range of u is contained in I. If $a, b \in J$, then*

$$\int_a^b f[u(x)]u'(x)\, dx = \int_{u(a)}^{u(b)} f(u)\, du. \tag{5.21}$$

PROOF. Let $c \in I$ and define

$$F(u) = \int_c^u f(t)\, dt. \tag{5.22}$$

From the Fundamental theorem of calculus (Theorem 5.7), we have

$$F'(u) = f(u).$$

Defining $G(x) = F[u(x)]$, we employ the Chain rule to obtain

$$G'(x) = F'[u(x)]u'(x) = f[u(x)]u'(x).$$

Since all functions under consideration are continuous, it follows that

$$\int_a^b f[u(x)]u'(x)\, dx = \int_a^b G'(x)\, dx$$

$$= G(b) - G(a)$$

$$= F[u(b)] - F[u(a)].$$

From Equation (5.22) above, we see that

$$F[u(b)] - F[u(a)] = \int_c^{u(b)} f(t)\, dt - \int_c^{u(a)} f(t)\, dt = \int_{u(a)}^{u(b)} f(t)\, dt,$$

and the theorem is proved.   □

PROBLEMS

1. (a) Suppose that f is continuous on $I = \{x: a \leqslant x \leqslant b\}$. Show that all upper and lower Darboux sums are Riemann sums.
   (b) Suppose that f is increasing on $I = \{x: a \leqslant x \leqslant b\}$. Show that all upper and lower Darboux sums are Riemann sums.
   (c) Give an example of a bounded function f defined on I in which a Darboux sum is not a Riemann sum.

2. Suppose that f is integrable on $I = \{x: a \leqslant x \leqslant b\}$, and suppose that $0 < m \leqslant f(x) \leqslant M$ for $x \in I$. Prove that $\int_a^b (1/f(x))\, dx$ exists.

3. Suppose that u, u', v, and v' are continuous on $I = \{x: a \leqslant x \leqslant b\}$. Establish the formula for **integration by parts**:

$$\int_a^b u(x)v'(x)\, dx = u(b)v(b) - u(a)v(a) - \int_a^b v(x)u'(x)\, dx.$$

4. Suppose u, v, w, u', v', and w' are continuous on $I = \{x: a \leqslant x \leqslant b\}$. Establish the extended integration by parts formula:

$$\int_a^b u(x)v(x)w'(x)\, dx = u(b)v(b)w(b) - u(a)v(a)w(a) - \int_a^b u(x)v'(x)w(x)\, dx$$

$$- \int_a^b u'(x)v(x)w(x)\, dx.$$

5. Give an example of a function f defined on $I = \{x: 0 \leqslant x \leqslant 1\}$ such that $|f|$ is integrable but f is not.

6. Show, from the definition, that the Riemann integral is unique (Theorem 5.9). [*Hint*: Assume that there are two different numbers A and A' satisfying Inequality (5.13), and then reach a contradiction.]

7. Prove the following extension of Theorem 5.13. Suppose that u is continuous for $a \leqslant x \leqslant b$, that u' is continuous only for $a < x < b$, and that u' tends to a limit both as $x \to a^+$ and $x \to b^-$. Define (assuming $u(a) < u(b)$)

$$f_0(u) = \begin{cases} f(u) & \text{for } u(a) \leqslant u \leqslant u(b), \\ f[u(a)] & \text{for } u \leqslant u(a), \\ f[u(b)] & \text{for } u \geqslant u(b), \end{cases}$$

$$u_0(x) = \begin{cases} u(x) & \text{for } a \leqslant x \leqslant b, \\ u(a) + u'(a)(x - a) & \text{for } x \leqslant a, \\ u(b) + u'(b)(x - b) & \text{for } x \geqslant b, \end{cases}$$

where $u'(a)$ and $u'(b)$ are the limiting values of $u'(x)$. Then Formula (5.21) holds.

8. State conditions on u, v, and f such that the following formula is valid:

$$\int_a^b f[u\{v(x)\}]u'[v(x)]v'(x)\, dx = \int_{u[v(a)]}^{u[v(b)]} f(u)\, du.$$

9. Suppose that f and g are integrable on $I = \{x: a \leqslant x \leqslant b\}$. Then f^2, g^2, and fg are integrable. (See Problem 14, Section 5.1.) Prove the **Cauchy–Schwarz inequality**

$$\left[\int_a^b f(x)g(x)\,dx \right]^2 \leqslant \left[\int_a^b f^2(x)\,dx \right]\left[\int_a^b g^2(x)\,dx \right].$$

[*Hint*: Set $\alpha = \int_a^b f^2(x)\,dx$, $\beta = \int_a^b f(x)g(x)\,dx$, and $\gamma = \int_a^b g^2(x)\,dx$ and observe that $\alpha z^2 + 2\beta z + \gamma \geqslant 0$ for all real numbers z.]

10. Given f on $I = \{x: a \leqslant x \leqslant b\}$. If $\int_a^b f^3(x)\,dx$ exists, does it follow that $\int_a^b f(x)\,dx$ exists?

5.3. The Logarithm and Exponential Functions

The logarithm function and the exponential function are undoubtedly familiar to the reader. In this section, we define these functions precisely and develop their principal properties.

Definitions. The **natural logarithm function,** denoted by log, is defined by the formula

$$\log x = \int_1^x \frac{1}{t}\,dt, \qquad x > 0.$$

Theorem 5.14. *Let $f: x \to \log x$ be defined for $x > 0$ and suppose a and b are positive numbers. The following ten statements hold:*

(i) $\log(ab) = \log a + \log b$.
(ii) $\log(a/b) = \log a - \log b$.
(iii) $\log 1 = 0$.
(iv) $\log(a^r) = r \log a$ *for every rational number r.*
(v) $f'(x) = 1/x$.
(vi) *f is increasing and continuous on $I = \{x: 0 < x < +\infty\}$.*
(vii) $1/2 \leqslant \log 2 \leqslant 1$.
(viii) $\log x \to +\infty$ *as $x \to +\infty$.*
(ix) $\log x \to -\infty$ *as $x \to 0^+$.*
(x) *The range of f is all of $\mathbb{R}^1$.*

PROOF. To prove (i), we write

$$\log(ab) = \int_1^{ab} \frac{1}{t}\,dt = \int_1^a \frac{1}{t}\,dt + \int_a^{ab} \frac{1}{t}\,dt.$$

Changing variables in the last integral on the right by letting $u = t/a$, we see that

$$\int_a^{ab} \frac{1}{t}\,dt = \int_1^b \frac{1}{u}\,du.$$

Hence $\log(ab) = \log a + \log b$.

To prove (ii), apply (i) to $a = b \cdot (a/b)$ getting

$$\log a = \log b + \log\left(\frac{a}{b}\right),$$

which yields (ii).

To verify (iii), simply set $a = b$ in (ii).

To establish (iv) we proceed step by step. If r is a positive integer, we get the result from (i) with $a = b$ and mathematical induction. For negative integers, write $a^{-n} = 1/a^n$ and employ (ii). Finally, if $r = p/q$ where p and q are integers, set $u = a^{1/q}$, and thus $u^q = a$. Hence $q \log u = \log a$. Since

$$a^r = a^{p/q} = u^p,$$

we find

$$\log(a^r) = \log(u^p) = p \log u = \frac{p}{q}\log a = r \log a.$$

Statement (v) is simply a statement of the Fundamental theorem of calculus (Theorem 5.7).

As for (vi), f is increasing since its derivative is positive; f is continuous since every differentiable function is continuous.

To establish the inequalities in (vii), we use upper and lower Darboux sums to show that

$$\frac{1}{2} \leqslant \int_1^2 \frac{1}{t}dt \leqslant 1.$$

The details are left to the reader. (A graph of $f: t \to 1/t$ makes the result evident. See Figure 5.7(a). A sketch of $f: x \to \log x$ is shown in Figure 5.7(b).)

To prove (viii), note that if $x > 2^n$, n a positive integer, then

$$\log x > \log(2^n) = n \log 2 \geqslant \tfrac{1}{2}n$$

(because of vi). Let M be any positive number. Choose n so that $n > 2M$; hence $\log x > M$ if $x > 2^n$, and (viii) is established. Inequality (ix) is shown similarly by choosing $x < 2^{-n}$.

The Intermediate-value theorem together with (viii) and (ix) allows us to deduce (x).   □

Since the logarithm function is increasing (item (vi) above), its inverse is a function; therefore, the following definition makes sense.

Definition. The inverse of the logarithm function is called the **exponential function** and is denoted by exp.

Theorem 5.15. *If $f: x \to \exp x$ is the exponential function, the following nine statements hold:*

(i) *f is continuous and increasing for all $x \in \mathbb{R}^1$; the range of f is $I = \{x: 0 < x < +\infty\}$.*

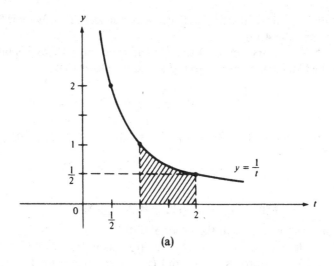

(a)

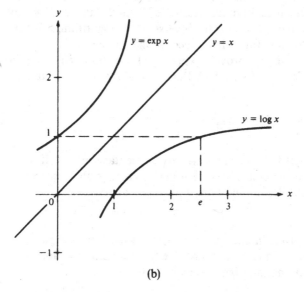

(b)

Figure 5.7. The logarithm and its inverse.

(ii) $f'(x) = \exp x$ *for all* x.
(iii) $\exp(x + y) = (\exp x) \cdot (\exp y)$.
(iv) $\exp(x - y) = (\exp x)/(\exp y)$.
 (v) $\exp(rx) = (\exp x)^r$, *for* r *rational*.
(vi) $f(x) \to +\infty$ *as* $x \to +\infty$.
(vii) $f(x) \to 0$ *as* $x \to -\infty$.
(viii) $\log(\exp x) = x$ *for all* $x \in \mathbb{R}^1$ *and* $\exp(\log x) = x$ *for all* $x > 0$.
 (ix) *If* $a > 0$ *and* r *is rational, then* $\exp(r \log a) = a^r$.

PROOF. Items (i) and (viii) are immediate consequences of the Inverse function theorem (Theorem 4.17).

To establish (ii), first set $y = \exp x$. From Theorem 4.18, exp is differentiable. By the Chain rule applied to $\log y = x$, it follows that

$$\frac{1}{y} y' = 1 \quad \text{or} \quad y' = y = \exp x.$$

To prove (iii), set $y_1 = \exp x_1$ and $y_2 = \exp x_2$. Then $x_1 = \log y_1$, $x_2 = \log y_2$, and

$$x_1 + x_2 = \log y_1 + \log y_2 = \log(y_1 y_2).$$

Hence

$$\exp(x_1 + x_2) = y_1 y_2 = (\exp x_1)(\exp x_2).$$

The proof of (iv) is similar to the proof of (iii).

The formula in (v) is obtained by induction (as in the proof of part (iv) of Theorem 5.14). The proofs of (vi) and (vii) follow from the corresponding results for the logarithm function. See Figure 5.7 in which we note that since the exponential function is the inverse of the logarithm, it is the reflection of the logarithm function with respect to the line $y = x$.

Item (ix) is simply proved: $\exp(r \log a) = \exp(\log(a^r)) = a^r$, the first equality coming from Theorem 5.14, Part (iv) and the second from Part (viii) of this theorem.   □

Expressions of the form

$$a^x, \quad a > 0,$$

for x rational have been defined by elementary means. If $x = p/q$, then we merely take the pth power of a and then take the qth root of the result. However, the definition of quantities such as

$$3^{\sqrt{2}}, \quad (\sqrt{7})^{\pi}$$

cannot be given in such an elementary way. For this purpose, we use the following technique, a standard one for extending the domain of a function from the rational numbers to the real numbers.

Definition. For $a > 0$, and $x \in \mathbb{R}^1$, we define

$$a^x = \exp(x \log a).$$

Observe that when x is rational this formula coincides with (ix) of Theorem 5.15 so that the definition is consistent with the basic idea of "raising to a power."

Theorem 5.16. *Define $f: x \to a^x$, $a > 0$. Then the following three statements hold:*

(i) *f is positive and continuous for all $x \in \mathbb{R}^1$.*

(ii) *If $a \neq 1$, then the range of f is $I = \{x: 0 < x < +\infty\}$.*
(iii) *If $a > 1$, then f is increasing; if $a < 1$, then f is decreasing.*

The proof of this theorem follows directly from the definition of a^x and is left to the reader.

Definitions. If $b > 0$ and $b \neq 1$, the function $\log_b$ (*called* **logarithm to the base** b) is defined as the inverse of the function $f: x \to b^x$. When $b = 10$ we call the function the **common logarithm**.

The next theorem describes many of the familiar properties of the logarithm to any base and of the function a^x. The proof is left to the reader.

Theorem 5.17. *Let $a > 0$, $b > 0$ be fixed. Define $f: x \to a^x$. Then the following six statements hold*:

(i) $a^x \cdot a^y = a^{x+y}$; $a^x/a^y = a^{x-y}$; $(a^x)^y = a^{xy}$; $(ab)^x = a^x b^x$; $(a/b)^x = a^x/b^x$.
(ii) $f'(x) = a^x \log a$.
(iii) *If $b \neq 1$, then* $\log_b x = \log x/\log b$.
(iv) *If $b \neq 1$, the function $g: x \to \log_b x$ is continuous for all $x \in I = \{x: 0 < x < +\infty\}$.*
(v) *If $a \neq 1$, $b \neq 1$, then* $\log_b x = (\log_a x) \cdot (\log_b a)$.
(vi) *If $b \neq 1$, $x > 0$, $y > 0$, then* $\log_b(xy) = \log_b x + \log_b y$; $\log_b(x/y) = \log_b x - \log_b y$; $\log_b(x^y) = y \log_b x$; $\log_b 1 = 0$.

Definition. $e = \exp 1$.

Theorem 5.18. *The following statements about the logarithm function and exponential function are valid*:

(i) $\log_e x = \log x$ *for all $x > 0$.*
(ii) $\exp x = e^x$ *for all x.*
(iii) *Given $f: x \to x^n$ with n an arbitrary real number, then $f'(x) = nx^{n-1}$.*
(iv) $\lim_{x \to 0^+} (1 + x)^{1/x} = e$.

Remarks. The proofs of these four statements are left to the reader. Statements (i) and (ii) are almost immediate consequences of the definitions. It is interesting to note that the formula in (iii) is almost always proved in elementary courses for n a *rational number*, although the formula is frequently stated as being valid for all real numbers n. The proof of (iii) simply uses the definition: $x^n = \exp(n \log x)$. Then the Chain rule is used to differentiate $\exp(n \log x)$, from which the result follows. Similarly (iv) is established by writing $(1 + x)^{1/x} = \exp[(1/x) \log(1 + x)]$. Since $\exp$ is a continuous function, we compute the quantity $\lim_{x \to 0^+} [(1/x) \log(1 + x)]$. The expression $(1 + x)^{1/x}$ may be used to give an approximate value for e if x is small.

PROBLEMS

1. Prove the three properties of the function $f: x \to a^x$ as stated in Theorem 5.16.

2. Show that the familiar "laws of exponents" are valid for the function $f: x \to a^x$. (Theorem 5.17, Part (i).)

3. Given $f: x \to a^x$, show that $f'(x) = a^x \log a$. (Theorem 5.17, Part (ii).)

4. Given $f: x \to x^n$ with n any real number; show that $f'(x) = nx^{n-1}$. (Theorem 5.18, Part (iii).)

5. Show that $2 < e < 4$.

6. Prove that $\lim_{x\to 0^+}(1 + x)^{1/x} = e$. (Theorem 5.18, Part (iv).)

7. If $x > -1$, show that $\log(1 + x) \leqslant x$.

8. Prove that $e^x \geqslant 1 + x$ for all x.

9. Show that $F: x \to [1 + (1/x)]^x$ is an increasing function.

10. Given $f: x \to \log x$. From the definition of derivative, we know that

$$\lim_{h\to 0} \frac{f(x + h) - f(x)}{h} = \frac{1}{x}.$$

Use this fact to deduce that $\lim_{h\to 0}(1 + h)^{1/h} = e$.

5.4. Jordan Content and Area

In this section, we develop a precise theory of area of bounded sets of points in $\mathbb{R}^1$ (i.e., sets which are interior to some rectangle). We may think of S as a reasonably simple region such as the set of points "inside and on the curve" in Figure 5.8, but what we do will apply equally well to any bounded set. Since

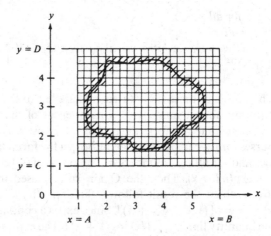

Figure 5.8. A grid in R^2.

S is bounded, we may assume that every point (x, y) of S satisfies the conditions

$$A < x < B \quad \text{and} \quad C < y < D$$

where A, B, C, and D are *integers*; i.e., (x, y) is interior to the rectangle bounded by the lines $x = A$, $x = B$, $y = C$, and $y = D$ (see Figure 5.8).

Now for each n, suppose that $\mathbb{R}^2$ is divided up into small squares of side $(\frac{1}{2})^n$ by drawing all the lines $x = i/2^n$ and $y = j/2^n$ where i and j are integers; a typical square consists of all points (x, y) for which

$$\frac{i-1}{2^n} \leqslant x \leqslant \frac{i}{2^n} \quad \text{and} \quad \frac{j-1}{2^n} \leqslant y \leqslant \frac{j}{2^n}. \tag{5.23}$$

Definitions. For a given n, these squares will be called the squares of the nth **grid**. For a given n, a square which lies entirely in S is an **inner square** for S; any square which contains at least one point of S is a **covering square** for S.

For each n, there are 4^n squares of the nth grid in each unit square and hence, since A, B, C, and D are integers, there are $4^n(B - A)(D - C)$ such squares in the rectangle

$$R = \{(x, y): A \leqslant x \leqslant B, C \leqslant y \leqslant D\}.$$

It is natural to consider the quantities.

$$A_n^-(S) = \frac{1}{4^n} \quad \text{times the number of inner squares for} \quad S$$

and

$$A_n^+(S) = \frac{1}{4^n} \quad \text{times the number of covering squares for} \quad S.$$

Evidently $A_n^-(S)$ represents the "area" (yet to be defined) of the union of all the inner squares and $A_n^+(S)$ represents that of the union of the covering squares.

Lemma 5.3. *Suppose that S is interior to the rectangle R above; i.e., if $(x, y) \in S$, then $A < x < B$ and $C < y < D$ and A, B, C, and D are integers. Then each covering square r for S is contained in R.*

PROOF. Suppose $(x, y) \in S \cap r$ where

$$\frac{i-1}{2^n} \leqslant x \leqslant \frac{i}{2^n} \quad \text{and} \quad \frac{j-1}{2^n} \leqslant y \leqslant \frac{j}{2^n} \quad \text{for some } i \text{ and } j.$$

Since $A < x < B$, it follows that $i/2^n \geqslant x > A$ or $i > 2^n \cdot A$. Therefore $i - 1 \geqslant 2^n \cdot A$ since i and A are integers. Likewise, the inequality $(i - 1)/2^n \leqslant x < B$ holds, so that $(i - 1) < 2^n \cdot B$ and therefore $i \leqslant 2^n \cdot B$. Consequently,

$$A \leqslant \frac{i-1}{2^n} < \frac{i}{2^n} \leqslant B.$$

In a similar way it can be shown that

$$C \leqslant \frac{j-1}{2^n} < \frac{j}{2^n} \leqslant D,$$

and so r is in R. $\square$

Concerning the quantities $A_n^-(S)$ and $A_n^+(S)$, we have the following results:

Theorem 5.19. *Suppose that S is interior to the rectangle R as in Lemma 5.3. Then*

(a) $0 \leqslant A_n^-(S) \leqslant A_n^+(S) \leqslant (B-A)(D-C)$ *for each n;*
(b) $A_{n+1}^-(S) \geqslant A_n^-(S)$ *for each n;*
(c) $A_{n+1}^+(S) \leqslant A_n^+(S)$ *for each n;*
(d) *the sequences $A_n^-(S)$ and $A_n^+(S)$ tend to limits, denoted by $A^-(S)$ and $A^+(S)$, respectively, and*

$$0 \leqslant A^-(S) \leqslant A^+(S) \leqslant (B-A)(D-C);$$

(e) *the quantities $A_n^-(S)$, $A_n^+(S)$, $A^-(S)$, and $A^+(S)$ are all independent of the size of the rectangle R so long as R contains S in its interior.*

PROOF. Part (a) follows from the lemma, the definitions, and the fact that there are exactly $4^n(B-A)(D-C)$ squares of the nth grid in R. Part (b) follows since if r is an inner square of the nth grid, the four squares of the $(n+1)$st grid which constitute r are all inner squares. Part (c) follows since if r' is a covering square of the $(n+1)$st grid, then r' contains a point (x, y) of S so that the square r of the nth grid which contains r' also contains (x, y) and so is a covering square for S. Part (d) follows from Parts (a), (b), and (c) since the sequences are bounded and monotone; Part (e) is evident. $\square$

Definitions. The number $A^-(S)$ and $A^+(S)$ are called the **inner** and **outer areas** *of S*, respectively. In case $A^-(S) = A^+(S)$, we say that S is a **figure** and define its **area** $A(S)$ as the common value of $A^-(S)$ and $A^+(S)$. The area of a set is also called the **Jordan content**.

Of course, if S is too complicated, it can happen that $A^-(S) < A^+(S)$. As an example, let S consist of all the points (x, y) where x and y are both rational numbers with $0 < x < 1$ and $0 < y < 1$. Its inner area is zero since no square consists entirely of such points and its outer area is 1 since every square of the nth grid in $R = \{(x, y): 0 \leqslant x \leqslant 1, 0 \leqslant y \leqslant 1\}$ contains such a point.

Since not every bounded set is a figure, it is important to prove theorems which will guarantee that a given set *is* a figure, for only then can we legitimately speak of its area. In order to state such theorems, we now introduce some additional terminology about sets in $\mathbb{R}^2$: If S_1 and S_2 are sets, the **difference** $S_1 - S_2$ (or $S_1 \sim S_2$) consists of all points in S_1 which are not in S_2. A point P_0 is an **interior point of S** if and only if it is the center of an actual

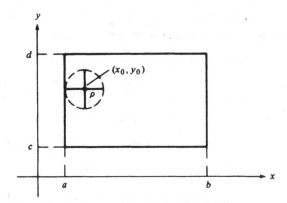

Figure 5.9. (x_0, y_0) is an interior point.

disk contained in S; the totality of interior points of S is denoted by $S^{(0)}$. If $P_1(x_1, y_1)$ and $P_2(x_2, y_2)$ are any two points of $\mathbb{R}^2$, then

$$\sqrt{(x_1 - x_2)^2 + (y_1 - y_2)^2}$$

is the **Euclidean distance** between P_1 and P_2. We denote it by $|P_1 P_2|$. A point $P_0(x_0, y_0)$ is said to be a **limit point** of S if and only if there is a sequence $\{P_n(x_n, y_n)\}$ of points of S such that $P_n \neq P_0$ for each n and $|P_0 P_n| = \sqrt{(x_n - x_0)^2 + (y_n - y_0)^2} \to 0$; the **union** of a set S and the set of all its limit points is called the **closure** of S and is denoted by $\bar{S}$. If a set S contains all its limit points, i.e., $S = \bar{S}$, then S is said to be **closed**. The **boundary** of a set S is the set $\bar{S} - S^{(0)}$.

Lemma 5.4. *Let* $R = \{(x, y): a \leqslant x \leqslant b \text{ and } c \leqslant y \leqslant d\}$ *be a rectangle in* $\mathbb{R}^2$. *Then* (i) R *is closed, and* (ii) $R^{(0)} = \{(x, y): a < x < b \text{ and } c < y < d\}$.

PROOF. The proof of Part (i) and the proof that $\{(x, y): a < x < b \text{ and } c < y < d\} \subset R^{(0)}$ are left to the reader (see Figure 5.9). To complete the proof of (ii), suppose (x_0, y_0) is interior to R. Then all points (x, y) such that $(x - x_0)^2 + (y - y_0)^2 \leqslant \rho^2$ belong to R for some $\rho > 0$. In particular, the points (x, y_0) with $x_0 - \rho \leqslant x \leqslant x_0 + \rho$ and the points (x_0, y) with $y_0 - \rho \leqslant y \leqslant y_0 + \rho$ (see Figure 5.9) all belong to R. Thus $a \leqslant x_0 - \rho < x_0 < x_0 + \rho \leqslant b$ and $c \leqslant y_0 - \rho < y_0 < y_0 + \rho \leqslant d$. That is $R^{(0)} \subset \{(x, y): a < x < b, c < y < d\}$. $\square$

We now prove several intuitively evident, elementary theorems about area.

Theorem 5.20. *Suppose that each of the sets* $S_1, \ldots, S_p$ *is bounded; some of them may intersect.*

(a) *If* $S_1 \subset S_2$, *then* $A^-(S_1) \leqslant A^-(S_2)$ *and* $A^+(S_1) \leqslant A^+(S_2)$.

(b) $A^+(S_1 \cup \cdots \cup S_p) \leqslant A^+(S_1) + \cdots + A^+(S_p)$.

(c) *If no two of the S_i have common interior points, then*

$$A^-(S_1 \cup \cdots \cup S_p) \geqslant A^-(S_1) + \cdots + A^-(S_p).$$

(d) *If each S_i is a figure, then $S_1 \cup \cdots \cup S_p$ is a figure and if no two of the S_i have common interior points, then*

$$A(S_1 \cup \cdots \cup S_p) = A(S_1) + \cdots + A(S_p).$$

(e) *If R is a rectangle as in Lemma 5.4, then R and $R^{(0)}$ are figures and*

$$A(R^{(0)}) = A(R) = (b - a)(d - c).$$

PROOF. (a) Each inner square for S_1 is an inner square for S_2, and each covering square for S_1 is one for S_2. Thus

$$A_n^-(S_1) \leqslant A_n^-(S_2) \quad \text{and} \quad A_n^+(S_1) \leqslant A_n^+(S_2) \quad \text{for each } n.$$

Part (a) then follows from Theorem 5.19 and the limit of inequalities.

The proof of Part (b) is left to the reader.

To prove Part (c), suppose that r is an inner square for some particular S_i. Then, of course, r is an inner square for the union. Moreover, r is *not* an inner square for any S_j with $j \neq i$ for if it were, every interior point of r would be an interior point of both S_i and S_j contrary to the hypotheses. Thus we must have

$$A_n^-(S_1 \cup \cdots \cup S_p) \geqslant A_n^-(S_1) + \cdots + A_n^-(S_p) \quad \text{for each } n. \tag{5.24}$$

(That the strict inequality may hold in Expression (5.24) is indicated in Figure 5.10.)

The proof of Part (d) is left to the reader.

We shall prove Part (e) for the rectangle R; the proof for $R^{(0)}$ is similar. For each n, let i_n be the unique integer such that

$$\frac{i_n - 1}{2^n} < a \leqslant \frac{i_n}{2^n}$$

and define

$$a_n = \frac{i_n - 1}{2^n}, \qquad a_n' = \frac{i_n}{2^n},$$

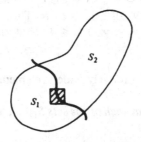

Figure 5.10. $A_n^-(S_1 \cup S_2) > A_n^-(S_1) + A_n^-(S_2)$.

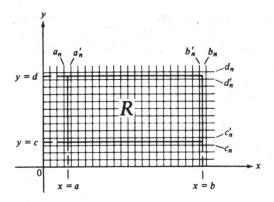

Figure 5.11. $R = \{(x, y): a \leqslant x \leqslant b, c \leqslant y \leqslant d\}$.

(see Figure 5.11). In a similar way, define b_n, b_n', c_n, c_n', d_n, and d_n' so that

$$a_n < a \leqslant a_n', \qquad b_n' \leqslant b < b_n, \qquad a_n' - a_n = b_n - b_n' = (\tfrac{1}{2})^n, \quad \text{etc.} \quad (5.25)$$

By counting the number of inner and covering squares, we obtain

$$A_n^-(R) = (b_n' - a_n')(d_n' - c_n') \quad \text{and} \quad A_n^+(R) = (b_n - a_n)(d_n - c_n). \quad (5.26)$$

From Expression (5.25), it follows that $a - (\tfrac{1}{2})^n \leqslant a_n < a$, etc., so that $a_n \to a$, $a_n' \to a$, etc., and Part (e) follows by passing to the limit in Equation (5.26). $\quad\square$

We define a **boundary square** of the nth grid for S to be a covering square for S which is not an inner square. The shaded squares in Figure 5.8 are exactly the boundary squares for the set S of that figure. The next lemma and theorem provide useful facts about the union and intersection of figures. Also, we observe the intuitively evident fact that the boundary of a figure must have zero area.

Lemma 5.5

(a) *A square r of the nth grid is a boundary square for $S \Leftrightarrow r$ contains a point of S and a point not in S.*
(b) *A set S is a figure $\Leftrightarrow$ the area of the union of its boundary squares of the nth grid $\to 0$ as $n \to \infty$.*

The proof is left to the reader.

Theorem 5.21. *Suppose that $S_1, \ldots, S_p$ are figures and $p \geqslant 2$. Then the union $S_1 \cup \cdots \cup S_p$, the intersection $S_1 \cap \cdots \cap S_p$, and the difference $S_1 - S_2$ are all figures. The boundary of a figure has area zero.*

PROOF. We shall prove that $S_1 \cup \cdots \cup S_p$ is a figure for the case $p = 2$. The general result follows by induction. Let r be a boundary square of the nth grid, for $S_1 \cup S_2$. Then r contains a point P in $S_1 \cup S_2$ and a point Q not in $S_1 \cup S_2$. Then P is in S_1 or S_2 or both and Q is not in either S_1 or S_2. If $P \in S_1$, then r is a boundary square for S_1; if $P \in S_2$, r is one for S_2, and if P is in both, then r is one for both. Thus the number of boundary squares for $S_1 \cup S_2$ is less than or equal to the number for S_1 plus the number for S_2. The result follows by dividing by 4^n and letting $n \to \infty$.

The proofs that $S_1 \cap \cdots \cap S_p$ and $S_1 - S_2$ are figures are similar to the above proof and are left to the reader. The proof that the boundary of a figure has zero area is also left to the reader.   ☐

PROBLEMS

1. Let R be the rectangle $R = \{(x, y): a \leqslant x \leqslant b, c \leqslant y \leqslant d\}$. Prove that R is closed (Lemma 5.4(i)).

2. Show that the interior of the rectangle R in Problem 1 is the set $R^0 = \{(x, y): a < x < b, c < y < d\}$ (Lemma 5.4(ii)).

3. Let S_1 and S_2 be bounded sets. Show that

$$A^+(S_1 \cup S_2) \leqslant A^+(S_1) + A^+(S_2).$$

4. If S_1 and S_2 are figures which have no common interior points, show that

$$A(S_1 \cup S_2) = A(S_1) + A(S_2).$$

5. If S_1 and S_2 are figures, show that

$$A(S_1 \cup S_2) + A(S_1 \cap S_2) = A(S_1) + A(S_2).$$

6. Show that a set S in $\mathbb{R}^2$ is a figure if and only if the area of the union of its boundary squares of the nth grid tends to zero as $n \to \infty$ (Lemma 5.5(b)).

7. If $S_1, S_2, \ldots, S_p$ are figures, show that $S_1 \cap S_2 \cap \cdots \cap S_p$ is a figure.

8. If S_1 and S_2 are figures, show that $S_1 - S_2$ is a figure.

9. Prove that the boundary of a figure S in $\mathbb{R}^2$ has area zero.

10. Let S be a set contained in a rectangle R in $\mathbb{R}^2$. Show that

$$A_-(S) = A(R) - A^+(R - S).$$

11. Prove the following theorem: *Suppose that f is defined and bounded on $[a, b]$, $c < f(x)$ for every x on $[a, b]$, and suppose $F_1 = \{(x, y): a \leqslant x \leqslant b, c \leqslant y < f(x)\}$ and $F_2 = \{(x, y): a \leqslant x \leqslant b, c \leqslant y \leqslant f(x)\}$. Then (see Figure 5.12)*

$$\underline{\int_a^b} [f(x) - c]\, dx \leqslant A^-(F_1) \leqslant A^-(F_2),$$

$$\overline{\int_a^b} [f(x) - c]\, dx \geqslant A^+(F_2) \geqslant A^+(F_1).$$

[Hint: *Define $g(x) = f(x) - c$. Let $\varepsilon > 0$ be given. Then there is a subdivision Δ such that $S_-(g, \Delta) > \int_a^b g(x)\, dx - \varepsilon$. Let $m_i = \inf g(x)$ on I_i. For each i, let $r_i =$*

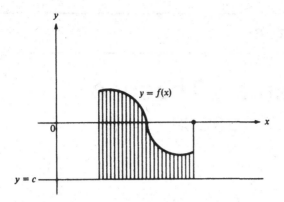

Figure 5.12

$\{(x, y): x \in I_i, c \leqslant y < m_i + c\}$. *Then* $r_1 \cup r_2 \cup \cdots \cup r_n \subset F_1 \subset F_2$. *Hence* $A^-(F_2) \geqslant$
$A^-(F_1) \geqslant A(r_1 \cup r_2 \cup \cdots \cup r_n) = S_-(g, \Delta) > \underline{\int}_a^b g(x)\, dx - \varepsilon.]$

12. Prove the following theorem: *Suppose f and g are integrable on $[a, b]$ and $f(x) \leqslant$ $g(x)$ on $[a, b]$. Suppose also that $S = \{(x, y): a \leqslant x \leqslant b, f(x) \leqslant y \leqslant g(x)\}$. Then S is a figure and*

$$A(S) = \int_a^b [g(x) - f(x)]\, dx.$$

CHAPTER 6

Elementary Theory of Metric Spaces

6.1. The Schwarz and Triangle Inequalities; Metric Spaces

In Chapters 2 through 5 we developed many properties of functions from $\mathbb{R}^1$ into $\mathbb{R}^1$ with the purpose of proving the basic theorems in differential and integral calculus of one variable. The next step in analysis is the establishment of the basic facts needed in proving the theorems of calculus in two and more variables. One way would be to prove extensions of the theorems of Chapters 2–5 for functions from $\mathbb{R}^2$ into $\mathbb{R}^1$, then for functions from $\mathbb{R}^3$ into $\mathbb{R}^1$, and so forth. However, all these results can be encompassed in one general theory obtained by introducing the concept of a metric space and by considering functions defined on one metric space with range in a second metric space. In this chapter we introduce the fundamentals of this theory and in the following two chapters the results are applied to differentiation and integration in Euclidean space in any number of dimensions.

We establish a simple version of the Schwarz inequality, one of the most useful inequalities in analysis.

Theorem 6.1 (Schwarz inequality). *Let* $x = (x_1, x_2, \ldots, x_N)$ *and* $y = (y_1, y_2, \ldots, y_N)$ *be elements of* $\mathbb{R}^N$. *Then*

$$\left| \sum_{i=1}^{N} x_i y_i \right| \leq \left(\sum_{i=1}^{N} x_i^2 \right)^{1/2} \left(\sum_{i=1}^{N} y_i^2 \right)^{1/2}. \tag{6.1}$$

The equality sign holds $\Leftrightarrow$ *either all the* x_i *are zero or there is a number* λ *such that* $y_i = \lambda x_i$ *for* $i = 1, 2, \ldots, N$.

PROOF. If every x_i is zero, then the equality sign in Expression (6.1) holds. Assume there is at least one x_i not zero. We form the function

$$f(\lambda) = \sum_{i=1}^{N} (y_i - \lambda x_i)^2,$$

which is nonnegative for all values of λ. We set

$$A = \sum_{i=1}^{N} x_i^2, \qquad B = \sum_{i=1}^{N} x_i y_i, \qquad C = \sum_{i=1}^{N} y_i^2.$$

Then clearly,

$$f(\lambda) = A\lambda^2 - 2B\lambda + C \geqslant 0, \qquad A > 0.$$

From elementary calculus it follows that the nonnegative minimum value of the quadratic function $f(\lambda)$ is

$$\frac{AC - B^2}{A}.$$

The statement $AC - B^2 \geqslant 0$ is equivalent to Expression (6.1). The equality sign in Expression (6.1) holds if $f(\lambda) = 0$ for some value of λ, say λ_1; in this case $y_i - \lambda_1 x_i = 0$ for every i. $\qquad\square$

We recall that in the Euclidean plane the length of any side of a triangle is less than the sum of the lengths of the other two sides. A generalization of this fact is known as the Triangle inequality. It is proved by means of a simple application of the Schwarz inequality.

Theorem 6.2 (Triangle inequality). *Let $x = (x_1, x_2, \ldots, x_N)$ and $y = (y_1, y_2, \ldots, y_N)$ be elements of $\mathbb{R}^N$. Then*

$$\sqrt{\sum_{i=1}^{N} (x_i + y_i)^2} \leqslant \sqrt{\sum_{i=1}^{N} x_i^2} + \sqrt{\sum_{i=1}^{N} y_i^2}. \tag{6.2}$$

The equality sign in Expression (6.2) holds $\Leftrightarrow$ either all the x_i are zero or there is a nonnegative number λ such that $y_i = \lambda x_i$ for $i = 1, 2, \ldots, N$.

PROOF. We have

$$\sum_{i=1}^{N} (x_i + y_i)^2 = \sum_{i=1}^{N} (x_i^2 + 2x_i y_i + y_i^2) = \sum_{i=1}^{N} x_i^2 + 2\sum_{i=1}^{N} x_i y_i + \sum_{i=1}^{N} y_i^2.$$

We apply the Schwarz inequality to the middle term on the right, obtaining

$$\sum_{i=1}^{N} (x_i + y_i)^2 \leqslant \sum_{i=1}^{N} x_i^2 + 2\sqrt{\sum_{i=1}^{N} x_i^2}\sqrt{\sum_{i=1}^{N} y_i^2} + \sum_{i=1}^{N} y_i^2,$$

and so

$$\sum_{i=1}^{N} (x_i + y_i)^2 \leqslant \left(\sqrt{\sum_{i=1}^{N} x_i^2} + \sqrt{\sum_{i=1}^{N} y_i^2}\right)^2.$$

Since the left side of this inequality is nonnegative, we may take the square root to obtain Expression (6.2).

If all x_i are zero, then equality in Expression (6.2) holds. Otherwise, equality holds if and only if it holds in the application of Theorem 6.1. The nonnegativity of λ is required to ensure that $2\sum_{i=1}^{N} x_i y_i$ has all nonnegative terms when we set $y_i = \lambda x_i$.   □

Corollary. *Let* $x = (x_1, x_2, \ldots, x_N)$, $y = (y_1, y_2, \ldots, y_N)$, *and* $z = (z_1, z_2, \ldots, z_N)$ *be elements of* $\mathbb{R}^N$. *Then*

$$\sqrt{\sum_{i=1}^{N} (x_i - z_i)^2} \leqslant \sqrt{\sum_{i=1}^{N} (x_i - y_i)^2} + \sqrt{\sum_{i=1}^{N} (y_i - z_i)^2}. \tag{6.3}$$

The equality sign in Expression (6.3) *holds* ⇔ *there is a number* r *with* $0 \leqslant r \leqslant 1$ *such that* $y_i = rx_i + (1 - r)z_i$ *for* $i = 1, 2, \ldots, N$.

PROOF. Setting $a_i = x_i - y_i$ and $b_i = y_i - z_i$ for $i = 1, 2, \ldots, N$, we see that Inequality (6.3) reduces to Inequality (6.2) for the elements $a = (a_1, \ldots, a_N)$ and $b = (b_1, b_2, \ldots, b_N)$. The number r is $\lambda_1/(1 + \lambda_1)$ in Theorem 6.1.   □

Remark. The Corollary to Theorem 6.2 is the familiar assertion that the sum of the lengths of any two sides of a triangle exceeds the length of the third side in Euclidean N-dimensional space. The equality sign in Expression (6.3) occurs when the three points all lie on the same line segment with y falling between x and z.

In this chapter we shall be concerned with *sets* or *collections* of elements, which may be chosen in any manner whatsoever. A set will be considered fully described whenever we can determine whether or not any given element is a member of the set.

Definition. Let S and T be sets. The **Cartesian product of S and T**, denoted $S \times T$, is the set of all ordered pairs (p, q) in which $p \in S$ and $q \in T$. The Cartesian product of any finite number of sets $S_1, S_2, \ldots, S_N$ is the set of ordered N-tuples $(p_1, p_2, \ldots, p_N)$ in which $p_i \in S_i$ for $i = 1, 2, \ldots, N$. We write $S_1 \times S_2 \times \cdots \times S_N$.

EXAMPLES. (1) The space $\mathbb{R}^N$ is the Cartesian product $\mathbb{R}^1 \times \mathbb{R}^1 \times \cdots \times \mathbb{R}^1$ (N factors).

(2) The Cartesian product of $I_1 = \{x: a \leqslant x \leqslant b\}$ and $I_2 = \{x: c \leqslant x \leqslant d\}$ is the rectangle $T = \{(x, y): a \leqslant x \leqslant b, c \leqslant y \leqslant d\}$. That is,

$$T = I_1 \times I_2.$$

(3) The Cartesian product of the circle $C = \{(x, y): x^2 + y^2 = 1\}$ with $\mathbb{R}^1$ yields a right circular cylinder U in $\mathbb{R}^3$:

$$U = \{(x, y, z): x^2 + y^2 = 1, -\infty < z < \infty\} = C \times \mathbb{R}^1.$$

Definition. Let S be a set and suppose d is a function with domain consisting of all pairs of points of S and with range in $\mathbb{R}^1$. That is, d is a function from $S \times S$ into $\mathbb{R}^1$. We say that S and the function d form a **metric space** when the function d satisfies the following conditions:

(i) $d(x, y) \geqslant 0$ for all $(x, y) \in S \times S$; and $d(x, y) = 0$ if and only if $x = y$.
(ii) $d(y, x) = d(x, y)$ for all $(x, y) \in S \times S$.
(iii) $d(x, z) \leqslant d(x, y) + d(y, z)$ for all x, y, z in S. (Triangle inequality.)

The function d satisfying Conditions (i), (ii), and (iii) is called the **metric** or **distance function** in S. Hence a metric space consists of the pair (S, d).

EXAMPLES OF METRIC SPACES. (1) In the space $\mathbb{R}^N$, choose

$$d(x, y) = \sqrt{\sum_{i=1}^{N} (x_i - y_i)^2}$$

where $x = (x_1, x_2, \ldots, x_N)$ and $y = (y_1, y_2, \ldots, y_N)$. The function d is a metric. Conditions (i) and (ii) are obvious, while (iii) is precisely the content of the Corollary to Theorem 6.2. The pair $(\mathbb{R}^N, d)$ is a metric space. This metric, known as the **Euclidean metric**, is the familiar one employed in two and three dimensional Euclidean geometry.

(2) In the space $\mathbb{R}^N$, choose

$$d_1(x, y) = \max_{1 \leqslant i \leqslant N} |x_i - y_i|$$

where $x = (x_1, x_2, \ldots, x_N)$ and $y = (y_1, y_2, \ldots, y_N)$. The reader can verify that d_1 is a metric. Therefore $(\mathbb{R}^N, d_1)$ is a metric space. We observe that this metric space is different from the space $(\mathbb{R}^N, d)$ exhibited in the first example.

(3) Let S be any set. We define

$$\bar{d}(x, y) = \begin{cases} 0 & \text{if } x = y, \\ 1 & \text{if } x \neq y. \end{cases}$$

Clearly, $\bar{d}$ is a metric and $(S, \bar{d})$ is a metric space. Thus *any* set may have a metric attached to it, and thereby become a metric space. When attaching a metric to a set, we say that the set or space is **metrized**. Of course, such a simple metric as $\bar{d}$ merely tells us whether or not two points coincide and is hardly useful.

(4) Let $\mathscr{C}$ be the collection of all continuous functions which have $I = \{x: 0 \leqslant x \leqslant 1\}$ for domain and have range in $\mathbb{R}^1$. For any two elements f, g in $\mathscr{C}$, define

$$d(f, g) = \max_{0 \leqslant x \leqslant 1} |f(x) - g(x)|.$$

It is not difficult to verify that d is a metric. Hence $(\mathscr{C}, d)$ is a metric space.

Examples 1, 2, and 3 above show that a given set may become a metric space in a variety of ways. Let S be a given set and suppose that (S, d_1) and

(S, d_2) are metric spaces. We define the metrics d_1 and d_2 as **equivalent** if and only if there are positive constants c and C such that

$$cd_1(x, y) \leqslant d_2(x, y) \leqslant Cd_1(x, y)$$

for all x, y in S. It is not difficult to show that the metrics in Examples 1 and 2 are equivalent. On the other hand, if the set S in Example 3 is taken to be $\mathbb{R}^N$, it is easy to show that $\bar{d}$ is not equivalent to either of the metrics in Examples 1 and 2.

PROBLEMS

1. Show that $(\mathbb{R}^N, d_1)$ is a metric space where

$$d_1(x, y) = \max_{1 \leqslant i \leqslant N} |x_i - y_i|,$$

$x = (x_1, x_2, \ldots, x_N)$, $y = (y_1, y_2, \ldots, y_N)$.

2. Suppose that (S, d) is a metric space. Show that (S, d') is a metric space where

$$d'(x, y) = \frac{d(x, y)}{1 + d(x, y)}.$$

   [*Hint*: Show first that $d(x, z) = \lambda[d(x, y) + d(y, z)]$ for some λ with $0 \leqslant \lambda \leqslant 1$.]

3. Given the metric spaces $(\mathbb{R}^N, d)$ and $(\mathbb{R}^N, d')$

$$d(x, y) = \sqrt{\sum_{i=1}^{N} (x_i - y_i)^2}$$

   and d' is defined as in Problem 2. Decide whether or not d and d' are equivalent.

4. Show that the metrics d and d_1 in Examples (1) and (2) above of metric spaces are equivalent.

5. Show that $(\mathbb{R}^N, d_2)$ is a metric space where

$$d_2(x, y) = \sum_{i=1}^{N} |x_i - y_i|,$$

$x = (x_1, x_2, \ldots, x_N)$, $y = (y_1, y_2, \ldots, y_N)$. Is d_2 equivalent to the metric d_1 given in Problem 1?

6. Let (x_1, x_2), (x_1', x_2') be points of $\mathbb{R}^2$. Show that $(\mathbb{R}^2, d_3)$ is a metric space where

$$d_3((x_1, x_2), (x_1', x_2')) = \begin{cases} |x_2| + |x_2'| + |x_1 - x_1'| & \text{if } x_1 \neq x_1', \\ |x_2 - x_2'| & \text{if } x_1 = x_1' \end{cases}.$$

7. For $x, y \in \mathbb{R}^1$, define $d_4(x, y) = |x - 3y|$. Is $(\mathbb{R}^1, d_4)$ a metric space?

8. Let $\mathscr{C}$ be the collection of continuous functions which have $I = \{x : 0 \leqslant x \leqslant 1\}$ for domain and have range in $\mathbb{R}^1$. Show that $(\mathscr{C}, d)$ is a metric space where

$$d(f, g) = \max_{0 \leqslant x \leqslant 1} |f(x) - g(x)|$$

for $f, g \in \mathscr{C}$.

9. Let $\mathscr{C}$ be the same collection described in Problem 8. Define

$$\bar{d}(f, g) = \int_0^1 |f(x) - g(x)| \, dx$$

for $f, g \in \mathscr{C}$. Show that $(\mathscr{C}, \bar{d})$ is a metric space.

10. Suppose $x = (x_1, x_2, \ldots, x_n, \ldots)$, $y = (y_1, y_2, \ldots, y_n, \ldots)$ are infinite sequences such that

$$\sum_{i=1}^{\infty} x_i^2 \quad \text{and} \quad \sum_{i=1}^{\infty} y_i^2$$

both converge.[1] Prove that the series $\sum_{i=1}^{\infty} x_i y_i$ converges and establish the **Schwarz inequality**.

$$\left| \sum_{i=1}^{\infty} x_i y_i \right| \le \left(\sum_{i=1}^{\infty} x_i^2 \right)^{1/2} \left(\sum_{i=1}^{\infty} y_i^2 \right)^{1/2}. \tag{6.4}$$

[*Hint*: Use Theorem 6.1 and the theorem on the limit of inequalities.]

11. State and prove conditions which must prevail when the equality sign holds in Formula (6.4) of Problem 10.

12. Suppose $x = (x_1, x_2, \ldots, x_n, \ldots)$ and $y = (y_1, y_2, \ldots, y_n, \ldots)$ are as in Problem 10. Show that $\sum_{i=1}^{\infty} (x_i + y_i)^2$ converges and prove the **Triangle inequality**

$$\left(\sum_{i=1}^{\infty} (x_i + y_i)^2 \right)^{1/2} \le \left(\sum_{i=1}^{\infty} x_i^2 \right)^{1/2} + \left(\sum_{i=1}^{\infty} y_i^2 \right)^{1/2}. \tag{6.5}$$

Show that the equality sign holds in Formula (6.5) if and only if either $x = 0$ or there is a nonnegative number λ such that $y_i = \lambda x_i$ for all i.

13. Let l_2 be the collection of all infinite sequences $x = (x_1, x_2, \ldots, x_n, \ldots)$ such that $\sum_{i=1}^{\infty} x_i^2$ converges. Define

$$d(x, y) = \sqrt{\sum_{i=1}^{\infty} (x_i - y_i)^2}$$

for $x, y \in l_2$. Show that (l_2, d) is a metric space. (The space (l_2, d) is called **real numerical Hilbert space**.) [*Hint*: Prove an extension of the Corollary to Theorem 6.2, using the result in Problem 12.]

14. A sequence $x_1, x_2, \ldots, x_n, \ldots$ is **bounded** if and only if there is a number m such that $|x_i| \le m$ for all i. Let M denote the collection of all bounded sequences, and define

$$d(x, y) = \sup_{1 \le i \le \infty} |x_i - y_i|.$$

Show that (M, d) is a metric space.

15. Let B be the collection of all absolutely convergent series. Define

$$d(x, y) = \sum_{i=1}^{\infty} |x_i - y_i|.$$

Show that (B, d) is a metric space.

---

[1] Problems 10–15 assume the reader has studied convergence and divergence of infinite series.

16. Let C be the subset of $\mathbb{R}^2$ consisting of pairs $(\cos \theta, \sin \theta)$ for $0 \leqslant \theta < 2\pi$. Define

$$d^*(p_1, p_2) = |\theta_1 - \theta_2|,$$

where $p_1 = (\cos \theta_1, \sin \theta_1)$, $p_2 = (\cos \theta_2, \sin \theta_2)$. Show that (C, d^*) is a metric space. Is d^* equivalent to the metric d_1 of Problem 1 applied to the subset C of $\mathbb{R}^2$?

17. Let S be a set and d a function from $S \times S$ into $\mathbb{R}^1$ with the properties:
    (i) $d(x, y) = 0$ if and only if $x = y$.
    (ii) $d(x, z) \leqslant d(x, y) + d(z, y)$ for all $x, y, z \in S$.
    Show that d is a metric and hence that (S, d) is a metric space.

6.2. Elements of Point Set Topology

In this section we shall develop some of the basic properties of metric spaces. From an intuitive point of view it is natural to think of a metric space as a Euclidean space of one, two, or three dimensions. While such a view is sometimes helpful for geometric arguments, it is important to recognize that the definitions and theorems apply to arbitrary metric spaces, many of which have geometric properties far removed from those of the ordinary Euclidean spaces.

For convenience we will use the letter S to denote a metric space with the understanding that a metric d is attached to S.

Definition. Let $p_1, p_2, \ldots, p_n, \ldots$ denote a sequence of elements of a metric space S. We use the symbol $\{p_n\}$ to denote such a sequence. Suppose $p_0 \in S$. We say that p_n **tends to** p_0 **as** n **tends to infinity** if and only if $d(p_n, p_0) \to 0$ as $n \to \infty$. The notations $p_n \to p_0$ and $\lim_{n \to \infty} p_n = p_0$ will be used.

Theorem 6.3 (Uniqueness of limits). *Suppose that* $\{p_n\}, \bar{p}, \bar{q}$ *are elements of* S, *a metric space. If* $p_n \to \bar{p}$ *and* $p_n \to \bar{q}$ *as* $n \to \infty$, *then* $\bar{p} = \bar{q}$.

This theorem is an extension of the corresponding simpler result for sequences of real numbers discussed in Section 2.5. The proof follows the lines of the proof of Theorem 2.1 and is left to the reader.

Definitions. Let p_0 be an element of S, a metric space, and suppose r is a positive number. The **open ball with center at** p_0 **and radius** r is the set $B(p_0, r)$ given by

$$B(p_0, r) = \{p \in S : d(p, p_0) < r\}.$$

The **closed ball with center at** p_0 **and radius** r is the set $\overline{B(p_0, r)}$ given by

$$\overline{B(p_0, r)} = \{p \in S : d(p, p_0) \leqslant r\}.$$

Figure 6.1 shows an open ball with center at $(0, 0)$ and radius 1 in the space $\mathbb{R}^2$ with metric $d(x, y) = (\sum_{i=1}^{2}(x_i - y_i)^2)^{1/2}$. Figure 6.2 shows an open

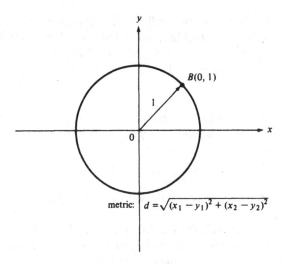

Figure 6.1. An open ball with center at $(0, 0)$ and radius 1. The circle forming the boundary of $B(0, 1)$ is not part of the open ball.

ball with center at $(0, 0)$ and radius 1 in the space $\mathbb{R}^2$ with metric $d(x, y) = \max_{i=1,2} |x_i - y_i|$.

Figures 6.1 and 6.2 illustrate the care which must be exercised in employing geometric reasoning for statements about a general metric space. For example, suppose the metric

$$d(x, y) = \begin{cases} 0 & \text{if } x = y, \\ 1 & \text{if } x \neq y, \end{cases}$$

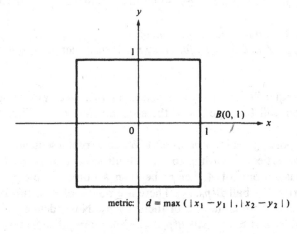

Figure 6.2. An open ball with center at $(0, 0)$ and radius 1. The four sides of the square forming the boundary of $B(0, 1)$ are not part of the open ball.

is attached to $\mathbb{R}^2$. Then any open ball of radius less than 1 contains only one point, the center; also, any ball of radius larger than 1 contains all of $\mathbb{R}^2$.

We note that in Figures 6.1 and 6.2 the closed ball $\bar{B}(0, 1)$ consists in each case of the union of the open ball and the boundary (circle or square) of the ball.

Definition. Let A be a set in a metric space S and suppose p_0 is an element of A. We say p_0 is an **isolated point** of A if and only if for some positive number r there is an open ball $B(p_0, r)$ such that

$$B(p_0, r) \cap A = \{p_0\},$$

where $\{p_0\}$ is the set consisting of the single element p_0.

For example, in $\mathbb{R}^1$ with the Euclidean metric the set $A = \{x : 1 \leqslant x \leqslant 2, x = 3, 4\}$ has the elements 3 and 4 as isolated points of A. None of the points in the interval $1 \leqslant x \leqslant 2$ is an isolated point of A.

Definition. A point p_0 is a **limit point**[2] **of a set** A if and only if every open ball $B(p_0, r)$ contains a point p of A which is distinct from p_0. Note that p_0 may or may not be an element of A.

For example, in $\mathbb{R}^1$ with the usual Euclidean metric defined by $d(x_1, x_2) = |x_1 - x_2|$, the set $C = \{x : 1 \leqslant x < 3\}$ has $x = 3$ as a limit point. In fact, every member of C is a limit point of C.

A set A in a metric space S is **closed** if and only if A contains all of its limit points.

A set A in a metric space S is **open** if and only if each point p_0 in A is the center of an open ball $B(p_0, r)$ which is contained in A. That is, $B(p_0, r) \subset A$. It is important to notice that the radius r may change from point to point in A.

Theorem 6.4. *A point p_0 is a limit point of a set A if and only if there is a sequence $\{p_n\}$ with $p_n \in A$ and $p_n \neq p_0$ for every n and such that $p_n \to p_0$ as $n \to \infty$.*

PROOF

(a) If a sequence $\{p_n\}$ of the theorem exists, then clearly every open ball with p_0 as center will have points of the sequence (all $\neq p_0$). Thus p_0 is a limit point.

(b) Suppose p_0 is a limit point of A. We construct a sequence $\{p_n\}$ with the desired properties. According to the definition of limit point, the open ball $B(p_0, \frac{1}{2})$ has a point of A. Let p_1 be such a point. Define $r_2 = d(p_0, p_1)/2$ and construct the ball $B(p_0, r_2)$. There are points of A in this ball (different from p_0), and we denote one of them by p_2. Next, define $r_3 = d(p_0, p_2)/2$ and choose $p_3 \in A$ in the ball $B(p_0, r_3)$. Continuing this process, we see that

---

[2] The term *cluster point* is also in common use.

$r_n < 1/2^n \to 0$ as $n \to \infty$. Since $d(p_0, p_n) < r_n$, it follows that $\{p_n\}$ is a sequence tending to p_0, and the proof is complete. $\qquad\qquad\qquad\qquad\qquad\qquad\qquad\qquad\square$

In Section 5.4 we defined a limit point in the special case of $\mathbb{R}^2$ with the Euclidean metric. We gave a definition in terms of a sequence $\{p_n\}$ rather than the one in terms of open balls. Theorem 6.4 shows that the two definitions are equivalent.

Theorem 6.5

(a) *An open ball is an open set.*
(b) *A closed ball is a closed set.*

PROOF

(a) Let $B(p_0, r)$ be an open ball. Suppose $q \in B(p_0, r)$. We must show that there is an open ball with center at q which is entirely in $B(p_0, r)$. Let $r_1 = d(p_0, q)$. We consider a ball of radius $\bar{r} = (r - r_1)/2$ with center at q. See Figure 6.3. Let q' be a typical point of $B(q, \bar{r})$. We shall show that $q' \in B(p_0, r)$. We can accomplish this by showing that $d(p_0, q') < r$. We have

$$d(p_0, q') \leqslant d(p_0, q) + d(q, q') < r_1 + \bar{r} = r_1 + \tfrac{1}{2}(r - r_1)$$

$$= \tfrac{1}{2}(r + r_1) < r.$$

(b) To show that $\overline{B(p_0, r)}$ is a closed set we must show that every limit point of $\overline{B(p_0, r)}$ belongs to $\overline{B(p_0, r)}$. If q is such a limit point, there is a sequence $p_1, p_2, \ldots, p_n, \ldots$ in $\overline{B(p_0, r)}$ such that $d(p_n, q) \to 0$. However, for each n, we have

$$d(p_n, p_0) \leqslant r.$$

Therefore

$$d(q, p_0) \leqslant d(q, p_n) + d(p_n, p_0) \leqslant d(q, p_n) + r.$$

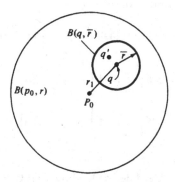

Figure 6.3. An open ball is an open set.

Letting $n \to \infty$, the Theorem on the limit of inequalities shows that $d(q, p_0)$ $\leqslant r$. That is, $q \in \overline{B(p_0, r)}$. $\square$

Remark. In the proof of Theorem 6.5, Figure 6.3 was used as an intuitive aid in following the argument. It is important to observe that the argument itself is entirely analytic and does not depend on any "geometric reasoning."

Let S be a space of elements, by which we mean that S is the totality of points (or elements) under consideration. A set of points is a collection of elements in S. We shall often deal with a set of sets, that is, a set whose elements are sets in S. We speak of such a set as a **family of sets** and denote it by a script letter such as $\mathscr{F}$. If the number of sets in the family is finite, we use the term *finite family* of sets and use subscripts to identify the members. For example, $\{A_1, A_2, \ldots, A_n\}$ where each A_i is a set in S comprise a finite family of sets.

Definitions. Let $\mathscr{F}$ be a family of sets, a typical member of $\mathscr{F}$ being denoted by the letter A. That is, A is a set of points in a space S. We define the **union of the sets in** $\mathscr{F}$, denoted $\bigcup_{A \in \mathscr{F}} A$, by the formula

$$\bigcup_{A \in \mathscr{F}} A = \{p: p \in S \text{ and } p \text{ is in at least one set of } \mathscr{F}\}.$$

We define the **intersection of the sets in** $\mathscr{F}$, denoted $\bigcap_{A \in \mathscr{F}} A$, by the formula

$$\bigcap_{A \in \mathscr{F}} A = \{p: p \in S \text{ and } p \text{ is in every set of } \mathscr{F}\}.$$

If A and B are sets in S, we define their **difference** $B - A$ by the formula

$$B - A = \{p: p \text{ is in } B \text{ and } p \text{ is not in } A\}.$$

If A is any set in S, we define the **complement of** A, denoted $\mathscr{C}(A)$, as the set $S - A$.

The following important identities, known as the **de Morgan formulas**, are useful in the study of families of sets.

Theorem 6.6 (de Morgan formulas). *Suppose that S is any space and $\mathscr{F}$ is a family of sets. Then*

(a)
$$\mathscr{C}\left[\bigcup_{A \in \mathscr{F}} A\right] = \bigcap_{A \in \mathscr{F}} \mathscr{C}(A).$$

(b)
$$\mathscr{C}\left[\bigcap_{A \in \mathscr{F}} A\right] = \bigcup_{A \in \mathscr{F}} \mathscr{C}(A).$$

PROOF

(a) We employ a standard device to show that two sets A and B are equal: we show that A is contained in B and that B is contained in A. For this purpose suppose that $p \in \mathscr{C}[\bigcup_{A \in \mathscr{F}} A]$. Then p is not in the union of the sets in $\mathscr{F}$, and hence p is not in any $A \in \mathscr{F}$. Thus p is in $\mathscr{C}(A)$ for every A in $\mathscr{F}$. That is, $p \in \bigcap_{A \in \mathscr{F}} \mathscr{C}(A)$. We have shown that $\mathscr{C}[\bigcup_{A \in \mathscr{F}} A] \subset \bigcap_{A \in \mathscr{F}} \mathscr{C}(A)$. Next, let $p \in \bigcap_{A \in \mathscr{F}} \mathscr{C}(A)$ and, by reversing the discussion we find that $\bigcap_{A \in \mathscr{F}} \mathscr{C}(A) \subset [\bigcup_{A \in \mathscr{F}} A]$. Hence Part (a) is proved. Part (b) is proved similarly and is left to the reader.   □

Remark. Observe that Theorem 6.6 applies for any space S whether or not it is metrized.

Theorem 6.7

(a) *The union of any family $\mathscr{F}$ of open sets in a metric space is open.*
(b) *Let $A_1, A_2, \ldots, A_n$ be a finite family of open sets. Then the intersection $\bigcap_{i=1}^{n} A_i$ is open.*

PROOF

(a) Suppose that $p \in \bigcup_{A \in \mathscr{F}} A$. Then p is in at least one set A in $\mathscr{F}$. Since A is open there is an open ball with center at p and radius r, denoted $B(p, r)$, which is entirely in A. Hence $B(p, r) \subset \bigcup_{A \in \mathscr{F}} A$. We have just shown that any point of $\bigcup_{A \in \mathscr{F}} A$ is the center of an open ball which is also in $\bigcup_{A \in \mathscr{F}} A$. Thus the union of any family of open sets is open.

(b) Suppose that $p \in \bigcap_{i=1}^{n} A_i$. Then for each i there is an open ball $B(p, r_i)$ which is entirely in A_i. Define $\bar{r} = \min(r_1, r_2, \ldots, r_n)$. The open ball $B(p, \bar{r})$ is in every A_i and hence is in $\bigcap_{i=1}^{n} A_i$. Thus the set $\bigcap_{i=1}^{n} A_i$ is open.   □

Remark. The result of Part (b) of Theorem 6.7 is false if the word *finite* is dropped from the hypothesis. To see this consider $\mathbb{R}^2$ with the Euclidean metric and define the infinite family of open sets $A_k, k = 1, 2, \ldots$, by the formula

$$A_k = \left\{ (x, y) : 0 \leqslant x^2 + y^2 < \frac{1}{k} \right\}.$$

Each A_k is an open ball (and therefore an open set), but the intersection $\bigcap_{k=1}^{\infty} A_k$ of this family consists of the single point $(0, 0)$, clearly not an open set (see Figure 6.4).

Theorem 6.8

(a) *Let A be any set in a metric space S. Then A is closed $\Leftrightarrow \mathscr{C}(A)$ is open.*
(b) *The space S is both open and closed.*
(c) *The null set is both open and closed.*

PROOF. We first assume A is closed and prove that $\mathscr{C}(A)$ is open. Let $p \in \mathscr{C}(A)$ and suppose there is no open ball about p lying entirely in $\mathscr{C}(A)$. We shall reach a contradiction. If there is no such ball then every ball about p contains

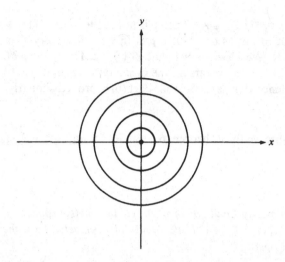

Figure 6.4. The intersection of an infinite family of open sets is not necessarily open.

points of A (necessarily different from p since $p \in \mathscr{C}(A)$). This fact asserts that p is a limit point of A. But since A is assumed closed, we conclude that p is in A, which contradicts the fact that $p \in \mathscr{C}(A)$. Thus there is an open ball about p in $\mathscr{C}(A)$, and so $\mathscr{C}(A)$ is open.

The proof that A is closed if $\mathscr{C}(A)$ is open is left to the reader.

Part (b) is a direct consequence of the definitions of open and closed sets. Part (c) follows immediately from Parts (a) and (b) since the null set is the complement of S.   □

Theorem 6.9

(a) *The intersection of any family $\mathscr{F}$ of closed sets is closed.*
(b) *Let $A_1, A_2, \ldots, A_n$ be a finite family of closed sets. Then the union $\bigcup_{i=1}^{n} A_i$ is closed.*

The proof of Theorem 6.9 is an immediate consequence of the de Morgan formulas (Theorem 6.6) and Theorems 6.7 and 6.8 The details are left to the reader.

It is important to observe that Part (b) of Theorem 6.9 is false if the word *finite* is dropped from the hypothesis. The reader may easily construct an infinite family of closed sets such that the union of the family is not a closed set.

Definition. Let $\{p_n\}$ be an infinite sequence of points. An infinite sequence of points $\{q_n\}$ is called a **subsequence** of $\{p_n\}$ if and only if there is an *increasing* sequence of positive integers $k_1, k_2, \ldots, k_n, \ldots$ such that for every n

$$q_n = p_{k_n}.$$

To illustrate subsequences, observe that the sequence

$$p_1, p_2, \ldots, p_n, \ldots$$

has subsequences such as

$$p_1, p_3, p_5, \ldots,$$

$$p_5, p_{10}, p_{15}, \ldots,$$

$$p_1, p_4, p_7, \ldots.$$

It is clear that *the subscripts* $\{k_n\}$ *of a subsequence have the property that* $k_n \geqslant n$ *for every n.* This fact may be proved by induction, as we noted in Section 3.3.

Theorem 6.10. *Suppose that* $\{p_n\}$ *is an infinite sequence in a metric space S, and that* $\{q_n\}$ *is a subsequence. If* $p_n \to p_0$ *as* $n \to \infty$, *then* $q_n \to p_0$ *as* $n \to \infty$.

(Every subsequence of a convergent sequence converges to the same limit.)

PROOF. Let $\varepsilon > 0$ be given. From the definition of convergence, there is a positive integer N such that $d(p_n, p_0) < \varepsilon$ for all $n > N$. However, for each n, we have $q_n = p_{k_n}$ with $k_n \geqslant n$. Thus

$$d(q_n, p_0) = d(p_{k_n}, p_0) < \varepsilon \quad \text{for all} \quad n > N.$$

Therefore $q_n \to p_0$ as $n \to \infty$.   □

Remark. Part (a) of Theorem 6.9 may be proved directly from the definition of a closed set. To do so, the result of Theorem 6.10 is used.

Definitions. A point p is an **interior point** of a set A in a metric space if and only if there is an $r > 0$ such that $B(p, r)$ is contained in A. The **interior of a set** A, denoted $A^{(0)}$, is the set of all interior points of A. For any set A, the **derived set** of A, denoted A', is the collection of limit points of A. The **closure** of A, denoted $\bar{A}$, is defined by $\bar{A} = A \cup A'$. The **boundary of a set** A, denoted ∂A, is defined by $\partial A = \bar{A} - A^{(0)}$.

EXAMPLE. Let A be the open ball in $\mathbb{R}^2$ (with the Euclidean metric) given by $A = \{(x, y): 0 \leqslant x^2 + y^2 < 1\}$. Then every point of A is an interior point, so that $A = A^{(0)}$. More generally, the property which characterizes any open set is the fact that every point is an interior point. The boundary of A is the circle $\partial A = \{(x, y): x^2 + y^2 = 1\}$. The closure of A is the closed ball $\bar{A} = \{(x, y): 0 \leqslant x^2 + y^2 \leqslant 1\}$. We also see that A', the derived set of A, is identical with $\bar{A}$.

Theorem 6.11. *Let A be any set in a metric space S. We have*

(a) A' *is a closed set.*
(b) $\bar{A}$ *is a closed set.*
(c) ∂A *is a closed set.*

(d) $A^{(0)}$ is an open set.
(e) If $A \subset B$ and B is closed, then $\bar{A} \subset B$.
(f) If $B \subset A$ and B is open, then $B \subset A^{(0)}$.
(g) A point is a closed set.
(h) A is open $\Leftrightarrow \mathscr{C}A$ is closed.

The proofs of these facts are left to the reader.

PROBLEMS

1. Suppose that $p_n \to p_0$ and $q_n \to q_0$ in a metric space S . If $d(p_n, q_n) < a$ for all n , show that $d(p_0, q_0) \leqslant a$.

2. If $p_n \to p_0$, $p_n \to q_0$ in a metric space S , show that $p_0 = q_0$ (Theorem 6.3).

3. Given $\mathbb{R}^2$ with the metric $d(x, y) = |y_1 - x_1| + |y_2 - x_2|$, $x = (x_1, x_2)$, $y = (y_1, y_2)$. Describe (and sketch) the ball with center at $(0, 0)$ and radius 1.

4. Given $\mathbb{R}^1$ with the metric $d(x, y) = |x - y|$. Find an example of a set which is neither open nor closed.

5. Given $\mathbb{R}^1$ with $d(x, y) = |x - y|$. Show that a finite set consists only of isolated points. Is it true that a set consisting only of isolated points must be finite?

6. Given an example of a set in $\mathbb{R}^1$ with exactly four limit points.

7. In $\mathbb{R}^1$ with the Euclidean metric, define $A = \{x : 0 \leqslant x \leqslant 1 \text{ and } x \text{ is a rational number}\}$. Describe the set $\bar{A}$.

8. Given $\mathbb{R}^2$ with the Euclidean metric. Show that the set $S = \{(x, y) : 0 < x^2 + y^2 < 1\}$ is open. Describe the sets $S^{(0)}$, S' , ∂S , $\bar{S}$, $\mathscr{C}(S)$.

9. If S is any space and $\mathscr{F}$ is a family of sets, show that

$$\mathscr{C}\left[\bigcap_{A \in \mathscr{F}} A\right] = \bigcup_{A \in \mathscr{F}} \mathscr{C}(A)$$

(Part (b) of the de Morgan formulas).

10. Let A , B , and C be arbitrary sets in a space S . Show that
    (a) $\mathscr{C}(A - B) = B \cup \mathscr{C}(A)$
    (b) $A - (A - B) = A \cap B$
    (c) $A \cap (B - C) = (A \cap B) - (A \cap C)$
    (d) $A \cup (B - A) = A \cup B$
    (e) $(A - C) \cup (B - C) = (A \cup B) - C$

11. In $\mathbb{R}^2$ with the Euclidean metric, find an infinite collection of open sets $\{A_n\}$ such that $\bigcap_n A_n$ is the closed ball $\bar{B}(0, 1)$.

12. In a metric space S , show that a set A is closed if $\mathscr{C}(A)$ is open (Theorem 6.8).

13. Show that in a metric space the intersection of any family $\mathscr{F}$ of closed sets is closed.

14. Let $A_1, A_2, \ldots, A_n$ be a finite family of closed sets in a metric space. Show that the union is a closed set.

15. Let $A_1, A_2, \ldots, A_n, \ldots$ be sets in a metric space. Define $B = \bigcup A_i$. Show that $\bar{B} \supset \bigcup \bar{A_i}$ and give an example to show that $\bar{B}$ may not equal $\bigcup \bar{A_i}$.

16. In $\mathbb{R}^1$ with the Euclidean metric, give an example of a sequence $\{p_n\}$ with two subsequences converging to different limits. Give an example of a sequence which has infinitely many subsequences converging to different limits.

17. Using only the definition of closed set prove that the intersection of any family of closed sets is closed (Theorem 6.9(a)).

18. Suppose that f is continuous on an interval $I = \{x : a \leqslant x \leqslant b\}$ with $f(x) > 0$ on I. Let $S = \{(x, y) \in \mathbb{R}^2 : a \leqslant x \leqslant b, 0 \leqslant y \leqslant f(x)\}$ (Euclidean metric).
    (a) Show that S is closed.
    (b) Find S' and $\bar{S}$.
    (c) Find $S^{(0)}$ and prove the result.
    (d) Find ∂S.

19. If A is a set in a metric space, show that A' and $\bar{A}$ are closed sets (Theorem 6.11(a), (b)).

20. If A is a set in a metric space, show that ∂A is closed and $A^{(0)}$ is open (Theorem 6.11(c), (d)).

21. If A and B are sets in a metric space, show that if $A \subset B$ and B is closed, then $\bar{A} \subset B$ (Theorem 6.11(e)).

22. If A and B are sets in a metric space, show that if $B \subset A$ and B is open, then $B \subset A^{(0)}$ (Theorem 6.11(f)).

23. If A is a set in a metric space, show that

$$A \text{ is open} \Leftrightarrow \mathscr{C}(A) \text{ is closed}$$

(Theorem 6.11(h)).

6.3. Countable and Uncountable Sets

If A and B are sets with a finite number of elements we can easily compare their sizes by pairing the elements of A with those of B. If, after completing the pairing process, there remain unpaired members of one of the sets, say of A, then A is said to be larger than B. The same situation holds for sets with infinitely many elements. While many pairs of infinite sets can be matched in a one-to-one way, it is a remarkable fact that there are infinite sets which cannot be paired with each other. The problem of the relative sizes of sets with infinitely many members has been studied extensively, and in this section we shall describe only the basic properties of the sizes of finite and infinite sets. These properties and the more sophisticated developments of the theory of infinite sets turn out to be useful tools in the development of analysis. The theorems we consider below pertain to sets in R^N although Theorems 6.12(b) and 6.13 hold for arbitrary sets.

Definitions. A set S is **denumerable** if and only if S can be put into one-to-one correspondence with the positive integers. A set S is **countable** if and only if S is either finite or denumerable.

Theorem 6.12

(a) *Any nonempty subset of the natural numbers is countable.*
(b) *Any subset of a countable set is countable.*

PROOF. To establish Part (a), let S be the given nonempty subset. By Theorem 1.30, S has a smallest element; we denote it by k_1. If $S = \{k_1\}$, then S has exactly one element. If the set $S - \{k_1\}$ is not empty, it has a smallest element which we denote k_2. We continue the process. If, for some integer n, the set $S - \{k_1, k_2, \ldots, k_n\}$ is empty, then S is finite and we are done. Otherwise, the sequence $k_1, k_2, \ldots, k_n, \ldots$ is infinite. We shall show that $S = \{k_1, k_2, \ldots, k_n, \ldots\}$. Suppose, on the contrary, that there is an element of S not among the k_n. Then there is a smallest such element which we denote by p. Now k_1 is the smallest element of S and so $k_1 \leqslant p$. Let T be those elements of $k_1, k_2, \ldots, k_n, \ldots$ with the property that $k_i \leqslant p$. Then T is a finite set and there is an integer i such that $k_i \leqslant p < k_{i+1}$ for some i. Since p is not equal to any of the k_j, it follows that $k_i < p < k_{i+1}$. By construction, the element k_{i+1} is the smallest element of S which is larger than $k_1, k_2, \ldots, k_j$. Since p is in S, we have a contradiction of the inequality $k_i < p < k_{i+1}$. Thus $S = \{k_1, k_2, \ldots, k_n, \ldots\}$.

Part (b) is a direct consequence of Part (a).   □

Theorem 6.13. *The union of a countable family of countable sets is countable.*

PROOF. Let $S_1, S_2, \ldots, S_m, \ldots$ be a countable family of countable sets. For each fixed m, the set S_m can be put into one-to-one correspondence with some or all of the pairs $(m, 1), (m, 2), \ldots, (m, n), \ldots$. Hence the union S can be put into one-to-one correspondence with a subset of the totality of ordered pairs (m, n) with $m, n = 1, 2, \ldots$. However, the totality of ordered pairs can be arranged in a single sequence. We first write the ordered pairs as shown in Figure 6.5.

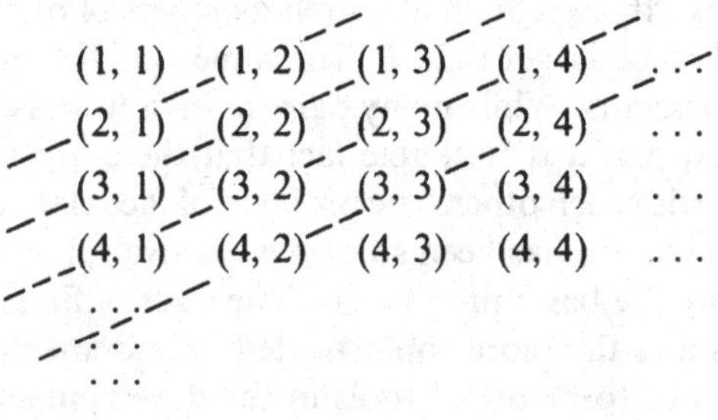

$$(1, 1) \quad (1, 2) \quad (1, 3) \quad (1, 4) \quad \cdots$$
$$(2, 1) \quad (2, 2) \quad (2, 3) \quad (2, 4) \quad \cdots$$
$$(3, 1) \quad (3, 2) \quad (3, 3) \quad (3, 4) \quad \cdots$$
$$(4, 1) \quad (4, 2) \quad (4, 3) \quad (4, 4) \quad \cdots$$
$$\vdots$$

Figure 6.5

Then we write the terms in diagonal order as follows:

$$(1, 1), (2, 1), (1, 2), (3, 1), (2, 2), (1, 3), (4, 1), \dots .$$

This sequence, is in one-to-one correspondence with the positive integers. The reader may verify that the ordered pair (p, q) in the above ordering corresponds to the natural number r given by

$$r = \frac{(p + q - 1)(p + q - 2)}{2} + q. \qquad \square$$

Theorem 6.13 shows that the set of all integers (positive, negative, and zero) is denumerable. Since the rational numbers can be put into one-to-one correspondence with a subset of ordered pairs (p, q) of integers (p is the numerator, q the denominator), we see that the rational numbers are denumerable.

The next theorem shows that the set of all real numbers is essentially larger than the set of all rational numbers. Any attempt to establish a one-to-one correspondence between these two sets must fail.

Theorem 6.14. $\mathbb{R}^1$ *is not countable.*

PROOF. It is sufficient to show that the interval $I = \{x : 0 < x < 1\}$ is not countable. Suppose I is countable. Then we may arrange the members of I in a sequence $x_1, x_2, \dots, x_n, \dots$. We shall show there is a number $x \in I$ not in this sequence, thus contradicting the fact that I is countable. Each x_i has a unique proper decimal development, as shown in Figure 6.6. Each d_{ij} is a digit between 0 and 9.

$$x_1 = 0.d_{11} \, d_{12} \, d_{13} \dots$$
$$x_2 = 0.d_{21} \, d_{22} \, d_{23} \dots$$
$$x_3 = 0.d_{31} \, d_{32} \, d_{33} \dots$$
$$\vdots \quad \vdots \quad \vdots \quad \vdots$$

Figure 6.6. Decimal expansion of $x_1, x_2, \dots, x_n, \dots$.

We now define a number x in I by the following decimal development:

$$x = 0.a_1 a_2 a_3 \dots . \qquad (6.6)$$

Each digit a_i is given by

$$a_i = \begin{cases} 4 & \text{if } d_{ii} \neq 4, \\ 5 & \text{if } d_{ii} = 4. \end{cases}$$

Then Expression (6.6) is a proper decimal expansion of the number x and it differs from x_n in the nth place. Hence $x \neq x_n$ for every n, and I is not countable.

$\square$

The proofs of the next two theorems are left to the reader. (See Problems 5, 6, and 7 at the end of this section and the hints given there.)

Theorem 6.15

(a) *Any family of disjoint open intervals in $\mathbb{R}^1$ is a countable family.*
(b) *Any family of disjoint open sets in $\mathbb{R}^N$ is a countable family.*

Theorem 6.16. *Let f be a monotone function defined on an interval I of $\mathbb{R}^1$. Then the set of points of discontinuity of f is countable.*

Definitions. An **open cell** in $\mathbb{R}^N$ is a set of the form

$$\{(x_1, x_2, \ldots, x_n): a_i < x_i < b_i, i = 1, 2, \ldots, N\}.$$

We suppose that $a_i < b_i$ for every i. Note that an open cell is a straightforward generalization of an open interval in $\mathbb{R}^1$. Similarly, a **closed cell** is a set of the form

$$\{(x_1, x_2, \ldots, x_N): a_i \leqslant x_i \leqslant b_i, i = 1, 2, \ldots, N\}.$$

Again, we assume that $a_i < b_i$ for every i.

Theorem 6.17. *An open set in $\mathbb{R}^1$ is the union of a countable family of disjoint open intervals.*

PROOF. Let G be the given open set and suppose $x_0 \in G$. Since G is open there is an open interval J which contains x_0 and such that $J \subset G$. Define $I(x_0)$ as the union of all such open intervals J. Then $I(x_0)$ is open and contained in G. Furthermore, since the union of a set of open intervals having a common point is an open interval, $I(x_0)$ is an open interval. Let x_1 and x_2 be distinct points of G. We shall show that $I(x_1)$ and $I(x_2)$ are disjoint or that $I(x_1) \equiv I(x_2)$. Suppose that $I(x_1) \cap I(x_2)$ is not empty. Then $I(x_1) \cup I(x_2)$ is an interval in G which contains both x_1 and x_2. Hence $I(x_1) \cup I(x_2)$ is contained in both $I(x_1)$ and $I(x_2)$. Therefore $I(x_1) \equiv I(x_2)$. Thus G is composed of the union of disjoint open intervals of the type $I(x_0)$ and by Theorem 6.15(a) this union is countable. $\qquad \square$

Theorem 6.18. *An open set in $\mathbb{R}^N$ is the union of a countable family of closed cells whose interiors are disjoint.*

PROOF. Let G be an open set in $\mathbb{R}^N$. Construct the hyperplanes (i.e. hyperplanes parallel to the coordinate hyperplanes, $x_i = 0$) given by

$$x_i = \frac{k_i}{2^n}, \qquad i = 1, 2, \ldots, N$$

where $k_1, k_2, \ldots, k_N$ are any integers and n is any nonnegative integer. For each such integer n, the inequalities

$$\frac{k_i - 1}{2^n} \leqslant x_i \leqslant \frac{k_i}{2^n}, \qquad i = 1, 2, \ldots, N, n = 1, 2, 3, \ldots,$$

define a **hypercube**, and the totality of all such hypercubes as the $\{k_i\}$ take on integer values determines the **nth grid** of $\mathbb{R}^N$. Let F_0 be the union of all the closed hypercubes of the 0th grid which are contained in G. Let G_1 be the union of all closed hypercubes of the first grid which are in G but such that no hypercube of G_1 is contained in F_0. Define $F_1 = F_0 \cup G_1$. We define G_2 as the union of all closed hypercubes of the second grid which are in G but such that no hypercube of G_2 is contained in F_1. Define $F_2 = F_1 \cup G_2$. We continue the process. Clearly, the sets $F_0, G_1, G_2, \ldots, G_k, \ldots$ are such that each set consists of a countable family of hypercubes and no two such cubes have any common interior points. Furthermore, for each n, the set F_n is the union of all hypercubes of the nth grid which are in G. We shall show that $G = \bigcup_n F_n$. To see this, let $p_0 \in G$. Since G is open there is a ball B with center p_0 and radius r such that $B \subset G$. For any n such that $2^{-n}\sqrt{N} < r$, there is a hypercube of the nth grid which contains p_0 and is entirely in B. Hence this hypercube is in G. Because p_0 is arbitrary, it follows that $G \subset \bigcup_n F_n$. Since each F_n is contained in G, the proof is complete. $\square$

Remark. The decomposition of open sets into the union of closed cells is not unique. See Problem 10 at the end of this section.

PROBLEMS

1. Show that the totality of **rational points** in $\mathbb{R}^2$ (that is, points both of whose coordinates are rational numbers) is denumerable.

2. Show that the totality of rational points in $\mathbb{R}^N$ is denumerable.

3. Define the set S as follows. The element x is in S if x is an infinite sequence of the form $(r_1, r_2, \ldots, r_n, 0, 0, \ldots, 0, \ldots)$. That is, from some n on, the sequence consists entirely of zeros and the nonzero entries are rational numbers. Show that S is denumerable.

4. An **algebraic number** is any number which is the root of some equation of the form

$$a_0 x^n + a_1 x^{n-1} + \cdots + a_{n-1} x + a_n = 0, \qquad a_0 \neq 0$$

   in which each a_i is an integer and n is a positive integer. The number n may vary from element to element. A **real algebraic number** is a real root of any such equation.
   (a) Show that the totality of real algebraic numbers is countable.
   (b) Show that the totality of algebraic numbers is countable.

5. Show that any family of pairwise disjoint open intervals in $\mathbb{R}^1$ is countable. [*Hint:* Set up a one-to-one correspondence between the disjoint intervals and a subset of the rational numbers. (Theorem 6.15 (a).)]

6. Show that any family of disjoint open sets in $\mathbb{R}^N$ is countable. (Theorem 6.15 (b).) See the hint in Problem 5.

7. Prove Theorem 6.16 [*Hint:* Show that the points of discontinuity can be put into one-to-one correspondence with a family of disjoint open intervals.]

8. Show that every infinite set contains a denumerable subset.

9. Let G be an open set in $\mathbb{R}^1$. Show that G can be represented as the union of open intervals with rational endpoints.

10. Let $G = \{(x, y): 0 < x < 1, 0 < y < 1\}$ be a subset of $\mathbb{R}^2$.
    (a) Give explicitly a family of closed cells (as in Theorem 6.18) in $\mathbb{R}^2$ whose union is G.
    (b) Show that the choices of the closed cells may be made in infinitely many ways.

11. Let $I = \{x: 0 < x < 1\}$. Let S be the totality of functions f having domain I and range contained in I. Prove that the set S cannot be put into one-to-one correspondence with any subset of $\mathbb{R}^1$.

12. Let $I = \{x: 0 < x < 1\}$ and $J = \{(x, y): 0 < x < 1, 0 < y < 1\}$. Show that the points of I and J can be put into one-to-one correspondence. Extend the result to I and $K = \{(x_1, \ldots, x_N): 0 < x_i < 1, i = 1, 2, \ldots, N\}$.

6.4. Compact Sets and the Heine–Borel Theorem

In Theorem 3.10 we established the Bolzano–Weierstrass theorem which states that *any bounded infinite sequence in $\mathbb{R}^1$ contains a convergent subsequence.* In this section we investigate the possibility of extending this theorem to metric spaces. We shall give an example below which shows that the theorem does not hold if the bounded sequence in $\mathbb{R}^1$ is replaced by a bounded sequence in a metric space S. However, the important notion of compactness, which we now define, leads to a modified extension of the Bolzano–Weierstrass theorem.

Definition. A set A in a metric space S is **compact** if and only if each sequence of points (p_n) in A contains a subsequence, say $\{q_n\}$, which converges to a point p_0 in A.

We see immediately from the definition of compactness and from the Bolzano–Weierstrass theorem that every closed, bounded set in $\mathbb{R}^1$ is compact. Also, it is important to notice that the subsequence $\{q_n\}$ not only converges to p_0 but that p_0 is in A

Definition. A set A in a metric space S is **bounded** if and only if A is contained in some ball $B(p, r)$ with $r > 0$.

Theorem 6.19. *A compact set A in a metric space S is bounded and closed.*

PROOF. Assume A is compact and not bounded. We shall reach a contradiction. Define the sequence $\{p_n\}$ as follows: choose $p_1 \in A$. Since A is not bounded, there is an element p_2 in A such that $d(p_1, p_2) > 1$. Continuing, for each n, there is an element p_n in A with $d(p_1, p_n) > n - 1$. Since A is compact, we can find a subsequence $\{q_n\}$ of $\{p_n\}$ such that q_n converges to an element $p_0 \in A$. From the definition of convergence, there is an integer N such that $d(q_n, p_0) < 1$ for all $n \geq N$. But

$$d(q_n, p_1) \leq d(q_n, p_0) + d(p_0, p_1) < 1 + d(p_0, p_1). \tag{6.7}$$

Since q_n is actually at least the nth member of the sequence $\{p_n\}$, it follows that $d(q_n, p_1) > n - 1$. For n sufficiently large, this inequality contradicts Inequality (6.7). Hence A is bounded.

Now assume that A is not closed. Then there is a limit point p_0 of A which is not in A. According to Theorem 6.4, there is a sequence $\{p_n\}$ of elements in A such that $p_n \to p_0$ as $n \to \infty$. But any subsequence of a convergent sequence converges to the same element. Hence the definition of compactness is contradicted unless $p_0 \in A$.   □

It is natural to ask whether or not the converse of Theorem 6.19 holds. We give an example of a bounded sequence of elements in a metric space which has no convergent subsequence. Thus any set containing this sequence, even if it is bounded and closed, cannot be compact. The space we choose for the example is the space $\mathscr{C}$ of continuous functions on $I = \{x : 0 \leq x \leq 1\}$ with range in $\mathbb{R}^1$ and with the metric

$$d(f, g) = \max_{0 \leq x \leq 1} |f(x) - g(x)|.$$

We consider the sequence f_n defined by

$$f_n(x) = \begin{cases} 0 & \text{for } 0 \leq x \leq 2^{-n-1}, \\ 2^{n+2}(x - 2^{-n-1}) & \text{for } 2^{-n-1} \leq x \leq 3 \cdot 2^{-n-2}, \\ -2^{n+2}(x - 2^{-n}) & \text{for } 3 \cdot 2^{-n-2} \leq x \leq 2^{-n}, \\ 0 & \text{for } 2^{-n-1} \leq x \leq 1. \end{cases}$$

Figure 6.7 shows a typical member of the sequence $\{f_n\}$. Since the "sawtooth" parts of two different functions f_n, f_m do not overlap, it is easily verified that

$$d(f_n, f_m) = 1 \quad \text{for } n \neq m.$$

Furthermore, $d(f_n, 0) = 1$ for every n. Hence we have a bounded sequence which cannot have a convergent subsequence since every element is at distance 1 from every other element. The set $\{f_n\}$ is not compact in $\mathscr{C}$.

For some metric spaces the property of being bounded and closed is equivalent to compactness. The next theorem establishes this equivalence for the spaces $\mathbb{R}^N$ with a Euclidean or equivalent metric.

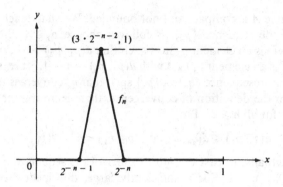

Figure 6.7. A bounded sequence with no convergent subsequence.

Theorem 6.20. *A bounded, closed set in $\mathbb{R}^N$ is compact.*

PROOF. Let A be bounded and closed. Then A lies in a cell

$$C = \{(x_1, x_2, \ldots, x_N) : a_i \leqslant x_i \leqslant b_i, \quad i = 1, 2, \ldots, N\}.$$

Let $\{p_n\}$ be any sequence in A. Writing $p_k = (p_k^1, p_k^2, \ldots, p_k^N)$, we have $a_i \leqslant p_k^i \leqslant b_i$, $i = 1, 2, \ldots, N$, $k = 1, 2, \ldots, n, \ldots$. Starting with $i = 1$, it is clear that $\{p_k^1\}$ is a bounded sequence of real numbers. By the Bolzano–Weierstrass theorem, $\{p_k^1\}$ has a convergent subsequence. Denote this convergent subsequence $\{p_{k'}^1\}$. Next let $i = 2$ and consider the subsequence $\{p_{k'}^2\}$ of $\{p_k^2\}$. Since $\{p_{k'}^2\}$ is bounded, it has a convergent subsequence which we denote $\{p_{k''}^2\}$. Observe that $\{p_{k''}^1\}$ being a subsequence of the convergent sequence $\{p_{k'}^1\}$ is also convergent. Proceeding to $i = 3$, we choose a subsequence of $\{p_{k''}^3\}$ which converges, and so on until $i = N$. We finally obtain a subsequence of $\{p_k\}$, which we denote by $\{q_k\}$ such that every component of q_k consists of a convergent sequence of real numbers. Since A is closed, q_k converges to an element c of $\mathbb{R}^N$ with c in A. The set A is compact. $\square$

The proof of the next theorem is left to the reader.

Theorem 6.21. *In any metric space a closed subset of a compact set is compact.*

Definition. A sequence $\{p_n\}$ in a metric space S is a **Cauchy sequence** if and only if for every $\varepsilon > 0$ there is a positive integer N such that $d(p_n, p_m) < \varepsilon$ whenever $n, m > N$.

Remarks. We know that a Cauchy sequence in $\mathbb{R}^1$ converges to a limit (Section 3.6). It is easy to see that a Cauchy sequence $\{p_n\}$ in $\mathbb{R}^N$ also converges to a limit. Writing $p_n = (p_n^1, p_n^2, \ldots, p_n^N)$, we observe that $\{p_k^i\}$, $k = 1, 2, \ldots$ is a Cauchy sequence of real numbers for each $i = 1, 2, \ldots, N$. Hence $p_k^i \to c^i$,

$i = 1, 2, \ldots, N$ and the element $c = (c^1, c^2, \ldots, c^N)$ is the limit of the sequence $\{p_k\}$. However, we cannot conclude that in a general metric space a Cauchy sequence always converges to an element in the space. In fact, the statement is false. To see this, consider the metric space (S, d) consisting of all rational numbers with the metric $d(a, b) = |a - b|$. Since a Cauchy sequence of rational numbers may converge to an irrational number, it follows that a Cauchy sequence in S may not converge to a limit in S.

The proofs of the following three theorems concerning Cauchy sequences and compact sets are left to the reader.

Theorem 6.22. *In a metric space S, if $p_n \to p_0$, then $\{p_n\}$ is Cauchy sequence.*

Theorem 6.23. *If A is a compact set in a metric space S and if $\{p_n\}$ is a Cauchy sequence in A, then there is a point $p_0 \in A$ such that $p_n \to p_0$ as $n \to \infty$.*

Theorem 6.24. *Let A be a compact set in a metric space S. Let $S_1, S_2, \ldots, S_n, \ldots$ be a sequence of nonempty closed subsets of A such that $S_n \supset S_{n+1}$ for each n. Then $\bigcap_{n=1}^{\infty} S_n$ is not empty.*

The next theorem shows that a compact set may be covered by a finite number of balls of *any* fixed radius. Of course, the number of balls needed to cover a set will increase as the common radius of the balls becomes smaller.

Theorem 6.25. *Suppose that A is a compact set in a metric space S. Then for each positive number δ, there is a finite number of points $p_1, p_2, \ldots, p_n$ in A such that $\bigcup_{i=1}^{n} B(p_1, \delta) \supset A$.*

PROOF. Assume A is not empty and choose a point p_1 in A. Let δ be given. If $A \subset B(p_1, \delta)$ the theorem is established. Otherwise choose $p_2 \in A - B(p_1, \delta)$. If $A \subset B(p_1, \delta) \cup B(p_2, \delta)$ the theorem is established. Otherwise choose $p_3 \in A - (B(p_1, \delta) \cup B(p_2, \delta))$. We continue the process. If A is covered in a finite number of steps the result is proved. Otherwise there is an infinite sequence $p_1, p_2, \ldots, p_k, \ldots$ such that each p_i is at distance at least δ from all previous elements p_j. That is, $d(p_i, p_j) \geqslant \delta$ for all $i \neq j$. Hence no subsequence of $\{p_i\}$ can converge to a point, contradicting the hypothesis that A is compact. Therefore the selection of $p_1, p_2, \ldots$ must stop after a finite number of steps, and the result is proved. $\square$

Let H be a set in a space S and suppose that $\mathscr{F}$ is a family of sets in S. The family $\mathscr{F}$ **covers** H if and only if every point of H is a point in at least one member of $\mathscr{F}$. Coverings of compact sets by a finite number of open balls, as given in Theorem 6.25, have a natural extension to coverings by any family of open sets. The next theorem, an important and useful one in analysis, gives one form of this result. It is a direct extension of Theorem 3.15.

Theorem 6.26 (Heine–Borel theorem). *Let A be a compact set in a metric space S. Suppose that $\mathcal{F} = \{G_\alpha\}$ is a family of open sets which covers A. Then a finite subfamily of $\mathcal{F}$ covers A.*

PROOF. Suppose no such subfamily exists. We shall reach a contradiction. According to Theorem 6.25 with $\delta = 1/2$, there is a finite number of points $p_{11}, p_{12}, \ldots, p_{1k_1}$ in A such that $A \subset \bigcup_{i=1}^{k_1} B(p_{1i}, 1/2)$. Then one of the sets

$$A \cap \overline{B(p_{11}, 1/2)}, \qquad A \cap \overline{B(p_{12}, 1/2)}, \qquad \ldots, \qquad A \cap \overline{B(p_{1k_1}, 1/2)}$$

is such that an infinite number of open sets of $\mathcal{F}$ are required to cover it. (Otherwise, the theorem would be proved.) Call this set $A \cap B(p_1, 1/2)$ and observe that it is a compact subset of A. Now apply Theorem 6.25 to this compact subset with $\delta = 1/2^2$. Then there is a finite number of points $p_{21}, p_{22}, \ldots, p_{2k_2}$ in $A \cap \overline{B(p_1, 1/2)}$ such that

$$A \cap \overline{B(p_1, 1/2)} \subset \bigcup_{i=1}^{k_2} B(p_{2i}, 2^{-2}).$$

One of the balls $B(p_{21}, 2^{-2}), \ldots, B(p_{2k_2}, 2^{-2})$ is such that an infinite number of open sets of $\mathcal{F}$ are required to cover the portion of A contained in that ball. Denote that ball by $B(p_2, 1/4)$. Continuing in this way we obtain a sequence of closed balls

$$\overline{B\left(p_1, \frac{1}{2}\right)}, \qquad \overline{B\left(p_2, \frac{1}{4}\right)}, \qquad \ldots, \qquad \overline{B\left(p_n, \frac{1}{2^n}\right)}, \ldots$$

such that $p_n \in A \cap B(p_{n-1}, 1/2^{n-1})$, and an infinite number of open sets of $\mathcal{F}$ are required to cover $A \cap \overline{B(p_n, 1/2^n)}$. Since $d(p_{n-1}, p_n) \leq 1/2^{n-1}$ for each n, the sequence $\{p_n\}$ is a Cauchy sequence and hence converges to some element p_0. Now since A is compact, we have $p_0 \in A$. Therefore p_0 lies in some open set G_0 of $\mathcal{F}$. Since $p_n \to p_0$, there is a sufficiently large value of n such that $\overline{B(p_n, 1/2^n)} \subset G_0$. However, this contradicts the fact that an infinite number of members of $\mathcal{F}$ are required to cover $B(p_n, 1/2^n) \cap A$. $\qquad\qquad\square$

Remarks. The compactness of A is essential in establishing the Heine–Borel theorem. It is not enough to assume, for example, that A is bounded. To see this, recall the set $\{f_n\}$ of functions in $\mathscr{C}$, the space of continuous functions, which are defined following Theorem 6.19. There we saw that $d(f_n, f_m) = 1$ for all $n \neq m$. Hence we can cover $\{f_n\}$ by a family $\mathcal{F}$ of balls $\{B(f_n, \frac{1}{2})\}$ and each ball will contain exactly one function of the set $\{f_n\}$. No finite subfamily can cover this set. Note that $\{f_n\}$ is a bounded set.

A simpler example is given by the set in $\mathbb{R}^1$ defined by $A = \{x : 0 < x < 1\}$. We cover A with the family $\mathcal{F}$ of open intervals defined by $I_a = \{x : a/2 < x < 1\}$ for each $a \in A$. It is not difficult to verify that no finite subcollection of $\{I_a\}$ can cover A.

The next theorem is a useful equivalent form of the Heine–Borel theorem. The statement and proof should be compared with that of Theorem 3.16.

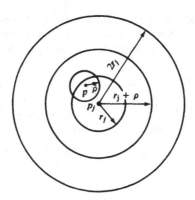

Figure 6.8. $B(p, \rho)$ is in one open set of $\mathscr{F}$.

Theorem 6.27 (Lebesgue lemma). *Let A be a compact set and $\mathscr{F} = \{G_a\}$ a family of open sets which covers A, as in Theorem 6.26. The there is a positive number ρ such that for every $p \in A$ the ball $B(p, \rho)$ is contained in one of the open sets of $\mathscr{F}$.*

PROOF. Since $\mathscr{F}$ covers A, each point p of A is the center of some ball of radius say $2r_p$ which is entirely in *one* open set of $\mathscr{F}$. Hence the ball $B(p, r_p)$ is also in the same open set. Denote by $\mathscr{G}$ the family of open balls $\{B(p, r_p)\}$ for all $p \in A$. By the Heine–Borel theorem, a finite subfamily of $\mathscr{G}$ which we denote

$$B(p_1, r_1), B(p_2, r_2), \ldots, B(p_k, r_k)$$

covers A. Let $\rho = \min_{1 \leqslant i \leqslant k} r_i$. Now if $p \in A$, then p is in some ball $B(p_j, r_j)$. We have (see Figure 6.8)

$$B(p, \rho) \subset B(p_j, r_j + \rho) \subset B(p_j, 2r_j).$$

The ball $B(p_j, 2r_j)$ was chosen so that it lies in a single open set of $\mathscr{F}$. Since p is an arbitrary point of A, the proof is complete. $\qquad\square$

The equivalent form of the Heine–Borel theorem given by the Lebesgue lemma is particularly useful in the proof of the basic theorem on the change of variables in a multiple integral. This topic is discussed in Chapter 14.

PROBLEMS

1. Show that the union of a compact set and a finite set is compact.

2. (a) Show that the intersection of any number of compact sets is compact.
   (b) Show that the union of any finite number of compact sets is compact.
   (c) Show that the union of an infinite number of compact sets may not be compact.

3. Consider the space $\mathbb{R}^2$ with the Euclidean metric. We define $A = \{(x, y) : 0 \leqslant x \leqslant 1, 0 \leqslant y \leqslant 1\}$, and we define $B = A - \{P(1, 1)\}$. Show that B can be covered by an infinite family $\mathscr{F}$ of open rectangles in such a way that no finite subfamily of $\mathscr{F}$ covers B.

4. Let l_2 be the space of sequences $x = \{x_1, x_2, \ldots, x_n, \ldots\}$ such that $\sum_{i=1}^{\infty} x_i^2$ converges. In this space, define $e_n = \{0, 0, \ldots, 0, 1, 0 \ldots\}$ in which all entries are zero except the nth which is 1. The distance between two points x and y is $d(x, y) = (\sum_{i=1}^{\infty} (x_i - y_i)^2)^{1/2}$. Show that the set $\{e_n\}$ is a bounded noncompact set in l_2.

5. In $\mathbb{R}^1$ we cover the set $A = \{x : 0 \leqslant x \leqslant 1\}$ by a family of open intervals defined as follows:

$$I_a = \left\{ x : \frac{a}{2} < x < 2 \right\}$$

for all a such that $0 < a \leqslant 1$. We define $I_0 = \{x : -1/5 \leqslant x \leqslant 1/5\}$. Let $\mathscr{F} = I_0 \cup \{I_a : 0 < a \leqslant 1\}$. For this family $\mathscr{F}$ find a value of ρ in the Lebesgue lemma.

6. Suppose that $\mathscr{F}$ is a family of open sets in $\mathbb{R}^2$ which covers the rectangle $R = \{(x, y) : a \leqslant x \leqslant b, c \leqslant y \leqslant d\}$. Show that R can be divided into a finite number of rectangles by lines parallel to the sides of R such that no two rectangles have common interior points and such that each closed rectangle is contained in one open set of $\mathscr{F}$.

7. Let $\mathscr{A}$ be the space of sequences $x = \{x_1, x_2, \ldots, x_n, \ldots\}$ in which only a finite number of the x_i are different from zero. In $\mathscr{A}$ define $d(x, y)$ by the formula

$$d(x, y) = \max_{1 \leqslant i < \infty} |x_i - y_i|.$$

(a) Show that $\mathscr{A}$ is a metric space.
(b) Find a closed bounded set in $\mathscr{A}$ which is not compact.

8. Suppose that $\mathscr{F}$ is a family of open sets in $\mathbb{R}^2$ which covers the circle $C = \{(x, y) : x^2 + y^2 = 1\}$. Show that there is a $\rho > 0$ such that $\mathscr{F}$ covers the set

$$A = \{(x, y) : (1 - \rho)^2 \leqslant x^2 + y^2 \leqslant (1 + \rho)^2\}.$$

9. Show that in a metric space a closed subset of a compact set is compact (Theorem 6.21).

10. Give an example of points in $\mathbb{R}^1$ which form a compact set and whose limit points form a countable set.

11. Show that in a metric space a convergent sequence is a Cauchy sequence (Theorem 6.22).

12. Let $\{p_n\}$ be a Cauchy sequence in a compact set A in a metric space. Show there is a point $p_0 \in A$ such that $p_n \to p_0$ as $n \to \infty$ (Theorem 6.23).

13. Let $S_1, \ldots, S_n, \ldots$ be a sequence of closed sets in the compact set A such that $S_n \supset S_{n+1}$, $n = 1, 2, \ldots$. Show that $\bigcap_{n=1}^{\infty} S_n$ is not empty.

14. In $\mathbb{R}^3$ with the Euclidean metric consider all points on the surface $x_1^2 + x_2^2 - x_3^2 = 1$. Is this set compact?

6.5. Functions on Compact Sets

In Chapter 3, we discussed the elementary properties of functions defined on $\mathbb{R}^1$ with range in $\mathbb{R}^1$. We were also concerned with functions whose domain consisted of a part of $\mathbb{R}^1$, usually an interval. In this section we develop properties of functions with domain all or part of an arbitrary metric space S and with range in $\mathbb{R}^1$. That is, we consider real-valued functions. We first take up the fundamental notions of limits and continuity.

Definition. Suppose that f is a function with domain A, a subset of a metric space S, and with range in $\mathbb{R}^1$; we write $f: A \to \mathbb{R}^1$. We say that $f(p)$ **tends to** l **as** p **tends to** p_0 **through points of** A if and only if (i) p_0 is a limit point of A, and (ii) for each $\varepsilon > 0$ there is a $\delta > 0$ such that

$$|f(p) - l| < \varepsilon \text{ for all } p \text{ in } A$$

with the property that $0 < d(p, p_0) < \delta$. We write

$$f(p) \to l \quad \text{as } p \to p_0, \qquad p \in A.$$

We shall also use the notation

$$\lim_{\substack{p \to p_0 \\ p \in A}} f(p) = l.$$

Observe that the above definition does not require $f(p_0)$ to be defined nor does p_0 have to belong to the set A. However, when both of these conditions hold, we are able to define continuity for real-valued functions.

Definitions. Let A be a subset of a metric space S, and suppose $f: A \to \mathbb{R}^1$ is given. Let $p_0 \in A$. We say that f **is continuous with respect to** A **at** p_0 if and only if (i) $f(p_0)$ is defined, and (ii) either p_0 is an isolated point of A or p_0 is a limit point of A and

$$f(p) \to f(p_0) \quad \text{as } p \to p_0, \qquad p \in A.$$

We say that f **is continuous on** A if and only if f is continuous with respect to A at every point of A.

Remarks. If the domain of f is the entire metric space S, then we say f is continuous at p_0, omitting the phrase "with respect to A". The definitions of limit and continuity for functions with domain in one metric space S_1 and range in another metric space S_2 are extensions of those given above. See Problem 9 at the end of this section. A fuller discussion is given in Section 6.7.

The theorems in Section 2.2 all have straightforward extensions to functions defined on subsets of a metric space with the exception of the theorem on composite functions (Theorem 2.7). To illustrate the extensions we first establish the analog of Theorem 2.1 on the uniqueness of limits.

Theorem 6.28 (Uniqueness of limits). *Let A be a subset of a metric space S and suppose $f: A \to \mathbb{R}^1$ is given. If*

$$f(p) \to L \quad as \ p \to p_0, \qquad p \in A,$$

and

$$f(p) \to M \quad as \ p \to p_0, \qquad p \in A,$$

then $L = M$.

PROOF. Suppose $L \neq M$. We assume $L < M$, the proof for $L > M$ being the same. Define $\varepsilon = (M - L)/2$. Then there are numbers $\delta_1 > 0$ and $\delta_2 > 0$ such that

$$|f(p) - L| < \varepsilon \quad \text{for all } p \in A \text{ for which } 0 < d(p, p_0) < \delta_1,$$
$$|f(p) - M| < \varepsilon \quad \text{for all } p \in A \text{ for which } 0 < d(p, p_0) < \delta_2. \tag{6.8}$$

Let δ be smaller of δ_1 and δ_2. Since p_0 is a limit point of A, there is a point $\bar{p} \in B(p_0, \delta)$ which is in A. We have from the inequalities in (6.8)

$$M - L \leqslant |M - f(\bar{p})| + |f(\bar{p}) - L| < \varepsilon + \varepsilon = 2\varepsilon,$$

a contradiction to $\varepsilon = (M - L)/2$. Hence $M = L$. $\qquad\qquad\qquad\qquad\square$

Note the similarity of the above proof to the proof of Theorem 2.1. The proofs of Theorems 2.2–2.6 and 2.8 to 2.10 may be generalized in exactly the same way. (See Problems 1–8 at the end of this section.)

The basic properties of functions defined on $\mathbb{R}^1$ discussed in Chapter 3 do not always have direct analogs for functions defined on a subset of a metric space. For example, the Intermediate-value theorem (Theorem 3.4) for functions defined on an interval of $\mathbb{R}^1$ does not carry over to real-valued functions defined on all or part of a metric space. However, when the domain of a real-valued function is a *compact* subset of a metric space, many of the theorems of Chapter 3 have natural extensions. For example, the following theorem is the analog of the Boundedness theorem (Theorem 3.12).

Theorem 6.29. *Let A be a compact subset of a metric space S. Suppose that $f: A \to \mathbb{R}^1$ is continuous on A. Then the range of f is bounded.*

PROOF. We assume the range is unbounded and reach a contradiction. Suppose that for each positive integer n, there is a $p_n \in A$ such that $|f(p_n)| > n$. Since A is compact, the sequence $\{p_n\} \subset A$ must have a convergent subsequence, say $\{q_n\}$, and $q_n \to \bar{p}$ with $\bar{p} \in A$. Since f is continuous on A, we have $f(q_n) \to f(\bar{p})$ as $n \to \infty$. Choosing $\varepsilon = 1$ in the definition of continuity on A and observing that $d(q_n, \bar{p}) \to 0$ as $n \to \infty$, we can state that there is an N_1 such that for $n > N_1$, we have

$$|f(q_n) - f(\bar{p})| < 1 \quad \text{whenever} \quad n > N_1.$$

We choose N_1 so large that $|f(\bar{p})| < N_1$. Now for $n > N_1$, we may write

$$|f(q_n)| = |f(q_n) - f(\bar{p}) + f(\bar{p})| \leqslant |f(q_n) - f(\bar{p})| + |f(\bar{p})| < 1 + |f(\bar{p})|.$$

Since q_n is at least the nth member of the sequence $\{p_n\}$, it follows that $|f(q_n)| > n$ for each n. Therefore,

$$n < |f(q_n)| < 1 + |f(\bar{p})| < 1 + N_1,$$

a contradiction for n sufficiently large. Hence the range of f is bounded. $\square$

Note the similarity of the above proof to that of Theorem 3.12. Also, observe the essential manner in which the compactness of A is employed. The result clearly does not hold if A is not compact. The next result is the analog of the Extreme-value theorem (Theorem 3.12). The proof is left to the reader.

Theorem 6.30. *Let A be a compact subset of a metric space S. Suppose that $f: A \to \mathbb{R}^1$ is continuous on A. Then the range of f contains its supremum and infimum.*

Definition. Let A be a subset of a metric space S, and suppose $f: A \to \mathbb{R}^1$ is given. We say that f is **uniformly continuous on a set B** if and only if (i) $B \subset A$, (ii) for each $\varepsilon > 0$ there is a $\delta > 0$ such that $|f(p) - f(q)| < \varepsilon$ whenever $p, q \in B$ and $d(p, q) < \delta$. (The quantity δ depends on ε but not on the particular points p, q in B.)

The Uniform continuity theorem (Theorem 3.13) has the following extension.

Theorem 6.31. *Suppose that $f: A \to \mathbb{R}^1$ is continuous on a compact set B, and $B \subset A$. Then f is uniformly continuous on B.*

The proof follows that of Theorem 3.13 and is left to the reader.

PROBLEMS

In Problems 1 through 7 a set A in a metric space S is given. All functions are real-valued (range in $\mathbb{R}^1$) and have domain A.

1. If c is a number and $f(p) = c$ for all $p \in A$, then show that for any limit point p_0 of A, we have

    $$\lim_{\substack{p \to p_0 \\ p \in A}} f(p) = c$$

    (Theorem on limit of a constant).

2. Suppose f and g are such that $f(p) = g(p)$ for all $p \in A - \{p_0\}$ where p_0 is a limit point of A, and suppose that $f(p) \to l$ as $p \to p_0, p \in A$. Show that $g(p) \to l$ as $p \to p_0$, $p \in A$ (Limit of equal functions).

3. Suppose that $f_1(p) \to l_1$ as $p \to p_0$, $p \in A$ and $f_2(p) \to l_2$ as $p \to p_0$, $p \in A$. Show that $f_1(p) + f_2(p) \to l_1 + l_2$ as $p \to p_0$, $p \in A$.

4. Under the hypotheses of Problem 3, show that $f_1(p) \cdot f_2(p) \to l_1 \cdot l_2$ as $p \to p_0$, $p \in A$.

5. Assume the hypotheses of Problem 3 hold and that $l_2 \neq 0$. Show that $f_1(p)/f_2(p) \to l_1/l_2$ as $p \to p_0$, $p \in A$.

6. Suppose that $f(p) \leqslant g(p)$ for all $p \in A$, that $f(p) \to L$ as $p \to p_0$, $p \in A$, and that $g(p) \to M$ as $p \to p_0$, $p \in A$. Show that $L \leqslant M$. Show that the same result holds if we assume $f(p) \leqslant g(p)$ in some open ball containing p_0.

7. State and prove an analog to Theorem 2.10 (Sandwiching theorem) for functions defined on a set A in a metric space with range in $\mathbb{R}^1$.

8. Prove the following extension of the composite function theorem (Theorem 2.7). Suppose that $f: \mathbb{R}^1 \to \mathbb{R}^1$ is continuous at L and that g is a function $g: S \to \mathbb{R}^1$, where S is a metric space. Assume that $g(p) \to L$ as $p \to p_0$, $p \in S$. Then $f[g(p)] \to f(L)$ as $p \to p_0$, $p \in S$.

9. Let S_1 and S_2 be two metric spaces. Suppose $f: S_1 \to S_2$ is given.
   (a) Define continuity at a point $p_0 \in S_1$.
   (b) Let A be a subset of S_1. Define continuity of f at p_0 with respect to A.
   (c) Define the statement: $f(p) \to q_0$ as $p \to p_0$ through points of A; that is
   $$\lim_{\substack{p \to p_0 \\ p \in A}} f(p) = q_0.$$

10. Let A be a subset of a metric space S and suppose we are given $f: A \to \mathbb{R}^2$ with the Euclidean metric in $\mathbb{R}^2$. Using the definition of limit obtained in Problem 9, prove a generalization of the theorem on uniqueness of limits (Theorem 6.28). Does the result hold if $\mathbb{R}^2$ is replaced by a general metric space S_2?

11. Let A be a subset of $\mathbb{R}^1$ and let $f: A \to \mathbb{R}^1$ be continuous on A. Show, by constructing an example, that if A is not an interval then the Intermediate-value theorem (Theorem 3.4) may not hold.

12. Suppose $f: A \to \mathbb{R}^1$ is continuous on A, a compact set in a metric space. Show that the range of f contains its supremum and infimum (Theorem 6.30).

13. Suppose that $f: A \to \mathbb{R}^1$ is continuous on a compact set B in A. Show that f is uniformly continuous on B (Theorem 6.31).

14. Let $A = \{(x, y): 0 \leqslant x \leqslant 1,\ 0 \leqslant y \leqslant 1\}$ be a subset of $\mathbb{R}^2$. We define $B = A - \{(1, 1)\}$. Show that the function $f: B \to \mathbb{R}^1$ given by
    $$f(x, y) = (1 - xy)^{-1}$$
    is continuous on B but not uniformly continuous on B.

15. Show how Theorem 6.29 may be proved by using the Heine–Borel theorem (Theorem 6.26).

16. Prove Theorem 6.31 by means of the Lebesgue lemma (Theorem 6.27).

6.6. Connected Sets

Intuitively, we think of a set of points as connected if any two points of the set can be joined by a path of points situated entirely within the set itself. Looked at another way, we may think of a set as connected if it cannot be divided into two or more "separate pieces." It is this latter characterization which we use for the definition of a connected set.

Definition. A set A in a metric space S is said to be **connected** if and only if A cannot be represented as the union of two nonempty disjoint sets neither of which contains a limit point of the other.

For example, the set A in $\mathbb{R}^1$ given by

$$A = \{x : 0 < x < 1 \text{ or } 1 < x < 2\}$$

is not connected. To see this, note that $A = B_1 \cup B_2$ where $B_1 = \{x : 0 < x < 1\}$ and $B_2 = \{x : 1 < x < 2\}$. The sets B_1 and B_2 are disjoint. While B_1 and B_2 have a common limit point ($x = 1$), the point itself is in neither set (Figure 6.9).

We give a second more interesting example by defining in $\mathbb{R}^2$ the sets $D_1 = \{(x, y) : (x - 1)^2 + y^2 \leqslant 1\}$ and $D_2 = \{(x, y) : (x + 1)^2 + y^2 < 1\}$. The set C given by $D_1 \cup D_2$ is connected. Although D_1 and D_2 are disjoint sets, observe that $P(0, 0)$, a member of D_1, is a limit point of D_2 (Figure 6.10). It can be shown that there is no way to decompose C into two nonempty sets neither of which contains a limit point of the other.

Notation. Let $f : X \to Y$ be a function from a metric space X into a metric space Y. Suppose that S is a subset of X. Then $f | S$ denotes **the function f restricted to S**. The **range of $f | S$** is denoted by $R(f|S)$.

The following theorem, an extension to metric spaces of Theorem 3.9, indicates the importance of the notation of connectedness.

Theorem 6.32 (Intermediate-value theorem). *Suppose f is a function from a metric space X into $\mathbb{R}^1$ which is continuous on a nonempty connected set S. Then $R(f|S)$ is an interval or a point.*

PROOF. If $f | S$ is constant, then $R(f|S)$ is a point. Otherwise, by virtue of Theorem 3.8, it is sufficient to show that if y_1 and $y_2 \in R(f|S)$ and c is between y_1 and y_2, then $c \in R(f|S)$. Suppose c is not in $R(f|S)$. Let S_1 be the subset of points $p \in S$ where $f(p) < c$ and let S_2 be the subset of points $p \in S$ where

$$0 < x < 1 \qquad\qquad 1 < x < 2$$

$$0 \qquad\qquad B_1 \qquad\qquad 1 \qquad\qquad B_2 \qquad\qquad 2$$

$A = B_1 \cup B_2$ is not connected

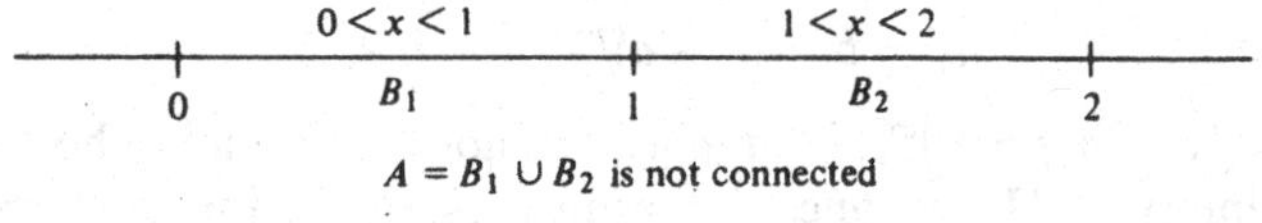

Figure 6.9

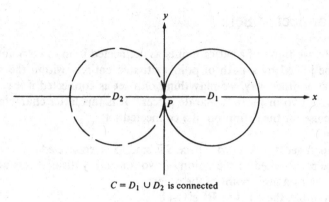

$C = D_1 \cup D_2$ is connected

Figure 6.10

$f(p) > c$. Then S_1 and S_2 are disjoint and also nonempty since $y_1, y_2 \in R(f|S)$, i.e., there are points p_i (in S_i) such that $f(p_i) = y_i$, $i = 1, 2$. Therefore, since S is connected, one of the S_i must contain a limit point of the other. On the other hand, suppose $p_0 \in S_1$ and let $\varepsilon = c - f(p_0)$. Then $\varepsilon > 0$ and there is a $\delta > 0$ such that $|f(p) - f(p_0)| < \varepsilon$ for all p in S with $d(p, p_0) < \delta$. For all such points, $f(p) < f(p_0) + \varepsilon = c$, so that $p \in S_1$. Thus $B(p_0, \delta)$ contains no points of S_2 and so p_0 is not a limit point of S_2. Similarly no point of S_2 is a limit point of S_1. This is a contradiction. Then c is in the range of f and so $R(f|S)$ is an interval. □

The proofs of the following two theorems are left to the reader (see Problems 1 and 2 at the end of the section).

Theorem 6.33. *Let $\mathscr{F}$ be any family of connected subsets of a metric space X such that any two members of $\mathscr{F}$ have a common point. Then the union $A = \bigcup\{S : S \in \mathscr{F}\}$ of all the sets in $\mathscr{F}$ is connected.*

Theorem 6.34. *If S is a connected subset of a metric space, then $\bar{S}$ is connected.*

Theorem 6.35. *Any interval I in $\mathbb{R}^1$ is a connected subset of $\mathbb{R}^1$.*

PROOF. It is sufficient to prove the theorem for a closed interval $I = [a, b]$ since any interval I is either already closed or is the union of an increasing sequence $\{[a_n, b_n]\}$ of closed intervals; that is, $[a_n, b_n] \subset [a_{n+1}, b_{n+1}]$ for each n. For example, $[a, b) = \bigcup_{n=1}^{\infty} [a_n, b_n]$, where $a_n = a$ and where

$$b_n = b - (b - a)/2^n, n = 1, 2, \ldots;$$

similarly, $[a, +\infty) = \bigcup_{n=1}^{\infty} [a, a + n]$. Analogous formulas hold for other types of intervals. Hence, once the result is established for a closed interval, Theorem 6.33 yields the result for any interval.

Suppose the interval $[a, b] = S_1 \cup S_2$ where S_1 and S_2 are disjoint and nonempty. Let $U_i = \sup S_i$, $i = 1, 2$; by renumbering the S_i, if necessary, we may assume $U_1 \leqslant U_2 \leqslant b$. Suppose $U_1 < b$. Then every x with $U_1 < x \leqslant b$ belongs to S_2 so $U_2 = b$. If $U_1 \in S_1$, it is certainly a limit point of S_2. If $U_1 \in S_2$, then it is not in S_1 and hence is a limit point of S_1 (see the Corollary to Theorem 3.5). If $U_1 = U_2 = b$, then $b \in S_1$ or S_2. If $b \in S_1$, it is a limit point of S_2 and vice-versa. Thus in all cases, one of the S_i contains a limit point of the other. Hence $[a, b]$ is connected. $\qquad\qquad\square$

Suppose f is a function from $\mathbb{R}^1$ to $\mathbb{R}^1$ which is continuous on an interval I. Then $f|I$, i.e. the set $\{(x, y) : x \in I \text{ and } y = f(x)\}$, is connected (the proof of this fact is left to the reader). For example, if we assume that the sine function is continuous and periodic on $\mathbb{R}^1$ and has range $J = \{y : -1 \leqslant y \leqslant 1\}$, we see that the following sets S_1 and S_2 are connected:

$$S_1 = \{(x, y) : x > 0, y = \sin(1/x)\},$$
$$S_2 = \{(x, y) : x < 0, y = \sin(1/x)\}.$$

(6.9)

Their union is not connected. However, if $-1 \leqslant \eta \leqslant 1$, the set

$$S_1 \cup S_2 \cup \{(x, y) : -1 \leqslant x \leqslant 1, y = \eta\}$$

is connected; but if $|\eta| > 1$, the set

$$S_1 \cup S_2 \cup \{(x, y) : -1 \leqslant x \leqslant 1, y = \eta\}$$

is not connected. The proofs of all these facts are left to the reader. Further theorems and examples concerning connected sets are given in the next section.

PROBLEMS

1. Prove Theorem 6.33. [*Hint:* Let $A = \sum_1 \cup \sum_2$ where $\sum_1$ and $\sum_2$ are disjoint and nonempty. Then prove by contradiction that some S of $\mathscr{F}$ has a nonempty intersection with both $\sum_1$ and $\sum_2$.]

2. Prove that if S is a connected subset of a metric space, then $\bar{S}$ is connected (Theorem 6.34).

3. Prove that if f is a function from $\mathbb{R}^1$ to $\mathbb{R}^1$ which is continuous on an interval I, then the set $\{(x, y) : x \in I, y = f(x)\}$ is a connected subset of $\mathbb{R}^2$.

In Problems 4, 5, and 6 assume that S_1 and S_2 are the sets defined in Equations (6.9); also assume the elementary properties of the sine function.

4. Prove that $S_1 \cup S_2$ is not connected.

5. Prove that $S_1 \cup S_2 \cup \{(x, y) : -1 \leqslant x \leqslant 1, y = \eta\}$ is connected if $|\eta| \leqslant 1$.

6. Prove that $S_1 \cup S_2 \cup \{(x, y) : -1 \leqslant x \leqslant 1, y = \eta\}$ is not connected if $|\eta| > 1$.

7. Prove that if A is a connected set in a metric space and $A \subset B \subset \bar{A}$, then B is connected.

8. Prove that if S is a closed subset of a metric space which is not connected then there exist closed, disjoint, nonempty subsets S_1 and S_2 of S such that $S = S_1 \cup S_2$.

9. Let A be the set of all rational numbers in $\mathbb{R}^1$. Show that A is not connected.

10. Let B be the set of points in $\mathbb{R}^2$ both of whose coordinates are rational. Show that B is not connected.

11. Let A and B be connected sets in a metric space with $A - B$ not connected and suppose $A - B = C_1 \cup C_2$ where $\bar{C}_1 \cap C_2 = C_1 \cap \bar{C}_2 = \varnothing$. Show that $B \cup C_1$ is connected.

12. Let $I = \{x : a \leqslant x \leqslant b\}$ and suppose that $f : I \to I$ is continuous. Show that f has a fixed point, i.e., show that the equation $f(x) = x$ has at least one solution. [*Hint:* Assume the result is false and form the function $F(x) = x - f(x)$. Note that $f(a) > a$ and $f(b) < b$, and that F is continuous under the hypothesis that $x \neq f(x)$ for all $x \in I$.]

13. In R^3 with the Euclidean metric, show that the set $S = \{(x_1, x_2, x_3) : x_1^2 + x_2^2 = x_3^2,$ $(x_1, x_2, x_3) + (0, 0, 0)\}$ is not connected.

6.7. Mappings from One Metric Space to Another

In Section 1.2 we defined a relation from $\mathbb{R}^1$ to $\mathbb{R}^1$, and we also described functions or mappings as special cases of relations. We now define relations and functions between arbitrary sets and show how some of the theorems developed for continuous real-valued functions may be extended to functions from one metric space to another.

Definitions. A **relation from a set** A **to a set** B is a collection of ordered pairs (p, q) in which $p \in A$ and $q \in B$. The set of all the elements p in the collection is called the **domain** of the relation and the set of all the elements q is called the **range**. If no two of the ordered pairs in the collection have the same first element, the relation is called a **mapping**. We also use the terms **function** and **transformation** to mean the same thing. If f is a mapping from A to B we write $f : A \to B$ and if p is in the domain, the $f(p)$ is the unique element q such that $(p, q) \in f$. If f is a mapping from A to B and the domain of f is all of A, we say that f is **on** A **into** B. If the range of f is all of B, we say the mapping is **onto** B. If A_1 is a subset of the domain of the mapping f, then **the image** $f(A_1)$ **under** f consists of all points q such that $f(p) = q$ for some $p \in A_1$. If B_1 is a subset of the range of f, the **inverse image** $f^{-1}(B_1)$ **under** f is the set of all p in the domain of f such that $f(p) \in B_1$. If T is a relation from A to B, the **inverse relation** T^{-1} is the set of all ordered pairs (q, p) such that $(p, q) \in T$.

Remarks. (i) The inverse relation of a mapping f is ordinarily not a mapping. However, if f is one-to-one, then the inverse f^{-1} will be a mapping also. (ii) We recall that the parametric equations of a curve in three-space are given by

$$x_1 = f(t), \qquad x_2 = g(t), \qquad x_3 = h(t).$$

These equations represent a mapping from $\mathbb{R}^1$ to $\mathbb{R}^3$. More generally a mapping f from $\mathbb{R}^N$ to $\mathbb{R}^M$ may be given explicitly by a set of M equations of the form

$$y_1 = f_1(x_1, x_2, \ldots, x_N),$$
$$y_2 = f_2(x_1, x_2, \ldots, x_N),$$
$$\vdots \qquad\qquad \vdots$$
$$y_M = f_M(x_1, x_2, \ldots, x_N).$$

Each f_i is a mapping from $\mathbb{R}^N$ to $\mathbb{R}^1$, and the collection represents a mapping from $\mathbb{R}^N$ to $\mathbb{R}^M = \mathbb{R}^1 \times \mathbb{R}^1 \times \cdots \times \mathbb{R}^1$ (M factors).

Suppose that the domain and range of a mapping are sets in metric spaces. Then we easily extend to such mappings the definitions of limit and continuity given in Chapter 3 for functions from $\mathbb{R}^1$ to $\mathbb{R}^1$. It turns out that many of the Theorems of Chapter 3 use only the fact that $\mathbb{R}^1$ is a metric space.

Definitions. Let (S_1, d_1) and (S_2, d_2) be metric spaces, and suppose that f is a mapping from S_1 into S_2. Let A be a subset of S_1. We say that $f(p)$ **tends to** q_0 **as p tends to** p_0 **for** $p \in A$ if and only if (i) p_0 is a limit point of A, and (ii) for every $\varepsilon > 0$ there is a $\delta > 0$ such that $d_2(f(p), q_0) < \varepsilon$ for all $p \in A$ such that $d_1(p, p_0) < \delta$. We say also that $f(p)$ **has the limit** q_0 as p tends to p_0 (after uniqueness of limits is proved). The symbols

$$f(p) \to q_0 \quad \text{as} \quad p \to p_0, \qquad p \in A,$$

and

$$\lim_{\substack{p \to p_0 \\ p \in A}} f(p) = q_0$$

are used. If A is the entire space S_1, we omit the phrase "for all $p \in A$" in the above definition. The mapping f is **continuous with respect to** A **at** p_0 if and only if (i) $f(p_0)$ is defined, and (ii) either p_0 is an isolated point of A or p_0 is a limit point of A and $f(p) \to f(p_0)$ as $p \to p_0$ for $p \in A$. A function f is **continuous on** A if and only if f is continuous with respect to A at each point of A. If A is all of S_1, we omit the phrase "with respect to A" in the definition of continuity.

It is important to observe that a function may be continuous with respect to a set A at some points of S_1 and yet not be continuous at these same points with respect to a larger set. For example, consider the function f from $\mathbb{R}^1$ into

$\mathbb{R}^1$ defined by

$$f(x) = \begin{cases} 1 & \text{if } x \text{ is rational,} \\ 0 & \text{if } x \text{ is irrational.} \end{cases}$$

Then f is not continuous at any point. However, if A denotes the set of rational numbers in $\mathbb{R}^1$, then f is continuous with respect to A at every rational number. This fact is obvious since f is constant on A. On the other hand, f is not continuous with respect to A at any irrational number although every irrational is a limit point of points in A.

The theorems in Section 6.5 on functions defined on a metric space into $\mathbb{R}^1$ have direct analogs to functions from one metric space to another. Some of these are stated in the problems at the end of this section. We illustrate the form such theorems take in the following version of the Composite function theorem.

Theorem 6.36 (Composite function theorem). *Suppose that f is a mapping on a metric space (S_1, d_1) into a metric space (S_2, d_2), and that g is a mapping on (S_2, d_2) into a metric space (S_3, d_3).*

(a) *Suppose that $f(p) \to q_0$ as $p \to p_0$ and that g is continuous at q_0. Then $g[f(p)] \to g(q_0)$ as $p \to p_0$.*

(b) *Suppose that f is continuous on S_1 and g is continuous on S_2. Then the composite function $g \circ f$ is continuous on S_1. (Here $g \circ f(p) = g[f(p)]$.)*

The proof is similar to the proof of the Composite function theorem for functions on $\mathbb{R}^1$ (Theorem 2.7).

The next theorem shows that continuous functions may be characterized in terms of the open sets in a metric space.

Theorem 6.37. *Let f be a mapping on a metric space (S_1, d_1) into a metric space (S_2, d_2). Then f is continuous on $S_1 \Leftrightarrow$ for every open set U in S_2, the set $f^{-1}(U)$ is an open set in S_1.*

PROOF

(a) Assume that f is continuous on S_1. Let U be an open set in S_2. We shall show that $f^{-1}(U)$ is an open set in S_1. Consider $p_0 \in f^{-1}(U)$. From the definition of continuity, for every $\varepsilon > 0$ there is a $\delta > 0$ such that $d_2(f(p), f(p_0)) < \varepsilon$ whenever $d_1(p, p_0) < \delta$. Select ε so small that the open ball $B(f(p_0), \varepsilon)$ is contained in U. Then $f^{-1}(B) \subset f^{-1}(U)$. Also, the open ball $B_1(p_0, \delta)$ which, from the definition of continuity is contained in $f^{-1}(B)$, is also in $f^{-1}(U)$. This shows that p_0 is an interior point of $f^{-1}(U)$. Therefore every point of $f^{-1}(U)$ is an interior point, and so $f^{-1}(U)$ is open.

(b) Assume that $f^{-1}(U)$ is open for every open set U in S_2. We shall show that f is continuous. Consider $p_0 \in S_1$ and let $\varepsilon > 0$ be given. The open ball $B(f(p_0), \varepsilon)$ in S_2 has an inverse image which is open and which contains p_0.

We choose δ so small that the open ball $B_1(p_0, \delta)$ is contained entirely in $f^{-1}(B)$. Then the definition of continuity is satisfied with this value of δ. $\square$

The statements of many theorems are greatly simplified when we introduce the notion of a "metric subspace" of a metric space. Such a space may be generated by any subset of a given metric space, and it enables us to regard the given subset as the universe of discourse instead of the whole space.

Definition. Suppose that (X, d) is a metric space and suppose $S \subset X$. The **metric subspace generated by** S is the metric space (S, d_S) where d_S is the restriction of the metric d to $(S \times S)$.

The proofs of the following facts are left to the reader.

Theorem 6.38

(a) *Suppose that f is a mapping from a metric space X into a metric space Y, that $S \subset X$, and that $P_0, P_n \in S$, $n = 1, 2, \ldots$; also, suppose that f is continuous with respect to S at P_0, that $f(P_n)$ is defined for each n, and that $P_n \to P_0$. Then $f(P_n) \to f(P_0)$.*

(b) *Suppose that f is a mapping from a metric space X with metric d into a metric space Y with metric d', that $S \subset X$, and that $P_0 \in S$. Then f is continuous with respect to S at $P_0 \Leftrightarrow f(P_0)$ is defined and, for each $\varepsilon > 0$, there is a $\delta > 0$ such that $d'(f(P), f(P_0)) < \varepsilon$ for all P in S for which $d(P, P_0) < \delta$.*

A partial converse to Theorem 6.38(a) is easily stated for the case that S is the whole space.

Theorem 6.39. *Suppose that f is a mapping from a metric space X onto a metric space Y, that $P_0 \in X$, and that $f(P_n) \to f(P_0)$ for every sequence $\{P_n\}$ in X such that $P_n \to P_0$. Then f is continuous at P_0.*

PROOF. Suppose f is not continuous at P_0. Then, from Theorem 6.38(b) with $S = X$, there is an $\varepsilon_0 > 0$ such that there is no δ satisfying the condition of that theorem. Thus, for each n, there is a P_n such that $d(P_0, P_n) < 1/n$ but $d'(f(P_n), f(P_0)) \geq \varepsilon_0$. Then $P_n \to P_0$ but $f(P_n)$ does not $\to f(P_0)$, a contradiction.
    $\square$

The following versions of the Composite function theorem hold.

Theorem 6.40 (Composite function theorem). *Suppose f is a continuous mapping from a metric space X into a metric space Y and g is a continuous mapping from Y into a metric space Z.*

(a) *If g is continuous at Q_0 and $f(P) \to Q_0$ as $P \to P_0$, then $g[f(P)] \to g(Q_0)$ as $P \to P_0$.*

(b) *If the domain of f is X, the domain of g is Y, if f is continuous on X and if g is continuous on Y, then $g \circ f$ is continuous on X. (Here $g \circ f$ denotes the composite map: $(g \circ f)(P) = g[f(P)]$.)*

The proofs are left to the reader.

Remarks

(i) If the domain of a function f is a subset of a metric space S_1, we may reword Theorem 6.37 appropriately. Consider the domain of f as a subspace, and then the inverse image of an open set in the range must be an open set in this subspace.

(ii) It is not true that continuous functions always map open sets onto open sets. For example, the function f from $\mathbb{R}^1$ into $\mathbb{R}^1$ given by $f(x) = x^2$ maps the open interval $I = \{x: -1 < x < 1\}$ onto the half-open interval $J = \{x: 0 \leqslant x < 1\}$.

(iii) The theorem corresponding to Theorem 6.37 for closed sets is a restatement of the definition of continuity: f is **continuous on** S_1 if and only if for every closed set A in S_2, the set $f^{-1}(A)$ is a closed set in S_1. Continuous functions do not necessarily map closed sets onto closed sets. For example, the function f from $\mathbb{R}^1$ into $\mathbb{R}^1$ given by $f(x) = 1/(1 + x^2)$ maps the closed set $A = \{x: 0 \leqslant x < \infty\}$ onto the half-open interval $I = \{x: 0 < x \leqslant 1\}$.

Definitions. Let (S, d) be a metric space and let A be a subset of S. Using the metric d, we know that we may consider (A, d) as a metric subspace. A set C contained in A is **open in** A if and only if C is open when considered as a set in the metric space (A, d). In a similar way, we define C is **closed in** A and C is **connected in** A.

For example, let A be a set of isolated points in $\mathbb{R}^1$. Then A, considered as a metric space, has the property that every subset is open. However, no nonempty isolated set is open in $\mathbb{R}^1$. Thus sets may be open when considered in reference to a subspace without being open in the entire space. Similar examples are easily obtained for closed sets.

Theorem 6.41. *Suppose that (A, d) is a metric subspace of the metric space (S, d). Then for any set C in A,*

(a) *C is open in $A \Leftrightarrow$ there is an open set G in S such that $C = G \cap A$.*
(b) *C is closed in $A \Leftrightarrow$ there is a closed set F in S such that $C = F \cap A$.*
(c) *If C is connected in A, then C is connected in S.*

PROOF

(a) Denote by $B(p, r)$ the open ball in S with center at p and radius r. If $p \in A$, denote by $B_A(p, r)$ the open ball in A with center at p and radius r. Clearly, $B_A(p, r) = A \cap B(p, r)$. Suppose C is open in A. Then each point $p \in C$

has an open ball $B_A(p, r_p)$ lying entirely in C. We have, taking the union of such balls,

$$C = \bigcup_{p \in C} B_A(p, r_p).$$

Since $B_A(p, r_p) = A \cap B(p, r_p)$, we may write

$$C = A \cap \left[\bigcup_{p \in C} B(p, r_p) \right].$$

But the union of any collection of open sets is open. Hence $C = A \cap G$ where $G = \bigcup_{p \in C} B(p, r_p)$ with G an open set in S.

Now suppose that $C = G \cap A$ where G is an open set in S. We wish to show that C is open in A. Each point $p \in C$ is the center of an open ball $B(p, r_p)$ contained in G. Since $B_A(p, r_p) = A \cap B(p, r_p)$, the ball $B_A(p, r_p)$ is contained in $A \cap G = C$. Therefore each point of C is an interior point with respect to A. Hence C is open in A.

The proofs of Parts (b) and (c) are Problems 7 and 8 at the end of this section.   □

Definition. A mapping f from a metric space (S_1, d_1) into a metric space (S_2, d_2) is **uniformly continuous on** A, a subset of S_1, if and only if (i) the domain of f contains A, (ii) for each $\varepsilon > 0$ there is a $\delta > 0$ such that $d_2(f(p), f(q)) < \varepsilon$ whenever $d_1(p, q) < \delta$ and $p, q \in A$, and (iii) the number δ is the same for all p, q in A.

Remarks. If a function is uniformly continuous on a set A, it is continuous at each point p in A *with respect to* A. It may not, however, be continuous at such points with respect to sets other than A. For example, the function f from $\mathbb{R}^1$ to $\mathbb{R}^1$, which is 1 if x is rational and 0 if x is irrational, is uniformly continuous on the set A of rational numbers and it is continuous with respect to A at each rational point. However, f is not continuous (with respect to $\mathbb{R}^1$) at any point.

Theorem 6.42. *Let A be a subset of a metric space (S_1, d_1), and suppose f is a continuous function on A into a metric space (S_2, d_2).*

(a) *If A is compact, then $f(A)$ is compact.*
(b) *If A is connected, then $f(A)$ is connected.*
(c) *If A is compact, then f is uniformly continuous on A.*
(d) *If A is compact and f is one-to-one, then f^{-1} is continuous.*

The proof of Theorem 6.42 is left to the reader. (These are Problems 12, 13, 14, and 15 at the end of this section.)

We now establish an important theorem about open connected sets in $\mathbb{R}^N$. It states that any two points in an open connected set G in $\mathbb{R}^N$ can be joined by a polygonal path which lies entirely in G.

Definition. Let $I = \{x : a \leqslant x \leqslant b\}$ be any interval and let z be a continuous mapping from I into $\mathbb{R}^N$. Then z is said to be **piecewise linear** if and only if there exist numbers $t_0, t_1, \ldots, t_k$ in I with $a = t_0 < t_1 < \cdots < t_k = b$ such that the mapping z is a linear function on each interval $I_j = \{x : t_{j-1} \leqslant x \leqslant t_j\}$, $j = 1, 2, \ldots, k$. That is, $z^j = (z_1^j, \ldots, z_N^j)$ has the form: $z_i^j = a_i^j x + b_i^j$, $i = 1, 2, \ldots, N$, $j = 1, 2, \ldots, k$. The a_i^j, b_i^j are numbers:

*Theorem 6.43.[3] *Let G be an open connected set in $\mathbb{R}^N$ and suppose that u and v are in G. Then there exists a piecewise linear mapping $z : I \to \mathbb{R}^N$ such that $z(a) = u$, $z(b) = v$ and the range of z is contained in G.*

PROOF. Let u be any fixed point of G, and define G_1 to be the set of all points v in G for which such a piecewise linear mapping exists. Clearly G_1 is not empty since, for any point in a sufficiently small ball with u as center, a *linear* mapping from u to that point can be constructed which is entirely in G. We show that G_1 is an open set. Let $v \in G_1$. Then there is a piecewise linear map $z : I \to \mathbb{R}^N$ with $z(a) = u$ and $z(b) = v$. Since G is open there is a ball $B(v, \rho)$ lying entirely in G. Suppose $w \in B(z, \rho)$; we choose c as any real number larger than b and denote $J = \{x : a \leqslant x \leqslant c\}$. Now we define the piecewise linear mapping $\xi : J \to \mathbb{R}^N$ so that $\xi(t) = z(t)$ for $t \in I$, and we let ξ be a linear map on the interval $J - I$ with $\xi(b) = v$ and $\xi(c) = w$. Then ξ is a piecewise linear mapping of the desired type on J. A change of scale transforms ξ into a piecewise linear map on I. Thus $w \in G_1$. We have shown that every point of $B(z, \rho)$ is in G_1 and so G_1 is open.

Now suppose $G - G_1$ is not empty. Since G_1 is open no point of G_1 is a limit point of $G - G_1$. Hence, since G is connected, there must be a point $v' \in G - G_1$ which is a limit point of G_1. Because G is open there is a ball $B(v', \rho)$ contained in G. Since v' is a limit point of G_1, there is a point w' in $B(v', \rho)$ which is in G_1 (see Figure 6.11). Then there is a piecewise linear

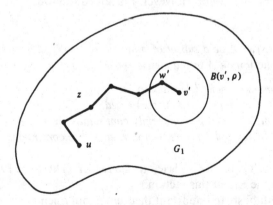

Figure 6.11. A piecewise linear map.

---

[3] Theorems with an asterisk may be omitted without loss of continuity.

mapping connecting u and w'. Now construct a linear map connecting w' and v'. This yields a piecewise linear mapping connecting u and v'. Thus $v' \in G_1$, a contradiction. Hence $G - G_1$ is empty and the result is established. $\quad\square$

*Theorem 6.44

(a) *A set S in a metric space is connected $\Leftrightarrow S$ is not the union of two nonempty disjoint subsets, each of which is open in S.*

(b) *A set S in a metric space is connected $\Leftrightarrow$ the only subsets of S which are both open and closed in S are S itself and the empty set, $\varnothing$.*

PROOF OF (a). Suppose that S is not connected. Then $S = S_1 \cup S_2$ where S_1 and S_2 are disjoint, nonempty, and neither contains a limit point of the other. Thus each point p of S_1 is the center of a ball $B(p, 2r_p)$ which contains no point of S_2; similarly, each point q of S_2 is the center of a ball $B(q, 2\rho_q)$ which contains no point of S_1. We define

$$G_1 = \bigcup_{p \in S_1} B(p, r_p), \qquad G_2 = \bigcup_{q \in S_2} B(q, \rho_p).$$

Clearly, G_1 and G_2 are open and nonempty; we show that they are disjoint. Assume there is a point $s \in G_1 \cap G_2$. Then for some $p \in S_1$ and $q \in S_2$, we must have $s \in B(p, r_p) \cap B(q, \rho_p)$. With d as the distance function, we have

$$d(s, p) < r_p \leqslant \tfrac{1}{2}d(p, S_2) \leqslant \tfrac{1}{2}d(p, q).$$

Also,

$$d(s, q) < \rho_p \leqslant \tfrac{1}{2}d(q, S_1) \leqslant \tfrac{1}{2}d(q, p).$$

Therefore, $d(p, q) \leqslant d(p, s) + d(s, q) < \tfrac{1}{2}d(p, q) + \tfrac{1}{2}d(p, q) = d(p, q)$, a contradiction. Hence there is no such point s; $G_1 \cap G_2 = \varnothing$. We observe that $S = (S \cap G_1) \cup (S \cap G_2)$ and so S is the union of two disjoint, nonempty subsets, each of which is open in S. We proved half of Part (a).

To establish the remainder of Part (a), suppose $S = T_1 \cup T_2$ where T_1 and T_2 are disjoint nonempty subsets of S, each of which is open in S. Since T_1 is open in S, there is an open set G_1 in the metric space such that $T_1 = G_1 \cap S$ (Theorem 6.41(a)). Then for each $p \in T_1$ there is a ball $B(p, r)$ contained in G_1 which contains no point of T_2. Hence T_1 contains no limit point of T_2. Similarly, T_2 contains no limit point of T_1. We have just shown that S is not connected, and Part (a) is established.

The proof of Part (b) is left to the reader. $\quad\square$

PROBLEMS

1. Suppose that (S_1, d_1) and (S_2, d_2) are metric spaces and that A is a subset of S_1. Let f be a mapping on A into S_2 which is continuous with respect to A at a point p_0. Suppose $p_1, p_2, \ldots, p_n, \ldots$ is a sequence of points of A such that $p_n \to p_0$ and $p_0 \in A$. Show that $f(p_n) \to f(p_0)$ as $p_n \to p_0$.

2. Show that if a function f is continuous with respect to A at any point p_0 it is continuous with respect to C at p_0 if C is any set contained in A.

3. Let f be a mapping from the metric space (S_1, d_1) into the metric space (S_2, d_2). State and prove a theorem on uniqueness of limits which is the analog of Theorem 6.28.

4. Prove Theorem 6.36.

5. Give an example of a continuous function from $\mathbb{R}^1$ into $\mathbb{R}^1$ which maps an open set onto a closed set.

6. Let (S, d) be a metric space and let A and C be subsets with $C \subset A$. Suppose A is closed. Show that if C is closed in A, then C is closed in S.

7. Prove Part (b) of Theorem 6.41.

8. Prove Part (c) of Theorem 6.41.

9. Let (S_1, d_1) and (S_2, d_2) be metric spaces and suppose that f is a mapping on S_1 into S_2. Show that f is continuous if and only if for every closed set A in S_2, the set $f^{-1}(A)$ is closed in S_1.

10. Given an example of a mapping from $\mathbb{R}^1$ into $\mathbb{R}^1$ which is continuous at every point of $\mathbb{R}^1$ but not uniformly continuous on $\mathbb{R}^1$.

11. Suppose that (S_1, d_1), (S_2, d_2), and (S_3, d_3) are metric spaces with f a uniformly continuous mapping on S_1 into S_2 and g a uniformly continuous mapping on S_2 into S_3. Show that $g \circ f$ is uniformly continuous on S_1.

12. Prove Part (a) of Theorem 6.42. [*Hint:* See Problem 1.]

13. Prove Part (b) of Theorem 6.42.

14. Prove Part (c) of Theorem 6.42.

15. Prove Part (d) of Theorem 6.42.

16. Let f and g be continuous mappings on (S_1, d_1) into (S_2, d_2). Let A be the set of points of S_1 such that $f(p) = g(p)$. Show that A is closed.

17. In $\mathbb{R}^3$, let S be the set given by

$$S = \{(x_1, x_2, x_3): 1 < x_1^2 + x_2^2 + x_3^2 < 4\}.$$

Give an explicit construction of a piecewise linear mapping z from $I = \{x: 0 \leqslant x \leqslant 1\}$ into $\mathbb{R}^3$ with range in S such that $z(0) = p_1$ and $z(1) = p_2$ where $p_1 = (-3/2, 0, 0)$, $p_2 = (3/2, 0, 0)$.

18. Prove Part (b) of Theorem 6.44.

19. Let A and B be compact sets in $\mathbb{R}^N$ such that $A \cap B = \varnothing$. Show that inf $d(p, q)$ where the infimum is taken for all $p \in A$, all $q \in B$ is positive.

CHAPTER 7

Differentiation in $\mathbb{R}^N$

7.1. Partial Derivatives and the Chain Rule

There are two principal extensions to $\mathbb{R}^N$ of the theory of differentiation of real-valued functions on $\mathbb{R}^1$. In this section, we develop the natural generalization of ordinary differentiation discussed in Chapter 4 to partial differentiation of functions from $\mathbb{R}^N$ to $\mathbb{R}^1$. In Section 7.3, we extend the ordinary derivative to the total derivative.

We shall use letters x, y, z, etc. to denote elements in $\mathbb{R}^N$. The components of an element x are designated by $(x_1, x_2, \ldots, x_N)$ and as usual, the Euclidean distance, given by the formula

$$d(x, y) = \sqrt{\sum_{i=1}^{N} (x_i - y_i)^2},$$

will be used. We also write $d(x, y) = |x - y|$ and $d(x, 0) = |x|$.

Definition. Let f be a function with domain an open set in $\mathbb{R}^N$ and range in $\mathbb{R}^1$. We define the N functions $f_{,i}$ with $i = 1, 2, \ldots, N$ by the formulas

$$f_{,i}(x_1, x_2, \ldots, x_N)$$
$$= \lim_{h \to 0} \frac{f(x_1, \ldots, x_i + h, \ldots, x_N) - f(x_1, \ldots, x_i, \ldots, x_N)}{h}$$

whenever the limit exists. The functions $f_{,1}, f_{,2}, \ldots, f_{,N}$ are called the **first partial derivatives** of f.

We assume the reader is familiar with the elementary processes of partial differentiation. The partial derivative with respect to the ith variable is computed by holding all other variables constant and calculating the ordinary derivative with respect to x_i. That is, to compute $f_{,i}$ at the value $a = (a_1, a_2, \ldots, a_N)$, form the function φ (from $\mathbb{R}^1$ to $\mathbb{R}^1$) by setting

$$\varphi(x_i) = f(a_1, a_2, \ldots, a_{i-1}, x_i, a_{i+1}, \ldots, a_N), \qquad (7.1)$$

and observe that

$$f_{,i}(a) = \varphi'(a_i). \qquad (7.2)$$

There are many notations for partial differentiation and, in addition to the one above, the most common ones are

$$D_i f, \qquad \frac{\partial f}{\partial x_i}, \qquad f_{x_i}.$$

Good notation is important in studying partial derivatives since the many letters and subscripts which occur can often lead to confusion. The two "best" symbols, especially when discussing partial derivatives of higher order, are $f_{,i}$ and $D_i f$, and we shall employ these most of the time.

We recall that the equations of a line segment in $\mathbb{R}^N$ connecting the points $a = (a_1, \ldots, a_N)$ and $a + h = (a_1 + h_1, \ldots, a_N + h_N)$ are given parametrically by the formulas (the parameter is t)

$$x_1 = a_1 + th_1, \qquad x_2 = a_2 + th_2, \ldots, x_N = a_N + th_N, \qquad 0 \le t \le 1.$$

The next theorem is the extension to functions from $\mathbb{R}^N$ to $\mathbb{R}^1$ of the Mean-value theorem (Theorem 4.12).

Theorem 7.1. *Let τ_i be the line segment in $\mathbb{R}^N$ connecting the points $(a_1, a_2, \ldots, a_i, \ldots, a_N)$ and $(a_1, a_2, \ldots, a_i + h_i, \ldots, a_N)$. Suppose f is a function from $\mathbb{R}^N$ into $\mathbb{R}^1$ with domain containing τ_i, and suppose that the domain of $f_{,i}$ contains τ_i. Then there is a real number ξ_i on the closed interval in $\mathbb{R}^1$ with endpoints a_i and $a_i + h_i$ such that*

$$f(a_1, \ldots, a_i + h_i, \ldots, a_N) - f(a_1, \ldots, a_i, \ldots, a_N) = h_i f_{,i}(a_1, \ldots, \xi_i, \ldots, a_N). \qquad (7.3)$$

PROOF. If $h_i = 0$, then Equation (7.3) holds. If $h_i \ne 0$, we use the notation of Expression (7.1) to write the left side of Equation (7.3) in the form

$$\varphi(a_i + h_i) - \varphi(a_i).$$

For this function on $\mathbb{R}^1$, we apply the Mean-value theorem (Theorem 4.12) to conclude that

$$\varphi(a_i + h_i) - \varphi(a_i) = h_i \varphi'(\xi_i).$$

This formula is a restatement of Equation (7.3).   □

Theorem 4.8, the Fundamental lemma of differentiation, has the following generalization for functions from $\mathbb{R}^N$ into $\mathbb{R}^1$. We state the theorem in the general case and prove it for $N = 2$.

Theorem 7.2 (Fundamental lemma of differentiation). *Suppose that the functions f and $f_{,1}, f_{,2}, \ldots, f_{,N}$ all have a domain in $\mathbb{R}^N$ which contains an open ball about the point $a = (a_1, a_2, \ldots, a_N)$. Suppose all the $f_{,i}$ $i = 1, 2, \ldots, N$, are continuous at a. Then*

(a) *f is continuous at a;*
(b) *there are functions $\varepsilon_1(x), \varepsilon_2(x), \ldots, \varepsilon_N(x)$, continuous at $x = 0$ such that $\varepsilon_1(0) = \varepsilon_2(0) = \cdots = \varepsilon_N(0) = 0$ and*

$$f(a + h) - f(a) = \sum_{i=1}^{N} [f_{,i}(a) + \varepsilon_i(h)] h_i \tag{7.4}$$

for $h = (h_1, h_2, \ldots, h_N)$ in some ball $B(0, r)$ in $\mathbb{R}^N$ of radius r and center at $h = 0$.

PROOF. For $N = 2$. We employ the identity

$$f(a_1 + h_1, a_2 + h_2) - f(a_1, a_2)$$
$$= [f(a_1 + h_1, a_2) - f(a_1, a_2)]$$
$$+ [f(a_1 + h_1, a_2 + h_2) - f(a_1 + h_1, a_2)]. \tag{7.5}$$

Since $f, f_{,1}$ and $f_{,2}$ are defined in an open ball about $a = (a_1, a_2)$, this identity is valid for h_1, h_2 in a sufficiently small ball (of radius, say, r) about $h_1 = h_2 = 0$. Applying Theorem 7.1 to the right side of Equation (7.5), we find there are numbers $\xi_1(h_1, h_2), \xi_2(h_1, h_2)$ on the closed intervals from a_1 to $a_1 + h_1$ and from a_2 to $a_2 + h_2$, respectively (see Figure 7.1), such that

$$f(a_1 + h_1, a_2 + h_2) - f(a_1, a_2) = f_{,1}(\xi_1, a_2) h_1 + f_{,2}(a_1 + h_1, \xi_2) h_2. \tag{7.6}$$

Equation (7.6) is valid for h_1, h_2 in the ball of radius r about $h_1 = h_2 = 0$.

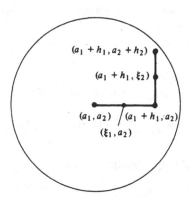

Figure 7.1

Now define

$$\varepsilon_1(h_1, h_2) = f_{,1}(\xi_1, a_2) - f_{,1}(a_1, a_2),$$

$$\varepsilon_2(h_1, h_2) = f_{,2}(a_1 + h_1, \xi_2) - f_{,2}(a_1, a_2). \tag{7.7}$$

We wish to show that ε_1 and ε_2 are continuous at $(0, 0)$. Let $\varepsilon > 0$ be given. Since $f_{,1}$ and $f_{,2}$ are continuous at (a_1, a_2), there is a $\delta > 0$ such that

$$|f_{,1}(x_1, x_2) - f_{,1}(a_1, a_2)| < \varepsilon \quad \text{and} \quad |f_{,2}(x_1, x_2) - f_{,2}(a_1, a_2)| < \varepsilon, \tag{7.8}$$

for (x_1, x_2) in a ball of radius δ with center at (a_1, a_2). Comparing Expressions (7.7) and (7.8), we see that if h_1 and h_2 are sufficiently small (so that (ξ_1, ξ_2) is close to (a_1, a_2)), then

$$|\varepsilon_1(h_1, h_2)| < \varepsilon \quad \text{and} \quad |\varepsilon_2(h_1, h_2)| < \varepsilon.$$

Moreover, $\varepsilon_1(0, 0) = 0$ and $\varepsilon_2(0, 0) = 0$. Substituting the values of ε_1 and ε_2 from Equations (7.7) into Equation (7.6) we obtain Part (b) of the theorem. The continuity of f at (a_1, a_2) follows directly from Equation (7.4).

The proof for $N > 2$ is similar. $\square$

The Chain rule for ordinary differentiation (Theorem 4.9) can be extended to provide a rule for taking partial derivatives of composite functions. We establish the result for a function $f: \mathbb{R}^N \to \mathbb{R}^1$ when it is composed with N functions $g^1, g^2, \ldots, g^N$, each of which is a mapping from $\mathbb{R}^M$ into $\mathbb{R}^1$. The integer M may be different from N. If $y = (y_1, y_2, \ldots, y_N)$ and $x = (x_1, x_2, \ldots, x_M)$ are elements of $\mathbb{R}^N$ and $\mathbb{R}^M$, respectively, then in customary terms, we wish to calculate the derivative of $H(x)$ (a function from $\mathbb{R}^M$ into $\mathbb{R}^1$) with respect to x_j, where $f = f(y_1, y_2, \ldots, y_N)$ and

$$H(x) = f[g^1(x), g^2(x), \ldots, g^N(x)]. \tag{7.9}$$

Theorem 7.3 (Chain rule). *Suppose that each of the functions $g^1, g^2, \ldots, g^N$ is a mapping from $\mathbb{R}^M$ into $\mathbb{R}^1$. For a fixed integer j between 1 and M, assume that $g^1_{,j}, g^2_{,j}, \ldots, g^N_{,j}$ are defined at some point $b = (b_1, b_2, \ldots, b_M)$. Suppose that f and its partial derivatives*

$$f_{,1}, f_{,2}, \ldots, f_{,N}$$

are continuous at the point $a = (g^1(b), g^2(b), \ldots, g^N(b))$. Form the function H as in Equation (7.9). Then the partial derivative of H with respect to x_j is given by

$$H_{,j}(b) = \sum_{i=1}^{N} f_{,i}[g^1(b), g^2(b), \ldots, g^N(b)]g^i_{,j}(b).$$

PROOF. Define the following functions from $\mathbb{R}^1$ into $\mathbb{R}^1$.

$$\varphi(x_j) = H(b_1, \ldots, b_{j-1}, x_j, b_{j+1}, \ldots, b_M)$$

$$\psi^i(x_j) = g^i(b_1, \ldots, b_{j-1}, x_j, b_{j+1}, \ldots, b_M), \qquad i = 1, 2, \ldots, N.$$

Then, according to Equation (7.9), we have

$$\varphi(x_j) = f[\psi^1(x_j), \psi^2(x_j), \ldots, \psi^N(x_j)].$$

With h denoting a real number, define

$$\Delta\varphi = \varphi(b_j + h) - \varphi(b_j)$$
$$\Delta\psi^i = \psi^i(b_j + h) - \psi^i(b_j), \qquad i = 1, 2, \ldots, N.$$

Since each ψ^i is continuous at b_j, it follows that $\Delta\psi^i \to 0$ as $h \to 0$. We apply Equation (7.4) of Theorem 7.2 to the function φ and use $\Delta\psi^i$ instead of h_i in that equation. We obtain

$$\Delta\varphi = \sum_{i=1}^{N} \{f_{,i}[\psi^1(b), \ldots, \psi^N(b)] + \varepsilon_i\}\Delta\psi^i. \tag{7.10}$$

In this formula, we have $\varepsilon_i = \varepsilon_i(\Delta\psi^1, \ldots, \Delta\psi^N)$ and $\varepsilon_i \to 0$ as $\Delta\psi^k \to 0$, $k = 1, 2, \ldots, N$.

Now write Equation (7.10) in the form

$$\frac{\Delta\varphi}{h} = \sum_{i=1}^{N} \{f_{,i}[g^1(b), \ldots, g^N(b)] + \varepsilon_i\}\frac{\Delta\psi^i}{h},$$

valid for $|h|$ sufficiently small. Letting h tend to 0, we get the statement of the Chain rule. $\qquad\qquad\square$

Definitions. Let D be a subset of $\mathbb{R}^N$ and suppose $f: D \to \mathbb{R}^1$ is a given function. For $x = (x_1, \ldots, x_N)$ and k a real number, we use the symbol kx to denote $(kx_1, kx_2, \ldots, kx_N)$. Then f is **homogeneous of degree** n if and only if (i) $kx \in D$ whenever $x \in D$ and $k \neq 0$, and (ii)

$$f(kx) = k^n f(x) \quad \text{for} \quad x \in D, k \neq 0.$$

The function f is **positively homogeneous of degree** n if and only if the two conditions above hold for all $k > 0$ and all $x \in D$.

Remarks. The quantity n need not be an integer. For example, the function $f: (x, y) \to x^{-1/3} + y^{-1/3}$ is homogeneous of degree $-1/3$. A function may be positively homogeneous but not homogeneous. The function $f: (x, y) \to \sqrt{x^2 + y^2}$ is positively homogeneous of degree 1 but not homogeneous.

Theorem 7.4 (Euler's theorem on homogeneous functions). *Suppose that* $f: \mathbb{R}^N \to \mathbb{R}^1$ *is positively homogeneous of degree n and suppose that* $f_{,1}, f_{,2}, \ldots,$ $f_{,N}$ *are continuous for* $a \neq 0$. *Then*

$$\sum_{i=1}^{N} a_i f_{,i}(a) = nf(a). \tag{7.11}$$

The proof is left to the reader.

PROBLEMS

1. Let D be an open connected set in $\mathbb{R}^N$ and suppose $f: D \to \mathbb{R}^1$ has the property that $f_{,1} = f_{,2} = \cdots = f_{,N} = 0$ for all $x \in D$. Show that $f \equiv$ constant in D.

2. Let $f: \mathbb{R}^N \to \mathbb{R}^1$ be given and suppose that $g^1, g^2, \ldots, g^N$ are N functions from $\mathbb{R}^M$ into $\mathbb{R}^1$. Let $h^1, h^2, \ldots, h^M$ be M functions from $\mathbb{R}^p$ into $\mathbb{R}^1$. Give a formula for the Chain rule for $H_{,i}(x)$, where

$$H(x) = f\{g^1[h^1(x), \ldots, h^M(x)], \quad g^2[h^1(x), \ldots, h^M(x)], \ldots\}.$$

3. Write a proof of Theorem 7.2 for $N = 3$.

4. Use the Chain rule to compute $H_{,2}(x)$ where H is given by (7.9) and $f(x) = x_1^2 + x_2^3 - 3x_3$, $g^1(x) = \sin 2x_1 + x_2 x_3$, $g^2(x) = \tan x_2 + 3x_3$, $g^3(x) = x_1 x_2 x_3$.

5. Use the Chain rule to compute $H_{,3}(x)$ where H is given by (7.9) and $f(x) = 2x_1 x_2 + x_3^2 - x_4^2$, $g^1(x) = \log(x_1 + x_2) - x_3^2$, $g^2(x) = x_1^2 + x_4^2$, $g^3(x) = x_1^2 x_3 + x_4$, $g^4(x) = \cos(x_1 + x_3) - 2x_4$.

6. Consider the function $f: \mathbb{R}^2 \to \mathbb{R}^1$ defined by

$$f(x_1, x_2) = \begin{cases} \dfrac{x_1 x_2}{x_1^2 + x_2^2}, & (x_1, x_2) \neq (0, 0), \\ 0, & x_1 = x_2 = 0. \end{cases}$$

(a) Show that f is not continuous at $x_1 = x_2 = 0$.
(b) Show that $f_{,1}$ and $f_{,2}$ exist at $x_1 = x_2 = 0$. Why does this fact not contradict Theorem 7.2?

7. Consider the function $f: \mathbb{R}^2 \to \mathbb{R}^1$ such that $f_{,1}$ and $f_{,2}$ exist and are bounded in a region about $x_1 = x_2 = 0$. Show that f is continuous at $(0, 0)$.

8. Given the function $f: \mathbb{R}^2 \to \mathbb{R}^1$ defined by

$$f(x_1, x_2) = \begin{cases} \dfrac{x_2^3}{x_1^2 + x_2^2}, & (x_1, x_2) \neq (0, 0), \\ 0 & x_1 = x_2 = 0. \end{cases}$$

Show that $f_{,1}$ and $f_{,2}$ are bounded near $(0, 0)$ and therefore (Problem 6) that f is continuous at $(0, 0)$.

9. If $f: \mathbb{R}^N \to \mathbb{R}^1$ is homogeneous of degree 0, show by a direct computation that f satisfies Euler's differential equation:

$$\sum_{i=1}^{N} x_i f_{,i} = 0.$$

10. Prove Theorem 7.4, Euler's theorem on homogeneous functions.

7.2. Taylor's Theorem; Maxima and Minima

Definitions. Let f be a function from $\mathbb{R}^N$ into $\mathbb{R}^1$. We define the **second partial derivative** $f_{,i,j}$ as the first partial derivative of $f_{,i}$ with respect to x_j. We define the **third partial derivative** $f_{,i,j,k}$ as the first partial derivative of $f_{,i,j}$. Fourth, fifth, and higher derivatives are defined similarly.

In computing second partial derivatives it is natural to ask whether the order of computation affects the result. That is, is it always true that $f_{,i,j} = f_{,j,i}$ for $i \neq j$? There are simple examples which show that the order of computation may lead to different results. (See Problem 3 at the end of this section.) The next theorem gives a sufficient condition which validates the interchange of order of partial differentiation.

Theorem 7.5. Let $f: \mathbb{R}^N \to \mathbb{R}^1$ be given and suppose that f, $f_{,i}$, $f_{,i,j}$, and $f_{,j,i}$ are all continuous at a point a. Then

$$f_{,i,j}(a) = f_{,j,i}(a).$$

PROOF. We establish the result for $N = 2$ with $i = 1$ and $j = 2$. The proof in the general case is exactly the same.

Writing $a = (a_1, a_2)$, we define

$$\Delta_2 f = f(a_1 + h, a_2 + h) - f(a_1 + h, a_2) - f(a_1, a_2 + h) + f(a_1, a_2). \quad (7.12)$$

We shall show that $\Delta_2 f / h^2$ tends to the limit $f_{,1,2}(a_1, a_2)$ and also that the same quantity tends to $f_{,2,1}$. Hence the two second derivatives must be equal. Define

$$\varphi(s) = f(a_1 + s, a_2 + h) - f(a_1 + s, a_2), \quad (7.13)$$

$$\psi(t) = f(a_1 + h, a_2 + t) - f(a_1, a_2 + t). \quad (7.14)$$

Then

$$\Delta_2 f = \varphi(h) - \varphi(0), \qquad \Delta_2 f = \psi(h) - \psi(0). \quad (7.15)$$

We apply the Mean-value theorem to Equations (7.15), getting

$$\Delta_2 f = \varphi'(s_1)h, \qquad \Delta_2 f = \psi'(t_1)h, \quad (7.16)$$

where s_1 and t_1 are numbers between 0 and h. From Equations (7.13) and (7.14), it follows that

$$\varphi'(s_1) = f_{,1}(a_1 + s_1, a_2 + h) - f_{,1}(a_1 + s_1, a_2), \quad (7.17)$$

$$\psi'(t_1) = f_{,2}(a_1 + h, a_2 + t_1) - f_{,2}(a_1, a_2 + t_1). \quad (7.18)$$

A second application of the Mean-value theorem to Equations (7.17) and (7.18) yields

$$\varphi'(s_1) = f_{,1,2}(a_1 + s_1, a_2 + t_2)h,$$

$$\psi'(t_1) = f_{,2,1}(a_1 + s_2, a_2 + t_1)h,$$

where s_2 and t_2 are numbers between 0 and h. Substituting these expressions in Equations (7.16) we find

$$\frac{1}{h^2}\Delta_2 f = f_{,1,2}(a_1 + s_1, a_2 + t_2) = f_{,2,1}(a_1 + s_2, a_2 + t_1).$$

Letting h tend to zero and observing that s_1, s_2, t_1, and t_2 all tend to zero with h, we obtain the desired result. $\square$

Definitions. A **multi-index** α is an element $(\alpha_1, \alpha_2, \ldots, \alpha_N)$ in $\mathbb{R}^N$ where each α_i is a nonnegative integer. The **order** of a multi-index, denoted by $|\alpha|$, is the nonnegative integer $\alpha_1 + \alpha_2 + \cdots + \alpha_N$. We extend the factorial symbol to multi-indices by defining $\alpha! = \alpha_1! \cdot \alpha_2! \ldots \alpha_N!$. If β is another multi-index $(\beta_1, \beta_2, \ldots, \beta_N)$, we define

$$\alpha + \beta = (\alpha_1 + \beta_1, \alpha_2 + \beta_2, \ldots, \alpha_N + \beta_N).$$

Let $x = (x_1, x_2, \ldots, x_N)$ be any element of $\mathbb{R}^N$. Then the monomial x^α is defined by the formula

$$x^\alpha = x_1^{\alpha_1} x_2^{\alpha_2} \ldots x_N^{\alpha_N}.$$

Clearly, the degree of x^α is $|\alpha|$. Any polynomial in $\mathbb{R}^N$ is a function f of the form

$$f(x) = \sum_{|\alpha| \leq n} c_\alpha x^\alpha \tag{7.19}$$

in which α is a multi-index, the c_α are constants, and the sum is taken over all multi-indices with order less than or equal to n, the degree of the polynomial.

EXAMPLE 1. Given $f(x_1, x_2) = x_1^3 + 3x_1 x_2^2 - 3x_1^2 - 3x_2^2 + 4$. Write f in the form of Equation (7.19).

Solution. We set $\alpha = (\alpha_1, \alpha_2)$ and consider multi-indices with $|\alpha| \leq 3$ since the polynomial is of degree 3. Then

$$f(x_1, x_2) = \sum_{|\alpha| \leq 3} c_{\alpha_1 \alpha_2} x_1^{\alpha_2} x_2^{\alpha_2}$$

where

$$c_{30} = 1, \quad c_{12} = 3, \quad c_{21} = c_{03} = 0, \quad c_{20} = -3,$$

$$c_{11} = 0, \quad c_{02} = -3, \quad c_{10} = c_{01} = 0, \quad c_{00} = 4. \qquad \square$$

Lemma 7.1 (Binomial theorem). *Suppose that x, $y \in \mathbb{R}^1$ and n is a positive integer. Then*

$$(x + y)^n = \sum_{j=0}^{n} \frac{n!}{j!(n-j)!} x^{n-j} y^j.$$

The proof is easily established by induction on n, and we leave the details to the reader.

The Multinomial theorem, an extension to several variables of the binomial theorem, is essential for the development of the Taylor expansion for functions of several variables.

Lemma 7.2 (Multinomial theorem). *Suppose that* $x = (x_1, x_2, \ldots, x_N)$ *is an element of* $\mathbb{R}^N$ *and that* n *is any positive integer. Then*

$$(x_1 + x_2 + \cdots + x_N)^n = \sum_{|\alpha|=n} \frac{n!}{\alpha!} x^{\alpha}$$

$$\equiv \sum_{\alpha_1 + \cdots + \alpha_N = n} \frac{n!}{\alpha_1! \ldots \alpha_N!} x_1^{\alpha_1} \ldots x_N^{\alpha_N}. \qquad (7.20)$$

PROOF. We fix the integer n and prove Equation (7.20) by induction on the integer N. For $N = 1$, Equation (7.20) becomes

$$x_1^n = \sum_{\alpha_1 = n} \frac{n!}{\alpha_1!} x_1^n = x_1^n,$$

which is true. Now, suppose that Equation (7.20) holds for $N = k$. We shall prove that it also holds for $N = k + 1$. To do this, we first observe that the binomial theorem yields

$$(x_1 + x_2 + \cdots + x_{k+1})^n = [(x_1 + \cdots + x_k) + x_{k+1}]^n$$

$$= \sum_{j=0}^{n} \frac{n!}{j!(n-j)!} x_{k+1}^j (x_1 + \cdots + x_k)^{n-j}. \qquad (7.21)$$

From the induction hypothesis, the right side of Equation (7.21) becomes

$$\sum_{j=0}^{n} \sum_{\alpha_1 + \cdots + \alpha_k = n-j} \frac{n!}{j!(n-j)!} \frac{(n-j)!}{\alpha_1! \ldots \alpha_k!} x_1^{\alpha_1} \ldots x_k^{\alpha_k} x_{k+1}^j. \qquad (7.22)$$

Setting $\alpha_{k+1} = j$ and cancelling $(n-j)!$, we find that Expression (7.22) becomes

$$\sum_{\alpha_1 + \cdots + \alpha_{k+1} = n} \frac{n!}{\alpha_1! \ldots \alpha_k! \alpha_{k+1}!} x_1^{\alpha_1} \ldots x_k^{\alpha_k} x_{k+1}^{\alpha_{k+1}}.$$

This last expression is the right side of Equation (7.20) with k replaced by $k + 1$. The induction is established. $\qquad \square$

Let G be an open set in $\mathbb{R}^N$ and let $f: G \to \mathbb{R}^1$ be a function with continuous second derivatives in G. We know that in the computation of second derivatives the order of differentiation may be interchanged. That is, $f_{,i,j} = f_{,j,i}$ for all i and j. We may also write $f_{,i,j} = D_j[D_i f]$ and, using the symbol $\circ$ for composite maps as described in Theorem 6.40, we have for two differential maps D_i and D_j

$$D_i \circ D_j(f) = D_i[D_j f].$$

If f has third and higher order derivatives, then

$$D_i \circ D_j \circ D_k(f) = D_i\{D_j[D_k(f)]\},$$

$$D_i \circ D_j \circ D_k \circ D_l(f) = D_i(D_j\{D_k[D_l(f)]\}),$$

and so on. We shall usually omit the symbol $\circ$, especially when the order of differentiation may be interchanged. A linear combination of differential maps is called a **differential operator**. If a, b, c, d are constants, the combination

$$aD_1 D_2 D_3 + bD_2 D_1 D_1 + cD_1 D_3 + dD_2$$

is an example of a differential operator. A differential operator acts on functions which are assumed to have continuous derivatives up to the required order on a fixed open set in $\mathbb{R}^N$. With any polynomial in $\mathbb{R}^N$ of the form

$$P(\xi_1, \xi_2, \ldots, \xi_N) = \sum_{|\alpha| \leqslant n} c_\alpha \xi^\alpha, \tag{7.23}$$

we associate the differential operator

$$P(D_1, D_2, \ldots, D_N) = \sum_{|\alpha| \leqslant n} c_\alpha D^\alpha. \tag{7.24}$$

If α is the multi-index $(\alpha_1, \alpha_2, \ldots, \alpha_N)$, then D^α is the operator given by $D^\alpha = D_1^{\alpha_1} D_2^{\alpha_2} \ldots D_N^{\alpha_N}$. That is, $D^\alpha f$ means that f is first differentiated with respect to x_N exactly α_N times; then it is differentiated α_{N-1} times with respect to x_{N-1}, and so on until all differentiations of f are completed. The **order** of the differential operator (7.24) is the degree of the polynomial (7.23).

By induction it is easy to see that every differential operator consisting of a linear combination of differential maps is of the form (7.24). The differentiations may be performed in any order. The polynomial $P(\xi_1, \xi_2, \ldots, \xi_N)$ in (7.23) is called the **auxiliary polynomial** of the operator (7.24). We illustrate the use of the notation with an example.

EXAMPLE 2. Suppose that $f: \mathbb{R}^2 \to \mathbb{R}^1$ is given by $f(x_1, x_2) = x_1 e^{x_1} \cos x_2$. Let $P(\xi_1, \xi_2) = \xi_1^4 + 2\xi_1^2 \xi_2^2 + \xi_2^4$ be a given polynomial. Show that $P(D_1, D_2)f = 0$.

Solution. We have

$$D_1(f) = \frac{\partial f}{\partial x_1} = (x_1 + 1)e^{x_1} \cos x_2; \qquad D_1^2(f) = D_1[D_1(f)];$$

$$= (x_1 + 2)e^{x_1} \cos x_2;$$

$$D_1^3(f) = (x_1 + 3)e^{x_1} \cos x_2; \qquad D_1^4(f) = (x_1 + 4)e^{x_1} \cos x_2;$$

$$D_2(f) = -x_1 e^{x_1} \sin x_2;$$

$$D_2^2(f) = -x_1 e^{x_1} \cos x_2; \qquad D_1^2 D_2^2(f) = -(x_1 + 2)e^{x_1} \cos x_2;$$

$$D_2^4(f) = x_1 e^{x_1} \cos x_2.$$

Therefore,

$$P(D_1, D_2)f = D_1^4(f) + 2D_1^2 D_2^2(f) + D_2^4(f) = 0. \qquad \square$$

We now introduce several definitions and simple facts concerning the algebra of linear operators with the purpose of applying them to linear differential operators as defined by (7.24). These operators are useful in the proof of Taylor's theorem and the second derivative test for maxima and minima of functions of several variables (Theorem 7.8 below).

Definitions. Let $L_1, L_2, \ldots, L_k$ be differential operators, each of which has the same domain and range in a Euclidean space. Let $c_1, c_2, \ldots, c_k$ be real numbers. The operator denoted by $c_1 L_1 + c_2 L_2 + \cdots + c_k L_k$ is the operator L such that $L(f) = c_1 L_1(f) + \cdots + c_k L_k(f)$ for all f in the domain of the L_i, $i = 1, 2, \ldots, k$. The operator $L_1 L_2$ is the operator L such that $L(f) = L_1\{L_2(f)\}$ for all f in the domain of L_2 and with the function $L_2(f)$ in the domain of L_1. The operators $L_1 L_2 L_3$, $L_1 L_2 L_3 L_4$, etc., are defined similarly.

Lemma 7.3. *Suppose that $L_1, L_2, \ldots, L_k$ are differential operators and $c_1, c_2, \ldots, c_k$ are real numbers. Let $P_i(\xi)$ be the auxiliary polynomial for L_i, $i = 1, 2, \ldots, k$. Then*

(i) *the auxiliary polynomial $P(\xi)$ for $c_1 L_1 + c_2 L_2 + \cdots + c_k L_k$ is $P(\xi) = c_1 P_1(\xi) + \cdots + c_k P_k(\xi)$;*

(ii) *the auxiliary polynomial for $L_1 L_2 \ldots L_k$ is $P_1(\xi) P_2(\xi) \ldots P_k(\xi)$.*

PROOF. Let n be the maximum order of all the operators L_i. Then we may write

$$L_i = \sum_{|\alpha| \leqslant n} b_{i\alpha} D^\alpha, \qquad i = 1, 2, \ldots, k.$$

(If some of the L_i are of order less than n, then the $b_{i\alpha}$ are zero for $|\alpha|$ beyond the order of the operator.) We have

$$c_1 L_1 + \cdots + c_k L_k = \sum_{j=1}^{k} c_j \sum_{|\alpha| \leqslant n} b_{j\alpha} D^\alpha,$$

and the auxiliary polynomial for L is

$$\sum_{j=1}^{k} c_j \sum_{|\alpha| \leqslant n} b_{j\alpha} \xi^\alpha = \sum_{j=1}^{k} c_j P_j(\xi).$$

We establish Part (ii) for two operators, the general result following by induction. Let

$$L_1 = \sum_{|\alpha| \leqslant m} c_\alpha D^\alpha, \qquad L_2 = \sum_{|\beta| \leqslant n} d_\beta D^\beta, \qquad L^1 = L_1 L_2.$$

Then

$$L^1(f) = L_1[L_2(f)] = \sum_{|\alpha| \leqslant m} c_\alpha D^\alpha \sum_{|\beta| \leqslant n} d_\beta D^\beta$$

$$P^1(\xi) = \sum_{|\alpha| \leqslant m} \sum_{|\beta| \leqslant n} c_\alpha d_\beta \xi^\alpha \xi^\beta = P_1(\xi) P_2(\xi). \qquad \square$$

Definitions. Let $f: \mathbb{R}^N \to \mathbb{R}^1$ be a given function, and suppose $a \in \mathbb{R}^N$ and $b \in \mathbb{R}^N$ with $|b| = 1$. The **directional derivative of f in the direction b at the point**

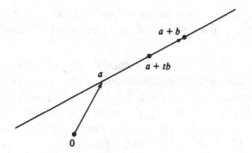

Figure 7.2. The line joining a and $a + b$.

a, denoted by $d_b f(a)$, is the number defined by

$$d_b f(a) = \lim_{t \to 0} \frac{f(a + tb) - f(a)}{t}. \tag{7.25}$$

Note that the difference quotient in the definition is taken by subtracting $f(a)$ from the value of f taken on the line segment in $\mathbb{R}^N$ joining a and $a + b$. See Figure 7.2. The **second directional derivative of f in the direction b at the point** a is simply $d_b[d_b f](a)$ and it is denoted by $(d_b)^2 f(a)$. The nth directional derivative, $(d_b)^n f(a)$, is defined similarly.

Lemma 7.4. *Suppose that $f: \mathbb{R}^N \to \mathbb{R}^1$ and all its partial derivatives up to and including order n are continuous in a ball $B(a, r)$. Then*

$$(d_b)^n f(a) = \sum_{|\alpha|=n} \frac{n!}{\alpha!} b^\alpha D^\alpha f(a). \tag{7.26}$$

Symbolically, the nth directional derivative is written

$$(d_b)^n = \sum_{|\alpha|=n} \frac{n!}{\alpha!} b^\alpha D^\alpha.$$

PROOF. For $n = 1$, we set $\phi(t) = f(a + tb)$. Then $d_b f(a) = \phi'(0)$. We use the Chain rule to compute ϕ' and, denoting $b = (b_1, b_2, \ldots, b_N)$, we find

$$d_b f(a) = b_1 D_1 f(a) + \cdots + b_N D_N f(a).$$

That is, $d_b = b_1 D_1 + \cdots + b_N D_N$. By induction, we obtain

$$(d_b^n) f = (b_1 D_1 + \cdots + b_N D_N)^n f.$$

Using the Multinomial theorem (Lemma 7.2), we get Equation (7.26). □

Theorem 7.6 (Taylor's theorem for functions from $\mathbb{R}^1$ to $\mathbb{R}^1$). *Suppose that $f: \mathbb{R}^1$ to $\mathbb{R}^1$ and all derivatives of f up to and including order $n + 1$ are continuous on an interval $I = \{x: |x - a| < r\}$. Then for each x on I, there is a* *number ξ on the open interval between a and x such that*

$$f(x) = \sum_{k=0}^{n} \frac{1}{k!} f^{(k)}(a)(x-a)^k + \frac{1}{(n+1)!} f^{(n+1)}(\xi)(x-a)^{n+1}. \qquad (7.27)$$

The proof of Theorem 7.6 makes use of Rolle's theorem (Theorem 4.11) and is deferred until Chapter 9. See Theorem 9.24. Observe that for $n = 0$, Formula (7.27) is the Mean-value theorem. (See Problem 6 at the end of this section and the hint given there.) The last term in (7.27) is called the **remainder**.

We now use Theorem 7.6 to establish Taylor's theorem for functions from a domain in $\mathbb{R}^N$ to $\mathbb{R}^1$.

Theorem 7.7 (Taylor's theorem with remainder). *Suppose that $f: \mathbb{R}^N \to \mathbb{R}^1$ and all its partial derivatives up to and including order $n+1$ are continuous on a ball $B(a, r)$. Then for each x in $B(a, r)$, there is a point ξ on the straight line segment from a to x such that*

$$f(x) = \sum_{|\alpha| \leqslant n} \frac{1}{\alpha!} D^\alpha f(a)(x-a)^\alpha + \sum_{|\alpha|=n+1} \frac{1}{\alpha!} D^\alpha f(\xi)(x-a)^\alpha. \qquad (7.28)$$

PROOF. If $x = a$ the result is obvious. If $x \neq a$, define $b = (b_1, b_2, \ldots, b_N)$ by

$$b_i = \frac{x_i - a_i}{|x - a|}.$$

We note that $|b| = 1$ and define $\phi(t) = f(a + tb)$. We observe that $\phi(0) = f(a)$ and $\phi(|x - a|) = f(a + |x - a| \cdot b) = f(x)$. By induction, it follows that ϕ has continuous derivatives up to and including order $n + 1$ since f does. Now we apply Taylor's theorem (Theorem 7.6) to ϕ, a function of one variable. Then

$$\phi(t) = \sum_{j=0}^{n} \frac{1}{j!} \phi^{(j)}(0) t^j + \frac{1}{(n+1)!} \phi^{(n+1)}(\tau) t^{n+1}, \qquad (7.29)$$

where τ is between 0 and t. From Lemma 7.4, it is clear that

$$\phi^{(j)}(t) = (d_b)^j f(a + tb) = \sum_{|\alpha|=j} \frac{j!}{\alpha!} b^\alpha D^\alpha f(a + tb). \qquad (7.30)$$

We set $t = |x - a|$ in Equation (7.29), getting

$$\phi(|x-a|) = f(x) = \sum_{j=0}^{n} \frac{1}{j!} \phi^{(j)}(0)|x-a|^j + \frac{1}{(n+1)!} \phi^{(n+1)}(\tau)|x-a|^{n+1}.$$

Inserting Equation (7.30) into this expression, we find

$$f(x) = \sum_{j=0}^{n} \frac{1}{j!} \left[\sum_{|\alpha|=j} \frac{j!}{\alpha!} b^\alpha D^\alpha f(a)|x-a|^j \right]$$

$$+ \frac{1}{(n+1)!} \left[\sum_{|\alpha|=n+1} \frac{(n+1)!}{\alpha!} b^\alpha D^\alpha(\xi)|x-a|^{n+1} \right].$$

By definition we have $x_i - a_i = b_i|x - a|$, and so $(x - a)^\alpha = b^\alpha|x - a|^{|\alpha|}$. Employing this fact in the above expression for $f(x)$, we obtain Equation (7.28). $\square$

The last terms in formula (7.28) are known as the **remainder**.

For functions from $\mathbb{R}^1$ into $\mathbb{R}^1$ the Second Derivative test is one of the most useful for identifying the maximum and minimum points on the graph of the function. We recall that a function f with two derivatives and with $f'(a) = 0$ has a relative maximum at a if its second derivative at a is negative. It has a relative minimum at a if $f''(a) > 0$. If $f''(a) = 0$, the test fails. With the aid of Taylor's theorem for functions from $\mathbb{R}^N$ into $\mathbb{R}^1$ we can establish the corresponding result for functions of N variables.

Definitions. Let $f: \mathbb{R}^N \to \mathbb{R}^1$ be given. The function f has a **local maximum at the value** a if and only if there is a ball $B(a, r)$ such that $f(x) - f(a) \leqslant 0$ for $x \in B(a, r)$. The function f has a **strict local maximum** at a if and only if $f(x) - f(a) < 0$ for $x \in B(a, r)$ except for $x = a$. The corresponding definitions for **local minimum** and **strict local minimum** reverse the inequality sign. If f has partial derivatives at a, we say that f has a **critical point at the value** a if and only if $D_i f(a) = 0$, $i = 1, 2, \ldots, N$.

Theorem 7.8 (Second derivative test). *Suppose that $f: \mathbb{R}^N \to \mathbb{R}^1$ and its partial derivatives up to and including order 2 are continuous in a ball $B(a, r)$. Suppose that f has a critical point at a. For $h = (h_1, h_2, \ldots, h_N)$, define $\Delta f(a, h) = f(a + h) - f(a)$; also define*

$$Q(h) = \sum_{|\alpha|=2} \frac{1}{\alpha!} D^\alpha f(a) h^\alpha = \frac{1}{2!} \sum_{i,j=1}^N D_i D_j f(a) h_i h_j. \tag{7.31}$$

(a) *If $Q(h) > 0$ for $h \neq 0$, then[1] f has a strict local minimum at a.*
(b) *If $Q(h) < 0$ for $h \neq 0$, then f has a strict local maximum at a.*
(c) *If $Q(h)$ has a positive maximum and a negative minimum, then $\Delta f(a, h)$ changes sign in any ball $B(a, \rho)$ such that $\rho < r$.*

PROOF. We establish Part (a), the proofs of Parts (b) and (c) being similar. Taylor's theorem with remainder (Theorem 7.7) for $n = 1$ and $x = a + h$ is

$$f(a + h) = f(a) + \sum_{|\alpha|=1} D^\alpha f(a) h^\alpha + \sum_{|\alpha|=2} \frac{1}{\alpha!} D^\alpha f(\xi) h^\alpha, \tag{7.32}$$

where ξ is on the straight line segment connecting a with $a + h$. Since f has a critical point at a, the first sum in Equation (7.32) is zero, and the

---

[1] If Q is a quadratic form such as $Q(h) = \sum_{i,j=1}^N a_{ij} h_i h_j$, then Q is **positive definite** if and only if $Q(h) > 0$ for all $h \neq 0$. Also, Q is **negative definite** when $-Q$ is positive definite.

second sum may be written

$$\Delta f(a, h) = \sum_{|\alpha|=2} \frac{1}{\alpha!} D^\alpha f(a) h^\alpha + \sum_{|\alpha|=2} \frac{1}{\alpha!} [D^\alpha f(\xi) - D^\alpha f(a)] h^\alpha. \qquad (7.33)$$

Setting $|h|^2 = h_1^2 + h_2^2 + \cdots + h_N^2$ and

$$\varepsilon(h) = \sum_{|\alpha|=2} \frac{1}{\alpha!} [D^\alpha f(\xi) - D^\alpha f(a)] \frac{h^\alpha}{|h|^2},$$

we find from Equation (7.33)

$$\Delta f(a, h) = Q(h) + |h|^2 \varepsilon(h). \qquad (7.34)$$

Because the second partial derivatives of f are continuous near a, it follows that $\varepsilon(h) \to 0$ as $h \to 0$. Also,

$$Q(h) = |h|^2 \sum_{i,j=1}^N D_i D_j f(a) \frac{h_i}{|h|} \cdot \frac{h_j}{|h|} \equiv |h|^2 Q_1(h).$$

The expression $Q_1(h)$ is continuous for h on the unit sphere in $\mathbb{R}^N$. According to the hypothesis in Part (a), Q_1, a continuous function, must have a positive minimum on the unit sphere which is a closed set. Denote this minimum by m. Hence,

$$Q(h) \geqslant |h|^2 m \quad \text{for all } h.$$

Now choose $|h|$ so small that $|\varepsilon(h)| < m/2$. Inserting the inequalities for $Q(h)$ and $|\varepsilon(h)|$ into Equation (7.34), we find

$$\Delta f(a, h) > \tfrac{1}{2}|h|^2 m$$

for $|h|$ sufficiently small and $h \neq 0$. We conclude that f has a strict local minimum at a. $\qquad \square$

Remarks. The quantity Q given in Expression (7.31) is a quadratic form. In linear algebra, we develop the fact that Q is positive definite if and only if the matrix $(A_{ij}) \equiv (D_i D_j f(a))$ has all positive eigenvalues. Also, Q is negative definite when all the eigenvalues of (A_{ij}) are negative. The quadratic form Q has a positive maximum and a negative minimum when the matrix has at least one positive and at least one negative eigenvalue. It is not necessary to find all the eigenvalues in order to determine the properties of Q. It is sufficient to "complete the square", as in elementary algebra, to determine which of the cases (a), (b), or (c) of Theorem 7.8 prevails.

PROBLEMS

1. Given $f: \mathbb{R}^3 \to \mathbb{R}^1$ defined by

$$f(x_1, x_2, x_3) = (x_1^2 + x_2^2 + x_3^2)^{-1/2}, \qquad (x_1, x_2, x_3) \neq (0, 0, 0).$$

Show that f satisfies the equation

$$f_{,1,1} + f_{,2,2} + f_{,3,3} = 0.$$

2. Given $f: \mathbb{R}^2 \to \mathbb{R}^1$ defined by $f(x_1, x_2) = x_1^4 - 2x_1^3 x_2 - x_1 x_2$ and given $L_1(D) = 2D_1 - 3D_2$, $L_2(D) = D_1 D_2$, show that $(L_1 L_2)(f) = (L_2 L_1)(f)$.

3. Given $f: \mathbb{R}^2 \to \mathbb{R}^1$ defined by

$$f(x_1, x_2) = \begin{cases} x_1 x_2 \dfrac{x_1^2 - x_2^2}{x_1^2 + x_2^2}, & x_1^2 + x_2^2 > 0, \\ 0, & x_1 = x_2 = 0. \end{cases}$$

Show that $f_{,1,2}(0, 0) = -1$ and $f_{,2,1}(0, 0) = 1$.

4. Write out the proof of Theorem 7.5 for a function f from $\mathbb{R}^N$ into $\mathbb{R}^1$.

5. Prove the Binomial theorem (Lemma 7.1).

6. Establish the Taylor expansion for functions from $\mathbb{R}^1$ into $\mathbb{R}^1$ (Theorem 7.6). [*Hint*: Make use of the function

$$\varphi(t) = f(x) - f(t) - \frac{f'(t)(x - t)}{1!} - \cdots - \frac{f^{(n)}(t)(x - t)^n}{n!} - R_n(a, x)\frac{(x - t)^{n+1}}{(x - a)^{n+1}}$$

where $R_n(a, x)$ is the remainder in the Taylor expansion of $f(x)$. Note that $\varphi(a) = \varphi(x) = 0$.]

7. Write out explicitly all the terms of the Taylor expansion for a function $f: \mathbb{R}^3 \to \mathbb{R}^1$ for the case $n = 2$.

8. Find the relative maxima and minima of the function $f: \mathbb{R}^2 \to \mathbb{R}^1$ given by

$$f(x_1, x_2) = x_1^3 + 3x_1 x_2^2 - 3x_1^2 - 3x_2^2 + 4.$$

9. Find the critical points of the function $f: \mathbb{R}^4 \to \mathbb{R}^1$ given by

$$f(x_1, x_2, x_3, x_4) = x_1^2 + x_2^2 + x_3^2 - x_4^2 - 2x_1 x_2 + 4x_1 x_3 + 3x_1 x_4$$
$$- 2x_2 x_4 + 4x_1 - 5x_2 + 7.$$

10. Write out proofs of Parts (b) and (c) of the Second derivative test (Theorem 7.8).

In each of Problems 11 through 13, determine whether $Q: \mathbb{R}^3 \to \mathbb{R}^1$ is positive definite, negative definite, or neither.

11. $Q(x_1, x_2, x_3) = x_1^2 + 5x_2^2 + 3x_3^2 - 4x_1 x_2 + 2x_1 x_3 - 2x_2 x_3$

12. $Q(x_1, x_2, x_3) = x_1^2 + 3x_2^2 + x_3^2 - 4x_1 x_2 + 2x_1 x_3 - 6x_2 x_3$

13. $Q(x_1, x_2, x_3) = -x_1^2 - 2x_2^2 - 4x_3^2 - 2x_1 x_2 - 2x_1 x_3$

7.3. The Derivative in $\mathbb{R}^N$

Each partial derivative of a function $f: \mathbb{R}^N \to \mathbb{R}^1$ is a mapping from $\mathbb{R}^N$ into $\mathbb{R}^1$. This generalization of the ordinary derivative, useful for many purposes, is unsatisfactory in that it singles out a particular direction in which the differentiation is performed. We now take up another extension of the ordinary derivative, one in which the difference quotient tends to a limit as $x \to a$ regardless of the direction of approach.

Let A be an open subset of $\mathbb{R}^N$, and suppose that f and g are functions from A into $\mathbb{R}^1$. We denote by $d(x, y)$ the Euclidean distance between two points x, y in $\mathbb{R}^N$.

Definition. The continuous functions f and g are **tangent at a point** $a \in A$ if and only if

$$\lim_{\substack{x \to a \\ x \neq a}} \frac{|f(x) - g(x)|}{d(x, a)} = 0.$$

We note that if two functions are tangent at a point a then, necessarily, $f(a) = g(a)$. Also, if $f, g,$ and h are functions with f and g tangent at a and with g and h tangent at a, then we verify simply that f and h are tangent at a. To see this, observe that

$$\frac{|f(x) - h(x)|}{d(x, a)} \leqslant \frac{|f(x) - g(x)|}{d(x, a)} + \frac{|g(x) - h(x)|}{d(x, a)}.$$

As $x \to a$, the right side tends to 0 and hence so must the left side.

Let $L: \mathbb{R}^N \to \mathbb{R}^1$ be a linear function; that is, L is of the form

$$L(x) = b_0 + \sum_{k=1}^{N} b_k x_k,$$

where $b_0, b_1, \ldots, b_N$ are real numbers. It may happen that a function f is tangent to a linear function at a point a. If so, this linear function is unique, as we now show.

Theorem 7.9. *Suppose that L_1 and L_2 are linear functions tangent to a function f at a point a. Then $L_1 \equiv L_2$.*

PROOF. It is convenient to write the linear functions L_1 and L_2 in the form

$$L_1(x) = c_0 + \sum_{k=1}^{N} c_k(x_k - a_k), \qquad L_2(x) = c_0' + \sum_{k=1}^{N} c_k'(x_k - a_k).$$

From the discussion above, it follows that $L_1(a) = L_2(a) = f(a)$. Hence, $c_0 = c_0'$. Also, for every $\varepsilon > 0$, we have

$$|L_1(x) - L_2(x)| \leqslant \varepsilon\, d(x, a) \tag{7.35}$$

for all x sufficiently close to a. For $z \in \mathbb{R}^N$, we use the notation $\|z\| = d(z, 0)$. Now, with δ a real number, choose

$$x - a = \frac{\delta z}{\|z\|}.$$

Then, for sufficiently small $|\delta|$, we find from Inequality (7.35) that

$$\left| \sum_{k=1}^{N} (c_k - c_k') \frac{\delta z_k}{\|z\|} \right| \leqslant \varepsilon |\delta| \frac{\|z\|}{\|z\|} = \varepsilon |\delta|.$$

Therefore,

$$\left| \sum_{k=1}^{N} (c_k - c_k') \frac{z_k}{\|z\|} \right| \leqslant \varepsilon,$$

and since this inequality holds for all positive ε, we must have $c_k = c_k'$, $k = 1, 2, \ldots, N$. $\square$

Definitions. Suppose that $f: A \to \mathbb{R}^1$ is given, where A is an open set in $\mathbb{R}^N$ containing the point a. The function f is **differentiable at** a if and only if there is a linear function $L(x) = f(a) + \sum_{k=1}^{N} c_k(x_k - a_k)$ which is tangent to f at a. A function f is **differentiable on a set** A in $\mathbb{R}^N$ if and only if it is differentiable at each point of A. The function L is called the **tangent linear function to** f **at the point** a. The function L is also called the **derivative** or **total derivative of** f at a. We use the symbol $f'(a)$ for the derivative of f at the point a in $\mathbb{R}^N$.

Theorem 7.10. *If f is differentiable at a point a, then f is continuous at a.*

The proof is left to the reader.

As we saw in Section 7.1, a function f may have partial derivatives without being continuous. An example of such a function is given in Problem 6 at the end of Section 7.1. Theorem 7.10 suggests that differentiability is a stronger condition than the existence of partial derivatives. The next theorem verifies this point.

Theorem 7.11. *If f is differentiable at a point a, then all first partial derivatives of f exist at a.*

PROOF. Let L be the tangent linear function to f at a. We write

$$L(x) = f(a) + \sum_{k=1}^{N} c_k(x_k - a_k).$$

From the definition of derivative, it follows that

$$\lim_{\substack{x \to a \\ x \neq a}} \frac{\left| f(x) - f(a) - \sum_{k=1}^{N} c_k(x_k - a_k) \right|}{d(x, a)} = 0. \tag{7.36}$$

For x we choose the element $a + h$ where $h = (0, 0, \ldots, 0, h_i, 0, \ldots, 0)$. Then Equation (7.36) becomes

$$\lim_{\substack{h_i \to 0 \\ h_i \neq 0}} \frac{|f(a + h) - f(a) - c_i h_i|}{|h_i|} = 0;$$

we may therefore write

$$\left| \frac{f(a + h) - f(a)}{h_i} - c_i \right| = \varepsilon(h_i), \tag{7.37}$$

where $\varepsilon(h_i) \to 0$ as $h_i \to 0$.

We recognize the left side of Equation (7.37) as the expression used in the definition of $f_{,i}(a)$. In fact, we conclude that

$$c_i = f_{,i}(a), \qquad i = 1, 2, \ldots, N. \qquad \square$$

A partial converse of Theorem 7.11 is also true.

Theorem 7.12. *Suppose that $f_{,i}, i = 1, 2, \ldots, N$ are continuous at a point a. Then f is differentiable at a.*

This result is most easily established by means of the Taylor expansion with remainder (Theorem 7.7) with $n = 0$. We leave the details for the reader.

Definitions. Suppose that f has all first partial derivatives at a point a in $\mathbb{R}^N$. The **gradient of f** is the element in $\mathbb{R}^N$ whose components are

$$(f_{,1}(a), f_{,2}(a), \ldots, f_{,N}(a)).$$

We denote the gradient of f by ∇f or grad f.

Suppose that A is a subset of $\mathbb{R}^N$ and $f \colon A \to \mathbb{R}^1$ is differentiable on A. Let $h = (h_1, h_2, \ldots, h_N)$ be an element of $\mathbb{R}^N$. We define the **total differential** df as that function from $A \times \mathbb{R}^N \to \mathbb{R}^1$, given by the formula

$$df(x, h) = \sum_{k=1}^{N} f_{,k}(x)h_k. \tag{7.38}$$

Using the inner or dot product notation for elements in $\mathbb{R}^N$, we may also write

$$df(x, h) = \nabla f(x) \cdot h.$$

Remarks. Equation (7.38) shows that the total differential (also called the **differential**) of a function f is linear in h and bears a close resemblance to the tangent linear function. The differential vanishes when $x = a$ and $h = 0$, while the tangent linear function has the value $f(a)$ at the corresponding point. The Chain rule (Theorem 7.3) takes a natural form for differentiable functions, as the following theorem shows.

Theorem 7.13 (Chain rule). *Suppose that each of the functions $g^1, g^2, \ldots, g^N$ is a mapping from $\mathbb{R}^M$ into $\mathbb{R}^1$ and that each g^i is differentiable at a point $b = (b_1, b_2, \ldots, b_M)$. Suppose that $f \colon \mathbb{R}^N \to \mathbb{R}^1$ is differentiable at the point $a = (g^1(b), g^2(b), \ldots, g^N(b))$. Form the function*

$$H(x) = f[g^1(x), g^2(x), \ldots, g^N(x)].$$

Then H is differentiable at b and

$$dH(b, h) = \sum_{i=1}^{N} f_{,i}[g(b)] \, dg^i(b, h).$$

The proof is similar to the proof of Theorem 7.3, and we leave it to the reader.

PROBLEMS

1. Let f and g be functions from $\mathbb{R}^N$ into $\mathbb{R}^1$. Show that if f and g are differentiable at a point a, then $f + g$ is differentiable at a. Also, αf is differentiable at a for every real number α.

2. Let $f: \mathbb{R}^3 \to \mathbb{R}^1$ be given by

$$f(x_1, x_2, x_3) = x_1 x_3 e^{x_1 x_2} + x_2 x_3 e^{x_1 x_3} + x_1 x_2 e^{x_2 x_3}.$$

Find ∇f and $df(x, h)$.

3. Let f and g be functions from $\mathbb{R}^N$ into $\mathbb{R}^1$. Show that if f and g are differentiable, and α and β are numbers, then

$$\nabla(\alpha f + \beta g) = \alpha \nabla f + \beta \nabla g,$$
$$\nabla(fg) = f\nabla g + g\nabla f.$$

4. Let $f: \mathbb{R}^N \to \mathbb{R}^1$ and $g: \mathbb{R}^1 \to \mathbb{R}^1$ be given. Assume that the range of f is contained in the domain of g. Show that, in such a case,

$$\nabla[g(f)] = g'\nabla f.$$

5. Show that if f is differentiable at a point a, then it is continuous at a. (Theorem 7.10.)

6. Suppose that all first partial derivatives of a function $f: \mathbb{R}^N \to \mathbb{R}^1$ are continuous at a point a. Show that f is differentiable at a. (Theorem 7.12.)

7. Given $f: \mathbb{R}^N \to \mathbb{R}^1$ and an element $b \in \mathbb{R}^N$ with $\|b\| = 1$. The **directional derivative of f in the direction b at the point a**, denoted by $d_b f$, is the function from $\mathbb{R}^N$ into $\mathbb{R}^1$ defined by

$$d_b f = \lim_{t \to 0} \frac{f(a + tb) - f(a)}{t}.$$

Suppose that all first partial derivatives of f are continuous at a point a. Show that the directional derivative of f in every direction exists at a and that

$$(d_b f)(a) = \nabla f(a) \cdot b.$$

8. Suppose that A is an open set in $\mathbb{R}^N$. Let $f: \mathbb{R}^N \to \mathbb{R}^1$ be differentiable at each point of A. Assume that the derivative of f, a mapping from A into $\mathbb{R}^1$, is also differentiable at each point of A. Then show that all second partial derivatives of f exist at each point of A and that

$$D_i D_j f = D_j D_i f$$

for all $i, j = 1, 2, \ldots, N$.

9. Prove Theorem 7.13.

10. Suppose f and g are functions from $\mathbb{R}^N$ into $\mathbb{R}^M$. We denote by d_N and d_M Euclidean distance in $\mathbb{R}^N$ and $\mathbb{R}^M$, respectively. We say that f **and** g **are tangent at a point** $a \in \mathbb{R}^N$ if and only if

$$\lim_{x \to a} \frac{d_M(f(x), g(x))}{d_N(x, a)} = 0.$$

Show that if f is tangent to a linear function $L: \mathbb{R}^N \to \mathbb{R}^M$ at a point a, then there can be only one such. [*Hint*: Write L as a system of linear equations and follow the proof of Theorem 7.9.]

11. (a) Let $f: \mathbb{R}^N \to \mathbb{R}^M$ be given. Using the result of Problem 10, define the **derivative of** f **at a point** a.
    (b) Let the components of f be denoted by $f^1, f^2, \ldots, f^M$. If f is differentiable at a, show that all partial derivatives of $f^1, f^2, \ldots, f^M$ exist at the point a.

12. Let $L: \mathbb{R}^N \to \mathbb{R}^M$ be a linear function. Show that $L = (L_1, L_2, \ldots, L_M)$ where $L_j = c_0^j + \sum_{k=1}^N c_k^j x_k$, $j = 1, 2, \ldots, M$.

13. Let A be a closed region in $\mathbb{R}^N$. Suppose that $f: \mathbb{R}^N \to \mathbb{R}^1$ is differentiable in a region containing A and that f has a maximum value at a point $a \in \partial A$. Show that $d_n f \leqslant 0$ at the point a where n is the inward pointing unit normal to ∂A at the point a.

CHAPTER 8

Integration in $\mathbb{R}^N$

8.1. Volume in $\mathbb{R}^N$

In order to construct the theory of Riemann integration in $\mathbb{R}^N$, we need to develop first a theory of volume for sets of points in $\mathbb{R}^N$. This development is a straightforward generalization of the theory of area (Jordan content) given in Section 5.4. We shall outline the main steps of the the the theory and leave the proofs of the theorems to the reader. We first recall the definitions of open and closed cells given earlier in Section 6.3.

Definitions. Let $a = (a_1, a_2, \ldots, a_N)$ and $b = (b_1, b_2, \ldots, b_N)$ be points in $\mathbb{R}^N$ with $a_i < b_i$ for $i = 1, 2, \ldots, N$. The set $R = \{x : a_i < x_i < b_i, i = 1, 2, \ldots, N\}$ is called an **open cell** in $\mathbb{R}^N$. A **closed cell** is the set $\{x : a_i \leqslant x_i \leqslant b_i, i = 1, 2, \ldots, N\}$. If the relations

$$b_1 - a_1 = b_2 - a_2 = \cdots = b_n - a_n$$

hold, then the cell is called a **hypercube**. Note that for $N = 2$ an open cell is the interior of a rectangle, and for $N = 3$ an open cell is the interior of a rectangular parallelepiped. For $N = 2$ and 3, hypercubes are squares and cubes, respectively.

Paralleling the theory given for $N = 2$, we divide all of $\mathbb{R}^N$ into closed hypercubes given by the inequalities

$$\frac{k_i - 1}{2^n} \leqslant x_i \leqslant \frac{k_i}{2^n}, \qquad i = 1, 2, \ldots, N, \tag{8.1}$$

where the k_i, $i = 1, 2, \ldots, N$, are *integers*. As we shall see below, the volume of each such hypercube is $1/2^{nN}$.

Let S be a bounded set in $\mathbb{R}^N$. We wish to define the volume of S. Suppose that S is contained in an open cell $R = \{x : A_i < x_i < B_i, i = 1, 2, \ldots, N\}$ where, for convenience, $A_1, \ldots, A_N, B_1, \ldots, B_N$ are *integers*. $\bar{R}$ denotes the closed cell having the same "sides" as R.

Definitions. Suppose that $\mathbb{R}^N$ is divided into hypercubes as given by Expression (8.1). Such a system of hypercubes divides $\mathbb{R}^N$ into a **grid**. Each such hypercube entirely contained in the point set S is called an **inner cube** for S of the nth grid. Hypercubes which contain at least one point of S but are not inner cubes are called **boundary cubes**. Hypercubes which contain no points of S are called **exterior cubes**.

A hypercube which has each edge of length 1 contains $(2^n)^N = 2^{Nn}$ hypercubes of the nth grid. Therefore the closed cell $\bar{R}$ containing S has

$$2^{Nn}(B_1 - A_1)(B_2 - A_2)\ldots(B_N - A_N)$$

hypercubes of the nth grid.

We define the quantities

$V_n^-(S) = 2^{-Nn}$ times the number of inner cubes for S,
$V_n^+(S) = V_n^-(S) + 2^{-Nn}$ times the number of boundary cubes for S.

The following lemma is the direct analog of Lemma 5.3.

Lemma 8.1. *Let* $R = \{x : A_i < x_i < B_i, i = 1, 2, \ldots, N\}$ *be an open cell where* $A_i, B_i, i = 1, 2, \ldots, N,$ *are integers. Suppose that* $S \subset R$. *Then, for every subdivision of* $\mathbb{R}^N$ *into hypercubes given by* (8.1), *all inner cubes and all boundary cubes of* S *are contained in* R.

Theorem 8.1. *Suppose that* R *and* S *are as in Lemma 8.1. Then*

(i) $0 \leqslant V_n^-(S) \leqslant V_n^+(S) \leqslant (B_1 - A_1)(B_2 - A_2)\ldots(B_N - A_N)$;
(ii) $V_n^-(S) \leqslant V_{n+1}^-(S)$ *for each* n;
(iii) $V_{n+1}^+(S) \leqslant V_n^+(S)$ *for each* n;
(iv) *the sequences* $\{V_n^-(S)\}, \{V_n^+(S)\}$ *tend to limits as* $n \to \infty$. *If the limits are denoted by*

$$V^-(S) = \lim_{n \to \infty} V_n^-(S) \quad and \quad V^+(S) = \lim_{n \to \infty} V_n^+(S),$$

then

$$0 \leqslant V^-(S) \leqslant V^+(S) \leqslant \prod_{i=1}^{N} (B_i - A_i);$$

(v) *the quantities* $V_n^-(S), V_n^+(S), V^-(S), V^+(S)$ *are independent of* R.

Note that Theorem 5.19 is exactly the same as Theorem 8.1 for the case $N = 2$. The proof of Theorem 8.1 parallels that of Theorem 5.19 and is left to the reader.

Definitions. The number $V^-(S)$ is called the **inner volume** of S and the number $V^+(S)$ is called the **outer volume** of S. A set of points S is said to have **volume** if and only if $V^-(S) = V^+(S)$. The volume of S, denoted $V(S)$ (sometimes denoted $V_N(S)$ if we wish to emphasize the number of dimensions) is this common value. A set of points in $\mathbb{R}^N$ which has a volume is called a **figure**.

Theorem 8.2. *Let S_1 and S_2 be bounded sets in $\mathbb{R}^N$. Then*

(i) *if S_1 is contained in S_2, then $V^-(S_1) \leqslant V^-(S_2)$ and $V^+(S_1) \leqslant V^+(S_2)$;*
(ii) $V^+(S_1 \cup S_2) \leqslant V^+(S_1) + V^+(S_2)$;
(iii) *if S_1 and S_2 have no common interior points, then*

$$V^-(S_1 \cup S_2) \geqslant V^-(S_1) + V^-(S_2);$$

(iv) *if S_1 and S_2 are figures and if S_1 and S_2 have no common interior points, then $S_1 \cup S_2$ is a figure and*

$$V(S_1 \cup S_2) = V(S_1) + V(S_2);$$

(v) *the open cell $R = \{x : a_i < x_i < b_i, i = 1, 2, \ldots, N\}$ is a figure and*

$$V(R) = \prod_{i=1}^{N} (b_i - a_i).$$

Also, $\bar{R}$ is a figure and $V(\bar{R}) = V(R)$.

The statement of Theorem 8.2 is the same as Theorem 5.20 for $N = 2$, and the proofs are similar.

Theorem 8.3. *Suppose that S_1 and S_2 are figures in $\mathbb{R}^N$. Then $S_1 \cup S_2$ is a figure. Also, $S_1 \cap S_2$ and $S_1 - S_2$ are figures. The boundary of any figure has volume zero.*

The proof of Theorem 8.3 follows in precise detail the proof of Theorem 5.21 and is left to the reader. Theorems 8.2 and 8.3 are easily extended to the case where $S_1, S_2, \ldots, S_k$ is any finite collection of sets in $\mathbb{R}^N$.

PROBLEMS

1. Give an example of a bounded set S in $\mathbb{R}^N$ for which $V^-(S) < V^+(S)$.

2. Consider the open cell $R = \{x : a_i < x_i < b_i\}$ in which $a_1, a_2, \ldots, a_N, b_1, b_2, \ldots, b_N$ are irrational numbers. Show that $V_n^-(R) < V_n^+(R)$ for every n.

3. Prove Lemma 8.1. Is the same result true if the $\{A_i\}$ and $\{B_i\}$ are rational numbers provided that the grid given by the inequalities (8.1) has n sufficiently large?

4. Give an example of a nonempty bounded set S in $\mathbb{R}^N$ with the property that $V_n^-(S) = V_n^+(S)$ for every n.

5. Prove Theorem 8.1.

6. State and prove for sets in $\mathbb{R}^N$ the analog of Lemma 5.3 of Section 5.4.

7. Give an example of two sets S_1 and S_2 in $\mathbb{R}^N$ such that neither S_1 nor S_2 is a figure but $S_1 \cup S_2$ is a figure and $S_1 \cap S_2$ is a figure. Is it possible that $S_1 - S_2$ is a figure?

8. Prove Theorem 8.2.

9. Prove Theorem 8.3.

10. Show that if $S_1, S_2, \ldots, S_k$ are figures in $\mathbb{R}^N$, then $S_1 \cup S_2 \cup \cdots \cup S_k$ and $S_1 \cap S_2 \cap \cdots \cap S_k$ are figures. If $S_1, \ldots, S_k, \ldots$ is an infinite collection of figures, is it true that $\bigcup_{n=1}^{\infty} S_n$ is a figure?

8.2. The Darboux Integral in $\mathbb{R}^N$

The development of the theory of integration in $\mathbb{R}^N$ for $N \geqslant 2$ parallels the one-dimensional case given in Section 5.1. For functions defined on an interval I in $\mathbb{R}^1$, we formed upper and lower sums by dividing I into a number of subintervals. In $\mathbb{R}^N$ we begin with a figure F, i.e., a bounded region which has volume and, in order to form upper and lower sums, we divide F into a number of subfigures. These subfigures are the generalizations of the subintervals in $\mathbb{R}^1$, and the limits of the upper and lower sums as the number of subfigures tends to infinity yield upper and lower integrals.

Definition. Let F be a bounded set in $\mathbb{R}^N$ which is a figure. A **subdivision** of F is a finite collection of figures, $\{F_1, F_2, \ldots, F_n\}$, no two of which have common interior points and whose union is F. That is,

$$F = F_1 \cup F_2 \cup \cdots \cup F_n, \qquad \text{Int}(F_i) \cap \text{Int}(F_j) = \varnothing \quad \text{for} \quad i \neq j.$$

We denote such a subdivision by the single letter Δ.

Let D be a set in $\mathbb{R}^N$ containing F and suppose that $f: D \to \mathbb{R}^1$ is bounded on F. Let Δ be a subdivision of F and set

$$M_i = \sup f \quad \text{on} \quad F_i, \qquad m_i = \inf f \quad \text{on} \quad F_i.$$

Definitions. The **upper sum** *of* f with respect to the subdivision Δ is defined by the formula

$$S^+(f, \Delta) = \sum_{i=1}^{n} M_i V(F_i),$$

where $V(F_i)$ is the volume in $\mathbb{R}^N$ of the figure F_i. Similarly, the **lower sum** *of* f is defined by

$$S_-(f, \Delta) = \sum_{i=1}^{n} m_i V(F_i).$$

Let Δ be a subdivision of a figure F. Then Δ', another subdivision of F, is

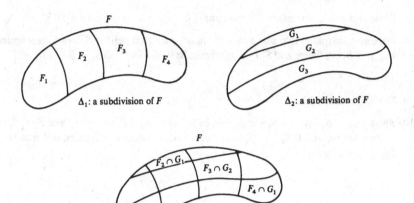

Figure 8.1

called a **refinement** *of* Δ if and only if every figure of Δ' is entirely contained in *one* of the figures of Δ. Suppose that $\Delta_1 = \{F_1, F_2, \ldots, F_n\}$ and $\Delta_2 = \{G_1, G_2, \ldots, G_m\}$ are two subdivisions of F. We say that Δ', the subdivision consisting of all nonempty figures of the form $F_i \cap G_j$, $i = 1, 2, \ldots, n$, $j = 1, 2, \ldots, m$, is the **common refinement** *of* Δ_1 *and* Δ_2. Note that Δ' is a refinement of both Δ_1 and Δ_2 (see Figure 8.1).

Theorem 8.4. *Let F be a figure in $\mathbb{R}^N$ and suppose that $f: F \to \mathbb{R}^1$ is bounded on F.*

(a) *If $m \leqslant f(x) \leqslant M$ for all $x \in F$ and if Δ is any subdivision of F, then*

$$mV(F) \leqslant S_-(f, \Delta) \leqslant S^+(f, \Delta) \leqslant MV(F).$$

(b) *If Δ' is a refinement of Δ, then*

$$S_-(f, \Delta) \leqslant S_-(f, \Delta') \quad and \quad S^+(f, \Delta') \leqslant S^+(f, \Delta).$$

(c) *If Δ_1 and Δ_2 are any two subdivisions of F, then*

$$S_-(f, \Delta_1) \leqslant S^+(f, \Delta_2).$$

PROOF

(a) The proof of Part (a) is identical to the proof of Part (a) in Theorem 5.1, the same theorem for functions from $\mathbb{R}^1$ into $\mathbb{R}^1$.

(b) Let $\Delta' = \{F_1', \ldots, F_m'\}$ be a refinement of $\Delta = \{F_1, F_2, \ldots, F_n\}$. We denote by $F_1', F_2', \ldots, F_k'$ the figures of Δ' contained in F_1. Then, using the symbols

$$m_1 = \inf f \quad in \quad F_1 \quad and \quad m_i' = \inf f \quad in \quad F_i', \qquad i = 1, 2, \ldots, k,$$

we have immediately $m_1 \leqslant m_i'$, $i = 1, 2, \ldots, k$, since each F_i' is a subset of F_1.

Because $V(F_1) = V(F_1') + \cdots + V(F_k')$, we have

$$m_1 V(F_1) \leqslant m_1'(F_1') + \cdots + m_k' V(F_k'). \tag{8.2}$$

The same type of inequality as (8.2) holds for $F_2, F_3, \ldots, F_n$. Summing these inequalities, we get $S_-(f, \Delta) \leqslant S_-(f, \Delta')$. The proof that $S^+(f, \Delta') \leqslant S^+(f, \Delta)$ is similar.

(c) If Δ_1 and Δ_2 are two subdivisions, let Δ' denote the common refinement. Then, from Parts (a) and (b), it follows that

$$S_-(f, \Delta_1) \leqslant S_-(f, \Delta') \leqslant S^+(f, \Delta') \leqslant S^+(f, \Delta_2). \qquad \square$$

Definitions. Let F be a figure in $\mathbb{R}^N$ and suppose that $f: F \to \mathbb{R}^1$ is bounded on F. The **upper integral** of f is defined by

$$\overline{\int_F} f \, dV = \text{g.l.b. } S^+(f, \Delta),$$

where the greatest lower bound is taken over all subdivisions Δ of F. The **lower integral** of f is

$$\underline{\int_F} f \, dV = \text{l.u.b. } S_-(f, \Delta),$$

where the least upper bound is taken over all possible subdivisions Δ if F. If

$$\overline{\int_F} f \, dV = \underline{\int_F} f \, dV,$$

then we say that f is **Darboux integrable**, or just **integrable**, on F and we designate the common value by

$$\int_F f \, dV.$$

When we wish to emphasize that the integral is N-dimensional, we write

$$\int_F f \, dV_N.$$

The following elementary results for integrals in $\mathbb{R}^N$ are the direct analogs of the corresponding theorems given in Chapter 5 for functions from $\mathbb{R}^1$ into $\mathbb{R}^1$. The proofs are the same except for the necessary alterations from intervals in $\mathbb{R}^1$ to figures in $\mathbb{R}^N$.

Theorem 8.5. *Let F be a figure in $\mathbb{R}^N$. Let f, f_1, f_2 be functions from F into $\mathbb{R}^1$ which are bounded.*

(a) *If $m \leqslant f(x) \leqslant M$ for $x \in F$, then*

$$mV(F) \leqslant \underline{\int_F} f \, dV \leqslant \overline{\int_F} f \, dV \leqslant MV(F).$$

(b) *If k is a positive number and $g = kf$, then*

$$\underline{\int_F} g \, dV = k \underline{\int_F} f \, dV \quad \text{and} \quad \overline{\int_F} g \, dV = k \overline{\int_F} f \, dV.$$

If k is negative, then

$$\underline{\int_F} g \, dV = k \overline{\int_F} f \, dV \quad \text{and} \quad \overline{\int_F} g \, dV = k \underline{\int_F} f \, dV.$$

(c) *The following inequalities hold:*

$$\underline{\int_F} (f_1 + f_2) \, dV \geqslant \underline{\int_F} f_1 \, dV + \underline{\int_F} f_2 \, dV;$$

$$\overline{\int_F} (f_1 + f_2) \, dV \leqslant \overline{\int_F} f_1 \, dV + \overline{\int_F} f_2 \, dV.$$

(d) *If $f_1(x) \leqslant f_2(x)$ for all $x \in F$, then*

$$\underline{\int_F} f_1(x) \, dV \leqslant \underline{\int_F} f_2(x) \, dV, \quad \overline{\int_F} f_1(x) \, dV \leqslant \overline{\int_F} f_2(x) \, dV.$$

(e) *Suppose G is another figure in $\mathbb{R}^N$ such that F and G have no common interior points. If f is defined and bounded on $F \cup G$, then*

$$\underline{\int_{F \cup G}} f \, dV = \underline{\int_F} f \, dV + \underline{\int_G} f \, dV;$$

$$\overline{\int_{F \cup G}} f \, dV = \overline{\int_F} f \, dV + \overline{\int_G} f \, dV.$$

If the functions considered in Theorem 8.5 are Darboux integrable, then Formulas (a)–(e) of that theorem are modified as in the Corollary to Theorem 5.3.

Theorem 8.6. *Let F be a figure in $\mathbb{R}^N$ and suppose that $f: F \to \mathbb{R}^1$ is bounded.*

(a) *f is Darboux integrable on F if and only if for each $\varepsilon > 0$ there is a subdivision Δ of F such that*

$$S^+(f, \Delta) - S_-(f, \Delta) < \varepsilon.$$

(b) *If f is uniformly continuous on F, then f is Darboux integrable on F.*
(c) *If f is Darboux integrable on F, then $|f|$ is also.*
(d) *If f_1, f_2 are each Darboux integrable on F, then $f_1 \cdot f_2$ is also.*
(e) *If f is Darboux integrable on F and H is a figure contained in F, then f is Darboux integrable on H.*

The proof of Theorem 8.6 follows the lines of the analogous theorem in $\mathbb{R}^1$. For positive functions from an interval I in $\mathbb{R}^1$ to $\mathbb{R}^1$, the Darboux integral gives a formula for finding the area under a curve. If F is a figure in $\mathbb{R}^N$, $N \geqslant 2$, and if $f: F \to \mathbb{R}^1$ is a nonnegative function, then the Darboux integral of f gives the $N + 1$-dimensional volume "under the hypersurface f." Theorem 8.8 below states the precise result. (See also Problems 11 and 12 in Section 5.4.)

The next theorem, which is used in the proof of Theorem 8.8, states that the volume of the direct product $F \times G$ of two figures F and G is the product of the volumes of F and G. This fact generalizes the simple formula which states that the volume of a rectangular box in $\mathbb{R}^N$ is the product of the lengths of the N one-dimensional edges of the box. Similarly, the volume of a right circular cylinder in $\mathbb{R}^3$ is the product of the area of the base (a circular disk in $\mathbb{R}^2$) and the height of the cylinder (length in $\mathbb{R}^1$ of the generator of the cylinder).

Theorem 8.7. *Let F be a figure in $\mathbb{R}^N$ and G a figure in $\mathbb{R}^M$. Form the set B in $\mathbb{R}^{N+M}$ given by $B = \{(x, y): x \in F, y \in G\}$. Then B is a figure and*

$$V_{N+M}(B) = V_N(F) \cdot V_M(G). \tag{8.3}$$

SKETCH OF PROOF. Divide the space $\mathbb{R}^{N+M}$ into hypercubes of side 2^{-n}. Then F is divided into hypercubes (N-dimensional) and G is divided into hypercubes (M-dimensional). Each inner hypercube of B is the direct product of an inner hypercube of F and one of G. Thus the totality of inner hypercubes of B is the product of the number of inner hypercubes of F with the number of inner hypercubes of G. Thus we have

$$V_n^-(B) = V_n^-(F) \cdot V_n^-(G).$$

Letting $n \to \infty$, we find $V^-(B) = V^-(F)V^-(G)$. A similar argument holds for outer volume. Since F and G are figures, the result follows.   □

Theorem 8.8. *Let F be a figure in $\mathbb{R}^N$ and suppose that $f: F \to \mathbb{R}^1$ is bounded on F. Let c be a constant such that $c < f(x)$ for all $x \in F$. Define the sets in $\mathbb{R}^{N+1}$*

$$G_1 = \{(x, y): x \in F, \quad y \in \mathbb{R}^1 \quad and \quad c \leqslant y < f(x)\},$$
$$G_2 = \{(x, y): x \in F, \quad y \in \mathbb{R}^1 \quad and \quad c \leqslant y \leqslant f(x)\}.$$

(a) *Then*

$$\underline{\int_F} [f(x) - c] \, dV_N = V_{N+1}^-(G_1) \leqslant V_{N+1}^-(G_2),$$

$$\overline{\int_F} [f(x) - c] \, dV_N = V_{N+1}^+(G_2) \geqslant V_{N+1}^+(G_1).$$

(b) *Suppose that f and $g: F \to \mathbb{R}^1$ are integrable on F and that $f(x) \leqslant g(x)$ for all $x \in F$. Define the set in $\mathbb{R}^{N+1}$*

$$G = \{(x, y): x \in F \quad and \quad f(x) \leqslant y \leqslant g(x)\}.$$

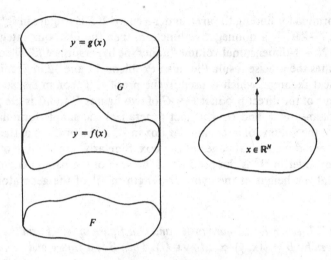

Figure 8.2. A set G in $\mathbb{R}^{N+1}$.

Then G is a figure in $\mathbb{R}^{N+1}$ *and*

$$V_{N+1}(G) = \int_F [g(x) - f(x)] \, dV_N.$$

PROOF. Since $G_2 \supset G_1$, the inequalities in Part (a) are immediate. The remainder of Part (a) follows from the definition of upper and lower Darboux integrals and the definition of inner and outer volume in $\mathbb{R}^{N+1}$.

Theorem 8.7 implies the result in Part (b) since each element of volume in $V_{N+1}(G)$ is the product of an N-dimensional hypercube and a one-dimensional length (see Figure 8.2). □

PROBLEMS

1. Let F be a figure in $\mathbb{R}^N$. Give an example of a function $f: F \to \mathbb{R}^N$ such that f is bounded on F and $\underline{\int}_F f \, dV \neq \overline{\int}_F f \, dV$.

2. Let F be a figure in $\mathbb{R}^N$ and suppose that $f: F \to \mathbb{R}^1$ is continuous on F. Let Δ be a subdivision of F. Suppose that $S_-(f, \Delta) = S_-(f, \Delta')$ for every refinement Δ' of Δ. What can be concluded about the function f?

3. Write a proof of Theorem 8.4, Part (a).

4. Suppose that F is a figure in $\mathbb{R}^N$ and that $f: F \to \mathbb{R}^1$ is uniformly continuous on F. In $\mathbb{R}^{N+1}$ define the set $A = \{(x, y): x \in F, y = f(x)\}$. Show that $V_{N+1}(A) = 0$.

5. Prove the analog of Lemma 5.1 for functions f from $\mathbb{R}^N$ into $\mathbb{R}^1$.

6. Suppose that F is a figure in $\mathbb{R}^N$ and that f is integrable over F. Show that

$$\left| \int_F f(x) \, dV \right| \leq \int_F |f(x)| \, dV.$$

7. Prove Theorem 8.5.

8. (a) Define

$$F = \{(x_1, x_2) : 0 \leqslant x_1^2 + x_2^2 \leqslant 1\}, \qquad G = \{(x_3, x_4) : 0 \leqslant x_3 \leqslant 1, 0 \leqslant x_4 \leqslant 1\}.$$

   Define in $\mathbb{R}^4$ the set $B = F \times G$. Show how Theorem 8.7 can be used to find $V(B)$.

   (b) Let F_1 be the unit ball in $\mathbb{R}^M$ and G_1 the unit hypercube in $\mathbb{R}^N$. Find $V_{M+N}(B)$ where $B = F_1 \times G_1$.

9. (a) Let F be a figure in $\mathbb{R}^N$ and suppose that $f : F \to \mathbb{R}^1$ is uniformly continuous on F. Show that f is Darboux integrable on F (Theorem 8.6, Part (b)).

   (b) Prove Theorem 8.6, Part (c).

   (c) Prove Theorem 8.6, Part (d).

   (d) Prove Theorem 8.6, Part (e).

10. Let F be a figure in $\mathbb{R}^N$ with $V(F) > 0$ and suppose that $f : F \to \mathbb{R}^1$ is continuous on F. Suppose that for every continuous function $g : F \to \mathbb{R}^1$ we have $\int_F fg \, dV = 0$. Prove that $f \equiv 0$ on F.

11. Write a complete proof of Theorem 8.7.

12. Let F be a figure and suppose that $f : F \to \mathbb{R}^1$ is integrable on F but not nonnegative. Interpret geometrically $\int_F f \, dV$ and $\int_F |f| \, dV$.

13. (a) Write a complete proof of Theorem 8.8(a).

    (b) Write a complete proof of Theorem 8.8(b).

14. Use Formula (8.3) of Theorem 8.7 to find the volume in $\mathbb{R}_5$ of the figure

$$F = \{(x_1, x_2, x_3, x_4, x_5) : x_1^2 + x_2^2 + x_3^2 \leqslant 1, x_4^2 + x_5^2 \leqslant 1\}.$$

8.3. The Riemann Integral in $\mathbb{R}^N$

The method of extending the Riemann integral from $\mathbb{R}^1$ to $\mathbb{R}^N$ is similar to the extension of the Darboux integral described in Section 8.2.

Definitions. Let A be a set in a metric space S with metric d. We define the **diameter** *of* A as the sup $d(x, y)$ where the supremum is taken over all x, y in A. The notation diam A is used for the diameter of A.

Suppose that F is a figure in $\mathbb{R}^N$ and that $\Delta = \{F_1, F_2, \ldots, F_n\}$ is a subdivision of F. The **mesh** *of* Δ, denoted $\|\Delta\|$, is the maximum of the diameters of $F_1, F_2, \ldots, F_n$.

Definition. Let f be a function from F, a figure in $\mathbb{R}^N$, into $\mathbb{R}^1$. Then f is **Riemann integrable** *on* F if and only if there is a number L with the following property: for each $\varepsilon > 0$ there is a $\delta > 0$ such that if Δ is any subdivision of F with $\|\Delta\| < \delta$, and $x^i \in F_i$, $i = 1, 2, \ldots, n$, then

$$\left| \sum_{i=1}^{n} f(x^i) V_N(F_i) - L \right| < \varepsilon.$$

This inequality must hold no matter how the x^i are chosen in the F_i. The number L is called the **Riemann integral of f over F**, and we use the symbol

$$\int_F f \, dV$$

for this value.

Theorem 8.9. *The Riemann integral of a function is unique.*

The proof is a simple consequence of the theorem on uniqueness of limits (Theorem 2.1) and is left to the reader.

Definition. A figure F in $\mathbb{R}^N$ is **regular** if and only if for every $\varepsilon > 0$, F possesses a subdivision $\{F_1, F_2, \ldots, F_n\}$ with mesh $\|\Delta\| < \varepsilon$ and such that $V_N(F_i) > 0$, $i = 1, 2, \ldots, n$.

Remarks. We recall that in $\mathbb{R}^1$ the Riemann integral of a function f is unaffected if the value of f is changed at a finite number of points. Similarly, in $\mathbb{R}^N$ the Riemann integral of a function f may be unaffected if the value of f is changed on certain sets which have N-dimensional volume zero. For example, consider the figure F consisting of the disk in $\mathbb{R}^2$ with a protruding line segment, as shown in Figure 8.3. The integral of a function f over F is unaffected by a change in the value of f on this protruding line segment. The figure F is not regular since a subdivision with mesh size smaller than the length of the protruding line segment must have at least one F_i with zero 2-dimensional area. For regular figures we have the following analog of Theorem 5.10.

Theorem 8.10. *Let F be a regular figure in $\mathbb{R}^N$ and suppose that $f: F \to \mathbb{R}^1$ is Riemann integrable on F. Then f is bounded on F.*

PROOF. Let L be the Riemann integral of f on F. From the definition of the integral, there is a $\delta > 0$ such that for any subdivision $\{F_1, F_2, \ldots, F_n\}$ with mesh $\|\Delta\| < \delta$ and for any $x^i \in F_i$, we have (if we take $\varepsilon = 1$)

$$\left| \sum_{i=1}^n f(x^i) V(F_i) - L \right| < 1.$$

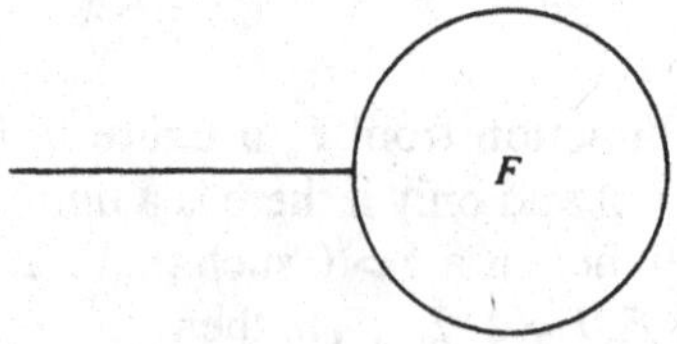

Figure 8.3. F includes a protruding line segment.

Therefore,

$$\left| \sum_{i=1}^{n} f(x^i) V(F_i) \right| \leq 1 + |L|. \tag{8.4}$$

Suppose that $\{F_1, F_2, \ldots, F_n\}$ is a particular subdivision in which $V(F_i) > 0$ for every i. Let x^1 be a point in F_1. Then from Inequality (8.4) we obtain

$$|f(x^1)| V(F_1) \leq 1 + |L| + \sum_{i=2}^{n} |f(x^i)| V(F_i),$$

or

$$|f(x^1)| \leq \frac{1}{V(F_1)} \left[1 + |L| + \sum_{i=2}^{n} |f(x^i)| V(F_i) \right]. \tag{8.5}$$

Since x^1 is any point in F_1, and since $x^2, x^3, \ldots, x^n$ may be kept fixed as x^1 varies over F_1, Inequality (8.5) shows that f is bounded on F_1. Similarly, f is bounded on $F_2, F_3, \ldots, F_n$ and so f is bounded on F. □

We recall that in $\mathbb{R}^1$ the Darboux and Riemann integrals are the same. The next two theorems state that the same result holds for integrals in $\mathbb{R}^N$.

Theorem 8.11. *Let F be a figure in $\mathbb{R}^N$ and suppose that $f: F \to \mathbb{R}^1$ is Riemann integrable on F. Then f is Darboux integrable on F and the two integrals are equal.*

The proof is similar to the proof of Theorem 5.11 for functions on $\mathbb{R}^1$ and is omitted.

The converse of Theorem 8.11 is contained in the following result.

Theorem 8.12. *Let F be figure in $\mathbb{R}^N$ and suppose that $f: F \to \mathbb{R}^1$ is bounded on F. Then*

(a) *for each $\varepsilon > 0$ there is a $\delta > 0$ such that*

$$S^+(f, \Delta) < \overline{\int_F} f \, dV + \varepsilon \quad and \quad S_-(f, \Delta) > \underline{\int_F} f \, dV - \varepsilon \tag{8.6}$$

for every subdivision of mesh $\|\Delta\| < \delta$;
(b) *if f is Darboux integrable on F, it is Riemann integrable on F and the integrals are equal.*

PROOF. Observe that (b) is an immediate consequence of (a) since each Riemann sum is between $S_-(f, \Delta)$ and $S^+(f, \Delta)$. Hence if $\overline{\int_F} f \, dV = \underline{\int_F} f \, dV$, this value is also the value of the Riemann integral.

We shall prove the first inequality in (8.6), the proof of the second being similar.

Since f is bounded, there is a number M such that $|f(x)| \leq M$ for all $x \in F$. Let $\varepsilon > 0$ be given. According to the definition of upper Darboux integral,

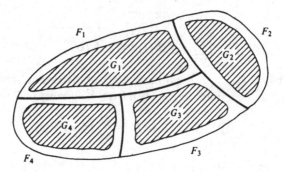

Figure 8.4. Volume of $F_i - G_i$ is small.

there is a subdivision $\Delta_0 = \{F_1, F_2, \ldots, F_m\}$ such that

$$S^+(f, \Delta_0) < \overline{\int_F} f \, dV + \frac{\varepsilon}{2}. \tag{8.7}$$

Define $F_i^{(0)}$ to be the set of interior points of F_i. It may happen that $F_i^{(0)}$ is empty for some values of i. For each $F_i^{(0)}$ which is not empty, we select a closed figure G_i contained in $F_i^{(0)}$ such that

$$V(F_i - G_i) < \frac{\varepsilon}{4Mm}, \qquad i = 1, 2, \ldots, m.$$

See Figure 8.4 where such a selection is shown with $m = 4$ for a figure in $\mathbb{R}^2$. It is not difficult to verify that such closed figures $G_1, G_2, \ldots, G_m$ can always be found. For example, each G_i may be chosen as the union of closed hypercubes interior to F_i for a sufficiently small grid size.

Since each set G_i is closed and is contained in $F_i^{(0)}$, there is a positive number δ such that every ball $B(x, \delta)$ with x in some G_i has the property that $B(x, \delta)$ is contained in the corresponding set $F_i^{(0)}$. Let Δ be any subdivision with mesh less than δ. We shall show that the first inequality in (8.6) holds for this subdivision.

We separate the figures of Δ into two classes: $J_1, J_2, \ldots, J_n$ are those figures of Δ containing points of some G_i; $K_1, K_2, \ldots, K_q$ are the remaining figures of Δ.

Denote by Δ' the common refinement of Δ and Δ_0. Because of the manner in which we chose δ, each J_i is contained entirely in some $F_k^{(0)}$. Therefore, J_1, $J_2, \ldots, J_n$ are figures in the refinement Δ'. The remaining figures of Δ' are composed of the sets $K_i \cap F_j$, $i = 1, 2, \ldots, q$; $j = 1, 2, \ldots, m$. We have the inequality

$$\sum_{k=1}^{q} V(K_k) < \sum_{i=1}^{m} V(F_i - G_i) < m \cdot \frac{\varepsilon}{4Mm} = \frac{\varepsilon}{4M}.$$

We introduce the notation

$$M_i = \sup_{x \in J_i} f(x), \qquad M_i' = \sup_{x \in K_i} f(x), \qquad M_{ij} = \sup_{x \in K_i \cap F_j} f(x).$$

Using the definitions of $S^+(f, \Delta)$ and $S^+(f, \Delta')$, we find

$$S^+(f, \Delta) = \sum_{i=1}^{n} M_i V(J_i) + \sum_{i=1}^{q} M_i' V(K_i),$$

$$S^+(f, \Delta') = \sum_{i=1}^{n} M_i V(J_i) + \sum_{i=1}^{q} \sum_{j=1}^{m} M_{ij} V(K_i \cap F_j).$$

Now it is clear that $V(K_i) = \sum_{j=1}^{m} V(K_i \cap F_j)$. Therefore, by subtraction, it follows that

$$S^+(f, \Delta) - S^+(f, \Delta') = \sum_{i=1}^{q} \sum_{j=1}^{m} (M_i' - M_{ij}) V(K_i \cap F_j)$$

$$\leqslant 2M \sum_{i=1}^{q} \sum_{j=1}^{m} V(K_i \cap F_j)$$

$$\leqslant 2M \sum_{i=1}^{q} V(K_i) < 2M \cdot \frac{\varepsilon}{4M} = \frac{\varepsilon}{2}. \qquad (8.8)$$

According to Part (b) of Theorem 8.4, we have

$$S^+(f, \Delta') \leqslant S^+(f, \Delta_0).$$

Combining this fact with Inequalities (8.7) and (8.8), we conclude that

$$S^+(f, \Delta) < \overline{\int_F} f \, dV + \varepsilon,$$

which is the first inequality in Part (a) of the Theorem. $\qquad\square$

The following result forms the basis for interchanging the order of integration in multiple integrals and for evaluating them.

Theorem 8.13. *Let F be a figure in $\mathbb{R}^M$ and G a figure in $\mathbb{R}^N$. Suppose that f is defined and bounded on the set $F \times G$ which is in $\mathbb{R}^{M+N}$. Then*

$$\overline{\int_{F \times G}} f \, dV_{M+N} \geqslant \overline{\int_F} \left[\overline{\int_G} f \, dV_N \right] dV_M;$$

$$\overline{\int_{F \times G}} f \, dV_{M+N} \geqslant \overline{\int_G} \left[\overline{\int_F} f \, dV_M \right] dV_N;$$

$$\underline{\int_{F \times G}} f \, dV_{M+N} \leqslant \underline{\int_F} \left[\underline{\int_G} f \, dV_N \right] dV_M;$$

$$\underline{\int_{F \times G}} f \, dV_{M+N} \leqslant \underline{\int_G} \left[\underline{\int_F} f \, dV_M \right] dV_N.$$

PROOF. Since the proofs of all the inequalities are similar, we shall prove only the first one. Let $\varepsilon > 0$ be given. Then there is a $\delta > 0$ such that if Δ is any subdivision of $F \times G$ with mesh less than δ, we have

$$S^+(f, \Delta) < \int_{F \times G}^{\overline{}} f \, dV_{M+N} + \varepsilon. \tag{8.9}$$

Now let $\{F_1, F_2, \ldots, F_m\}$ and $\{G_1, G_2, \ldots, G_n\}$ be subdivisions of F and G, respectively, each with mesh size less than $\delta/\sqrt{2}$. Then the subdivision of $F \times G$ consisting of $F_i \times G_j$, $i = 1, 2, \ldots, m$; $j = 1, 2, \ldots, n$, is a subdivision of $F \times G$ with mesh less than δ (see Theorem 8.7). We define

$$M_{ij} = \sup f(x, y) \quad \text{for} \quad (x, y) \in F_i \times G_j.$$

Then, using Theorem 8.7, we find

$$S^+(f, \Delta) = \sum_{i=1}^{m} \sum_{j=1}^{n} M_{ij} V_{M+N}(F_i \times G_j) = \sum_{i=1}^{m} \sum_{j=1}^{n} M_{ij} V_M(F_i) V_N(G_j).$$

For each $x \in F_i$, we have

$$\sum_{j=1}^{n} M_{ij} V_N(G_j) \geqslant \int_{G}^{\overline{}} f(x, y) \, dV_N.$$

We now define

$$M_i = \sum_{j=1}^{n} M_{ij} V_N(G_j),$$

and we see that

$$S^+(f, \Delta) = \sum_{i=1}^{m} M_i V_M(F_i) \geqslant \int_{F}^{\overline{}} \left[\int_{G}^{\overline{}} f(x, y) \, dV_N \right] dV_M.$$

Combining this inequality with (8.9) and observing that ε is arbitrary, we obtain the desired result. $\qquad\qquad\qquad\qquad\qquad\qquad\qquad\qquad\qquad\square$

The customary theorems on the equality of multiple integrals and iterated integrals are a consequence of Theorem 8.13. For example, if a function f is defined in a figure B in $\mathbb{R}^2$, we extend the definition of f to a rectangle R in $\mathbb{R}^2$ which contains B by setting $f = 0$ for (x, y) outside of B. Then f is integrable over B if and only if it is integrable over R, and the two integrals are equal. Since R is the product of two intervals in $\mathbb{R}^1$, we may apply Theorem 8.13. More generally, we have the following result.

Corollary to Theorem 8.13. *Suppose that F is a figure in $\mathbb{R}^M$ and that G_x is a figure in $\mathbb{R}^N$ for each $x \in F$. Define $B = \{(x, y) : x \in F, y \in G_x$ for each such $x\}$. Let $f : B \to \mathbb{R}^1$ be integrable over B and suppose that f is integrable over G_x for each $x \in F$. Then*

$$\phi(x) = \int_{G_x} f(x, y) \, dV_N$$

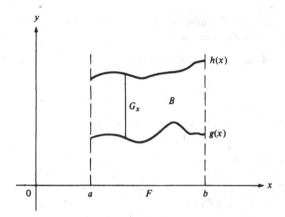

Figure 8.5. $B = \{(x, y) : a \leqslant x \leqslant b, y \in G_x\}$.

is integrable over F and

$$\int_B f(x, y)\, dV_{M+N} = \int_F \left[\int_{G_x} f(x, y)\, dV_N \right] dV_M.$$

Remark. Figure 8.5 shows a simple illustration of the corollary for functions on $\mathbb{R}^2$. Let $B = \{(x, y): a \leqslant x \leqslant b, g(x) \leqslant y \leqslant h(x)\}$. Then $F = \{x: a \leqslant x \leqslant b\}$ and $G_x = \{y: g(x) \leqslant y \leqslant h(x)$ for each $x\}$. The corollary states that

$$\int_B f(x, y)\, dA = \int_a^b \left[\int_{g(x)}^{h(x)} f(x, y)\, dy \right] dx.$$

PROBLEMS

1. Prove that the Riemann integral of a function in $\mathbb{R}^N$ is unique (Theorem 8.9).

2. Show that the union of two regular figures is regular. Give examples to show that the intersection and difference of regular figures may not be regular.

3. Suppose that F is a figure in $\mathbb{R}^N$ for $N \geqslant 2$ and that $f: F \to \mathbb{R}^1$ is Riemann integrable on F. Show that f may not be bounded on F. (Compare with Theorem 8.10.)

4. Let f be a nonnegative unbounded function defined on a figure F in $\mathbb{R}^N$ into $\mathbb{R}^1$. Define

$$f_n = \begin{cases} f & \text{if} |f| \leqslant n, \\ 0 & \text{if} |f| > n, \end{cases}$$

and suppose that f_n is Riemann integrable for each n. Define the **Riemann integral for unbounded functions** by the formula $\int_F f\, dV = \lim_{n \to \infty} \int_F f_n\, dV$ when the limit exists. Letting $f(x, y) = 1/(x^2 + y^2)^\alpha$ and $F = \{(x, y): 0 \leqslant x^2 + y^2 \leqslant 1\}$ in $\mathbb{R}^2$, show that $\int_F f(x, y)\, dV$ exists for $\alpha < 1$ and that it does not for $\alpha \geqslant 1$.

5. Let F be a figure in $\mathbb{R}^N$. Show that if $f: F \to \mathbb{R}^1$ is Riemann integrable then it is Darboux integrable (Theorem 8.11).

6. Suppose f is defined in the square $S = \{(x, y): 0 \leqslant x \leqslant 1, 0 \leqslant y \leqslant 1\}$ by the formula

$$f(x, y) = \begin{cases} 1 & \text{if } x \text{ is irrational,} \\ 4y^3 & \text{if } x \text{ is rational.} \end{cases}$$

(a) Show that $\int_0^1 (\int_0^1 f(x, y)\, dy)\, dx$ exists and has the value 1.
(b) Show that $\int_S f(x, y)\, dV$ does not exist.

7. Suppose that in the square $S = \{(x, y): 0 \leqslant x \leqslant 1, 0 \leqslant y \leqslant 1\}$ we define the set A as the set of points (x, y) such that x and y are rational and that, when they are represented in the form of $x = p_1/q_1, y = p_2/q_2$ (lowest terms), then $q_1 = q_2$. Suppose that $f: S \to \mathbb{R}^1$ is given by

$$f(x, y) = \begin{cases} 0 & \text{if } (x, y) \in A, \\ 1 & \text{if } (x, y) \in S - A. \end{cases}$$

(a) Show that $\int_0^1 (\int_0^1 f(x, y)\, dy)\, dx$ and $\int_0^1 (\int_0^1 f(x, y)\, dx)\, dy$, both exist and have the value 1.
(b) Show that $\int_S f(x, y)\, dV$ does not exist.

8. Write a proof of the Corollary to Theorem 8.13.

9. Let F be a regular figure in $\mathbb{R}^N$. Suppose $f: F \to \mathbb{R}^1$ and $g: F \to \mathbb{R}^1$ are Riemann integrable on F. Show that fg if Riemann integrable on F.

CHAPTER 9

Infinite Sequences and Infinite Series

9.1. Tests for Convergence and Divergence

It is customary to use expressions such as

$$u_1 + u_2 + \cdots + u_n + \cdots \quad \text{and} \quad \sum_{n=1}^{\infty} u_n \qquad (9.1)$$

to represent infinite series. The u_i are called the **terms** of the series, and the quantities

$$s_n = u_1 + u_2 + \cdots + u_n, \qquad n = 1, 2, \ldots,$$

are called the **partial sums** of the series. The symbols in (9.1) not only define an infinite series but also are used as an expression for the sum of the series when it converges. To avoid this ambiguity we define an infinite series in terms of ordered pairs.

Definitions. An **infinite series** is an ordered pair $(\{u_n\}, \{s_n\})$ of infinite sequences in which $s_n = u_1 + u_2 + \cdots + u_n$ for each n. The u_n are called the **terms** of the series and the s_n are called the **partial sums**. If there is a number s such that $s_n \to s$ as $n \to \infty$, we say the series is **convergent** and that the **sum** of the series is s. If the s_n do not tend to a limit we say that the series is **divergent**.

It is clear that an infinite series is uniquely determined by the sequence $\{u_n\}$ of its terms. There is almost never any confusion in using the symbols in (9.1) for an infinite series. While the definition in terms of ordered pairs is satisfactory from the logical point of view, it does require a cumbersome notation. Rather than have unwieldy proofs which may obscure their essential features, we shall use the standard symbols of $u_1 + u_2 + \cdots + u_n + \cdots$ and $\sum_{n=1}^{\infty} u_n$ to denote infinite series. The context will always show whether the expression represents the series itself or the sum of the terms.

Theorem 9.1. *If the series $\sum_{n=1}^{\infty} u_n$ converges, then $u_n \to 0$ as $n \to \infty$.*

PROOF. For all $n > 1$, we have $u_n = s_n - s_{n-1}$. If s denotes the sum of the series, then $s_n \to s$ and $s_{n-1} \to s$ as $n \to \infty$. Hence $u_n \to s - s = 0$ as $n \to \infty$. $\qquad\square$

Remarks. If the terms u_n of an infinite series tend to zero, it does not necessarily follow that the series $\sum_{n=1}^{\infty} u_n$ converges. See the Corollary to Theorem 9.5 which illustrates this point.

Let

$$u_1 + u_2 + \cdots + u_n + \cdots$$

be a given series. Then a new series may be obtained by deleting a finite number of terms at the beginning. It is clear that *the new series will be convergent if and only if the original series is.*

Theorem 9.2. *Let $\sum_{n=1}^{\infty} u_n$, $\sum_{n=1}^{\infty} v_n$ be given series and let $c \neq 0$ be a constant.*

(a) *If $\sum_{n=1}^{\infty} u_n$, $\sum_{n=1}^{\infty} v_n$ are convergent, then $\sum_{n=1}^{\infty} (u_n + v_n)$, $\sum_{n=1}^{\infty} (u_n - v_n)$, and $\sum_{n=1}^{\infty} cu_n$ are convergent series. Also,*

$$\sum_{n=1}^{\infty} (u_n \pm v_n) = \sum_{n=1}^{\infty} u_n \pm \sum_{n=1}^{\infty} v_n, \qquad \sum_{n=1}^{\infty} cu_n = c \sum_{n=1}^{\infty} u_n.$$

(b) *If $\sum_{n=1}^{\infty} u_n$ diverges, then $\sum_{n=1}^{\infty} cu_n$ diverges.*

PROOF. For each positive integer n, we have

$$\sum_{k=1}^{n} (u_k \pm v_k) = \sum_{k=1}^{n} u_k \pm \sum_{k=1}^{n} v_k, \qquad \sum_{k=1}^{n} (cu_k) = c \sum_{k=1}^{n} u_k.$$

Then Part (a) follows from the theorems on limits. (See Section 2.5.) To prove Part (b), we have only to observe that if $\sum_{n=1}^{\infty} cu_n$ converges, then so does $\sum_{n=1}^{\infty} (1/c)(cu_n) = \sum_{n=1}^{\infty} u_n$. $\qquad\square$

A series of the form

$$a + ar + ar^2 + \cdots + ar^n + \cdots$$

is called a **geometric series**. The number r is the **common ratio**.

Theorem 9.3. *A geometric series with $a \neq 0$ converges if $|r| < 1$ and diverges if $|r| \geqslant 1$. In the convergent case, we have*

$$\sum_{n=1}^{\infty} ar^{n-1} = \frac{a}{1-r}.$$

PROOF. We verify easily that for each positive integer n

$$s_n = a + ar + \cdots + ar^{n-1} = \frac{a}{1-r} - \frac{ar^n}{1-r}.$$

If $|r| < 1$, then $r^n \to 0$ as $n \to \infty$ (see Section 2.5, Problem 6). Hence $s_n \to a/(1-r)$. If $|r| \geqslant 1$, then u_n does not tend to zero. According to Theorem 9.1, the series cannot converge.   □

Theorem 9.4 (Comparison test). *Suppose that $u_n \geqslant 0$ for all n.*

(a) *If $u_n \leqslant a_n$ for all n and $\sum_{n=1}^{\infty} a_n$ converges, then $\sum_{n=1}^{\infty} u_n$ converges and $\sum_{n=1}^{\infty} u_n \leqslant \sum_{n=1}^{\infty} a_n$.*
(b) *Let $a_n \geqslant 0$ for all n. If $\sum_{n=1}^{\infty} a_n$ diverges and $u_n \geqslant a_n$ for all n, then $\sum_{n=1}^{\infty} u_n$ diverges.*

The proof is left to the reader. (See Problems 11 and 12 at the end of this section.)

Let f be continuous on $[a, \infty)$. We define

$$\int_a^{\infty} f(x)\, dx = \lim_{b \to +\infty} \int_a^b f(x)\, dx \tag{9.2}$$

when the limit on the right exists. The term **improper integral** is used when the range of integration is infinite. We say the improper integral **converges** when the limit in (9.2) exists; otherwise the integral **diverges**.

Theorem 9.5 (Integral test). *Suppose that f is continuous, nonnegative, and nonincreasing on $[1, \infty)$. Suppose that $\sum_{n=1}^{\infty} u_n$ is a series with $u_n = f(n)$, $n = 1, 2, \ldots$ Then*

(a) *$\sum_{n=1}^{\infty} u_n$ converges if $\int_1^{\infty} f(x)\, dx$ converges; and*
(b) *$\sum_{n=1}^{\infty} u_n$ diverges if $\int_1^{\infty} f(x)\, dx$ diverges.*

PROOF. Since f is positive and nonincreasing, we have (see Figure 9.1) for $n \geqslant 2$,

$$\sum_{j=2}^{n} u_j \leqslant \int_1^n f(x)\, dx \leqslant \sum_{j=1}^{n-1} u_j.$$

We define

$$F(X) = \int_1^X f(x)\, dx.$$

If this integral converges, then $F(X)$ is a nondecreasing function which tends to a limit, and so $F(n)$ is a bounded nondecreasing sequence. Thus denoting $s_n = \sum_{j=2}^{n} u_j$, we see that $s_n \leqslant F(n)$, and s_n tends to a limit. Part (a) is now established. If the integral diverges, then $F(X) \to +\infty$ as $X \to \infty$. Therefore $F(n) \to +\infty$. Since $F(n) \leqslant \sum_{j=1}^{n-1} u_j$, we conclude that the series diverges.   □

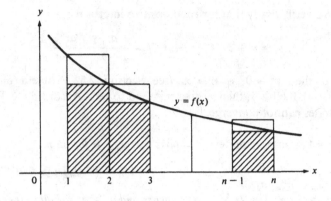

Figure 9.1. The integral test for convergence.

Corollary to Theorem 9.5. *The series*

$$\sum_{n=1}^{\infty} \frac{1}{n^p},$$

known as the p-series, converges if $p > 1$ and diverges if $p \leqslant 1$.

The proof of the Corollary is an immediate consequence of the integral test with $f(x) = 1/x^p$. The details are left to the reader. For $0 < p \leqslant 1$, the terms of the p-series tend to zero as $n \to \infty$ although the series diverges. This fact shows that the converse of Theorem 9.1 is false. The hypothesis that $u_n \to 0$ as $n \to \infty$ does *not* imply the convergence of $\sum_{n=1}^{\infty} u_n$.

EXAMPLE. Test the series

$$\sum_{n=1}^{\infty} \frac{1}{(n + 2) \log(n + 2)}$$

for convergence or divergence.

Solution. We define $f(x) = 1/(x + 2) \log(x + 2)$ and observe that f is positive and nonincreasing with $f(n) = 1/(n + 2) \log(n + 2)$. We have

$$\int_1^a \frac{dx}{(x + 2) \log(x + 2)} = \int_3^{a+2} \frac{du}{u \log u}$$

$$= \int_3^{a+2} \frac{d(\log u)}{\log u}$$

$$= \log[\log(a + 2)] - \log \log 3.$$

Since $\log[\log(a + 2)] \to +\infty$ as $a \to +\infty$, the integral diverges. Therefore the series does. $\square$

PROBLEMS

In each of Problems 1 through 10 test for convergence or divergence.

1. $\displaystyle\sum_{n=1}^{\infty} \frac{1}{\sqrt{n^3}}$

2. $\displaystyle\sum_{n=1}^{\infty} \frac{1}{\sqrt[3]{n^2}}$

3. $\displaystyle\sum_{n=1}^{\infty} \frac{1}{n(n+2)}$

4. $\displaystyle\sum_{n=1}^{\infty} \frac{1}{n \cdot 2^n}$

5. $\displaystyle\sum_{n=1}^{\infty} \frac{n+1}{n^3}$

6. $\displaystyle\sum_{n=1}^{\infty} \frac{1}{2n+3}$

7. $\displaystyle\sum_{n=1}^{\infty} \frac{2n+3}{n^3}$

8. $\displaystyle\sum_{n=1}^{\infty} \frac{\log n}{n^{3/2}}$

9. $\displaystyle\sum_{n=1}^{\infty} \frac{n}{e^n}$

10. $\displaystyle\sum_{n=1}^{\infty} \frac{n^p}{n!}, p > 0$ constant

11. Prove Theorem 9.4(a). [*Hint*: Use the fact that a bounded, nondecreasing sequence tends to a limit (Axiom of continuity).]

12. Prove Theorem 9.4(b).

13. For what values of p does the series $\sum_{n=1}^{\infty} \log n/n^p$ converge?

14. For what values of p does the series $\sum_{n=2}^{\infty} (\log n)^p/n$ converge?

15. Prove the Corollary to Theorem 9.5.

16. Prove the **Limit comparison theorem**:
Suppose that $a_n \geqslant 0, b_n \geqslant 0, n = 1, 2, \ldots,$ and that

$$\lim_{n \to \infty} \frac{a_n}{b_n} = L > 0.$$

Then either $\sum_{n=1}^{\infty} a_n$ and $\sum_{n=1}^{\infty} b_n$ both converge on both diverge. [*Hint*. For sufficiently large n, we have $\frac{1}{2}L < a_n/b_n < \frac{3}{2}L$. Now use the comparison test (Theorem 9.4) and the fact that the early terms of a series do not affect convergence.]

17. Use the result of Problem 16 to test for convergence:

$$\sum_{n=1}^{\infty} \frac{2n^2 + n + 2}{5n^3 + 3n}.$$

Take

$$a_n = \frac{2n^2 + n + 2}{5n^3 + 3n}, \qquad b_n = \frac{1}{n}.$$

18. Use the result of Problem 16 to test for convergence:

$$\sum_{n=1}^{\infty} \frac{1}{\sqrt[3]{n^2 + 5}}.$$

Take

$$a_n = \frac{1}{\sqrt[3]{n^2 + 5}}, \qquad b_n = \frac{1}{n^{2/3}}.$$

9.2. Series of Positive and Negative Terms; Power Series

When all the terms of a series are nonnegative, the Comparison test (Theorem 9.4) is a useful tool for testing the convergence or divergence of an infinite series. (See also the Limit comparison test in Problem 16 at the end of Section 9.1.) We now show that the same test may be used when a series has both positive and negative terms.

Definition. A series $\sum_{n=1}^{\infty} u_n$ which is such that $\sum_{n=1}^{\infty} |u_n|$ converges is said to be **absolutely convergent**. However, if $\sum_{n=1}^{\infty} u_n$ converges and $\sum_{n=1}^{\infty} |u_n|$ diverges, then the series $\sum_{n=1}^{\infty} u_n$ is said to be **conditionally convergent**.

The following theorem shows that if a series is absolutely convergent, then the series itself converges.

Theorem 9.6. *If $\sum_{n=1}^{\infty} |u_n|$ converges, then $\sum_{n=1}^{\infty} u_n$ converges and*

$$\left| \sum_{n=1}^{\infty} u_n \right| \leqslant \sum_{n=1}^{\infty} |u_n|.$$

PROOF. For $n = 1, 2, \ldots$, we define

$$v_n = \frac{|u_n| + u_n}{2}, \qquad w_n = \frac{|u_n| - u_n}{2}.$$

Then we have

$$u_n = v_n - w_n, \qquad |u_n| = v_n + w_n,$$

and

$$0 \leqslant v_n \leqslant |u_n|, \qquad 0 \leqslant w_n \leqslant |u_n|.$$

Both $\sum_{n=1}^{\infty} v_n$ and $\sum_{n=1}^{\infty} w_n$ converge by the Comparison test (Theorem 9.4). Therefore $\sum_{n=1}^{\infty} (v_n - w_n) = \sum_{n=1}^{\infty} u_n$ converges. Also,

$$\left| \sum_{n=1}^{\infty} u_n \right| = \left| \sum_{n=1}^{\infty} (v_n - w_n) \right| \leqslant \sum_{n=1}^{\infty} (v_n + w_n) = \sum_{n=1}^{\infty} |u_n|. \qquad \square$$

Remark. A series may be conditionally convergent and not absolutely convergent. The next theorem shows that $\sum_{n=1}^{\infty} (-1)^n (1/n)$ is convergent. However, the series $\sum_{n=1}^{\infty} 1/n$, a p-series with $p = 1$, is divergent; hence $\sum_{n=1}^{\infty} (-1)^n (1/n)$ is a conditionally convergent series.

Theorem 9.7 (Alternating series theorem). *Suppose that the numbers u_n, $n = 1, 2, \ldots$, satisfy the conditions:*

(i) *the u_n are alternately positive and negative;*
(ii) *$|u_{n+1}| < |u_n|$ for every n; and*
(iii) *$\lim_{n \to \infty} u_n = 0$.*

Then $\sum_{n=1}^{\infty} u_n$ is convergent. Furthermore, if the sum is denoted by s, then s lies between the partial sums s_n and s_{n+1} for each n.

PROOF. Assume that u_1 is positive. If not, consider the series beginning with u_2, since discarding one term does not affect convergence. With $u_1 > 0$, we have clearly

$$u_{2n-1} > 0 \quad \text{and} \quad u_{2n} < 0 \quad \text{for all } n.$$

We now write

$$s_{2n} = (u_1 + u_2) + (u_3 + u_4) + \cdots + (u_{2n-1} + u_{2n}).$$

Since by (ii) above $|u_{2k}| < u_{2k-1}$ for each k, each term in parenthesis is positive, and so s_{2n} increases with n. Also,

$$s_{2n} = u_1 + (u_2 + u_3) + (u_4 + u_5) + \cdots + (u_{2n-2} + u_{2n-1}) + u_{2n}.$$

The terms in parentheses are negative and $u_{2n} < 0$. Therefore $s_{2n} < u_1$ for all n. Hence s_{2n} is a bounded, increasing sequence and therefore convergent to a number, say s; also, $s_{2n} \leqslant s$ for each n. By observing that $s_{2n-1} = s_{2n} - u_{2n}$, we have $s_{2n-1} > s_{2n}$ for all n. In particular $s_{2n-1} > s_2 = u_1 + u_2$ and so s_{2n-1} is bounded from below. Also,

$$s_{2n+1} = s_{2n-1} + (u_{2n} + u_{2n+1}) < s_{2n-1}.$$

Therefore s_{2n+1} is a decreasing sequence which tends to a limit. Since $u_{2n} \to 0$ as $n \to \infty$, we see that s_{2n} and s_{2n+1} tend to the same limit s. Because s_{2n-1} is decreasing, it follows that $s_{2n-1} \geqslant s$ for every n. Thus $s \leqslant s_p$ for odd p and $s \geqslant s_p$ for even p. □

The next test is one of the most useful for deciding absolute convergence of series.

Theorem 9.8 (Ratio test). *Suppose that $u_n \neq 0, n = 1, 2, \ldots,$ and that*

$$\lim_{n \to \infty} \left| \frac{u_{n+1}}{u_n} \right| = \rho \quad \text{or} \quad \left| \frac{u_{n+1}}{u_n} \right| \to +\infty \quad \text{as} \quad n \to \infty.$$

Then

(i) *if $\rho < 1$, the series $\sum_{n=1}^{\infty} u_n$ converges absolutely;*
(ii) *if $\rho > 1$ or $|u_{n+1}/u_n| \to +\infty$ as $n \to \infty$, the series diverges;*
(iii) *if $\rho = 1$, the test gives no information.*

PROOF

(i) Suppose that $\rho < 1$. Choose any ρ' such that $\rho < \rho' < 1$. Since $|u_{n+1}/u_n|$ converges to ρ, there exists an integer N such that $|u_{n+1}/u_n| < \rho'$ for all $n \geqslant N$. That is,

$$|u_{n+1}| < \rho' |u_n| \quad \text{for} \quad n \geqslant N.$$

By induction, we have

$$|u_n| \leqslant (\rho')^{n-N}|u_N| \quad \text{for all} \quad n \geqslant N.$$

The series $\sum_{n=N}^{\infty} |u_n|$ converges by the Comparison test, using the geometric series $\sum_{n=0}^{\infty} (\rho')^n$. The original series converges absolutely since the addition of the finite sum $\sum_{n=1}^{N-1} |u_n|$ does not affect convergence.

(ii) If $\rho > 1$ or $|u_{n+1}/u_n| \to +\infty$, then there is an integer N such that $|u_{n+1}/u_n| > 1$ for all $n \geqslant N$. Then $|u_n| > |u_N|$ for all $n \geqslant N$. Hence u_n does not approach zero as $n \to \infty$. By Theorem 9.1 the series cannot converge.

(iii) The p-series for *all* values of p yields the limit $\rho = 1$. Since the p-series converges for $p > 1$ and diverges for $p \leqslant 1$, the Ratio test can yield no information when $p = 1$.   □

EXAMPLE 1. Test for conditional and absolute convergence:

$$\sum_{n=1}^{\infty} \frac{(-1)^n n}{3^n}.$$

Solution. We use the Ratio test. Set $u_n = (-1)^n n/3^n$. Then

$$\left| \frac{u_{n+1}}{u_n} \right| = \frac{n+1}{3^{n+1}} \cdot \frac{3^n}{n} = \frac{1}{3}\left(\frac{n+1}{n} \right) = \frac{1}{3}\left(\frac{1 + (1/n)}{1} \right).$$

Therefore

$$\lim_{n \to \infty} \left| \frac{u_{n+1}}{u_n} \right| = \frac{1}{3} = \rho.$$

The series converges absolutely.   □

The next theorem provides a useful test for many series.

Theorem 9.9 (Root test). *Let $\sum_{n=1}^{\infty} u_n$ be a series with either*

$$\lim_{n \to \infty} (|u_n|)^{1/n} = \rho \quad \text{or} \quad \lim_{n \to \infty} (|u_n|)^{1/n} = +\infty.$$

Then

(i) *if $\rho < 1$, the series $\sum_{n=1}^{\infty} u_n$ converges absolutely;*
(ii) *if $\rho > 1$, or if $|u_n|^{1/n} \to +\infty$, the series diverges;*
(iii) *if $\rho = 1$, the test gives no information.*

PROOF

(i) Suppose $\rho < 1$: choose $\varepsilon > 0$ so small that $\rho + \varepsilon < 1$ as well. Since $|u_n|^{1/n} \to \rho$, it follows that $|u_n|^{1/n} < \rho + \varepsilon$ for all $n \geqslant N$ if N is sufficiently large. Therefore $|u_n| < (\rho + \varepsilon)^n$ for all $n \geqslant N$. We observe that

$$\sum_{n=1}^{\infty} (\rho + \varepsilon)^n$$

is a convergent geometric series since $\rho + \varepsilon < 1$. The Comparison test (Theorem 9.4) shows that $\sum_{n=1}^{\infty} |u_n|$ converges. Hence (i) is established.

(ii) Suppose $\rho > 1$ or $|u_n|^{1/n} \to +\infty$. Choose $\varepsilon > 0$ so small that $(\rho - \varepsilon) > 1$. Therefore in Case (ii) $\rho - \varepsilon < |u_n|^{1/n}$ for all sufficiently large n. We conclude that

$$\lim_{n \to \infty} |u_n| \neq 0$$

and hence both $\sum_{n=1}^{\infty} |u_n|$ and $\sum_{n=1}^{\infty} u_n$ are divergent series.   □

EXAMPLE 2. Test for convergence:

$$\sum_{n=2}^{\infty} \frac{1}{(\log n)^n}.$$

Solution. We have

$$|u_n|^{1/n} = \frac{1}{\log n} \to 0 \quad \text{as} \quad n \to \infty.$$

The series converges by the Root test.   □

A **power series** is a series of the form

$$c_0 + c_1(x - a) + c_2(x - a)^2 + \cdots + c_n(x - a)^n + \cdots,$$

in which a and c_i, $i = 0, 1, 2, \ldots$, are constants. If a particular value is given to x, then the above expression is an infinite series of numbers which can be examined for convergence or divergence. For those values of x in $\mathbb{R}^1$ which yield a convergent power series, a function is defined whose range is the actual sum of the series. Denoting this function by f, we write

$$f: x \to \sum_{n=0}^{\infty} c_0(x - a)^n.$$

It will be established later that most of the elementary functions such as the trigonometric, logarithmic, and exponential functions have power series expansions. In fact, power series may be used for the definition of many of the functions studied thus far. For example, the function $\log x$ may be defined by a power series rather than by an integral, as in Section 5.3. If a power series definition is used, the various properties of functions, such as those given in Theorems 5.14 and 5.15 are usually more difficult to establish.

We first state a lemma and then prove two theorems which establish the basic properties of power series.

Lemma 9.1. *If the series $\sum_{n=1}^{\infty} u_n$ converges, then there is a number M such that $|u_n| \leq M$ for all n.*

The proof is left to the reader. (See Problem 28 at the end of the Section.)

Theorem 9.10. *If the series $\sum_{n=0}^{\infty} c_n(x - a)^n$ converges for $x = x_1$ where $x_1 \neq a$, then the series converges absolutely for all x such that $|x - a| < |x_1 - a|$. Furthermore, there is a number M such that*

$$|c_n(x - a)^n| \leqslant M\left(\frac{|x - a|}{|x_1 - a|}\right)^n \quad \text{for} \quad |x - a| \leqslant |x_1 - a| \quad \text{and for all } n.$$

$$(9.3)$$

PROOF. By Lemma 9.1 there is a number M such that

$$|c_n(x_1 - a)^n| \leqslant M \quad \text{for all } n.$$

Then (9.3) follows since

$$|c_n(x - a)^n| = |c_n(x_1 - a)^n| \cdot \left|\frac{(x - a)^n}{(x_1 - a)^n}\right| \leqslant M \frac{|x - a|^n}{|x_1 - a|^n}.$$

We deduce the convergence of the series at $|x - a|$ by comparison with the geometric series, the terms of which are the right side of the inequality in (9.3). □

Theorem 9.11. *Let $\sum_{n=0}^{\infty} c_n(x - a)^n$ be a given power series. Then either*

(i) *the series converges only for $x = a$; or*
(ii) *the series converges for all values of x; or*
(iii) *there is a number R such that the series converges for $|x - a| < R$ and diverges for $|x - a| > R$.*

PROOF. There are simple examples of series which show that (i) and (ii) may happen. To prove (iii), suppose there is a number $x_1 \neq a$ for which the series converges and a number $x_2 \neq a$ for which it diverges. By Theorem 9.10 we must have $|x_1 - a| \leqslant |x_2 - a|$, for if $|x_2 - a| < |x_1 - a|$ the series would converge for $x = x_2$. Define the set S of real numbers

$$S = \{\rho: \text{the series converges for } |x - a| < \rho\},$$

and denote $R = \sup S$. Now suppose $|x' - a| < R$. Then there is a $\rho \in S$ such that $|x' - a| < \rho < R$. By Theorem 9.10, the series converges for $x = x'$. Hence the series converges for all x such that $|x - a| < R$. Now suppose that $|x'' - a| = \rho' > R$. If the series converges for x'' then $\rho' \in S$ and we contradict the fact that $R = \sup S$. Therefore the series diverges for $|x - a| > R$, completing the proof. □

EXAMPLE 3. Find the values of x for which the series

$$\sum_{n=1}^{\infty} \frac{(-1)^n(x - 1)^n}{2^n n^2}$$

converges.

Solution. We apply the Ratio test:

$$\left|\frac{u_{n+1}}{u_n}\right| = \frac{1}{2}|x - 1|\frac{n^2}{(n + 1)^2}$$

and

$$\lim_{n\to\infty}\left|\frac{u_{n+1}}{u_n}\right| = \frac{1}{2}|x - 1|.$$

Therefore the series converges for $\frac{1}{2}|x - 1| < 1$ or for $-1 < x < 3$. By noting that for $x = -1$ and $x = 3$, the series is a *p*-series[1] with $p = 2$, we conclude that the series converges for all x in the interval $-1 \leqslant x \leqslant 3$ and it diverges for all other values of x. □

PROBLEMS

In each of Problems 1 through 16, test the series for convergence or divergence. If the series is convergent, determine whether it is absolutely or conditionally convergent.

1. $\displaystyle\sum_{n=1}^{\infty}\frac{(-1)^{n-1}(10)^n}{n!}$

2. $\displaystyle\sum_{n=1}^{\infty}\frac{(-1)^{n-1}n!}{(10)^n}$

3. $\displaystyle\sum_{n=1}^{\infty}\frac{(-1)^{n-1}(n - 1)}{n^2 + 1}$

4. $\displaystyle\sum_{n=1}^{\infty}\frac{(-1)^{n-1}}{n(n - 1/2)}$

5. $\displaystyle\sum_{n=1}^{\infty}\frac{(-1)^n(4/3)^n}{n^4}$

6. $\displaystyle\sum_{n=1}^{\infty}\frac{(-1)^{n-1}n!}{1\cdot 3\cdot 5\cdots(2n - 1)}$

7. $\displaystyle\sum_{n=1}^{\infty}\frac{(-1)^n(n + 1)}{\sqrt[n]{n}}$

8. $\displaystyle\sum_{n=1}^{\infty}\frac{(-1)^n\cdot 2\cdot 4\cdot 6\cdots 2n}{1\cdot 4\cdot 7\cdots(3n - 2)}$

9. $\displaystyle\sum_{n=1}^{\infty}\frac{(-1)^n(2n^2 - 3n + 2)}{n^3}$

10. $\displaystyle\sum_{n=1}^{\infty}\frac{(-1)^{n+1}}{(n + 1)\log(n + 1)}$

11. $\displaystyle\sum_{n=1}^{\infty}\frac{(-1)^{n+1}\log(n + 1)}{n + 1}$

12. $\displaystyle\sum_{n=1}^{\infty}\frac{(-1)^{n-1}\log n}{n^2}$

13. $\displaystyle\sum_{n=1}^{\infty}\frac{(-2)^n}{n^3}$

14. $\displaystyle\sum_{n=1}^{\infty}\frac{n!}{e^n}$

15. $\displaystyle\sum_{n=1}^{\infty}(-1)^n 3^{-n}$

16. $\displaystyle\sum_{n=1}^{\infty}\left(\frac{n}{2n + 1}\right)^n$

In each of Problems 17 through 24, find all the values of x for which the given power series converges.

17. $\displaystyle\sum_{n=0}^{\infty}(n + 1)x^n$

18. $\displaystyle\sum_{n=1}^{\infty}\frac{(x - 2)^n}{\sqrt[n]{n}}$

---

[1] Actually, for $x = 3$ the absolute values of the terms form a *p*-series.

19. $\displaystyle\sum_{n=0}^{\infty} \frac{(3/2)^n x^n}{n+1}$

20. $\displaystyle\sum_{n=1}^{\infty} \frac{n(x+2)^n}{2^n}$

21. $\displaystyle\sum_{n=1}^{\infty} \frac{n!(x-3)^n}{1\cdot 3\cdots(2n-1)}$

22. $\displaystyle\sum_{n=0}^{\infty} \frac{(2n^2+2n+1)x^n}{2^n(n+1)^3}$

23. $\displaystyle\sum_{n=1}^{\infty} \frac{\log(n+1)2^n(x+1)^n}{n+1}$

24. $\displaystyle\sum_{n=1}^{\infty} \frac{(-1)^{n-1}(\log n)2^n x^n}{3^n n^2}$

25. If $\sum_{n=1}^{\infty} u_n$ is a convergent series of positive terms, show that $\sum_{n=1}^{\infty} u_n^p$ is convergent for every $p > 1$.

26. Find the interval of convergence of the binomial series

$$1 + \sum_{n=1}^{\infty} \frac{m(m-1)\ldots(m-n+1)}{n!} x^n \qquad m \text{ a constant.}$$

*27. Let $\sum_{n=1}^{\infty} u_n$ be a conditionally convergent series with terms satisfying the conditions of Theorem 9.7. Let A be any real number. Show that by rearranging the terms of the series, the sum will be a number in the interval $(A - 1, A + 1)$.

28. Prove Lemma 9.1. [*Hint:* Use Theorem 9.1 and the fact that a finite number of terms has (at least) one which is largest in absolute value.]

29. Give examples of Cases (i) and (ii) in Theorem 9.11.

9.3. Uniform Convergence of Sequences

Let $\{f_n\}$ be a sequence of functions with each function having a domain containing an interval I of $\mathbb{R}^1$ and with range in $\mathbb{R}^1$. The convergence of such a sequence may be examined at each value x in I. The concept of uniform convergence, one which determines the nature of the convergence of the sequence for all x in I, has many applications in analysis. Of special interest is Theorem 9.13 which states that if all the f_n are continuous then the limit function must be also.

Definition. We say the sequence $\{f_n\}$ **converges uniformly on the interval** I to the function f if and only if for each $\varepsilon > 0$ there is a number N *independent of* x such that

$$|f_n(x) - f(x)| < \varepsilon \quad \text{for all} \quad x \in I \quad \text{and all} \quad n > N. \qquad (9.4)$$

Uniform convergence differs from ordinary pointwise convergence in that the integer N *does not depend on* x, although naturally it depends on ε.

The geometric meaning of uniform convergence is illustrated in Figure 9.2. Condition (9.4) states that if ε is any positive number, then for $n > N$ the graph of $y = f_n(x)$ lies entirely below the graph of $f(x) + \varepsilon$ and entirely above the graph of $f(x) - \varepsilon$.

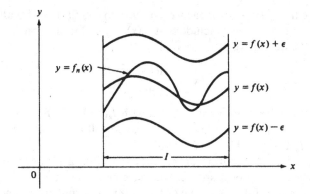

Figure 9.2. Illustrating uniform convergence.

It can happen that a sequence $\{f_n(x)\}$ converges to $f(x)$ for each x on an interval I but that the convergence is not uniform. For example, consider the functions

$$f_n: x \to \frac{2nx}{1 + n^2x^2}, \qquad I = \{x: 0 \leqslant x \leqslant 1\}.$$

The graphs of f_n for $n = 1, 2, 3, 4$ are shown in Figure 9.3. If $x \neq 0$, we write

$$f_n(x) = \frac{2x/n}{x^2 + (1/n^2)}$$

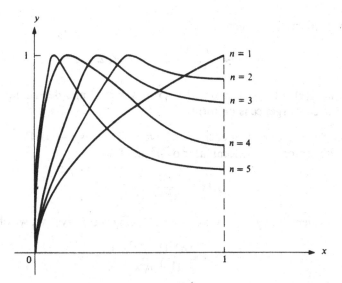

Figure 9.3. Illustrating nonuniform convergence to $f(x) \equiv 0$.

and observe that $f_n(x) \to 0$ for each $x > 0$. Moreover, $f_n(0) = 0$ for all n. Hence, setting $f(x) \equiv 0$ on I, we conclude that $f_n(x) \to f(x)$ for all x on I.

Taking the derivative of f_n, we find

$$f_n'(x) = \frac{2n(1 - n^2x^2)}{(1 + n^2x^2)^2}.$$

Therefore f_n has a maximum at $x = 1/n$ with $f_n(1/n) = 1$. Thus if $\varepsilon < 1$, there is no number N such that $|f_n(x) - f(x)| < \varepsilon$ for all $n > N$ and *all* x on I; in particular $|f_n(1/n) - f(1/n)| = 1$ for all n.

The definition of uniform convergence is seldom a practical method for deciding whether or not a specific sequence converges uniformly. The next theorem gives a useful and simple criterion for uniform convergence.

Theorem 9.12. *Suppose that f_n, $n = 1, 2, 3 \ldots$, and f are continuous on $I = \{x: a \leqslant x \leqslant b\}$. Then the sequence f_n converges uniformly to f on I if and only if the maximum value ε_n of $|f_n(x) - f(x)|$ converges to zero as $n \to \infty$.*

PROOF

(a) First, suppose the convergence is uniform. Let $\varepsilon > 0$ be given. Then there is an N such that $|f_n(x) - f(x)| < \varepsilon$ for all $n > N$ and all x on I. Since $|f_n(x) - f(x)|$ is continuous on I, it takes on its maximum value at some point x_n. Then $\varepsilon_n = |f_n(x_n) - f(x_n)|$. Hence $\varepsilon_n < \varepsilon$ for all $n > N$. Since there is an N for each $\varepsilon > 0$, it follows that $\varepsilon_n \to 0$ as $n \to \infty$.

(b) Suppose that $\varepsilon_n \to 0$ as $n \to \infty$. Then for each $\varepsilon > 0$ there is an N such that $\varepsilon_n < \varepsilon$ for all $n > N$. But, then, since ε_n is the maximum of $|f_n(x) - f(x)|$, we have $|f_n(x) - f(x)| \leqslant \varepsilon_n < \varepsilon$ for all $n > N$ and all x on I.   □

EXAMPLE. Given the sequence

$$f_n: x \to \frac{n^{3/2}x}{1 + n^2x^2}, \qquad n = 1, 2, \ldots,$$

show that $f_n(x) \to 0$ for each x on $I = \{x: 0 \leqslant x \leqslant 1\}$ and determine whether or not the convergence is uniform.

Solution. Since $f_n(0) = 0$ for each n, we have $f_n(0) \to 0$ as $n \to \infty$. For $x > 0$, we divide numerator and denominator by n^2 and find

$$f_n(x) = \frac{x/\sqrt{n}}{x^2 + (1/n^2)},$$

and it is evident that $f_n(x) \to 0$ as $n \to \infty$. Taking the derivative, we obtain

$$f_n'(x) = \frac{n^{3/2}(1 - n^2x^2)}{(1 + n^2x^2)^2}.$$

It is clear that $f_n'(x) = 0$ for $x = 1/n$; also $f_n(1/n) = \sqrt{n}/2$. Therefore $\varepsilon_n =$

$f_n(1/n)$ does *not* converge to 0 as $n \to \infty$ and hence the convergence of f_n is not uniform. $\qquad \square$

Although Theorem 9.12 is useful it cannot be applied unless the limit function f is known.

The importance of uniform convergence with regard to continuous functions is illustrated in the next theorem.

Theorem 9.13. *Suppose that* f_n, $n = 1, 2, \ldots$, *is a sequence of continuous functions on an interval I and that $\{f_n\}$ converges uniformly to f on I. Then f is continuous on I.*

PROOF. Suppose $\varepsilon > 0$ is given. Then there is an N such that

$$|f_n(x) - f(x)| < \frac{\varepsilon}{3} \quad \text{for all} \quad x \quad \text{on} \quad I \quad \text{and all} \quad n > N. \tag{9.5}$$

Let x_0 be any point in I. Since f_{N+1} is continuous on I, there is a $\delta > 0$ such that

$$|f_{N+1}(x) - f_{N+1}(x_0)| < \frac{\varepsilon}{3} \quad \text{for all} \quad x \in I \quad \text{such that} \quad |x - x_0| < \delta. \tag{9.6}$$

Also, by means of (9.5) and (9.6) we find

$$|f(x) - f(x_0)| \leq |f(x) - f_{N+1}(x)| + |f_{N+1}(x) - f_{N+1}(x_0)|$$

$$+ |f_{N+1}(x_0) - f(x_0)| < \frac{\varepsilon}{3} + \frac{\varepsilon}{3} + \frac{\varepsilon}{3} = \varepsilon,$$

which is valid for all x on I such that $|x - x_0| < \delta$. Thus f is continuous at x_0, an arbitrary point of I. $\qquad \square$

The next result shows that a uniformly convergent sequence of continuous functions may be integrated term-by-term.

Theorem 9.14 (Integration of uniformly convergent sequences). *Suppose that each f_n, $n = 1, 2, \ldots$, is continuous on the bounded interval I and that $\{f_n\}$ converges uniformly to f on I. Let $c \in I$ and define*

$$F_n(x) = \int_c^x f_n(t)\, dt.$$

Then f is continuous on I and F_n converges uniformly to the function

$$F(x) = \int_c^x f(t)\, dt.$$

PROOF. That f is continuous on I follows from Theorem 9.13. Let L be the length of I. For any $\varepsilon > 0$ it follows from the uniform convergence of $\{f_n\}$ that there is an N such that

$$|f_n(t) - f(t)| < \frac{\varepsilon}{L} \quad \text{for all} \quad n > N \quad \text{and all} \quad t \quad \text{on} \quad I.$$

We conclude that

$$|F_n(x) - F(x)| = \left| \int_c^x [f_n(t) - f(t)] \, dt \right|$$

$$\leqslant \left| \int_c^x |f_n(t) - f(t)| \, dt \right|$$

$$\leqslant \frac{\varepsilon}{L} |x - c| \leqslant \varepsilon$$

for all $n > N$ and all $x \in I$. Hence $\{F_n\}$ converges uniformly on I. $\qquad\square$

The next theorem illustrates when we can draw conclusions about the term-by-term differentiation of convergent sequences.

Theorem 9.15. *Suppose that $\{f_n\}$ is a sequence of functions each having one continuous derivative on an open interval I. Suppose that $f_n(x)$ converges to $f(x)$ for each x on I and that the sequence f_n' converges uniformly to g on I. Then g is continuous on I and $f'(x) = g(x)$ for all x on I.*

PROOF. That g is continuous on I follows from Theorem 9.13. Let c be any point on I. For each n and each x on I, we have

$$\int_c^x f_n'(t) \, dt = f_n(x) - f_n(c).$$

Since $\{f_n'\}$ converges uniformly to g and $f_n(x)$ converges to $f(x)$ for each x on I, we may apply Theorem 9.14 to get

$$\int_c^x g(t) \, dt = f(x) - f(c). \tag{9.7}$$

The result follows by differentiating (9.7). $\qquad\square$

Many of the results on uniform convergence of sequences of functions from $\mathbb{R}^1$ to $\mathbb{R}^1$ generalize directly to functions defined on a set A in a metric space S and with range in $\mathbb{R}^1$.

Definition. Let A be a set in a metric space S and suppose that $f_n : S \to \mathbb{R}^1$, $n = 1, 2, \ldots$, is a sequence of functions. The sequence $\{f_n\}$ **converges uniformly** on A to a function $f : A \to \mathbb{R}^1$ if and only if for every $\varepsilon > 0$ there is a number N such that

$$|f_n(x) - f(x)| < \varepsilon \quad \text{for all} \quad x \in A \quad \text{and all} \quad n > N.$$

The next theorem is an extension of Theorem 9.13.

Theorem 9.16. *Suppose that each f_n is continuous on a set A in a metric space S where $f_n: S \to \mathbb{R}^1$. If $\{f_n\}$ converges uniformly to f on A, then f is continuous on A.*

The proof is similar to the proof of Theorem 9.13 and is left to the reader.

Theorem 9.14 has a generalization to functions defined in N-dimensional Euclidean space.

Theorem 9.17. *Let F be a closed figure in $\mathbb{R}^N$ and suppose that $f_n: F \to \mathbb{R}^1$, $n = 1, 2, \ldots$, is a sequence of continuous functions which converges uniformly to f on the figure F. Then f is continuous on F and*

$$\int_F f \, dV_N = \lim_{n \to \infty} \int_F f_n \, dV_N.$$

The proof is similar to the proof of Theorem 9.14 and is left to the reader.

Let $\{f_n\}$ be a sequence defined on a bounded open set G in $\mathbb{R}^N$ with range in $\mathbb{R}^1$. Writing $f_n(x_1, x_2, \ldots, x_N)$ for the value of f_n at $(x_1, x_2, \ldots, x_N)$, we recall that

$$f_{n,k} \quad \text{and} \quad \frac{\partial}{\partial x_k} f_n(x_1, x_2, \ldots, x_N)$$

are symbols for the partial derivative of f_n with respect to x_k.

Theorem 9.18. *Let k be an integer such that $1 \leqslant k \leqslant N$. Suppose that for each n, the functions $f_n: G \to \mathbb{R}^1$ and $f_{n,k}$ are continuous on G, a bounded open set in $\mathbb{R}^N$. Suppose that $\{f_n(x)\}$ converges to $f(x)$ for each $x \in G$ and that $\{f_{n,k}\}$ converges uniformly to a function g on G. Then g is continuous on G and*

$$f_{,k}(x) = g(x) \quad \text{for all} \quad x \in G.$$

The proof is almost identical to the proof of Theorem 9.15 and is omitted.

If a sequence of continuous functions $\{f_n\}$ converges at every point to a continuous function f, it is not necessarily true that

$$\int f_n \, dV \to \int f \, dV \quad \text{as} \quad n \to \infty.$$

Simple convergence at every point is not sufficient as the following example shows. We form the sequence (for $n = 2, 3, 4, \ldots$)

$$f_n: x \to \begin{cases} n^2 x, & 0 \leqslant x \leqslant \dfrac{1}{n}, \\[2mm] -n^2 x^2 + 2n, & \dfrac{1}{n} \leqslant x \leqslant \dfrac{2}{n}, \\[2mm] 0, & \dfrac{2}{n} \leqslant x \leqslant 1. \end{cases}$$

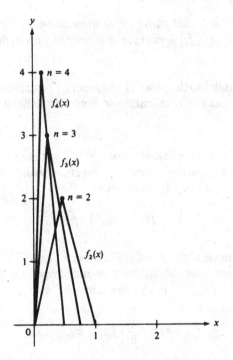

Figure 9.4. f_n converges to f but $\int_0^1 f_n$ does not converge to $\int_0^1 f$.

It is easy to see (Figure 9.4) that $f_n(x) \to f(x) \equiv 0$ for each $x \in I = \{x: 0 \leqslant x \leqslant 1\}$, but that

$$1 = \int_0^1 f_n(x)\, dx \nrightarrow \int_0^1 f(x)\, dx = 0.$$

Theorems 9.16, 9.17, and 9.18 may be generalized to sequences of functions with domain in a metric space (S_1, d_1) and range in another metric space (S_2, d_2). See Problem 19 at the end of this section.

PROBLEMS

In each of Problems 1 through 10 show that the sequence $\{f_n(x)\}$ converges to $f(x)$ for each x on I and determine whether or not the convergence is uniform.

1. $f_n: x \to \dfrac{2x}{1 + nx}, \quad f(x) \equiv 0, \quad I = \{x: 0 \leqslant x \leqslant 1\}.$

2. $f_n: x \to \dfrac{\cos nx}{\sqrt{n}}, \quad f(x) \equiv 0, \quad I = \{x: 0 \leqslant x \leqslant 1\}.$

3. $f_n: x \to \dfrac{n^3 x}{1 + n^4 x}, \quad f(x) \equiv 0, \quad I = \{x: 0 \leqslant x \leqslant 1\}.$

4. $f_n: x \to \dfrac{n^3 x}{1 + n^4 x^2}$, $f(x) \equiv 0$, $I = \{x: a \leqslant x < \infty, a > 0\}$.

5. $f_n: x \to \dfrac{nx^2}{1 + nx}$, $f(x) = x$, $I = \{x: 0 \leqslant x \leqslant 1\}$.

6. $f_n: x \to \dfrac{1}{\sqrt{x}} + \dfrac{1}{n}\cos(1/nx)$, $f(x) = 1/\sqrt{x}$, $I = \{x: 0 < x \leqslant 2\}$.

7. $f_n: x \to \dfrac{\sin nx}{2nx}$, $f(x) \equiv 0$, $I = \{x: 0 < x < \infty\}$.

8. $f_n: x \to x^n(1 - x)\sqrt{n}$, $f(x) \equiv 0$, $I = \{x: 0 \leqslant x \leqslant 1\}$.

9. $f_n: x \to \dfrac{1 - x^n}{1 - x}$, $f(x) = \dfrac{1}{1 - x}$, $I = \left\{x: -\dfrac{1}{2} \leqslant x \leqslant \dfrac{1}{2}\right\}$.

10. $f_n: x \to nxe^{-nx^2}$, $f(x) \equiv 0$, $I = \{x: 0 \leqslant x \leqslant 1\}$.

11. Show that the sequence $f_n: x \to x^n$ converges for each $x \in I = \{x: 0 \leqslant x \leqslant 1\}$ but that the convergence is not uniform.

12. Given that $f_n(x) = (n + 2)(n + 1)x^n(1 - x)$ and that $f(x) \equiv 0$ for x on $I = \{x: 0 \leqslant x \leqslant 1\}$. Show that $f_n(x) \to f(x)$ as $n \to \infty$ for each $x \in I$. Determine whether or not $\int_0^1 f_n(x)\, dx \to \int_0^1 f(x)\, dx$ as $n \to \infty$.

13. Prove Theorem 9.16.

14. Prove Theorem 9.17.

15. Prove Theorem 9.18.

16. Give an example of a sequence of functions $\{f_n\}$ defined on the set $A = \{(x, y): 0 \leqslant x \leqslant 1, 0 \leqslant y \leqslant 1\}$ such that f_n converges to a function f at each point $(x, y) \in A$ but $\int_A f_n(x, y)\, dV \nrightarrow \int_A f(x, y)\, dV$.

17. Suppose that $\{f_n\}$ converges uniformly to f and $\{g_n\}$ converges uniformly to g on a set A in a metric space S. Show that $\{f_n + g_n\}$ converges uniformly to $f + g$.

18. (a) Suppose $\{f_n\}$ and $\{g_n\}$ are bounded sequences each of which converges uniformly on a set A in a metric space S to functions f and g, respectively. Show that the sequence $\{f_n g_n\}$ converges uniformly to fg on A.
    (b) Give an example of sequences $\{f_n\}$ and $\{g_n\}$ which converge uniformly but are such that $\{f_n g_n\}$ does not converge uniformly.

19. Formulate a definition of uniform convergence of a sequence $\{f_n\}$ of mappings from a set A in a metric space (S_1, d_1) into a metric space (S_2, d_2). Prove that if $\{f_n\}$ are continuous and converge uniformly to a mapping f, then f is continuous.

20. Prove the following generalization of Theorem 9.17. Suppose that F is a figure in $\mathbb{R}^N$ and that $f_n: F \to \mathbb{R}^1$ is integrable on F for $n = 1, 2, \ldots$ If $\{f_n\}$ converges uniformly to f on F, then f is integrable over F and

$$\int_F f\, dV_N = \lim_{n \to \infty} \int_F f_n\, dV_N.$$

[*Hint*: First prove that f is integrable.]

9.4. Uniform Convergence of Series; Power Series

Let $u_k(x)$, $k = 1, 2, \ldots$, be functions defined on a set A in a metric space S with range in $\mathbb{R}^1$.

Definition. The infinite series $\sum_{k=1}^{\infty} u_k(x)$ has the partial sums $s_n(x) = \sum_{k=1}^{n} u_k(x)$. The series is said to **converge uniformly on a set A to a function s** if and only if the sequence of partial sums $\{s_n\}$ converges uniformly to s on the set A.

The above definition shows that theorems on uniform convergence of infinite series may be reduced to corresponding results for uniform convergence of sequences.

Theorem 9.19 (Analog of Theorem 9.13). *Suppose that u_n, $n = 1, 2, \ldots$, are continuous on a set A in a metric space S and that $\sum_{n=1}^{\infty} u_n(x)$ converges uniformly on A to a function $s(x)$. Then s is continuous on A.*

The next theorem is the analog for series of a generalization of Theorem 9.14. In this connection see Problem 20 of Section 9.3.

Theorem 9.20 (Term-by-term integration of infinite series)

(a) *Let $u_n(x)$, $n = 1, 2, \ldots$, be functions whose domain is a bounded interval I in $\mathbb{R}^1$ with range in $\mathbb{R}^1$. Suppose that each u_n is integrable on I and that $\sum_{n=1}^{\infty} u_n(x)$ converges uniformly on I to $s(x)$. Then s is integrable on I. If c is in I and U_n, S are defined by*

$$U_n(x) = \int_c^x u_n(t)\, dt, \qquad S(x) = \int_c^x s(t)\, dt,$$

then $\sum_{n=1}^{\infty} U_n(x)$ converges uniformly to $S(x)$ on I.

(b) *(See Theorem 9.17.) Let $u_n(x)$, $n = 1, 2, \ldots$, be defined on a figure F in $\mathbb{R}^N$ with range in $\mathbb{R}^1$. Suppose that each u_n is integrable on F and that $\sum_{n=1}^{\infty} u_n(x)$ converges uniformly on F to $s(x)$. Then s is integrable on F and*

$$\sum_{n=1}^{\infty} \int_F u_n\, dV_N = \int_F s\, dV_N.$$

The next result is the analog for series of Theorem 9.18.

Theorem 9.21 (Term-by-term differentiation of infinite series). *Let k be an integer with $1 \leqslant k \leqslant N$. Suppose that u_n and $u_{n,k}$, $n = 1, 2, \ldots$, are continuous functions defined on an open set G in $\mathbb{R}^N$ (range in $\mathbb{R}^1$). Suppose that the series $\sum_{n=1}^{\infty} u_n(x)$ converges for each $x \in G$ to $s(x)$ and that the series $\sum_{n=1}^{\infty} u_{n,k}(x)$ converges uniformly on G to $t(x)$. Then*

$$s_{,k}(x) = t(x) \quad \text{for all } x \text{ in } G.$$

The proofs of Theorems 9.19, 9.20, and 9.21 are similar to the proofs of the corresponding theorems for infinite sequences.

The following theorem gives a useful indirect test for uniform convergence. It is important to observe that the test can be applied without any knowledge of the sum of the series.

Theorem 9.22 (Weierstrass M-test). *Let $u_n(x)$, $n = 1, 2, \ldots$, be defined on a set A in a metric space S with range in $\mathbb{R}^1$. Suppose that $|u_n(x)| \leqslant M_n$ for all n and for all $x \in A$. Suppose that the series of constants $\sum_{n=1}^{\infty} M_n$ converges. Then $\sum_{n=1}^{\infty} u_n(x)$ and $\sum_{n=1}^{\infty} |u_n(x)|$ converge uniformly on A.*

PROOF. By the Comparison test (Theorem 9.4) we know that $\sum_{n=1}^{\infty} |u_n(x)|$ converges for each x. Set

$$t_n(x) = \sum_{k=1}^{n} |u_k(x)| \quad \text{and} \quad t(x) = \sum_{k=1}^{\infty} |u_k(x)|.$$

From Theorem 9.6 it follows that $\sum_{n=1}^{\infty} u_n(x)$ converges. Set

$$s_n(x) = \sum_{k=1}^{n} u_k(x), \qquad s(x) = \sum_{k=1}^{\infty} u_k(x).$$

Then

$$|s_n(x) - s(x)| = \left| \sum_{k=n+1}^{\infty} u_k(x) \right|$$

$$\leqslant \sum_{k=n+1}^{\infty} |u_k(x)| = |t(x) - t_n(x)|$$

$$\leqslant \sum_{k=n+1}^{\infty} M_k.$$

We define $S = \sum_{k=1}^{\infty} M_k$, $S_n = \sum_{k=1}^{n} M_k$. Then $\sum_{k=n+1}^{\infty} M_k = S - S_n$; since $S - S_n \to 0$ as $n \to \infty$ independently of x, we conclude that the convergence of $\{s_n\}$ and $\{t_n\}$ are uniform. $\qquad\square$

The next theorem on the uniform convergence of power series is a direct consequence of the Weierstrass M-test.

Theorem 9.23. *Suppose that the series $\sum_{n=0}^{\infty} c_n(x - a)^n$ converges for $x = x_1$ with $x_1 \neq a$. Then the series converges uniformly on $I = \{x: a - h \leqslant x \leqslant a + h\}$ for each $h < |x_1 - a|$. Also, there is a number M such that*

$$|c_n(x - a)^n| \leqslant M \cdot \left(\frac{h}{|x_1 - a|} \right)^n \tag{9.8}$$

for

$$|x - a| \leqslant h < |x_1 - a|.$$

PROOF. Inequality (9.8) is a direct consequence of the inequality stated in Theorem 9.10. The series $\sum_{n=0}^{\infty} M(h/|x_1 - a|)^n$ is a geometric series of con-

stants which converges. Therefore the uniform convergence of $\sum_{n=0}^{\infty} c_n(x - a)^n$
follows from the Weierstrass M-test. $\square$

Remarks

(i) The number h in Theorem 9.23 must be strictly less than $|x_1 - a|$ in
order that the series converge uniformly on I. To see this consider the series

$$\sum_{n=1}^{\infty} \frac{(-1)^n x^n}{n}$$

which converges for $x = 1$ but diverges for $x = -1$. If this series were to
converge uniformly for $|x| < 1$, then it would converge uniformly for $|x| \leqslant 1$.
(The reader may establish this fact.)

(ii) If the series in Theorem 9.23 converges absolutely for $x = x_1$, then
we may choose $h = |x_1 - a|$ and the series converges uniformly on $I = \{x : |x - a| \leqslant |x_1 - a|\}$. To see this we observe that for $|x - a| \leqslant |x_1 - a|$, we
have

$$|c_n(x - a)^n| \leqslant |c_n(x_1 - a)^n|.$$

EXAMPLE 1. Given the series

$$\sum_{n=0}^{\infty} (n + 1)x^n, \tag{9.9}$$

find all values of h such that the series converges uniformly on $I = \{x : |x| \leqslant h\}$.

Solution. For $|x| \leqslant h$, we have $|(n + 1)x^n| \leqslant (n + 1)h^n$. By the Ratio test, the
series $\sum_{n=0}^{\infty} (n + 1)h^n$ converges when $h < 1$. Therefore the series (9.9) con-
verges uniformly on $I = \{x : |x| \leqslant h\}$ if $h < 1$. The series (9.9) does not con-
verge for $x = \pm 1$, and hence there is uniform convergence if and only if $h < 1$.
                                                                                     $\square$

EXAMPLE 2. Given series

$$\sum_{n=1}^{\infty} \frac{x^n}{n^2}, \tag{9.10}$$

find all values of h such that the series converges uniformly on $I = \{x : |x| \leqslant h\}$.

Solution. For $|x| \leqslant h$, we have $|x^n/n^2| \leqslant h^n/n^2$. The p-series $\sum_{n=1}^{\infty} 1/n^2$ con-
verges and, by the Comparison test, the series $\sum_{n=1}^{\infty} h^n/n^2$ converges if $h \leqslant 1$.
By the Ratio test, the series (9.10) diverges if $x > 1$. We conclude that the series
(9.10) converges uniformly on $I = \{x : |x| \leqslant 1\}$. $\square$

Lemma 9.2. *Suppose that the series*

$$f(x) = \sum_{n=0}^{\infty} c_n(x - a)^n \tag{9.11}$$

converges for $|x - a| < R$ *with* $R > 0$. *Then* f *and* f' *are continuous on*

$|x - a| < R$ and

$$f'(x) = \sum_{n=1}^{\infty} nc_n(x - a)^{n-1} \quad for \quad |x - a| < R. \tag{9.12}$$

PROOF. Choose x_1 such that $|x_1 - a| < R$ and then choose h so that $0 < h < |x_1 - a|$. By Theorem 9.23, the series (9.11) converges uniformly on $I = \{x: |x - a| \leqslant h\}$, and there is a number M such that

$$|c_n(x - a)^n| \leqslant M\left(\frac{h}{|x_1 - a|}\right)^n \quad for \quad |x - a| \leqslant h.$$

According to Theorem 9.19, the function f is continuous on I. Also, we find

$$|nc_n(x - a)^{n-1}| \leqslant n|c_n|h^{n-1} \leqslant \frac{nM}{h}\left(\frac{h}{|x_1 - a|}\right)^n \equiv u_n.$$

From the Ratio test, the series $\sum_{n=1}^{\infty} u_n$ converges, so that the series (9.12) converges uniformly on I. Hence by Theorem 9.19, the function f' is continuous on I. Since h may be chosen as any positive number less than R, we conclude that f and f' are continuous for $|x - a| < R$.   □

With the aid of the above lemma, we obtain the following theorem on term-by-term differentiation and integration of power series.

Theorem 9.24. *Let f be given by*

$$f(x) = \sum_{n=0}^{\infty} c_n(x - a)^n \quad for \quad |x - a| < R \quad with \quad R > 0. \tag{9.13}$$

(i) *Then f possesses derivatives of all orders. For each positive integer m, the derivative $f^{(m)}(x)$ is given for $|x - a| < R$ by the term-by-term differentiation of (9.13) m times.*

(ii) *If F is defined for $|x - a| < R$ by*

$$F(x) = \int_a^x f(t)\, dt,$$

then F is given by the series obtained by term-by-term integration of the series (9.13).

(iii) *The constants c_n are given by*

$$c_n = \frac{f^{(n)}(a)}{n!}. \tag{9.14}$$

PROOF. The proof of (i) is obtained by induction using Lemma 9.2. We obtain (ii) by application of Theorem 9.20. To establish (iii) differentiate the series (9.13) n times and set $x = a$ in the resulting expression for $f^{(n)}(x)$.   □

Combining expressions (9.13) and (9.14), we see that any function f defined by a power series with $R > 0$ has the form

$$f(x) = \sum_{n=0}^{\infty} \frac{f^{(n)}(a)}{n!}(x-a)^n. \qquad (9.15)$$

A function that has continuous derivatives of all orders in the neighborhood of some point is said to be **infinitely differentiable**. It may happen that a function f is infinitely differentiable at some point a but is not representable by a power series such as (9.15). An example of such an f is given by

$$f: x \to \begin{cases} e^{-1/x^2} & \text{if } x \neq 0, \\ 0 & \text{if } x = 0. \end{cases}$$

By differentiation, we find that for $x \neq 0$

$$f'(x) = 2x^{-3}e^{-1/x^2},$$
$$f''(x) = x^{-6}(4 - 6x^2)e^{-1/x^2},$$
$$\vdots$$
$$f^{(n)}(x) = x^{-3n}P_n(x)e^{-1/x^2},$$

where P_n is a polynomial. By l'Hôpital's rule, it is not hard to verify that $f^{(n)}(x) \to 0$ as $x \to 0$ for $n = 1, 2, \ldots$ Therefore by Theorem 4.15, we know that $f^{(n)}(0) = 0$ for every n, and so f has continuous derivatives of all orders in a neighborhood of 0. The series (9.15) for f is identically zero, but the function f is not, and therefore the power series with $a = 0$ does not represent the function.

Definitions. Let $f: I \to \mathbb{R}^1$ be infinitely differentiable at a point $a \in I$ and suppose that the series (9.15) has a positive radius of convergence. Then f is said to be **analytic at** a. A function f is **analytic** on a domain if and only if it is analytic at each point of its domain.

A function must have properties in addition to infinite differentiability in order to be representable by a power series and therefore analytic. The principal tool for establishing the validity of power series expansions is given by Taylor's theorem with remainder.

Theorem 9.25 (Taylor's theorem with remainder). *Suppose that f and its first n derivatives are continuous on an interval containing $I = \{x: a \leq x \leq b\}$. Suppose that $f^{(n+1)}(x)$ exists for each x between a and b. Then there is a ξ with $a < \xi < b$ such that*

$$f(b) = \sum_{j=0}^{n} \frac{f^{(j)}(a)}{j!}(b-a)^j + R_n,$$

where

$$R_n = \frac{f^{(n+1)}(\xi)(b-a)^{n+1}}{(n+1)!}. \tag{9.16}$$

PROOF. We define R_n by the equation

$$f(b) = \sum_{j=0}^{n} \frac{f^{(j)}(a)}{j!}(b-a)^j + R_n,$$

that is,

$$R_n \equiv f(b) - \sum_{j=0}^{n} \frac{f^{(j)}(a)}{j!}(b-a)^j.$$

We wish to find the form R_n takes. For this purpose define ϕ for x on I by the formula

$$\phi(x) = f(b) - \sum_{j=0}^{n} \frac{f^{(j)}(x)}{j!}(b-x)^j - R_n \frac{(b-x)^{n+1}}{(b-a)^{n+1}}.$$

Then ϕ is continuous on I and $\phi'(x)$ exists for each x between a and b. A simple calculation shows that $\phi(a) = \phi(b) = 0$. By Rolle's theorem there is a number ξ with $a < \xi < b$ such that $\phi'(\xi) = 0$. We compute

$$\phi'(x) = -f'(x) + \sum_{j=1}^{n} \frac{f^{(j)}(x)}{(j-1)!}(b-x)^{j-1} - \sum_{j=1}^{n} \frac{f^{(j+1)}(x)}{j!}(b-x)^j$$

$$+ (n+1)R_n \frac{(b-x)^n}{(b-a)^{n+1}}.$$

Replacing j by $j-1$ in the second sum above, we find

$$\phi'(x) = -f'(x) + f'(x) + \sum_{j=2}^{n} \frac{f^{(j)}(x)}{(j-1)!}(b-x)^{j-1} - \sum_{j=2}^{n} \frac{f^{(j)}(x)}{(j-1)!}(b-x)^{j-1}$$

$$- \frac{f^{(n+1)}(x)}{n!}(b-x)^n + (n+1)R_n \frac{(b-x)^n}{(b-a)^{n+1}}$$

$$= -\frac{(n+1)(b-x)^n}{(b-a)^{n+1}}\left[\frac{f^{(n+1)}(x)(b-a)^{n+1}}{(n+1)!} - R_n\right].$$

The formula for R_n given in (9.16) is obtained by setting $x = \xi$ in the above expression and using the fact that $\phi'(\xi) = 0$. ∎

Remark. In the proof of Theorem 9.25 we assumed that $a < b$. However, the argument is unchanged if $b < a$, and hence (9.16) holds for any a, b contained in the interval where the hypotheses of the theorem are valid.

Using Theorem 9.25 we can now establish the validity of the Taylor expansion for many of the functions studied in elementary calculus.

Theorem 9.26. *For any values of a and x, the expansion*

$$e^x = e^a \sum_{n=0}^{\infty} \frac{(x - a)^n}{n!} \qquad (9.17)$$

is valid.

PROOF. We apply Taylor's theorem to $f(x) = e^x$. We have $f^{(n)}(x) = e^x$ for $n = 1, 2, \ldots$, and, setting $b = x$ in (9.16), we find

$$e^x = e^a \sum_{j=0}^{n} \frac{(x - a)^j}{j!} + R_n(a, x), \qquad R_n = \frac{e^{\xi}(x - a)^{n+1}}{(n + 1)!},$$

where ξ is between a and x. If $x > a$, then $a < \xi < x$ and so $e^{\xi} < e^x$; if $x < a$, then $x < \xi < a$ and $e^{\xi} < e^a$. Hence

$$R_n(a, x) \leqslant C(a, x)\frac{|x - a|^{n+1}}{(n + 1)!} \quad \text{where} \quad C(a, x) = \begin{cases} e^x & \text{if } x \geqslant a, \\ e^a & \text{if } x \leqslant a. \end{cases} \qquad (9.18)$$

Note that $C(a, x)$ is *independent of n*. By the Ratio test, the series in (9.17) converges for all x. The form of R_n in (9.18) shows that $R_n \to 0$ as $n \to \infty$. Therefore the series (9.17) converges to e^x for each x and a. $\qquad\square$

Lemma 9.3. *If $f(x) = \sin x$, then $f^{(k)}(x) = \sin(x + k\pi/2)$. If $f(x) = \cos x$, then $f^{(k)}(x) = \cos(x + k\pi/2)$.*

These facts are easily proved by induction.

Theorem 9.27. *For all a and x the following expansions are valid:*

$$\sin x = \sum_{n=0}^{\infty} \frac{\sin(a + n\pi/2)}{n!}(x - a)^n,$$

$$\cos x = \sum_{n=0}^{\infty} \frac{\cos(a + n\pi/2)}{n!}(x - a)^n.$$

The proof is left to the reader.

Theorem 9.28. *The following expansions are valid for $|x| < 1$:*

$$\log(1 + x) = \sum_{n=1}^{\infty} \frac{(-1)^{n+1}x^n}{n},$$

$$\arctan x = \sum_{n=1}^{\infty} \frac{(-1)^{n-1}x^{2n-1}}{2n - 1}.$$

The proof is left to the reader.

Theorem 9.29 (Simple substitutions in power series)

(a) If $f(u) = \sum_{n=0}^{\infty} c_n(u - b)^n$ for $|u - b| < R$ with $R > 0$, and if $b = kc + d$ with $k \neq 0$, then

$$f(kx + d) = \sum_{n=0}^{\infty} c_n k^n (x - c)^n \quad \text{for} \quad |x - c| < \frac{R}{|k|}.$$

(b) If $f(u) = \sum_{n=0}^{\infty} c_n(u - b)^n$ for $|u - b| < R$, then for every fixed positive integer k,

$$f[(x - c)^k + b] = \sum_{n=0}^{\infty} c_n(x - c)^{kn} \quad \text{for} \quad |x - c| < R^{1/k}.$$

PROOF

(a) If $u = kx + d$ and $b = kc + d$, then $u - b = k(x - c)$ and $c_n(u - b)^n = c_n k^n (x - c)^n$. Also, $|u - b| < R$ if and only if $|x - c| < R/|k|$.

(b) To prove Part (b) observe that the appropriate substitution is $u - b = (x - c)^k$. $\qquad\square$

A useful special case of Part (b) in Theorem 9.29 occurs when $b = c = 0$. Then we find that

$$f(x^k) = \sum_{n=0}^{\infty} c_n x^{kn}.$$

For example, from (9.17) we see that

$$e^{x^2} = \sum_{n=0}^{\infty} \frac{x^{2n}}{n!},$$

valid for all x.

In many applications of Taylor's theorem with remainder, it is important to obtain specific bounds on the remainder term. The following theorem which gives the remainder R_n in an integral form is often useful in obtaining precise estimates.

Theorem 9.30 (Taylor's theorem with integral form of the remainder). *Suppose that f and its derivatives of order up to $n + 1$ are continuous on an interval I containing a. Then for each $x \in I$,*

$$f(x) = \sum_{j=0}^{n} \frac{f^{(j)}(a)(x - a)^j}{j!} + R_n(a, x), \qquad (9.19)$$

where

$$R_n(a, x) = \int_a^x \frac{(x - t)^n}{n!} f^{(n+1)}(t)\, dt. \qquad (9.20)$$

PROOF. For each $x \in I$, by integrating a derivative, we have

$$f(x) = f(a) + \int_a^x f'(t) \, dt.$$

We integrate by parts in the above integral by setting

$$u = f'(t), \qquad v = -(x - t),$$
$$du = f''(t) \, dt, \qquad dv = dt.$$

We obtain

$$f(x) = f(a) - [(x - t)f'(t)]_a^x + \int_a^x (x - t)f''(t) \, dt$$

$$= f(a) + f'(a)(x - a) + \int_a^x (x - t)f''(t) \, dt.$$

We repeat the integration by parts in the integral above by setting

$$u = f''(t), \qquad v = -\frac{(x - t)^2}{2},$$

$$du = f'''(t) \, dt, \qquad dv = (x - t) \, dt.$$

The result is

$$f(x) = f(a) + f'(a)\frac{(x - a)}{1!} + f''(a)\frac{(x - a)^2}{2!} + \int_a^x f'''(t)\frac{(x - t)^2}{2!} \, dt.$$

We repeat the process and apply mathematical induction in the general case to obtain Formulas (9.19) and (9.20). $\qquad \qquad \square$

The "binomial formula" usually stated without proof in elementary courses is a direct corollary to Taylor's theorem.

Theorem 9.31 (Binomial series theorem). *For each $m \in \mathbb{R}^1$, the following formula holds:*

$$(1 + x)^m = 1 + \sum_{n=1}^{\infty} \frac{m(m - 1)\ldots(m - n + 1)}{n!} x^n \quad \text{for} \quad |x| < 1. \quad (9.21)$$

PROOF. We apply Theorem 9.30 with $a = 0$ and $f(x) = (1 + x)^m$. The result is

$$(1 + x)^m = 1 + \sum_{n=1}^{k} \frac{m(m - 1)\ldots(m - n + 1)}{n!} x^n + R_k(0, x),$$

where

$$R_k(0, x) = \int_0^x \frac{(x - t)^k}{k!} m(m - 1)\ldots(m - k)(1 + t)^{m-k-1} \, dt.$$

We wish to show that $R_k \to 0$ as $k \to \infty$ if $|x| < 1$. For this purpose we define

$$C_m(x) = \begin{cases} (1 + x)^{m-1} & \text{if} \quad m \geq 1, x \geq 0, \text{ or } m \leq 1, x \leq 0, \\ 1 & \text{if} \quad m \leq 1, x \geq 0, \text{ or } m \geq 1, x \leq 0. \end{cases}$$

Then we notice at once that $(1 + t)^{m-1} \leq C_m(x)$ for all t between 0 and x. Therefore

$$|R_k(0, x)| \leq C_m(x) \frac{|m(m - 1)\ldots(m - k)|}{k!} \int_0^x \left|\frac{x - t}{1 + t}\right|^k dt.$$

We define

$$u_k(x) = C_m(x) \frac{|m(m - 1)\ldots(m - k)|}{k!} |x|^{k+1},$$

and set $t = xs$ in the above integral. Hence

$$|R_k(0, x)| \leq u_k(x) \int_0^1 \frac{(1 - s)^k}{(1 + xs)^k} ds.$$

This last integral is bounded by 1 for each x such that $-1 \leq x \leq 1$, and hence $|R_k(0, x)| \leq u_k(x)$. By the Ratio test it is not difficult to verify that the series $\sum_{n=0}^\infty u_k(x)$ converges for $|x| < 1$. Therefore $u_k(x) \to 0$ as $k \to \infty$, and so $R_k(0, x) \to 0$ for $|x| < 1$ and all m. $\qquad \square$

Remarks

(i) If m is a positive integer then clearly the series (9.21) is finite, consisting of exactly $m + 1$ nonzero terms.

(ii) It is important to use the integral form for R_k in Theorem 9.31. If the form of R_k given in Theorem 9.25 is used, we find that

$$R_k(0, x) = \frac{m(m - 1)\ldots(m - k)}{k!}(1 + \xi)^{m-k-1}x^k$$

with ξ between 0 and x. If x is between -1 and $-1/2$ the best that can be said is

$$|R_k(0, x)| \leq C_m(x) \frac{m(m - 1)\ldots(m - k)}{k!} \left(\frac{x}{1 - |x|}\right)^k.$$

The right side does not tend to 0 as $k \to \infty$.

Problems

In each of Problems 1 through 10, determine the values of h for which the given series converges uniformly on the interval I.

1. $\sum_{n=1}^\infty \frac{x^n}{n}, \quad I = \{x: |x| \leq h\}.$

2. $\displaystyle\sum_{n=1}^{\infty} \frac{x^n}{\sqrt[n]{n}}, \quad I = \{x: |x| \leqslant h\}.$

3. $\displaystyle\sum_{n=1}^{\infty} n(x-1)^n, \quad I = \{x: |x-1| \leqslant h\}.$

4. $\displaystyle\sum_{n=0}^{\infty} \frac{(5x)^n}{n!}, \quad I = \{x: |x| \leqslant h\}.$

5. $\displaystyle\sum_{n=0}^{\infty} \frac{(-1)^n 2^n x^n}{3^n(n+1)}, \quad I = \{x: |x| \leqslant h\}.$

6. $\displaystyle\sum_{n=0}^{\infty} \frac{(n!)^2 (x-1)^n}{(2n)!}, \quad I = \{x: |x-1| \leqslant h\}.$

7. $\displaystyle\sum_{n=1}^{\infty} \frac{x^n}{(n+1)\log(n+1)}, \quad I = \{x: |x| \leqslant h\}.$

8. $\displaystyle\sum_{n=1}^{\infty} \frac{(\log n)2^n x^n}{3^n n\sqrt{n}}, \quad I = \{x: |x| \leqslant h\}.$

9. $\displaystyle\sum_{n=1}^{\infty} x^n(1-x), \quad I = \{x: |x| \leqslant h\}.$

10. $\displaystyle\sum_{n=1}^{\infty} \frac{x^2}{(1+nx^2)\sqrt{n}}, \quad I = \{x: |x| \leqslant h\}.$

11. Given that $\sum_{n=1}^{\infty} |a_n|$ converges, show that $\sum_{n=1}^{\infty} a_n \cos nx$ converges uniformly for all x.

12. Given that $\sum_{n=1}^{\infty} n|b_n|$ converges. Let $f(x) = \sum_{n=1}^{\infty} b_n \sin nx$. Show that $f'(x) = \sum_{n=1}^{\infty} nb_n \cos nx$ and that both series converge uniformly for all x.

13. Show that if $\sum_{n=0}^{\infty} c_n(x-a)^n$ converges uniformly for $|x-a| < h$, then it converges uniformly for $|x-a| \leqslant h$.

14. Prove Theorem 9.19.

15. Prove Theorem 9.20.

16. Prove Theorem 9.21.

17. Let

$$f(x) = \begin{cases} e^{-1/x^2} & \text{if } x \neq 0, \\ 0 & \text{if } x = 0. \end{cases}$$

Verify by l'Hôpital's rule that $f^{(n)}(x) \to 0$ as $x \to 0$ for each $n = 1, 2, \ldots$.

18. Prove Lemma 9.3.

19. Prove Theorem 9.27.

20. Prove Theorem 9.28.

21. Use induction on n to write a complete proof of Theorem 9.30.

In each of Problems 22 through 29 find the Taylor expansion about $a = 0$ and prove its validity.

22. $f: x \to (1 - x^2)^{-1/2}$.

23. $f: x \to (1 + x^2)^{-1/2}$.

24. $f: x \to (1 - x)^{-2}$.

25. $f: x \to (1 - x)^{-3}$.

26. $f: x \to \arcsin x$.

27. $f: x \to (1 - x^2)^{-3}$.

28. $f: x \to \arctan(x^2)$.

29. $f: x \to \arcsin(x^3)$.

In each of Problems 30 through 35, compute $f(x)$ to five decimals of accuracy. Use Taylor's theorem to guarantee the result.

30. $e^{0.2}$.

31. $\sin(0.5)$.

32. $\log(0.9)$.

33. $(0.94)\sqrt{3}$.

34. $(15)^{1/4}$.

35. $(63)^{1/6}$.

In each of Problems 36 through 38 compute the integral to five decimals of accuracy.

36. $\displaystyle\int_0^{1/2} \exp(x^2)\, dx$.

37. $\displaystyle\int_0^1 \frac{1 - \cos x}{x}\, dx$.

38. $\displaystyle\int_0^{0.3} \frac{dx}{\sqrt{1 + x^3}}$.

9.5. Unordered Sums[2]

Consider the double sequence

$$U_{mn} = \cos^m\left(\frac{1}{n}\right), \qquad m, n = 1, 2, \ldots$$

There is no natural way to obtain the limit of such a sequence as *both* m and n tend to infinity. It is easy to verify that

$$\lim_{n\to\infty}\left[\lim_{m\to\infty} \cos^m\left(\frac{1}{n}\right)\right] = 0 \quad \text{and} \quad \lim_{m\to\infty}\left[\lim_{n\to\infty} \cos^m\left(\frac{1}{n}\right)\right] = 1.$$

It also may happen that other limiting values result if m and n tend to infinity simultaneously and in such a way that they bear a specific relationship to each other. The situation becomes even more complicated for triple sequences, quadruple sequences, and so forth.

Rather than discuss multiple sequences on a case-by-case basis, we shall develop a single comprehensive theory, one which unifies the treatment of convergence questions for all multiple sequences and series.

---

[2] Sections 9.5, 9.6, and 9.7 may be omitted without loss of continuity.

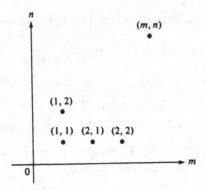

Figure 9.5. A set S in $\mathbb{R}^2$.

We consider functions defined from a set S into $\mathbb{R}^1$. Usually we shall think of S as a lattice in $\mathbb{R}^2$, $\mathbb{R}^3$ or $\mathbb{R}^N$ consisting of the points which have integer coordinates. For example, S might be all number pairs (m, n) where m and n are positive integers (Figure 9.5); or, S may be all number triples of the form (l, m, n) where l, m and n are nonnegative integers. We shall also consider sets S which consist of a finite number of elements, such as the set (Figure 9.6) $S = \{(m, n): 1 \leqslant m \leqslant 4, 2 \leqslant n \leqslant 5, m, n \text{ integers}\}$. However, S is not restricted to points on a lattice. The statements in this section remain valid for arbitrary sets S, countable or uncountable.

Definition. Let S be a finite set and suppose $f: S \to \mathbb{R}^1$ is a given function. We define the **sum of the values of** $f(p)$ **as** p **ranges over** S to be the quantity

$$\sum_{i=1}^{n} f(p_i)$$

where $p_1, \ldots, p_n$ is *any* arrangement of the elements of S. Since the number of

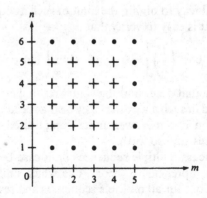

Figure 9.6. A finite set S.

terms in the sum is finite, the value of the sum does not depend on the ordering of the elements of S. We also denote this sum by

$$\sum_{p \in S} f(p). \tag{9.22}$$

In the most common situations the function f is designated by means of subscripts. For example, if $\{u_{mn}\}$, $m = 1, 2, 3$; $n = 1, 2, \ldots, 7$, are real numbers, then f is the function with domain $S = \{(m, n): 1 \leqslant m \leqslant 3, 1 \leqslant n \leqslant 7\}$ and with values u_{mn} in $\mathbb{R}^1$. We write

$$\sum_{m=1}^{3} \sum_{n=1}^{7} u_{mn}$$

instead of the sum (9.22).

Definition. Let S be an infinite set and suppose that $f: S \to \mathbb{R}^1$ is a given function. We say that f **has the sum** s **over** S if and only if for every $\varepsilon > 0$ there is a finite subset S' of S such that

$$\left| \sum_{p \in S''} f(p) - s \right| < \varepsilon$$

for *every* finite subset S'' of S which contains S'. We also say that f **has a sum over** S, without indicating the value s.

Remarks

(i) The subset S' in the definition depends, in general, on the value of ε.

(ii) We shall use the symbol

$$\sum_{p \in S} f(p) \tag{9.23}$$

to denote the sum over a finite or an infinite set S.

(iii) There is a strong inclination to compare the above definition with the definition of convergence of an infinite series: for every $\varepsilon > 0$ there is an integer n such that $|\sum_{n=1}^{m} a_n - s| < \varepsilon$ for every integer $m > n$. We might compare the set $1, 2, \ldots, n$ to S' and the set $1, 2, \ldots, m$ to S''. However, in defining the sum over a set S, nothing is said about how the finite set S' is chosen; it is not necessarily the *first* n integers if, say, S is the collection of natural numbers. The definition given for a sum over S allows us to choose a finite set of elements *arbitrarily* in the collection S. For this reason we refer to (9.23) as an **unordered sum**.

The next theorem shows that the sum over an infinite set can have at most one value s.

Theorem 9.32. *Let S be an infinite set and suppose that $f: S \to \mathbb{R}^1$ is a given function. Then f can have at most one sum s over S.*

PROOF. Assume that f has two sums s_1, s_2 over S with $s_1 < s_2$. We shall reach a contradiction. We set $\varepsilon = \frac{1}{2}(s_2 - s_1)$. Then there are finite sets S_1', S_2' such that

$$\left| \sum_{p \in S''} f(p) - s_1 \right| < \varepsilon \quad \text{and} \quad \left| \sum_{p \in \bar{S}''} f(p) - s_2 \right| < \varepsilon$$

for all finite sets S'', $\bar{S}''$ such that $S_1' \subset S''$ and $S_2' \subset \bar{S}''$. Let S^* be any finite set which contains $S_1' \cup S_2'$. Then

$$\sum_{p \in S^*} f(p) < s_1 + \varepsilon \quad \text{and} \quad s_2 - \varepsilon < \sum_{p \in S^*} f(p)$$

for this set S^*. We conclude that $s_2 - \varepsilon < s_1 + \varepsilon$, which, according to the definition of ε, is impossible. $\qquad\square$

Theorem 9.33. *Suppose that f_1 and f_2 each has a sum over a set S and that c_1, c_2 are real numbers. Then $c_1 f_1 + c_2 f_2$ has a sum over S, and*

$$\sum_{p \in S} (c_1 f_1(p) + c_2 f_2(p)) = c_1 \sum_{p \in S} f_1(p) + c_2 \sum_{p \in S} f_2(p).$$

PROOF. Let $\varepsilon > 0$ be given. Then there are finite sets S_1', S_2' such that

$$\left| \sum_{p \in S''} f_i(p) - \sum_{p \in S} f_i(p) \right| < \frac{\varepsilon}{1 + |c_1| + |c_2|} \quad \text{for} \quad i = 1, 2,$$

where S'' is any finite set containing S_1' and S_2'. We set $S' = S_1' \cup S_2'$. Then we have for such S'' containing S'

$$\left| \sum_{p \in S''} (c_1 f_1 + c_2 f_2) - \left(c_1 \sum_{p \in S} f_1 + c_2 \sum_{p \in S} f_2 \right) \right| \leq \sum_{i=1}^{2} |c_i| \left| \sum_{p \in S''} f_i(p) - \sum_{p \in S} f_i(p) \right|$$

$$< \frac{(|c_1| + |c_2|)\varepsilon}{1 + |c_1| + |c_2|} < \varepsilon. \qquad\square$$

Corollary. *Suppose that $f_1, f_2, \ldots, f_n$ each has a sum over S and that $c_1, c_2, \ldots, c_n \in \mathbb{R}^1$. Then $c_1 f_1 + c_2 f_2 + \cdots + c_n f_n$ has a sum and*

$$\sum_{p \in S} (c_1 f_1 + c_2 f_2 + \cdots + c_n f_n) = \sum_{i=1}^{n} c_i \sum_{p \in S} f_i.$$

We now show that if $f: S \to \mathbb{R}^1$ has a sum over S, then the sum of f over every finite subset of S is uniformly bounded.

Theorem 9.34. *Suppose that S is infinite and $f: S \to \mathbb{R}^1$ has a sum over S. Then there is a number M such that*

$$\left| \sum_{p \in S'} f(p) \right| \leq M$$

for every finite set S' contained in S.

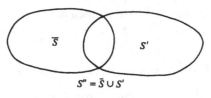

$$S'' = \bar{S} \cup S'$$

Figure 9.7

PROOF. Let $s = \sum_{p \in S} f(p)$ and choose $\varepsilon = 1$. Then there is a finite set S' such that

$$\left| \sum_{p \in S''} f(p) - s \right| < 1 \qquad (9.24)$$

for every finite set S'' which contains S'. From (9.24) we have

$$\left| \sum_{p \in S''} f(p) \right| < 1 + |s|. \qquad (9.25)$$

Now suppose that $\bar{S}$ is any finite subset of S. Define $S'' = \bar{S} \cup S'$ and observe that (9.25) is valid for this set S''. Also,

$$\bar{S} = (\bar{S} \cup S') - (S' - \bar{S})$$

(see Figure 9.7). We denote $A = \sum_{p \in S'} |f(p)|$. Therefore

$$\sum_{p \in \bar{S}} f(p) = \sum_{p \in S''} f(p) - \sum_{p \in S' - \bar{S}} f(p)$$

and

$$\left| \sum_{p \in \bar{S}} f(p) \right| \leqslant \left| \sum_{p \in S''} f(p) \right| + \sum_{p \in S'} |f(p)| \leqslant 1 + |s| + A.$$

We choose $M = 1 + |s| + A$, and since $\bar{S}$ is an arbitrary finite set, the result follows. $\qquad \square$

Theorem 9.35. Let $f: S \to \mathbb{R}^1$ be given and suppose that $f(p) \geqslant 0$ for all $p \in S$. Then f has a sum over S if and only if $\sum_{p \in S'} f(p)$ is uniformly bounded for all finite sets S' contained in S.

PROOF. If f has a sum over S, then the uniform boundedness results from Theorem 9.33. Now suppose the finite sums are uniformly bounded. We define

$$A = \sup_{S^*} \sum_{p \in S^*} f(p), \qquad (9.26)$$

where the supremum is taken over all finite sets S^* contained in S. Let $\varepsilon > 0$ be given. From the definition of supremum, there is a finite set S' such that

$\sum_{p \in S'} f(p) > A - \varepsilon$. Since $f(p) \geqslant 0$, we then have

$$A \geqslant \sum_{p \in S''} f(p) > A - \varepsilon \qquad (9.27)$$

for every finite set S'' containing S'. Since ε is arbitrary the result follows. □

From (9.26) and (9.27) in the above proof it is clear that when f has a sum over S, it is given by the number

$$\sum_{p \in S} f(p) = A.$$

Corollary. *Let* $f \colon S \to \mathbb{R}^1$, $g \colon S \to \mathbb{R}^1$ *be such that* $0 \leqslant f(p) \leqslant g(p)$ *for all* $p \in S$. *If* g *has a sum over* S, *then* f *does also and*

$$\sum_{p \in S} f(p) \leqslant \sum_{p \in S} g(p).$$

EXAMPLE 1. Let S be the set of pairs of all positive integers (m, n), $m, n = 1, 2, \ldots$. We define $f \colon S \to \mathbb{R}^1$ by $f(m, n) = 1/m^2 n^2$. Show that f has a sum over S.

Solution. Let S' be any finite subset of S and denote by m' and n' the largest integers in S'. Then

$$\sum_{(m,n) \in S'} \frac{1}{m^2 n^2} \leqslant \sum_{m=1}^{m'} \sum_{n=1}^{n'} \frac{1}{m^2 n^2}.$$

Since the series $\sum_{n=1}^{\infty} 1/n^2$ is convergent (in fact, to the value $\pi^2/6$), we find

$$\sum_{(m,n) \in S'} \frac{1}{m^2 n^2} \leqslant \frac{\pi^4}{36} = A$$

for all finite sets S'. Thus the finite sums are uniformly bounded and, according to Theorem 9.35, f has a sum over S. □

EXAMPLE 2. Let S be as in Example 1, and define $f \colon S \to \mathbb{R}^1$ by $f(m, n) = 1/(m^4 + n^4)$. Show that f has a sum over S.

Solution. We have $2m^2 n^2 \leqslant m^4 + n^4$ or $1/(m^4 + n^4) \leqslant 1/(2m^2 n^2)$ for all m and n. Set $g(m, n) = 1/2m^2 n^2$ and, by Example 1, g has a sum over S. We now employ the Corollary to Theorem 9.35 to conclude that f has a sum over S. □

Definitions. Let $f \colon S \to \mathbb{R}^1$ be given. Define the sets $S^+ = \{p \colon p \in S, f(p) \geqslant 0\}$; $S^- \{p \colon p \in S, f(p) < 0\}$. Also, denote $f^+(p) = \frac{1}{2}[|f(p)| + f(p)]$ and $f^-(p) = \frac{1}{2}[|f(p)| - f(p)]$.

From these definitions, it follows at once that

$$0 \leqslant f^+(p) \leqslant |f(p)|, \qquad 0 \leqslant f^-(p) \leqslant |f(p)|,$$
$$f^+ + f^- = |f|, \qquad f^+ - f^- = f.$$

Also,

$$f^+(p) = f(p) \quad \text{on } S^+, \qquad f^-(p) = 0 \quad \text{on } S^+,$$
$$f^-(p) = f(p) \quad \text{on } S^-, \qquad f^+(p) = 0 \quad \text{on } S^-.$$

In the study of infinite series of positive and negative terms, we saw in Theorem 9.6 that if a series converges absolutely then it converges conditionally. On the other hand, there are series which are conditionally convergent but not absolutely convergent. The series $\sum_{n=1}^{\infty} (-1)^n/n$ is such an example. The definition we have chosen for convergence of unordered sums is much stronger than the one selected for ordinary series. In fact, the next theorem shows that a function f has a sum over a set S if and only if the absolute value of f has a sum over S. We shall see later in this section that convergence of an unordered sum reduces, in the case of a series such as $\sum_{n=1}^{\infty} a_n$, to a type of convergence called **unconditional**, that is, a series which has the property that every possible rearrangement of the terms of the series yields a convergent series with the same sum. Unconditional convergence and absolute convergence are equivalent for single and multiple series of numbers.

Theorem 9.36. *f has a sum over S if and only if $|f|$ has a sum over S. When these sums over S exist, then*

$$\left| \sum_{p \in S} f(p) \right| \leqslant \sum_{p \in S} |f(p)|. \tag{9.28}$$

PROOF. Suppose $|f|$ has a sum over S. Then, by the Corollary to Theorem 9.35, f^+ and f^- have sums over S. Using Theorem 9.33, we find

$$\sum_{p \in S} f(p) = \sum_{p \in S} f^+(p) - \sum_{p \in S} f^-(p),$$
$$\sum_{p \in S} |f(p)| = \sum_{p \in S} f^+(p) - \sum_{p \in S} f^-(p),$$

so that (9.28) holds.

Now suppose that f has a sum over S. If we suppose that f^+ does not have a sum over S then, according to Theorem 9.34, for any number M there is a finite subset S' such that $\sum_{p \in S'} f^+(p) > M$. Since $f^+(p) = 0$ on S^-, we have

$$M < \sum_{p \in S'} f^+(p) = \sum_{p \in S'-S^-} f^+(p) = \sum_{p \in S' \cap S^+} f^+(p)$$
$$= \sum_{p \in S' \cap S^+} f(p).$$

Since M is arbitrary, Theorem 9.34 shows that f does not have a sum over S. Our supposition is false and we conclude that f^+ must have a sum over S. Consequently, $f^- = f^+ - f$ also has a sum over S and, finally, $|f| = f^+ + f^-$ has a sum over S. ☐

The proof of the following simple lemma is left to the reader.

Lemma 9.4. *Suppose that $f: S \to \mathbb{R}^1$ has a sum over S. If U is any set contained in S, then f has a sum over U.*

In problems involving infinite series, it is frequently useful to change the set S over which the sum is calculated. For example the two series

$$\sum_{n=1}^{\infty} \frac{1}{n^2}, \qquad \sum_{n=0}^{\infty} \frac{1}{(n+1)^2}$$

clearly have the same sum. The first sum is over the set $\{1, 2, \ldots\}$ and the second is over the set $\{0, 1, 2, \ldots\}$. When such a change from one series to another is made, the quantity being summed must be changed accordingly. In the above case the term $1/n^2$ is transformed to $1/(n+1)^2$. The next theorem shows that such changes are justified under quite general circumstances.

Theorem 9.37. *Let S_0 and S_1 be sets and $T: S_0 \to S_1$ a one-to-one mapping of S_0 onto S_1. Suppose that f has a sum over S_0 and that $g: S_1 \to \mathbb{R}^1$ is defined for each $q \in S_1$ by the formula $g(q) = f(p)$ with $T(p) = q$. Then g has a sum over S_1 and*

$$\sum_{q \in S_1} g(q) = \sum_{p \in S_0} f(p).$$

PROOF. Let the sum of f over S_0 be denoted by s. Then for every $\varepsilon > 0$ there is a finite set $S_0' \subset S_0$ such that

$$\left| \sum_{p \in S_0''} f(p) - s \right| < \varepsilon$$

for every finite set S_0'' containing S_0'. We denote $S_1' = T(S_0')$, $S_1'' = T(S_0'')$. Then by the definition of g, we have

$$\left| \sum_{q \in S_1''} g(q) - s \right| < \varepsilon$$

for every finite set S_1'' containing S_1'. Hence $\sum_{q \in S_1} g(q) = s$.   □

We recall that the symbol $\mathbb{N}$ is used to denote the set of all positive integers. We also use the symbol $\mathbb{N}_0$ to designate the set of all nonnegative integers, i.e. $\mathbb{N}_0 = \mathbb{N} \cup \{0\}$.

Theorem 9.38. *Let $f: \mathbb{N} \to \mathbb{R}^1$ be given. For each $n \in \mathbb{N}$, we let $a_n = f(n)$. Then f has a sum over $\mathbb{N}$ if and only if $\sum_{n=1}^{\infty} a_n$ converges absolutely. Also, the sums are equal:*

$$\sum_{n \in \mathbb{N}} a_n = \sum_{n=1}^{\infty} a_n.$$

PROOF. Suppose that f has a sum over $\mathbb{N}$. Then $|f|, f^+$, and f^- also have sums over $\mathbb{N}$. Let $s^+ = \sum_{n \in \mathbb{N}} a_n^+$ where $a_n^+ = \frac{1}{2}(|a_n| + a_n)$, and suppose that $\varepsilon > 0$ is given. Then there is a finite set N' such that

$$s^+ - \varepsilon < \sum_{n \in N'} a_n^+ \leqslant s^+.$$

We let n' be the largest number in N'. Then

$$s^+ - \varepsilon < \sum_{n=1}^{n''} a_n^+ \leqslant s^+$$

for all $n'' \geqslant n'$. Thus $\sum_{n=1}^{\infty} a_n^+$ converges. Similarly $\sum_{n=1}^{\infty} a_n^-$ converges where $a_n^- = \frac{1}{2}(|a_n| - a_n)$. We conclude that $\sum_{n=1}^{\infty} |a_n| = \sum_{n=1}^{\infty} (a_n^+ + a_n^-)$ converges.

Now suppose that $\sum_{n=1}^{\infty} |a_n|$ converges to s. Let N' be any finite subset of $\mathbb{N}$ with largest number n'. Then

$$\sum_{n \in N'} |a_n| \leqslant \sum_{n=1}^{n'} |a_n| \leqslant s.$$

Hence $|f|$ has a sum over $\mathbb{N}$, and by Theorem 9.36, so does f. $\square$

Corollary. *Suppose that the sequence $\sum_{n=1}^{\infty} a_n$ is absolutely convergent. Let $\{k_n\}$ be any sequence of positive integers. Then $\sum_{n=1}^{\infty} a_{k_n}$ is convergent. If the set $\{k_n\}$ consists of all of $\mathbb{N}$, then*

$$\sum_{n=1}^{\infty} a_{k_n} = \sum_{n=1}^{\infty} a_n.$$

If an infinite series has the property that every reordering of the terms yields a convergent series, then we say that the series is **unconditionally convergent**. The above Corollary shows that for the set $\mathbb{N}$, unconditional convergence and absolute convergence are equivalent. On the other hand, if a series is only conditionally convergent, then a reordering of the terms may yield a divergent series, as the following example shows.

EXAMPLE 3. Let $f: \mathbb{N} \to \mathbb{R}^1$ be given by $f(n) = (-1)^n/n$. Show that f does not have a sum over $\mathbb{N}$.

Solution. This fact is a direct result of Theorem 9.38 since the series $\sum_{n=1}^{\infty} (-1)^n/n$ is not absolutely convergent. We know that the two series

$$\sum_{n=1}^{\infty} \frac{1}{2n} \quad \text{and} \quad \sum_{n=1}^{\infty} \frac{-1}{2n-1},$$

being harmonic series, are divergent. Let r be any real number. Then by selecting terms first from the series with even-numbered denominators, then from the series with odd-numbered denominators, then from the even-numbered ones, and so forth, it is possible to obtain a rearrangement of f so that the sum of the rearranged series is r. The details of this process are left to the reader. See Problem 11 at the end of this section and the hint given there. $\square$

PROBLEMS

1. Let S be the set of all integers, positive, negative and zero. We define $f: S \to \mathbb{R}^1$ so that $f(n) = n/2^{|n|}$. Show that f has a sum over S.

2. Let S be defined as in Problem 1. Suppose that $g: S \to \mathbb{R}^1$ is given by $g(n) = \sqrt{|n|}/e^n$. Does g have a sum over S?

3. Let $S = \{x: x \text{ is rational}, 0 < x < 1\}$. Let $f: S \to \mathbb{R}^1$ be given by the formula

$$f\left(\frac{p}{q}\right) = \frac{1}{q}$$

where p, q are integers reduced to lowest terms of the rational number $r = p/q$. Does f have a sum over S?

4. Let $S = \{(m, n): m, n \text{ integers}, 0 < m < \infty, 0 < n < \infty\}$. Define $f: S \to \mathbb{R}^1$ so that $f(m, n) = 1/m^{3/2}n^{5/2}$. Show that f has a sum over S.

5. Let S be defined as in Problem 4. If $g: S \to \mathbb{R}^1$ is given by $g(m, n) = 1/mn^3$, does g have a sum over S?

6. Let $S = \{(l, m, n): l, m, n \text{ integers}, 0 < l < \infty, 0 < m < \infty, 0 < n < \infty\}$. Define $f: S \to \mathbb{R}^1$ by the formula $f(l, m, n) = 1/(l^8 + 2m^8 + n^8)$. Show that f has a sum over S.

7. Let $S = \{(m, n): m, n \text{ integers}, 0 < m < \infty, 0 < n < \infty, m \neq n\}$. Define $f: S \to \mathbb{R}^1$ by the formula $f(m, n) = 1/(m^2 - n^2)$. Show that f does not have a sum over S.

8. Prove Lemma 9.4.

9. Let S_0 and S_1 be sets and $T: S_0 \to S_1$ a mapping of S_0 onto S_1. Let $f: S_0 \to \mathbb{R}^1$ have a sum over S_0. For each $q \in S_1$, choose a single element p of the set $\{T^{-1}(q)\}$ and define the function $g: S_1 \to \mathbb{R}^1$ by the formula $g(q) = f(p)$. Show that g has a sum over S_1.

10. Write out a complete proof of the Corollary to Theorem 9.38.

11. Suppose that $\sum_{n=1}^{\infty} a_n$ is conditionally convergent. Let r be any real number. Show that there is a rearrangement $\sum_{n=1}^{\infty} b_n$ of the infinite series $\sum_{n=1}^{\infty} a_n$ such that $\sum_{n=1}^{\infty} b_n = r$. [*Hint:* Divide the terms of the series $\sum_{n=1}^{\infty} a_n$ into terms a'_n and a''_n in which $a'_n \geqslant 0$ and $a''_n < 0$. Then each of the series $\sum_{n=1}^{\infty} a'_n$ and $\sum_{n=1}^{\infty} a''_n$ must diverge. (Otherwise the original series would be absolutely convergent). Assume that the terms are rearranged so that $a'_{n+1} \leqslant a'_n$ and $-a''_{n+1} \leqslant -a''_n$ for each n. Clearly $a'_n \to 0$ and $a''_n \to 0$ as $n \to \infty$ as otherwise the original series would be divergent. Now choose enough terms of the first sequence so that r is just exceeded. Then choose enough terms of the second series so that the total sum falls just below r. Then choose terms of the first series to exceed r and continue. Show that this process must yield a sequence converging to r.]

9.6. The Comparison Test for Unordered Sums; Uniform Convergence

We learned in the study of the convergence of series that the comparison test is one of the most useful for deciding both convergence and divergence. The following analog for unordered sums is a useful test in many situations.

Theorem 9.39 (Comparison test). *Let $f: S \to \mathbb{R}^1$ and $F: S \to \mathbb{R}^1$ be given.*

(i) *If F has a sum over S and $|f(p)| \leqslant |F(p)|$ for all $p \in S$, then f has a sum over S.*

(ii) *If F does not have a sum over S and $|f(p)| \geqslant |F(p)|$ for all $p \in S$, then f does not have a sum over S.*

The proof is left to the reader (see Theorem 9.4).

EXAMPLE 1. Set $S = \{(m, n): m, n \text{ integers } 0 < m < \infty, 0 < n < \infty\}$ be given. Show that $f(m, n) = (-1)^{m+n}/(m^3 + n^3)$ has a sum over S.

Solution. We observe that $2m^{3/2}n^{3/2} \leqslant (m^3 + n^3)$ for all $(m, n) \in S$. We define $F(m, n) = 1/m^{3/2}n^{3/2}$, and it is clear that $\sum_{m=1}^{\infty}\sum_{n=1}^{\infty} 1/m^{3/2}n^{3/2}$ converges (two p-series with $p = 3/2$). Therefore $\sum_{(m,n) \in S} F(m, n)$ converges. Since $|f(m, n)| \leqslant |F(m, n)|$ for all $(m, n) \in S$, the result follows. □

EXAMPLE 2. Let S be as in Example 1. Show that $f(m, n) = 1/(m^2 + n^2)$ does not have a sum over S.

Solution. We define

$$F(m, n) = \begin{cases} \dfrac{1}{2n^2} & \text{for } m \leqslant n, \\[2mm] \dfrac{1}{2m^2} & \text{for } m > n. \end{cases}$$

Then, clearly $f(m, n) \geqslant F(m, n) > 0$ for all m, n. Let $S_{n'} = \{(m, n): 0 < m \leqslant n', 0 < n \leqslant n', m, n \text{ integers}\}$. Then

$$\sum_{(m,n) \in S_{n'}} F(m, n) > \sum_{n=1}^{n'}\sum_{m=1}^{n'} \frac{1}{n^2} = n' \sum_{n=1}^{n'} \frac{1}{n^2}$$

$$> \sum_{n=1}^{n'} \frac{1}{n}.$$

Thus, if M is any positive number, we can choose n' so large that

$$\sum_{(m,n) \in S_{n'}} F(m, n) > M.$$

Therefore F does not have a sum over S and, by the Comparison test, neither does f. □

The next two theorems are almost direct consequences of results already established. We leave the details of the proofs to the reader.

Theorem 9.40. *If $f: S \to \mathbb{R}^1$ and $g: S \to \mathbb{R}^1$ have sums over S, and $f(p) \leqslant g(p)$ for all $p \in S$, then*

$$\sum_{p \in S} f(p) \leqslant \sum_{p \in S} g(p).$$

Theorem 9.41 (Sandwiching theorem). *Let $f: S \to \mathbb{R}^1, g: S \to \mathbb{R}^1$, and $h: S \to \mathbb{R}^1$ be given. Suppose that f and h each has a sum over S and that $f(p) \leqslant g(p) \leqslant h(p)$ for all $p \in S$. Then g has a sum over S.*

Although the sets S may be quite arbitrary, we now show that if f has a sum over S then the set of points of S where f does not vanish is severely restricted.

Theorem 9.42. *Suppose that $f: S \to \mathbb{R}^1$ has a sum over S. Let $S^* = \{p \in S: f(p) \neq 0\}$. Then S^* is countable.*

PROOF. For $n = 2, 3, \ldots$, define

$$S_n = \left\{p \in S: \frac{1}{n} \leqslant |f(p)| < \frac{1}{n-1}\right\}.$$

The reader may complete the proof by showing that $S^* = \bigcup_{n=2}^{\infty} S_n$ and that each S_n must be a finite set. $\square$

Let A be a set in a metric space T and suppose that S is an arbitrary set. Let f be defined for $x \in A$ and $p \in S$ with values in $\mathbb{R}^1$; that is, $f: A \times S \to \mathbb{R}^1$. We write $f(x; p)$ for the function values of f. For each x in A, the function f may or may not have a sum over S; when f has a sum, we denote it by $s(x)$. For unordered sums we make the following definition, one which corresponds to uniform convergence for infinite series.

Definition. Let $f: A \times S \to \mathbb{R}^1$ be given. We say that f **has a sum uniformly on A over S** if and only if f has a sum over S for each $x \in A$ and, for every $\varepsilon > 0$ there is a finite set S' contained in S such that

$$\left| \sum_{p \in S''} f(x; p) - s(x) \right| < \varepsilon$$

for every finite set S'' containing S'; furthermore the set S' is *independent of x*. When the set A is clearly understood we say briefly that f has a sum uniformly over S.

The next result, an extension of Theorem 9.19, shows that if an unordered sum has all continuous terms and if the "convergence" is uniform, then the sum is a continuous function.

Theorem 9.43. *Suppose that for each p in S, the function $f: A \times S \to \mathbb{R}^1$ is continuous on the set A in the metric space T. Suppose that f has a sum uniformly over S. Then the function $s(x)$ defined by*

$$s(x) = \sum_{p \in S} f(x; p), \qquad x \in A,$$

is continuous on A.

PROOF. Let $x_0 \in A$ and $\varepsilon > 0$ be given. We wish to show that s is continuous at x_0. By hypothesis, there is a finite set S' such that

$$\left| \sum_{p \in S''} f(x_0; p) - s(x_0) \right| < \frac{\varepsilon}{3} \quad \text{and} \quad \left| \sum_{p \in S''} f(x; p) - s(x) \right| < \frac{\varepsilon}{3}$$

for any finite set S'' containing S'. We fix S'' and write

$$|s(x) - s(x_0)| \leq \left| s(x) - \sum_{p \in S''} f(x; p) \right| + \left| \sum_{p \in S''} [f(x; p) - f(x_0; p)] \right|$$

$$+ \left| \sum_{p \in S''} f(x_0; p) - s(x_0) \right|.$$

Now the hypothesis that f is continuous is employed to yield the result.   □

The proof of Theorem 9.43 should be compared to the proof of Theorem 9.13.

Theorem 9.44

(a) (Term-by-term integration of an unordered sum). *Let $I = \{x : a \leq x \leq b\}$ be a finite interval, and suppose that $f : I \times S \to \mathbb{R}^1$ has a sum on I uniformly over S. Suppose that for each $p \in S$, the function f is integrable on I. Define $s(x) = \sum_{p \in S} f(x; p)$. Then $s(x)$ is integrable on I and*

$$\int_a^b s(x) \, dx = \sum_{p \in S} \int_a^b f(x; p) \, dx.$$

(b) *Let F be a figure in $\mathbb{R}^1$ and suppose that $f : F \times S \to \mathbb{R}^1$ is integrable over F and has a sum uniformly over S, denoted by $s(x)$. Then s is integrable over F and*

$$\int_F s \, dV_N = \sum_{p \in S} \int_F f(x; p) \, dV_N.$$

The proofs of Parts (a) and (b) are similar to the proofs of Theorems 9.14 and 9.17, respectively, and are left to the reader.

The next theorem shows when unordered sums may be differentiated term-by-term.

Theorem 9.45. *Let G be an open set in $\mathbb{R}^N$ and suppose that $f : G \times S \to \mathbb{R}^1$ is continuous and has partial derivatives $f_{,k}(x; p)$, $k = 1, 2, \ldots, N$, which are continuous in G. Suppose that $f(x; p)$ has a sum over S denoted by $s(x)$, and that $f_{,k}(x; p)$ has a sum uniformly over S denoted by $t(x)$. Then*

$$s_{,k}(x) = t(x) \quad \text{for all} \quad x \in G.$$

The proof is similar to that of Theorem 9.15.

Theorem 9.46 (Weierstrass M-test). *Let A be a set in a metric space T and suppose that $u\colon A \times S \to \mathbb{R}^1$ satisfies $|u(x; p)| \leqslant f(p)$ for each $p \in S$. If $f(p)$ has a sum over S, then $u(x; p)$ and $|u(x; p)|$ each has a sum uniformly on A over S.*

The proof is similar to that of Theorem 9.22 and is left to the reader.

PROBLEMS

1. Let $S = \{(m, n)\colon m,\ n$ integers, $0 < m < \infty,\ 0 < n < \infty\}$. Given $f(m, n) = 1/(m^{5/2} + n^{5/2})$, show that f has a sum over S.

2. Let S be as in Problem 1 and define $f(m, n) = 1/(m + n^3)$. Show that f does not have a sum over S.

3. Let $S = \{(l, m, n)\colon l, m, n$ integers, $0 < l < \infty, 0 < m < \infty, 0 < n < \infty\}$. Define $f(l, m, n) = 1/(l^2 + m^2 + n^4)$. Does f have a sum over S?

4. Prove Theorem 9.39.

5. Prove Theorem 9.40.

6. Prove Theorem 9.41.

7. Complete the proof of Theorem 9.42.

8. Let $S = \{x\colon 0 < x < 1\}$ and suppose that $f\colon S \to \mathbb{R}^1$ has a range which contains the set $A = \{x\colon 0 < x < 1, x$ is rational$\}$. Show that f does not have a sum over S.

9. Let $I = \{x\colon a \leqslant x \leqslant b\}$ with $0 < a < b < 1$ and $S = \{n\colon -\infty < n < \infty, n$ an integer $\neq 0\}$. Suppose that $f\colon I \times S \to \mathbb{R}^1$ is given by $f(x; n) = x^n/n$. Does f have a sum uniformly on I over S?

10. Let $A = \{(x, y)\colon a \leqslant x \leqslant b, c \leqslant x \leqslant d\}$ with $0 < a < b < 1, 0 < c < d < 1$, and let $S = \{(m, n)\colon m, n$ integers $-\infty < m < \infty, -\infty < n < \infty, m \neq 0, n \neq 0\}$. Define $f\colon A \times S \to \mathbb{R}^1$ by $f(x, y; m, n) = x^m y^n/m^2 n^2$. Does f have a sum uniformly on A over S?

11. Prove Theorem 9.44(a).

12. Prove Theorem 9.44(b).

13. Let $S = \{n\colon n$ an integer, $+\infty < n < \infty\}$ and let $I = \{x\colon 0 \leqslant x < \infty\}$. Define $f(x, n) = e^{-nx}/(1 + n^4)$. Can the unordered sum $\sum_{n \in S} f(x, n)$ be differentiated term-by-term?

14. Prove Theorem 9.45.

15. Prove Theorem 9.46.

9.7. Multiple Sequences and Series

In Section 9.5 we introduced double sequences informally, and we showed how they are related to unordered sums. In this section we shall treat convergence problems for multiple sequences and series.

Definitions. We denote the collection of all pairs of positive integers by $\mathbb{N} \times \mathbb{N}$, and the collection of all pairs of nonnegative integers by $\mathbb{N}_0 \times \mathbb{N}_0$. A **double sequence** is a function u from $\mathbb{N}_0 \times \mathbb{N}_0$ into $\mathbb{R}^1$. The values of u are denoted by $u(m, n)$, $m, n = 0, 1, 2, \ldots$, or, more customarily, by u_{mn}. Sometimes the domain of a double sequence is $\mathbb{N} \times \mathbb{N}$ or a subset of $\mathbb{N} \times \mathbb{N}$. We say a **double sequence converges to** a **as** (m, n) **tends to infinity** if and only if for every $\varepsilon > 0$ there is a positive integer P such that $|u_{mn} - a| < \varepsilon$ for all (m, n) such that $m > P$ and $n > P$. We write $\{u_{mn}\} \to a$ as $(m, n) \to \infty$.

In analogy with the formal definition of an infinite series given in Section 9.1, we define a **double series** formally as an ordered pair $(\{u_{mn}\}, \{s_{mn}\})$ of double sequences such that

$$s_{mn} = \sum_{i=0}^{m} \sum_{j=0}^{n} u_{ij}.$$

The u_{mn}, $m, n = 0, 1, 2, \ldots$, are called the **terms** of the double series $(\{u_{mn}\}, \{s_{mn}\})$, and the s_{mn} are called its **partial sums**. The double series is said to be **convergent** if and only if there is a number s such that $s_{mn} \to s$ as $(m, n) \to \infty$. Otherwise, the double series is **divergent**. We will most often write a double series in the form

$$\sum_{i,j=0}^{\infty} u_{ij}$$

rather than use the ordered pair symbol.

The following theorem is a direct consequence of the definition of convergence and the theorems on limits.

Theorem 9.47

(a) *If the double series $\sum_{m,n=0}^{\infty} u_{mn}$, $\sum_{m,n=0}^{\infty} v_{mn}$ are both convergent and c and d are any numbers, then the series $\sum_{m,n=0}^{\infty} (cu_{mn} + dv_{mn})$ is convergent and*

$$\sum_{m,n=0}^{\infty} (cu_{mn} + dv_{mn}) = c \sum_{m,n=0}^{\infty} u_{mn} + d \sum_{m,n=0}^{\infty} v_{mn}.$$

(b) *If the series $\sum_{m,n=0}^{\infty} u_{mn}$ is divergent and if $c \neq 0$, then $\sum_{m,n=0}^{\infty} cu_{mn}$ is divergent.*

The next theorem shows the relationship between the convergence of a double series and an unordered sum.

Theorem 9.48. *Suppose that the terms u_{mn} of a double sequence $u: \mathbb{N}_0 \times \mathbb{N}_0 \to \mathbb{R}^1$ are all nonnegative. Then the double series $\sum_{m,n=0}^{\infty} u_{mn}$ is convergent if and only if the function u has a sum over $S = \mathbb{N}_0 \times \mathbb{N}_0$. Furthermore,*

$$\sum_{m,n=0}^{\infty} u_{mn} = \sum_{(m,n) \in S} u_{mn}. \tag{9.29}$$

PROOF

(a) Suppose that u has a sum over S which we denote by s. Let $\varepsilon > 0$ be given. Then there is a finite subset S' of S such that

$$\sum_{(m,n)\in S'} u_{mn} > s - \varepsilon.$$

Let P be a positive integer such that $P > m$ and $P > n$ for all $(m, n) \in S'$. Since $u_{ij} \geqslant 0$ for all i, j, it follows that

$$s - \varepsilon < \sum_{(m,n)\in S'} u_{mn}$$

$$\leqslant \sum_{i=1}^{m_0} \sum_{j=1}^{n_0} u_{ij}$$

$$= s_{m_0 n_0} \quad \text{if } m_0 \geqslant P \quad \text{and } n_0 \geqslant P.$$

Therefore $s - \varepsilon \leqslant s_{m_0 n_0} \leqslant s$ and so the double series converges.

(b) Suppose the double series converges to a sum s. Let S' be any finite subset of S and let $\varepsilon > 0$ be given. Then there is a positive integer P such that $P \geqslant m$ and $P \geqslant n$ for $(m, n) \in S'$, and furthermore the inequality $|s_{mn} - s| < \varepsilon$ holds whenever $m \geqslant P$ and $n \geqslant P$. Hence

$$\sum_{(m,n)\in S'} u_{mn} \leqslant \sum_{m=1}^{P} \sum_{n=1}^{P} u_{mn} = s_{pp} \leqslant s + \varepsilon.$$

Since ε and S' are arbitrary, u has a sum over S and its value is s. $\qquad\square$

It is clear that corresponding to Theorems 9.47 and 9.48 there are theorems for triple series, quadruple series and, in fact, for multiple series of any order.

Theorem 9.49. *Let* $u \colon \mathbb{N}_0 \times \mathbb{N}_0 \to \mathbb{R}^1$ *be given and suppose that its terms are* u_{mn}. *Then the double series* $\sum_{m,n=0}^{\infty} u_{mn}$ *is absolutely convergent if and only if the function u has a sum over $S = \mathbb{N}_0 \times \mathbb{N}_0$. In the case of convergence, Equation (9.29) holds.*

The above theorem is a direct consequence of Theorem 9.48 and the related theorems on unordered sums. We leave the details of the proof to the reader. Theorems 9.48 and 9.49 make the theory of absolutely convergent double series a special case of the theory of unordered sums given in Section 9.6. The next result is a special case of a general theorem on unordered sums (Theorem 9.51 below).

Theorem 9.50. *Let* $u \colon \mathbb{N}_0 \times \mathbb{N}_0 \to \mathbb{R}^1$ *be given with its terms denoted by* u_{mn}. *Suppose that* $\sum_{m,n=0}^{\infty} u_{mn}$ *is absolutely convergent. Then the single series*

$$\sum_{n=0}^{\infty} u_{mn}$$

is absolutely convergent for each $m = 0, 1, 2, \ldots$, *and the series* $\sum_{m=0}^{\infty} u_{mn}$ *is* *absolutely convergent for each $n = 0, 1, 2, \ldots$. If we define*

$$U_m = \sum_{n=0}^{\infty} u_{mn}, \qquad V_n = \sum_{m=0}^{\infty} u_{mn}, \qquad W_r = \sum_{m+n=r} u_{mn}$$

then the series $\sum_{m=0}^{\infty} U_m$, $\sum_{n=0}^{\infty} V_n$, $\sum_{r=0}^{\infty} W_r$ are all absolutely convergent and have the same value as $\sum_{m,n=0}^{\infty} u_{mn}$.

PROOF. We first show that $\sum_{n=0}^{\infty} u_{mn}$ is convergent for each m when $u_{mn} \geqslant 0$. By Theorem 9.49, we know that u has a sum over $S = \mathbb{N}_0 \times \mathbb{N}_0$. Since the partial sum $\sum_{n=0}^{P} u_{mn}$ is a finite subset of S, these partial sums are uniformly bounded. Hence the series is convergent for each m. Now consider the partial sum $\sum_{m=0}^{M} U_m$. For any $\varepsilon > 0$ and any positive number M, there is an integer P such that

$$\sum_{n=0}^{P} u_{mn} > \sum_{n=0}^{\infty} u_{mn} - \varepsilon/M.$$

Hence for any positive integer M, it follows that

$$\sum_{m=0}^{M} U_m = \sum_{m=0}^{M} \sum_{n=0}^{\infty} u_{mn} < \sum_{m=0}^{M} \sum_{n=0}^{P} u_{mn} + \varepsilon.$$

Since u has a sum over S, we conclude that the series $\sum_{m=0}^{\infty} U_m$ converges. Now for any $\varepsilon > 0$ there is a finite set $S' \subset S$ such that

$$\sum_{(m,n) \in S} u_{mn} > \sum_{(m,n) \in S'} u_{mn} - \varepsilon.$$

Also, there is an integer M' which we can take larger than the greatest value of m in S', such that

$$\sum_{m=0}^{\infty} U_m > \sum_{m=0}^{M'} U_m = \sum_{m=0}^{M'} \sum_{n=0}^{\infty} u_{mn} > \sum_{(m,n) \in S'} u_{mn} - \varepsilon.$$

Since ε is arbitrary it follows that $\sum_{m=0}^{\infty} U_m = \sum_{m,n=0}^{\infty} u_{mn}$. We treat the case where u_{mn} may be positive or negative by using the hypothesis of absolute convergence and separating the terms of the series into positive and negative parts. As in previous proofs, we apply the above argument to each of the series obtained in this way. The proofs for $\sum_{n=0}^{\infty} V_n$, $\sum_{r=0}^{\infty} W_r$ are similar. □

Theorem 9.50 is a special case of a general theorem on unordered sums which may be considered as a generalized associative law. This result allows us to extend the above theorem to multiple series of any order.

Theorem 9.51. *Let S' be a set. To every x in S' we associate a set denoted S_x. Define $S = \{(x, y): x \in S', y \in S_x\}$. Suppose that $f: S \to \mathbb{R}^1$ has a sum over S. Then f has a sum over S_x for each $x \in S'$. If we define $g(x) = \sum_{y \in S_x} f(x, y)$, then g has a sum over S' and*

$$\sum_{x \in S'} g(x) = \sum_{x \in S'} \left[\sum_{y \in S_x} f(x, y) \right] = \sum_{(x,y) \in S} f(x, y).$$

The proof follows the pattern of the proof of Theorem 9.50, and we leave the details to the reader.

We illustrate Theorem 9.51 with an example in triple series. Let $S' = \mathbb{N} \times \mathbb{N}$ and for each $x \in S'$, let $S_x = \mathbb{N}_0$. Then $S = \mathbb{N} \times \mathbb{N} \times \mathbb{N}_0$. Suppose $f: S \to \mathbb{R}^1$ has a sum over S; we denote the terms of f by u_{lmn}, $(l, m, n) \in S$. Theorem 9.51 then shows that

$$g(x) \equiv \sum_{y \in S_x} f(x, y) = \sum_{n=0}^{\infty} u_{lmn}$$

is convergent for all $(l, m) \in S'$. Also, if we denote the terms of $g(x)$ by v_{lm}, $(l, m) \in S'$, then

$$\sum_{l, m=1}^{\infty} v_{lm} = \sum_{l, m=1}^{\infty} \sum_{n=0}^{\infty} u_{lmn} = \sum_{(l, m, n) \in S} u_{lmn}.$$

The next theorem is a partial converse of Theorem 9.51.

Theorem 9.52. *Suppose that S', S, and S_x are as in Theorem 9.51. Let $f: S \to \mathbb{R}^1$ be given. Suppose that f has a sum over S_x for each $x \in S'$. Define $g(x) = \sum_{y \in S_x} |f(x, y)|$ and suppose that g has a sum over S'. Then f has a sum over S.*

We leave the proof of this and the following result to the reader.

Theorem 9.53 (Multiplication of unordered sums). *Let $f: S' \to \mathbb{R}^1$ and $g: S'' \to \mathbb{R}^1$ be given. Define $S = S' \times S''$ and $h(x, y) = f(x) \cdot g(y)$ for $x \in S'$, $y \in S''$. Suppose that f has a sum over S' and g has a sum over S''. Then h has a sum over S and*

$$\sum_{(x, y) \in S} h(x, y) = \left[\sum_{x \in S'} f(x) \right] \cdot \left[\sum_{y \in S''} g(y) \right].$$

We now show how the theorems of this section may be used to establish rules for the multiplication of power series. Suppose the series

$$f(x) = a_0 + a_1 x + \cdots + a_n x^n + \cdots,$$

$$g(x) = b_0 + b_1 x + \cdots + b_n x^n + \cdots,$$

are convergent for $|x| < R$. Without considering questions of convergence, we multiply the two series by following the rules for the multiplication of polynomials. The result is

$$b_0 f(x) = a_0 b_0 + a_1 b_0 x + \cdots + a_n b_0 x^n + \cdots,$$

$$b_1 x f(x) = \qquad a_0 b_1 x + \cdots + a_{n-1} b_1 x^n + \cdots,$$

$$\vdots$$

$$b_n x^n f(x) = \qquad\qquad\qquad a_0 b_n x^n + \cdots.$$

Adding, we obtain the series

$$a_0 b_0 + (a_0 b_1 + a_1 b_0)x + \cdots + (a_0 b_n + a_1 b_{n-1} + \cdots + a_n b_0)x^n + \cdots.$$

In summation notation, the product becomes

$$\sum_{i=0}^{\infty} a_i x^i \sum_{j=0}^{\infty} b_j x^j = \sum_{i=0}^{\infty} \sum_{j=0}^{\infty} a_i b_j x^{i+j}.$$

Setting $i + j = n$ and collecting terms, we obtain for the right side

$$\sum_{n=0}^{\infty} \left(\sum_{k=0}^{n} a_k b_{n-k} \right) x^n.$$

We shall show that the above expansion actually represents $f(x) \cdot g(x)$.

Definition. Let $\sum_{i=0}^{\infty} a_i (x - c)^i$ and $\sum_{j=0}^{\infty} b_j (x - c)^j$ be given power series. Then the series

$$\sum_{n=0}^{\infty} \left(\sum_{k=0}^{n} a_k b_{n-k} \right) (x - c)^n \qquad (9.30)$$

is called the **Cauchy product** of the two given series.

Theorem 9.54. *Suppose that the series*

$$f(x) = \sum_{n=0}^{\infty} c_n (x - c)^n, \qquad g(x) = \sum_{n=0}^{\infty} d_n (x - c)^n,$$

converge for $|x - c| < R$ with $R > 0$. Then for $|x - c| < R$ the product $f(x) \cdot g(x)$ is given by the Cauchy product of the two series.

PROOF. Since the two given series are absolutely convergent for $|x - c| < R$, it follows that

$$f(x) = \sum_{i \in \mathbb{N}_0} c_i (x - c)^i, \qquad g(x) = \sum_{j \in \mathbb{N}_0} d_j (x - c)^j.$$

In Theorem 9.53, we take $S' = \mathbb{N}_0$ and $S'' = \mathbb{N}_0$. Then the function h is defined for $(i, j) \in \mathbb{N}_0 \times \mathbb{N}_0$ by $h_{ij} = c_i d_j (x - c)^{i+j}$. Hence for each fixed x such that $|x - c| < R$, we have

$$f(x) \cdot g(x) = \sum_{(i,j) \in \mathbb{N}_0 \times \mathbb{N}_0} c_i d_j (x - c)^{i+j}.$$

Now for each $n \in \mathbb{N}_0$, define $S_n = \{(i, j) \in \mathbb{N}_0 \times \mathbb{N}_0 : i + j = n\}$. Then from the generalized associative law (Theorem 9.51) we find that the function

$$u_n(x) = \sum_{(i,j) \in S_n} c_i d_j (x - c)^n$$

has a sum over $\mathbb{N}_0$. Therefore

$$f(x) \cdot g(x) = \sum_{n \in \mathbb{N}_0} \left(\sum_{(i,j) \in S_n} c_i d_j \right) (x - c)^n$$

for $|x - a| < R$. This formula is of the form (9.30). $\square$

Suppose that g is represented by a power series for $|x - c| < R$. Then it is a consequence of Theorem 9.54 that if b is any real number, the expression $[g(x) - b]^n$ is given for $|x - c| < R$ by the power series obtained by using the series for g and computing successive Cauchy products. In this way we get a convergent power series expansion for the composition of two function. If

$$f(x) = \sum_{n=0}^{\infty} a_n(x - b)^n \quad \text{for} \quad |x - b| < R_0 \quad \text{with} \quad R_0 > 0,$$

then we form the series

$$f(g) = \sum_{n \in \mathbb{N}_0} a_n(g - b)^n. \tag{9.31}$$

We now substitute successive Cauchy products for $[g(x) - b]^n$ into (9.31). The resulting series converges provided that $|g(c) - b| < R_0$, and that $|x - c|$ is sufficiently small. Thus $f[g(x)]$ may be represented by a convergent power series in $(x - c)$.

Theorem 9.53 is useful for the extension to multinomial series of the results on double series.

If the terms u_{mn} of a double series are of the form $c_{mn}(x - a)^m(y - b)^n$ where c_{mn}, a, b, x, y are in $\mathbb{R}^1$ then we call

$$\sum_{m,n=0}^{\infty} c_{mn}(x - a)^m(y - b)^n$$

a **double power series**. **Triple, quadruple,** and **n-tuple** power series are defined similarly. The proofs of the next two theorems follow directly from previous results on convergence.

Theorem 9.55. *Suppose that the double power series $\sum_{m,n=0}^{\infty} c_{mn}x^my^n$ converges absolutely for $x = x_1$, $y = y_1$, and that $x_1 \neq 0$, $y_1 \neq 0$. Define $R = \{(x, y): |x| \leqslant |x_1|, |y| \leqslant |y_1|\}$. Then the series is absolutely convergent on R, and the function with terms $c_{mn}x^my^n$ has a sum uniformly on R over $\mathbb{N}_0 \times \mathbb{N}_0$.*

We may make the substitution $x = x' - a$ and $y = y' - b$ in Theorem 9.55 to obtain the analogous result for general double power series.

Theorem 9.56 (Term-by-term differentiation of double power series). *Suppose that the double power series*

$$f(x, y) = \sum_{m,n=0}^{\infty} c_{mn}(x - a)^m(y - b)^n \tag{9.32}$$

is absolutely convergent on $S = \{(x, y): |x - a| < r, |y - b| < s\}$. Then f has partial derivatives of all orders on S. These derivatives may be obtained by differentiating the series (9.32) term by term. Each differentiated series is absolutely convergent on S. In particular,

$$c_{mn} = \left| \frac{1}{m!n!} \frac{\partial^{m+n} f}{\partial x^m \partial y^n} \right|_{x=a, y=b}$$

EXAMPLE 1. Find the terms of the double power series expansion to degree 3 where $f(x, y) = e^{xy}$, $a = 1$ $b = 0$.

Solution. We have $f(1, 0) = 1$. By computation, $f_x(1, 0) = 0$, $f_y(1, 0) = 1$, $f_{xx}(1, 0) = 0$, $f_{xy}(1, 0) = 1$, $f_{yy}(1, 0) = 1$, $f_{xxx}(1, 0) = f_{xxy}(1, 0) = 0$, $f_{xyy}(1, 0) = 2$, $f_{yyy}(1, 0) = 1$. Therefore

$$e^{xy} = 1 + y + (x - 1)y + \frac{1}{2!}y^2 + (x - 1)y^2 + \frac{1}{3!}y^3 + \cdots . \qquad \square$$

EXAMPLE 2. Show that for all x, y,

$$\sum_{k=0}^{\infty} \frac{(x + y)^k}{k!} = \left(\sum_{m=0}^{\infty} \frac{x^m}{m!} \right) \left(\sum_{n=0}^{\infty} \frac{x^n}{n!} \right),$$

and hence that $e^{x+y} = e^x e^y$.

Solution. According to Theorem 9.53, we have for all x, y,

$$\left(\sum_{m=0}^{\infty} \frac{x^m}{m!} \right) \left(\sum_{n=0}^{\infty} \frac{y^n}{n!} \right) = \sum_{m,n=0}^{\infty} \frac{x^m y^n}{m!n!}.$$

Now we employ Theorem 9.50 to obtain (by setting $n = p - m$)

$$\sum_{m,n=0}^{\infty} \frac{x^m y^n}{m!n!} = \sum_{p=0}^{\infty} \sum_{m=0}^{p} \frac{x^m y^{p-m}}{m!(p - m)!} = \sum_{p=0}^{\infty} \frac{1}{p!} \left(\sum_{m=0}^{p} \frac{p! x^m y^{p-m}}{m!(p - m)!} \right)$$

$$= \sum_{p=0}^{\infty} \frac{(x + y)^p}{p!}. \qquad \square$$

PROBLEMS

In each of Problems 1 through 6, find the terms up to degree 3 of the double series expansion (9.32) of $f(x, y)$ as given. Take $a = b = 0$ in all cases.

1. $f(x, y) = e^x \cos y$ 2. $f(x, y) = (1 - x - 2y + x^2)^{-1}$

3. $f(x, y) = e^{-x} \sec y$ 4. $f(x, y) = e^{-2x} \log(1 + y)$

5. $f(x, y) = \cos(xy)$ 6. $f(x, y) = (1 + x + y)^{-1/2}$

7. State and prove the analogue of Theorem 9.47 for triple series.

8. Let $S = \mathbb{N} \times \mathbb{N}$ and suppose that $u: S \to \mathbb{R}^1$ is given by $u_{mn} = 1/mn^4$. Show that the double series $\sum_{m,n=1}^{\infty} 1/mn^4$ is not convergent and hence that u does not have a sum over S.

9. Prove Theorem 9.49.

10. Write the details of the proof of Theorem 9.50 for $W_r = \sum_{m+n=r} u_{mn}$.

11. Prove Theorem 9.51. Then let $S' = \mathbb{N}$ and $S_x = \mathbb{N} \times \mathbb{N}$ for all x, and show that a theorem analogous to Theorem 9.50 can be derived for triple series.

12. Let $\mathbb{Z}$ denote the set of all integers: positive, negative, and zero. Apply Theorem 9.51 with $S' = \mathbb{N}_0$ and $S_x = \mathbb{Z}$ for all x. Show how a theorem on convergence of "double series" of the form $\sum_{m=0}^{\infty} \sum_{n=-\infty}^{\infty} u_{mn}$ may be obtained.

13. Prove Theorem 9.52.

14. Suppose that in Theorem 9.52 we choose $S' = \mathbb{N}$, $S_x = \mathbb{N}$. We define $f(m, n) = (-1)^{m+n}/(m^2 + n^2)$. Are the hypotheses of the theorem satisfied?

15. Prove Theorem 9.53.

16. Use Theorem 9.53 to obtain a double power series expansion of $e^x \cos y$ about $(0, 0)$.

17. Prove Theorem 9.55.

18. State and prove a theorem on the term-by-term integration of double power series.

19. Prove Theorem 9.56.

20. Use power series expansions to show that $\cos(x + y) = \cos x \cos y - \sin x \sin y$.

CHAPTER 10

Fourier Series

10.1. Expansions of Periodic Functions

In the study of power series in Chapter 9, we saw that an analytic function f can be represented by a power series

$$f(x) = \sum_{n=0}^{\infty} c_n(x - a)^n \qquad (10.1)$$

for all values of x within the radius of convergence of the series. We recall that f has derivatives of all orders and that the coefficients c_n in (10.1) are given by $f^{(n)}(a)/n!$. In this chapter we shall be interested in series expansions of functions *which may not be smooth*. That is, we shall consider functions which may have only a finite number of derivatives at some points and which may be discontinuous at others. Of course, in such cases it is not possible to have expansions in powers of $(x - a)$ such as (10.1). To obtain representations of unsmooth functions we turn to expansions in terms of trigonometric functions such as

$$1, \cos x, \cos 2x, \ldots, \cos nx, \ldots,$$

$$\sin x, \sin 2x, \ldots, \sin nx, \ldots.$$

A **trigonometric series** is one of the form

$$\frac{1}{2}a_0 + \sum_{n=1}^{\infty} (a_n \cos nx + b_n \sin nx) \qquad (10.2)$$

in which the coefficients $\{a_n\}$ and $\{b_n\}$ are constants. Let f be a real-valued function defined on $I = \{x: -\pi \leqslant x \leqslant \pi\}$. The coefficients a_n and b_n, $n = 0, 1, 2\ldots$, are to be determined in such a way that f is represented by (10.2). To do so we make use of the so-called *orthogonality relations* of the trigonometric functions:

$$\int_{-\pi}^{\pi} \cos mx \cos nx \, dx = \int_{-\pi}^{\pi} \sin mx \sin nx \, dx = \begin{cases} \pi & \text{if } m = n, \\ 0 & \text{if } m \neq n, \end{cases}$$

and

$$\int_{-\pi}^{\pi} \cos mx \sin nx \, dx = 0 \quad \text{for } m, n = 1, 2, \ldots .$$

These are easily verified by elementary methods of integration. With the help of these formulas, explicit expressions will be found for the coefficients a_n, b_n in a trigonometric expansion such as (10.2).

Theorem 10.1. *Let f be continuous on $I = \{x: -\pi \leqslant x \leqslant \pi\}$. Suppose that the series*

$$\frac{a_0}{2} + \sum_{n=1}^{\infty} (a_n \cos nx + b_n \sin nx) \tag{10.3}$$

converges uniformly to f for all $x \in I$. Then

$$a_n = \frac{1}{\pi} \int_{-\pi}^{\pi} f(t) \cos nt \, dt, \qquad n = 0, 1, 2, \ldots, \tag{10.4}$$

$$b_n = \frac{1}{\pi} \int_{-\pi}^{\pi} f(t) \sin nt \, dt, \qquad n = 1, 2, \ldots . \tag{10.5}$$

PROOF. We define the partial sums

$$s_k(x) = \frac{1}{2} a_0 + \sum_{m=1}^{k} (a_m \cos mx + b_m \sin mx).$$

Since the sequence $s_k(x)$ converges uniformly to $f(x)$, it follows that $s_k(x) \cos nx$ converges uniformly to $f(x) \cos nx$ as $k \to \infty$ for each fixed n. We merely observe that

$$|s_k(x) \cos nx - f(x) \cos nx| = |s_k(x) - f(x)| \cdot |\cos nx| \leqslant |s_k(x) - f(x)|.$$

Similarly, $s_k \sin nx$ converges uniformly to $f(x) \sin nx$ for each fixed n. Therefore, for each fixed n

$$f(x) \cos nx = \frac{a_0}{2} \cos nx + \sum_{m=1}^{\infty} (a_m \cos mx \cos nx + b_m \sin mx \cos nx).$$

This uniformly convergent series may be integrated term-by-term between $-\pi$ and π to yield

$$\int_{-\pi}^{\pi} f(x) \cos nx \, dx = \pi a_n.$$

Similarly, by repeating the argument for $f(x) \sin nx$, we get Formula (10.5). $\square$

The numbers a_n and b_n are called the **Fourier coefficients of** f. When the a_n and b_n are given by (10.4) and (10.5), the trigonometrics series (10.3) is called the **Fourier series** of the function f.

Let f be any integrable function defined on $I = \{x: -\pi \leqslant x \leqslant \pi\}$. Then the coefficients a_n and b_n may be computed according to (10.4) and (10.5). However, there is no assurance that the Fourier series (10.3) will converge to f if f is an arbitrary integrable function. In general, we write

$$f(x) \sim \frac{1}{2}a_0 + \sum_{n=1}^{\infty} (a_n \cos nx + b_n \sin nx)$$

to indicate that the series on the right may or may not converge to f at some points $x \in I$. One of the central problems in the study of Fourier series concerns the identification of large classes of functions with the property that the Fourier series of these functions actually converge to the appropriate values.

Definition. A function f defined on $I = \{x: a \leqslant x \leqslant b\}$ is said to be **piecewise continuous on** I if and only if (i) there is a subdivision

$$a = x_0 < x_1 < x_2 < \cdots < x_n = b$$

such that f is continuous on each subinterval $I_k = \{x: x_{k-1} < x < x_k\}$, and (ii) at each of the subdivision points $x_0, x_1, \ldots, x_n$ both one-sided limits of f exist.

Thus, a piecewise continuous function has a finite number of points of discontinuity which occur at $x_0, x_1, \ldots, x_n$. At each such point the limits

$$\lim_{x \to x_k-} f(x) \quad \text{and} \quad \lim_{x \to x_k+} f(x)$$

exist, and we denote them by $f(x_k -)$ and $f(x_k +)$, respectively. The quantity $f(x_k +) - f(x_k -)$ is called the **jump of** f **at** x_k. The coefficients a_n and b_n, given by integrals, are unaffected if the values of f are changed at any finite number of points. Hence, two functions f_1 and f_2 which differ only at a finite number of points have the same Fourier series. Let f be a given piecewise continuous function. We say that f is **standardized** if its values at points of discontinuity are given by

$$f(x_i) = \frac{1}{2}[f(x_i +) + f(x_i -)].$$

Standardizing a piecewise continuous function does not change the Fourier coefficients. See Figure 10.1 for an example of a standardized function. For convenience in the study of Fourier series we shall usually assume that piecewise continuous functions have been standardized.

Definitions. A function f is **piecewise smooth on** $I = \{x: a \leqslant x \leqslant b\}$ if and only if (i) f is piecewise continuous, and (ii) f' exists and is piecewise continuous

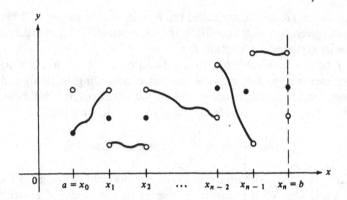

Figure 10.1. A standardized function.

on each subinterval $I_k = \{x: x_{k-1} < x < x_k\}$, $k = 1, 2, \ldots, n$. A function f is called **smooth on** I if and only if f and f' are continuous on I.

We postpone to Section 10.3 questions of convergence of Fourier series and consider now the formal process of determining the Fourier series for a variety of functions. Let f be a piecewise continuous function on $I = \{x: -\pi \leqslant x \leqslant \pi\}$. The **periodic extension** $\tilde{f}$ of f is defined by the formula

$$\tilde{f}(x) = \begin{cases} f(x) & \text{for } -\pi \leqslant x < \pi, \\ \tilde{f}(x - 2\pi) & \text{for } x \in \mathbb{R}^1. \end{cases}$$

Then we *standardize* $\tilde{f}$ at $-\pi$, π and all other points of discontinuity to that $\tilde{f}$ is defined for $-\infty < x < \infty$.

EXAMPLE 1. Find the Fourier series of the function

$$f(x) = x, \qquad x \in I = \{x: -\pi \leqslant x \leqslant \pi\}.$$

Solution. We form $\tilde{f}$, the periodic extension of f, and we standardize $\tilde{f}$ (see Figure 10.2). From (10.4) and (10.5), we find

$$a_n = \frac{1}{\pi} \int_{-\pi}^{\pi} x \cos nx \, dx, \qquad b_n = \frac{1}{\pi} \int_{-\pi}^{\pi} x \sin nx \, dx.$$

Integrating by parts, we obtain $a_0 = 0$ and

$$a_n = 0, \qquad b_n = (-1)^{n-1} \frac{2}{n}, \qquad n = 1, 2, \ldots.$$

Therefore

$$f(x) \sim 2\left[\sin x - \frac{\sin 2x}{2} + \frac{\sin 3x}{3} - \cdots \right], \qquad -\pi < x < \pi.$$

The series on the right represents $\tilde{f}$ for all values of x. We shall see later that

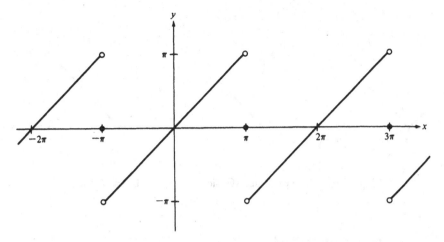

Figure 10.2. The periodic extension of $f(x) = x$, $-\pi \leqslant x \leqslant \pi$.

the series actually converges to x for $-\pi < x < \pi$, and clearly the series has
the value 0 for $x = \pm\pi$. □

In computing Fourier coefficients, we may frequently save labor by using
the integration properties of even and odd functions. A function f is said to
be **even** if

$$f(-x) = f(x)$$

for all x. Note that $\cos nx$ is an even function for every n. A function g is **odd** if

$$g(-x) = -g(x)$$

for all x. The functions $\sin nx$ are odd for all n. If c is any number, and if f is
even and g is odd, then

$$\int_{-c}^{c} f(x)\, dx = 2 \int_{0}^{c} f(x)\, dx, \qquad \int_{-c}^{c} g(x)\, dx = 0.$$

The product of two even functions is even, the product of two odd functions
is even, and the product of an even an odd function is odd.

We observe in Example 1 that $f(x) = x$ is odd. Therefore $f(x) \cos nx$ is odd
and so we can conclude *without computation* that $a_n = 0$ for all n.

EXAMPLE 2. Find the Fourier series for the function

$$f(x) = |x|, \qquad x \in I = \{x: -\pi \leqslant x \leqslant \pi\}.$$

Solution. We form the periodic extension of f as shown in Figure 10.3. Since
f is even, the functions $f(x) \sin nx$ are odd. Hence $b_n = 0$ for all n. Also, $a_0 = \pi$
and

$$a_n = \frac{2}{\pi} \int_{0}^{\pi} f(x) \cos nx\, dx = \frac{2}{\pi} \int_{0}^{\pi} x \cos nx\, dx.$$

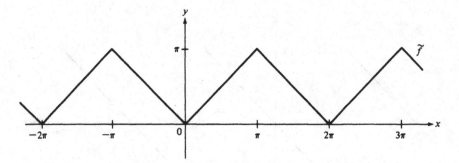

Figure 10.3. The periodic extension of $f(x) = |x|$, $-\pi \leqslant x \leqslant \pi$.

Integrating by parts, we find

$$a_n = \frac{2}{n^2\pi}[\cos n\pi - 1] = \frac{2}{n^2\pi}[(-1)^n - 1] = \begin{cases} \dfrac{-4}{(2k+1)^2\pi}, & n = 2k+1, \\ 0, & n = 2k+2, \end{cases}$$

for $k = 0, 1, 2, \ldots$. Therefore the Fourier series for f is given by

$$|x| \sim \frac{\pi}{2} - \frac{4}{\pi}\left[\frac{\cos x}{1^2} + \frac{\cos 3x}{3^2} + \cdots\right.$$
$$\left. + \frac{\cos(2k+1)x}{(2k+1)^2} + \cdots\right], \qquad -\pi \leqslant x \leqslant \pi.$$

Assuming for the moment that the above series actually converges to $|x|$, we set $x = 0$ and obtain the remarkable formula

$$\frac{1}{8}\pi^2 = \left[\frac{1}{1^2} + \frac{1}{3^2} + \frac{1}{5^2} + \frac{1}{7^2} + \cdots\right].$$

This formula is correct and may actually be used to compute π to any desired degree of accuracy. However, vastly superior series expansions are available for calculating π numerically. $\quad\square$

Problems

In each of Problems 1 though 10, find the Fourier series for the given function f.

1. $f(x) = \begin{cases} 0 & \text{for } x \in I_1 = \{x: -\pi \leqslant x < 0\}, \\ 1 & \text{for } x \in I_2 = \{x: 0 \leqslant x \leqslant \pi\}. \end{cases}$

2. $f(x) = \begin{cases} 0 & \text{for } x \in I_1 = \{x: -\pi \leqslant x < \pi/2\}, \\ 1 & \text{for } x \in I_2 = \{x: \pi/2 \leqslant x \leqslant \pi\}. \end{cases}$

3. $f(x) = x^2$ for $x \in I = \{x: -\pi \leqslant x \leqslant \pi\}$.

4. $f(x) = \begin{cases} 0 & \text{for } x \in I_1 = \{x: -\pi \leqslant x < 0\}, \\ x & \text{for } x \in I_2 = \{x: 0 \leqslant x \leqslant \pi\}. \end{cases}$

5. $f(x) = |\cos x|$ for $x \in I = \{x: -\pi \leqslant x \leqslant \pi\}$.

6. $f(x) = x^3$ for $x \in I = \{x: -\pi \leqslant x \leqslant \pi\}$.

7. $f(x) = e^{2x}$ for $x \in I = \{x: -\pi \leqslant x \leqslant \pi\}$.

8. $f(x) = \begin{cases} 0 & \text{for } x \in I_1 = \{x: -\pi \leqslant x < 0\}, \\ \sin x & \text{for } x \in I_2 = \{x: 0 \leqslant x \leqslant \pi\}. \end{cases}$

9. $f(x) = \sin^2 x$ for $x \in I = \{x: -\pi \leqslant x \leqslant \pi\}$.

10. $f(x) = x \sin x$ for $x \in I = \{x: -\pi \leqslant x \leqslant \pi\}$.

11. $f(x) = \begin{cases} -1 & \text{for } x \in I_1 = \{x: -\pi \leqslant x \leqslant 0\}, \\ 1 & \text{for } x \in I_2 = \{x: 0 < x \leqslant \pi\}. \end{cases}$

12. $f(x) = x^3$ for $x \in I = \{x: -\pi \leqslant x < \pi\}$.

13. $f(x) = \cos^3 x$ for $x \in I = \{x: -\pi < x < \pi\}$.

14. $f(x) = \sin^2 2x$ for $x \in I = \{x: -\pi < x < \pi\}$.

15. Given $f(x) = \begin{cases} \cos x & \text{for } x \in I_1 = \{x: -\pi < x \leqslant 0\}, \\ -\cos x & \text{for } x \in I_2 = \{x: 0 < x \leqslant \pi\}. \end{cases}$

Sketch $\tilde{f}$, the periodic extension of f, for x in $I_3 = \{x: -3\pi \leqslant x \leqslant 3\pi\}$.

16. Given $f(x) = x + \sin x$ for $x \in I = \{x: -\pi < x \leqslant \pi\}$. Sketch $\tilde{f}$, the periodic extension of f, for x in $I_1 = \{x: -3\pi \leqslant x < 3\pi\}$.

17. Verify the formulas

$$\int_{-\pi}^{\pi} \cos mx \cos nx \, dx = \int_{-\pi}^{\pi} \sin mx \sin nx \, dx = \begin{cases} \pi & \text{if } m = n, \\ 0 & \text{if } m \neq n. \end{cases}$$

18. (a) Find the Fourier series for

$$f(x) = \begin{cases} -\pi/4 & \text{for } x \in I_1 = \{x: -\pi \leqslant x < 0\}, \\ \pi/4 & \text{for } x \in I_2 = \{x: 0 \leqslant x \leqslant \pi\}. \end{cases}$$

(b) Assuming the series in Part (a) converges to f (as standardized), show that

(i) $\dfrac{\pi}{4} = 1 - \dfrac{1}{3} + \dfrac{1}{5} - \dfrac{1}{7} + \cdots$.

(ii) $\dfrac{\pi}{3} = 1 + \dfrac{1}{5} - \dfrac{1}{7} - \dfrac{1}{11} + \dfrac{1}{13} + \dfrac{1}{17} - \cdots$.

(iii) $\dfrac{\sqrt{3}}{6}\pi = 1 - \dfrac{1}{5} + \dfrac{1}{7} - \dfrac{1}{11} + \dfrac{1}{13} - \dfrac{1}{17} + \cdots$.

19. (a) Find the Fourier series for

$$f(x) = x + x^2 \quad \text{for} \quad x \in I = \{x: -\pi \leqslant x \leqslant \pi\}.$$

(b) Assuming the series in Part (a) converges to f (as standardized), show that $\pi^2/6 = \sum_{n=1}^{\infty} 1/n^2$.

10.2. Sine Series and Cosine Series; Change of Interval

Suppose that we wish to find the Fourier series of a function f which has domain $J = \{x: 0 \leqslant x \leqslant \pi\}$. Since the Fourier coefficients a_n and b_n are given in terms of integrals from $-\pi$ to π, we must somehow change the domain of f to $I = \{x: -\pi \leqslant x \leqslant \pi\}$. We can do this simply by defining f arbitrarily on the subinterval $I' = \{x: -\pi \leqslant x < 0\}$. Since we are interested in f only on J, properties of convergence of the series on I' are irrelevant. For example, we may set $f \equiv 0$ on I'. However, one choice which is useful for many purposes consists in defining f as an *even* function on I. Since $b_n = 0$, $n = 1, 2, \ldots$, for even functions, the Fourier series will have cosine terms only. We call such a series a **cosine series** and, as the original function has domain J, the Fourier expansion is called a **half-range series**.

A function f defined on J may be extended to I as an odd function. Then $a_n = 0$ for $n = 0, 1, 2, \ldots$, and the resulting series is called a **sine series**. We illustrate the process of obtaining cosine and sine series with two examples.

EXAMPLE 1. Given the function

$$f(x) = \begin{cases} 0 & \text{for } I_1 = \{x: 0 \leqslant x < \pi/2\}, \\ 1 & \text{for } I_2 = \{x: \pi/2 \leqslant x \leqslant \pi\}, \end{cases}$$

find the cosine series for f.

Solution. We extend f as an even function, as shown in Figure 10.4. Then the function is standardized so that $\tilde{f}(\pi/2) = \tilde{f}(3\pi/2) = \tilde{f}(-\pi/2) = \frac{1}{2}$. Since $\tilde{f}$ is even, we have $b_n = 0$, $n = 1, 2, \ldots$ Also,

$$a_n = \frac{2}{\pi} \int_0^\pi f(x) \cos nx \, dx, \qquad n = 0, 1, 2, \ldots.$$

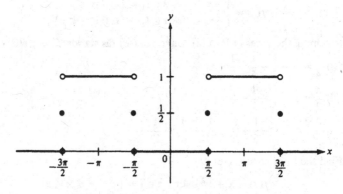

Figure 10.4. Extension as an even function.

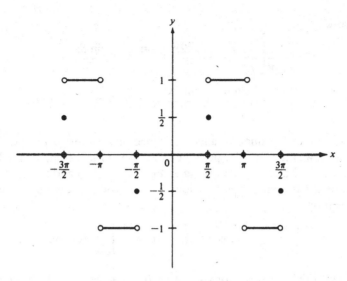

Figure 10.5. Extension as an odd function.

A simple calculation yields

$$a_0 = 1,$$

$$a_n = \frac{2}{\pi} \int_{\pi/2}^{\pi} \cos nx \, dx = \begin{cases} 0 & \text{if } n \text{ is even,} \\ \dfrac{2(-1)^{k+1}}{(2k+1)\pi} & \text{if } n = 2k+1, \, k = 0, 1, 2, \dots. \end{cases}$$

Therefore

$$\tilde{f}(x) \sim \frac{1}{2} - \frac{2}{\pi}\left[\frac{\cos x}{1} - \frac{\cos 3x}{3} + \frac{\cos 5x}{5} - \cdots \right], \qquad x \in I_1 \cup I_2. \qquad \square$$

EXAMPLE 2. Find the sine series for the function f for Example 1.

Solution. We extend f as an odd function as shown in Figure 10.5. The standardized function has $\tilde{f}(-3\pi/2) = \tilde{f}(\pi/2) = \cdots = \frac{1}{2}$, $\tilde{f}(-\pi) = \tilde{f}(\pi) = \cdots = 0$, and $\tilde{f}(-\pi/2) = \tilde{f}(3\pi/2) = \cdots = -\frac{1}{2}$. Then $a_n = 0$ for $n = 0, 1, 2, \dots$ and

$$b_n = \frac{2}{\pi} \int_0^{\pi} f(x) \sin nx \, dx = \frac{2}{\pi} \int_{\pi/2}^{\pi} \sin nx \, dx.$$

Hence

$$b_n = \begin{cases} \dfrac{2}{n\pi} & \text{if } n \text{ is odd,} \\ \dfrac{2}{k\pi}[(-1)^k - 1] & \text{if } n = 2k \text{ and } k = 1, 2, \dots. \end{cases}$$

Therefore

$$\tilde{f}(x) \sim \frac{2}{\pi}\left[\frac{\sin x}{1} - 2\frac{\sin 2x}{2} + \frac{\sin 3x}{3} + \frac{\sin 5x}{5}\right.$$

$$\left. - \frac{2\sin 6x}{6} + \frac{\sin 7x}{7} + \cdots\right] \quad \text{for} \quad x \in I_1 \cup I_2. \qquad \square$$

If f is a piecewise smooth function defined on an interval $J = \{x: c - \pi < x < c + \pi\}$, we may form the periodic extension of f and compute the Fourier coefficients $\{a_n\}$ and $\{b_n\}$ according to Formulas (10.4) and (10.5). It is clear that since the trigonometric functions have period 2π, these coefficients are also given by

$$a_n = \frac{1}{\pi}\int_{c-\pi}^{c+\pi} f(x)\cos nx\, dx, \qquad b_n = \frac{1}{\pi}\int_{c-\pi}^{c+\pi} f(x)\sin nx\, dx.$$

EXAMPLE 3. Given $f(x) = x$ for $x \in I = \{x: 0 \leqslant x \leqslant 2\pi\}$, find the Fourier series for f.

Solution. We extend f to be periodic and standardized, as shown in Figure 10.6. We compute the coefficients

$$a_0 = \frac{1}{\pi}\int_0^{2\pi} x\, dx = 2\pi, \qquad a_n = \frac{1}{\pi}\int_0^{2\pi} x\cos nx\, dx = 0, \qquad n = 1, 2, \ldots.$$

Also,

$$b_n = \frac{1}{\pi}\int_0^{2\pi} x\sin nx\, dx = -\frac{2}{n}, \qquad n = 1, 2, \ldots.$$

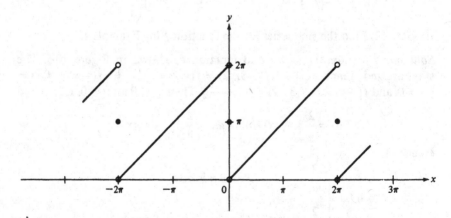

Figure 10.6. Periodic extension of $f(x) = x$, $0 \leqslant x \leqslant 2\pi$.

Therefore

$$x \sim \pi - 2 \sum_{n=1}^{\infty} \frac{\sin nx}{n} \quad \text{for} \quad 0 < x < 2\pi.$$ □

A function f which is piecewise smooth on an interval $I = \{x: -L \leqslant x \leqslant L\}$ for some number $L > 0$ can be represented by a **modified Fourier series**. We introduce a change of variable and define y and $g(y)$ by the relations

$$y = \frac{\pi x}{L}, \quad f(x) = f\left(\frac{Ly}{\pi}\right) = g(y) = g\left(\frac{\pi x}{L}\right).$$

The transformation maps I onto $I' = \{x: -\pi \leqslant x \leqslant \pi\}$, and then g is a piecewise smooth function on I'. Therefore

$$g(y) \sim \frac{a_0}{2} + \sum_{n=1}^{\infty} (a_n \cos ny + b_n \sin ny), \quad y \in I', \quad (10.6)$$

with

$$a_n = \frac{1}{\pi} \int_{-\pi}^{\pi} g(y) \cos ny \, dy, \quad b_n = \frac{1}{\pi} \int_{-\pi}^{\pi} g(y) \sin ny \, dy.$$

Returning to the variable x and the function f, we get the formulas for the coefficients a_n, b_n of the modified series which correspond to those in Theorem 10.1:

$$a_n = \frac{1}{L} \int_{-L}^{L} f(x) \cos \frac{n\pi x}{L} \, dx, \quad b_n = \frac{1}{L} \int_{-L}^{L} f(x) \sin \frac{n\pi x}{L} \, dx.$$

Series (10.6) becomes

$$f(x) \sim \frac{a_0}{2} + \sum_{n=1}^{\infty} \left(a_n \cos \frac{n\pi x}{L} + b_n \sin \frac{n\pi x}{L}\right), \quad x \in I.$$

EXAMPLE 4. Given

$$f(x) = \begin{cases} x + 1 & \text{for } x \in I_1 = \{x: -1 \leqslant x < 0\}, \\ x - 1 & \text{for } x \in I_2 = \{x: 0 \leqslant x \leqslant 1\}, \end{cases}$$

find the Fourier series of f on $I = I_1 \cup I_2$.

Solution. The graph of the (standardized) periodic extension of f is shown in Figure 10.7. Observe that f is odd and so $a_n = 0$ for $n = 0, 1, 2, \ldots$ Also,

$$b_n = 2 \int_0^1 (x - 1) \sin n\pi x \, dx = -\frac{2}{n\pi}, \quad n = 1, 2, \ldots.$$

We find

$$f(x) \sim -\frac{2}{\pi} \sum_{n=1}^{\infty} \frac{\sin n\pi x}{n} \quad \text{for} \quad x \in I.$$ □

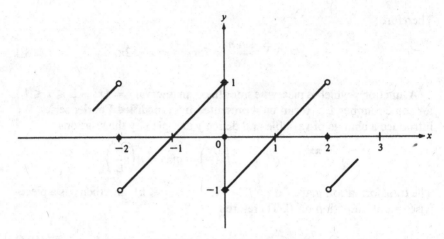

Figure 10.7. A standardized periodic extension.

PROBLEMS

In each of Problems 1 through 4 expand each function f in a cosine series. Sketch the standardized extension of f.

1. $f(x) = \begin{cases} 1 & \text{for } x \in I_1 = \{x: 0 \leqslant x < \pi/2\}, \\ 0 & \text{for } x \in I_2 = \{x: \pi/2 \leqslant x \leqslant \pi\}. \end{cases}$

2. $f(x) = \sin x$ for $x \in I = \{x: 0 \leqslant x \leqslant \pi\}$.

3. $f(x) = x$ for $x \in I = \{x: 0 \leqslant x \leqslant \pi\}$.

4. $f(x) = x^3$ for $x \in I = \{x: 0 \leqslant x \leqslant \pi\}$.

In each of Problems 5 through 8 expand each function f in a sine series. Sketch the standardized extension of f.

5. $f(x) = \begin{cases} 1 & \text{for } x \in I_1 = \{x: 0 \leqslant x < \pi/2\}, \\ -1 & \text{for } x \in I_2 = \{x: \pi/2 \leqslant x \leqslant \pi\}. \end{cases}$

6. $f(x) = \cos x$ for $x \in I = \{x: 0 \leqslant x \leqslant \pi\}$.

7. $f(x) = x$ for $x \in I = \{x: 0 \leqslant x \leqslant \pi\}$.

8. $f(x) = x^3$ for $x \in I = \{x: 0 \leqslant x \leqslant \pi\}$.

In each of Problems 9 through 12 find the Fourier series of the function f on the interval $I = \{x: -L < x < L\}$.

9. $f(x) = \begin{cases} 1 & \text{for } x \in I_1 = \{x: -2 \leqslant x < 0\}, \\ -1 & \text{for } x \in I_2 = \{x: 0 \leqslant x \leqslant 2\}. \end{cases}$

10. $f(x) = \begin{cases} 0 & \text{for } x \in I_1 = \{x: -2 \leqslant x < 0\}, \\ x & \text{for } x \in I_2 = \{x: 0 \leqslant x \leqslant 2\}. \end{cases}$

11. $f(x) = x^2$ for $x \in I = \{x: -1 \leqslant x \leqslant 1\}$.

12. $f(x) = 1 - |x|$ for $x \in I = \{x: -1 \leqslant x \leqslant 1\}$.

13. Suppose f is an odd function for $x \in I = \{x: -L < x < L\}$. In addition, suppose $f(L - x) = f(x)$ for $x \in I$. Prove that the Fourier coefficients $b_n = 0$ for all even n.

14. Suppose f is an odd function for $x \in I = \{x: -L < x < L\}$, and that $f(L - x) = -f(x)$ for $x \in I$. Prove that the Fourier coefficients $b_n = 0$ for all odd n.

10.3. Convergence Theorems

It is important to establish simple criteria which determine when a Fourier series converges. In this section we show how to obtain large classes of functions with the property that for each value of x in the domain of a function f, the Fourier series converges to $f(x)$.

Before establishing the next result which is useful throughout the study of Fourier series, we exhibit two simple facts about integrals. The first states that there are functions f which are not integrable but are such that f^2 is integrable. To see this we set $I = \{x: 0 \leqslant x \leqslant 1\}$ and define

$$f(x) = \begin{cases} 1 & \text{if } x \in I \text{ and } x \text{ is rational,} \\ -1 & \text{if } x \in I \text{ and } x \text{ is irrational.} \end{cases}$$

Then every upper Darboux sum is 1 and every lower Darboux sum is -1. Thus f is not integrable. On the other hand, $f^2(x) \equiv 1$ for all $x \in I$, and f^2 is integrable. The second fact states that if f is integrable on $I = \{x: a \leqslant x \leqslant b\}$, then f^2 is also integrable. To see this we observe that f is bounded on I and define M to be the l.u.b.$_{x \in I} f(x)$. Then using Darboux sums with the usual notation

$$S^+(f^2, \Delta) - S^-(f^2, \Delta) = \sum_{i=1}^{n} (M_i^2 - m_i^2) \Delta_i x$$

$$= \sum_{i=1}^{n} (M_i + m_i)(M_i - m_i) \Delta_i x$$

$$\leqslant 2M \sum_{i=1}^{n} (M_i - m_i) \Delta_i x.$$

Since f is integrable on I, the sum on the right tends to zero as $n \to \infty$ and $\|\Delta\| \to 0$. Hence f^2 is integrable.

We note at this point that in the study of integrals for unbounded functions which we take up in Chapter 11, it may happen that a function f (unbounded) may be integrable while f^2 is not integrable.

Theorem 10.2 (Bessel's inequality). *Suppose that f is integrable on $I = \{x: -\pi \leqslant x \leqslant \pi\}$. Let*

$$\frac{1}{2}a_0 + \sum_{n=1}^{\infty} (a_n \cos nx + b_n \sin nx)$$

be the Fourier series of f. Then

$$\frac{1}{2}a_0^2 + \sum_{n=1}^{\infty} (a_n^2 + b_n^2) \leqslant \frac{1}{\pi} \int_{-\pi}^{\pi} f^2(x)\,dx \quad \text{(Bessel's inequality)}. \quad (10.7)$$

PROOF. Denote the nth partial sum of the Fourier series by $s_n(x)$; that is,

$$s_n(x) = \frac{1}{2}a_0 + \sum_{k=1}^{n} (a_k \cos kx + b_k \sin kx).$$

Now write

$$\int_{-\pi}^{\pi} [f(t) - s_n(t)]^2\,dt = \int_{-\pi}^{\pi} f^2(t)\,dt - 2\int_{-\pi}^{\pi} f(t)s_n(t)\,dt + \int_{-\pi}^{\pi} s_n^2(t)\,dt. \quad (10.8)$$

From the definition of the Fourier coefficients, it follows that

$$\frac{1}{2}a_0^2 + \sum_{k=1}^{n} (a_k^2 + b_k^2) = \frac{1}{\pi} \int_{-\pi}^{\pi} f(t)s_n(t)\,dt. \quad (10.9)$$

Also, by multiplying out the terms of $s_n^2(t)$ and taking into account the orthogonality relations of the trigonometric functions, we verify that

$$\int_{-\pi}^{\pi} s_n^2(t)\,dt = \int_{-\pi}^{\pi} f(t)s_n(t)\,dt. \quad (10.10)$$

Therefore (10.8), (10.9), and (10.10) may be combined to give

$$0 \leqslant \int_{-\pi}^{\pi} [f(t) - s_n(t)]^2\,dt = \int_{-\pi}^{\pi} f^2(t)\,dt - \pi\left\{\frac{1}{2}a_0^2 + \sum_{k=1}^{n} (a_k^2 + b_k^2)\right\}. \quad (10.11)$$

Since f^2 is integrable we may let n tend to infinity in (10.11) and obtain

$$\frac{1}{2}a_0^2 + \sum_{k=1}^{\infty} (a_k^2 + b_k^2) \leqslant \frac{1}{\pi} \int_{-\pi}^{\pi} f^2(t)\,dt < \infty. \qquad \square$$

Bessel's inequality shows that a_n and b_n tend to zero as $n \to \infty$ for any function whose square is integrable on $I = \{x: -\pi \leqslant x \leqslant \pi\}$.

An expression which occurs frequently in the study of convergence of Fourier series is the **Dirichlet kernel** D_n defined by

$$D_n: x \to \frac{\sin(n + \frac{1}{2})x}{2 \sin \frac{1}{2}x}.$$

By using the trigonometric identity

$$2 \sin \tfrac{1}{2}x \cos kx = \sin(k + \tfrac{1}{2})x + \sin(k - \tfrac{1}{2})x,$$

it is easy to verify that

$$\frac{1}{2} + \sum_{k=1}^{n} \cos kx \equiv \frac{\sin(n + \frac{1}{2})x}{2 \sin \frac{1}{2}x} = D_n(x). \quad (10.12)$$

Thus the Dirichlet kernel has the following properties:

(i) $D_n(x)$ is an even function of x.
(ii) $\int_{-\pi}^{\pi} D_n(x)\, dx = \pi$.
(iii) D_n has period 2π.

Lemma 10.1. *Suppose that s_n is the nth partial sum of the Fourier series of a piecewise continuous function f with period 2π. Then*

$$s_n(x) - \frac{1}{2}[f(x+) + f(x-)] = \frac{1}{\pi} \int_0^{\pi} [f(x+u) - f(x+)]D_n(u)\, du$$

$$+ \frac{1}{\pi} \int_0^{\pi} [f(x-u) - f(x-)]D_n(u)\, du. \quad (10.13)$$

PROOF. Since $s_n(x) = \frac{1}{2}a_0 + \sum_{k=1}^{n}(a_k \cos kx + b_k \sin kx)$, we may insert the formulas for a_k and b_k to get

$$s_n(x) = \frac{1}{\pi} \int_{-\pi}^{\pi} f(t)\left[\frac{1}{2} + \sum_{k=1}^{n} (\cos kt \cos kx + \sin kt \sin kx)\right] dt$$

$$= \frac{1}{\pi} \int_{-\pi}^{\pi} f(t)\left[\frac{1}{2} + \sum_{k=1}^{n} \cos k(t-x)\right] dt.$$

We set $t = x + u$ in the above integral, obtaining

$$s_n(x) = \frac{1}{\pi} \int_{-\pi-x}^{\pi-x} f(x+u)\left[\frac{1}{2} + \sum_{k=1}^{n} \cos ku\right] du = \frac{1}{\pi} \int_{-\pi-x}^{\pi-x} f(x+u)D_n(u)\, du.$$

Since D_n and f are periodic with period 2π, the interval of integration may be changed to $I = \{u: -\pi < u < \pi\}$. Therefore

$$s_n(x) = \frac{1}{\pi} \int_{-\pi}^{0} f(x+v)D_n(v)\, dv + \frac{1}{\pi} \int_0^{\pi} f(x+u)D_n(u)\, du.$$

We replace v by $-u$ in the first integral (recalling that $D_n(-v) = D_n(v)$), and hence

$$s_n(x) = \frac{1}{\pi} \int_0^{\pi} [f(x+u) + f(x-u)]D_n(u)\, du.$$

Then taking Property (ii) of D_n into account, we obtain (10.13).   □

Theorem 10.3. *Suppose that f is piecewise smooth, standardized, and periodic with period 2π. Then the Fourier series of f converges to $f(x)$ for each x.*

PROOF. For each value of x, we shall show that $s_n(x) - f(x) \to 0$ as $n \to \infty$. We write (10.13) of Lemma 10.1 in the form

$$s_n(x) - f(x) = \frac{1}{\pi} \int_0^{\pi} [f(x+u) - f(x+)]D_n(u)\, du$$

$$+ \frac{1}{\pi} \int_0^\pi [f(x - u) - f(x -)] D_n(u) \, du$$

$$\equiv S_n(x) + T_n(x).$$

From the definition of $D_n(u)$, it follows that

$$T_n(x) = \frac{1}{\pi} \int_0^\pi \frac{f(x - u) - f(x -)}{2 \sin \frac{1}{2}u} \sin\left(n + \frac{1}{2}\right) u \, du$$

$$= \frac{1}{\pi} \int_0^\pi \frac{f(x - u) - f(x -)}{2 \sin \frac{1}{2}u} \left(\sin nu \cos \frac{1}{2}u + \cos nu \sin \frac{1}{2}u\right) du. \quad (10.14)$$

Define g_1 and g_2 by the formulas

$$g_1(x, u) = \frac{f(x - u) - f(x -)}{2 \sin \frac{1}{2}u} \cos \frac{1}{2}u,$$

$$g_2(x, u) = \tfrac{1}{2}(f(x - u) - f(x -)).$$

Then g_1 and g_2 are piecewise smooth except possibly for g_1 at $u = 0$. However, l'Hôpital's rule, shows that

$$g_1(x, 0 +) = -f'(x -),$$

and so g_1 is piecewise smooth everywhere. We now write (10.14) in the form

$$T_n(x) = \frac{1}{\pi} \int_0^\pi (g_1(x, u) \sin nu + g_2(x, u) \cos nu) \, du,$$

and we see that the right side is the nth Fourier coefficient of the sine series for $\frac{1}{2}g_1$ plus the nth coefficient of the cosine series for $\frac{1}{2}g_2$. According to Bessel's inequality, these coefficients tend to 0 as $n \to \infty$. Hence $T_n(x) \to 0$ as $n \to \infty$. Similarly, $S_n(x) \to 0$ as $n \to \infty$. □

A criterion for convergence which is less restrictive than Theorem 10.3 can be obtained with the aid of the following lemma, which has a number of important applications.

Theorem 10.4 (Riemann–Lebesgue lemma). *Let* f *be defined on* $I = \{x: -\pi \leqslant x \leqslant \pi\}$, *and suppose that* $|f|$ *is integrable on* I. *Let* $\{a_n\}$, $\{b_n\}$ *be the Fourier coefficients of* f. *Then* a_n, $b_n \to 0$ *as* $n \to \infty$.

PROOF. The main consideration occurs when $|f|$ is integrable in an improper sense.[1] (Otherwise f^2 is also integrable and Bessel's inequality yields the result.)

[1] For a discussion of improper integrals see Section 11.2. See, also, the example before Theorem 10.2.

For any positive number N we define

$$f_N(x) = \begin{cases} f(x) & \text{if } |f(x)| \leqslant N, \\ 0 & \text{if } |f(x)| > N. \end{cases}$$

Then from the definition of an improper integral, for any $\varepsilon > 0$, there is an N sufficiently large so that

$$\int_{-\pi}^{\pi} |f(x) - f_N(x)| \, dx < \frac{1}{2}\varepsilon.$$

Since $|f_N(x)|$ is bounded by N it follows that

$$\int_{-\pi}^{\pi} |f_N(x)|^2 \, dx \leqslant N \int_{-\pi}^{\pi} |f_N(x)| \, dx \leqslant N \int_{-\pi}^{\pi} |f(x)| \, dx < \infty.$$

Therefore by Bessel's inequality the Fourier coefficients for $f_N(x)$ tend to 0 as $n \to \infty$. That is, for any $\varepsilon > 0$, there is an n_0 such that

$$\left| \int_{-\pi}^{\pi} f_N(x) \cos nx \, dx \right| < \frac{1}{2}\varepsilon, \quad \left| \int_{-\pi}^{\pi} f_N(x) \sin nx \, dx \right| < \frac{1}{2}\varepsilon \quad \text{for} \quad n > n_0.$$

Consequently, for $n > n_0$ and N sufficiently large

$$\left| \int_{-\pi}^{\pi} f(x) \cos nx \, dx \right|$$
$$\leqslant \left| \int_{-\pi}^{\pi} f_N(x) \cos nx \, dx \right| + \left| \int_{-\pi}^{\pi} (f(x) - f_N(x)) \cos nx \, dx \right|$$
$$\leqslant \frac{1}{2}\varepsilon + \int_{-\pi}^{\pi} |f(x) - f_N(x)| \cdot |\cos nx| \, dx < \frac{1}{2}\varepsilon + \frac{1}{2}\varepsilon = \varepsilon.$$

The result for the sine coefficients is obtained in the same way. $\quad\square$

With the aid of the Riemann–Lebesgue lemma we establish the following criterion for convergence of Fourier series which is somewhat more general than Theorem 10.3.

Theorem 10.5 (Dini's test). *Suppose that f is standardized, periodic with period 2π, and that the integrals*

$$\int_{-\pi}^{\pi} \left| \frac{f(x+u) - f(x+)}{\sin \frac{1}{2}u} \right| \, du, \quad \int_{-\pi}^{\pi} \left| \frac{f(x-u) - f(x-)}{\sin \frac{1}{2}u} \right| \, du, \quad (10.15)$$

are finite for some value of s. Then the Fourier series of f converges to $f(x)$.

PROOF. We repeat the proof of Theorem 10.3 until Equation (10.14). Then because the integrals in (10.15) are finite we can apply the Riemann–Lebesgue lemma to the functions g_1 and g_2 defined in the proof of Theorem 10.3. Since f is assumed standardized, we conclude that $f(x) - s_n(x) \to 0$ as $n \to \infty$. $\quad\square$

Remarks

(i) Since the quantity $|(\sin \frac{1}{2}u)/u|$ is bounded, the Conditions (10.15) may be replaced by

$$\int_{-\pi}^{\pi} \left| \frac{f(x+u) - f(x+)}{u} \right| du < \infty,$$

$$\int_{-\pi}^{\pi} \left| \frac{f(x-u) - f(x-)}{u} \right| du < \infty. \tag{10.16}$$

(ii) It is important to have simple criteria on f itself, rather than a condition such as (10.16) which is usually difficult to verify. Observe that if f *has a bounded derivative*, then the integrals in (10.16) are certainly finite, in fact bounded by

$$2\pi \max_{-\pi \leqslant x \leqslant \pi} |f'(x)|.$$

(iii) A function is **Hölder continuous** if and only if for each x, there are constants M and α with $0 < \alpha \leqslant 1$ such that

$$|f(x) - f(y)| \leqslant M|x - y|^\alpha$$

for all y in $I = \{y: -\pi \leqslant y \leqslant \pi\}$. If f is Hölder continuous, then the integrals in (10.16) are bounded by an integral of the form

$$A \int_{-\pi}^{\pi} |u|^{\alpha-1} \, du.$$

Such an integral is finite so long as α is positive, and we note that any Hölder continuous function satisfies Condition (10.16).

It is useful to know when Fourier series can be differentiated and integrated term by term. For this purpose the following lemma is useful.

Lemma 10.2. *Suppose that f has period 2π and is piecewise smooth. Then its Fourier coefficients $a_n, b_n, n = 1, 2, \ldots$, satisfy the inequalities*

$$|a_n| \leqslant \frac{C}{n}, \qquad |b_n| \leqslant \frac{C}{n}, \qquad n = 1, 2, \ldots,$$

where C is a constant which depends only on f.

PROOF. Suppose the jumps of f occur at $-\pi = x_0 < x_1 < \cdots < x_{r-1} < x_r = \pi$. Then

$$a_n = \frac{1}{\pi} \int_{-\pi}^{\pi} f(t) \cos nt \, dt = \frac{1}{\pi} \sum_{i=1}^{r} \int_{x_{i-1}}^{x_i} f(t) \cos nt \, dt.$$

We may integrate by parts, and obtain

$$a_n = \frac{1}{\pi} \sum_{i=1}^{r} \left[\frac{f(t) \sin nt}{n} \right]_{x_{i-1}}^{x_i} - \frac{1}{\pi} \sum_{i=1}^{r} \frac{1}{n} \int_{x_{i-1}}^{x_i} f'(x) \sin nt \, dt.$$

Since f and f' are bounded, we obtain at once the estimate for a_n. The result for b_n is similar.   □

Corollary. *Suppose that f and its first $p - 2$ derivatives are periodic with period 2π, and $f^{(p-1)}$ is piecewise smooth. Then the Fourier coefficients a_n, b_n of f satisfy the inequalities*

$$|a_n| \leqslant \frac{C}{n^p}, \qquad |b_n| \leqslant \frac{C}{n^p}, \qquad n = 1, 2, \ldots,$$

where C does not depend on n.

To establish the Corollary we follow the method in the proof of Lemma 10.2, integrating by parts p times. The Corollary shows that *the more derivatives a function possesses, the more rapidly its Fourier series converges.*

Theorem 10.6 (Term-by-term differentiation of Fourier series). *Suppose that f is continuous everywhere and periodic with period 2π. Suppose that f' is piecewise smooth and standardized. Then*

(i) *The series obtained by differentiating the Fourier series for f term by term converges at every point to $f'(x)$.*
(ii) *The Fourier series of f converges uniformly to $f(x)$ for all x.*

PROOF. Let the jumps of f' occur at $-\pi = x_0 < x_1 < \cdots < x_{r-1} < x_r = \pi$. Define

$$g(x) = \int_{-\pi}^{x} f'(t)\, dt$$

and observe that g is continuous. Also, since $g' - f' \equiv 0$ for $x_{i-1} < x < x_i$, $i = 1, 2, \ldots, r$, the function $g - f$ must be constant on each subinterval. Since g and f are both continuous, $g - f$ is identically constant. Denote the Fourier coefficients of f' by A_n, B_n. For $n = 1, 2, \ldots$, we write

$$a_n = \frac{1}{\pi} \sum_{i=1}^{r} \int_{x_{i-1}}^{x_i} f(t) \cos nt\, dt.$$

We integrate by parts, obtaining

$$a_n = \frac{1}{\pi} \sum_{i=1}^{r} \frac{f(x_i) \sin nx_i - f(x_{i-1}) \sin nx_{i-1}}{n} - \frac{1}{n\pi} \sum_{i=1}^{r} \int_{x_{i-1}}^{x_i} f'(t) \sin nt\, dt$$

$$= \frac{-1}{n\pi} \int_{-\pi}^{\pi} f'(t) \sin nt\, dt = -\frac{B_n}{n}.$$

Similarly, we find

$$b_n = \frac{A_n}{n}.$$

Differentiation of the series

$$f(x) = \frac{1}{2}a_0 + \sum_{n=1}^{\infty} (a_n \cos nx + b_n \sin nx),$$

term by term yields the Fourier series for f'. The fact that the Fourier series for f' converges follows from Theorem 10.3.

To show that the Fourier series for f converges uniformly, we apply the Corollary to Lemma 10.2 to obtain

$$\left| \sum_{n=1}^{\infty} a_n \cos nx + b_n \sin nx \right| \leqslant 2C \sum_{n=1}^{\infty} \frac{1}{n^2}.$$

Since the series of constants on the right converges, the Fourier series converges uniformly.   □

Theorem 10.7 (Term-by-term integration of Fourier series). *Suppose that f is piecewise smooth everywhere and periodic with period 2π. Assume that the Fourier coefficient a_0 is zero, and define*

$$F(x) = \int_{-\pi}^{x} f(t)\, dt.$$

Then the Fourier series for F is obtained by integrating term by term the Fourier series for f, except for the constant term A_0 which is given by

$$A_0 = -\frac{1}{\pi} \int_{-\pi}^{\pi} xf(x)\, dx.$$

PROOF. The condition $a_0 = 0$ is required in order that F have period 2π. The relationship between the Fourier series for F and that for f now follows from Theorem 10.6. To find A_0, observe that

$$A_0 = \frac{1}{\pi} \int_{-\pi}^{\pi} \int_{-\pi}^{x} f(t)\, dt\, dx = \frac{1}{\pi} \int_{-\pi}^{\pi} f(t) \left[\int_{t}^{\pi} dx \right] dt$$

$$= \frac{1}{\pi} \int_{-\pi}^{\pi} (\pi - t) f(t)\, dt = -\frac{1}{\pi} \int_{-\pi}^{\pi} tf(t)\, dt.$$ □

Remarks

(i) If a function f does not have the property that $a_0 = 0$, we define $g(x) = f(x) - \frac{1}{2}a_0$, to which Theorem 10.7 will apply.

(ii) Theorem 10.7 does not require the uniform convergence of the derivative series $F'(x) = f(x)$. In general, integrated series will have better rates of convergence than the series itself.

EXAMPLE. In Example 1 of Section 10.1, we established the expansion

$$f(x) = x = 2 \sum_{n=1}^{\infty} \frac{(-1)^{n-1} \sin nx}{n}, \qquad x \in I = \{x: -\pi < x < \pi\}.$$

Use this result to find the Fourier series for $F: x \to x^2$ on I.

Solution. We have $a_0 = 0$ in the expansion of f and the function is piecewise smooth everywhere. Therefore Theorem 10.7 applies. Set

$$F_1'(x) = 4 \sum_{n=1}^{\infty} \frac{(-1)^{n-1} \sin nx}{n}$$

where $F_1(x) = x^2 - \pi^2$. Then

$$F(x) = \frac{\pi^2}{3} + 4 \sum_{n=1}^{\infty} \frac{(-1)^n \cos nx}{n^2}. \qquad \square$$

PROBLEMS

1. Find the Fourier expansion of $f: x \to (1/3)(\pi^2 x - x^3)$ on $I = \{x: -\pi \leqslant x \leqslant \pi\}$ and show that $\sum_{n=1}^{\infty} n^{-6} = \pi^6/945$.

2. Use Theorem 10.7 to find the Fourier expansion for $f: x \to |x|$ on $I = \{x: -\pi \leqslant x \leqslant \pi\}$.

3. Find the Fourier expansion for f given by

$$f: x \to \begin{cases} \frac{1}{2}(x^2 - \pi x), & 0 \leqslant x \leqslant \pi, \\ -\frac{1}{2}(x^2 + \pi x), & -\pi \leqslant x \leqslant 0. \end{cases}$$

4. Using the result of Problem 1, find the Fourier series of $f: x \to (1/12)(\pi^2 - x^2)^2$ on $I = \{x: -\pi \leqslant x \leqslant \pi\}$.

5. Find the Fourier series of the functions f and F given by

$$f: x \to |\sin x|, \qquad -\pi \leqslant x \leqslant \pi,$$

$$F: x \to \begin{cases} -1 + \cos x - \dfrac{2}{\pi}x & \text{for } I_1 = \{x: -\pi \leqslant x \leqslant 0\}, \\ \\ 1 - \cos x - \dfrac{2}{\pi}x & \text{for } I_2 = \{x: 0 \leqslant x \leqslant \pi\}. \end{cases}$$

6. Find the Fourier series for f given by

$$f: x \to \begin{cases} -(\pi + x) & \text{for } I_1 = \{x: -\pi \leqslant x \leqslant -\frac{1}{2}\pi\}, \\ x & \text{for } I_2 = \{x: -\frac{1}{2}\pi \leqslant x \leqslant \frac{1}{2}\pi\}, \\ \pi - x & \text{for } I_3 = \{x: \frac{1}{2}\pi \leqslant x \leqslant \pi\}. \end{cases}$$

7. Suppose that f, periodic with period 2π, possesses continuous derivatives of all orders for $-\infty < x < \infty$. Let a_n, b_n be the Fourier coefficients of f. What can be said about the ratios a_n/n^k, b_n/n^k as $n \to \infty$ where k is a positive integer?

8. Given the function $f: x \rightarrow x|x|$ for $x \in I = \{x: -\pi < x < \pi\}$. How many continuous derivatives does f have on I?

9. Sketch the graph of $D_n(x)$ for $n = 4$ and $x \in I = \{x: -\pi < x < \pi\}$.

10. Find the value of $\lim_{n \to \infty} \int_0^{\pi/6} D_n(x)\, dx$.

11. Find the maximum value of $D_n(x)$ for $x \in I = \{x: -\pi < x < \pi\}$.

12. Given

$$f(x) = \begin{cases} 0 & \text{for } x \in I_1 = \{x: -\pi < x < 0\}, \\ \sin x & \text{for } x \in I_2 = \{x: 0 < x < \pi\}. \end{cases}$$

(a) Find the Fourier series for f.

(b) By integrating the result in (a) obtain the Fourier series for

$$F(x) = \begin{cases} 0 & \text{for } x \in I_1, \\ 1 - \cos x & \text{for } x \in I_2. \end{cases}$$

CHAPTER 11

Functions Defined by Integrals; Improper Integrals

11.1. The Derivative of a Function Defined by an Integral; the Leibniz Rule

The solutions of problems in differential equations, especially those which arise in physics and engineering, are frequently given in terms of integrals. Most often either the integrand of the integral representing the solution is unbounded or the domain of integration is an unbounded set. In this chapter we develop rules for deciding when it is possible to interchange the processes of differentiation and integration—commonly known as differentiation under the integral sign. When the integrand becomes infinite at one or more points or when the interval of integration is infinite, a study of the convergence of the integral is needed in order to determine whether or not the differentiation process is allowable. We establish the required theorems for bounded functions and domains in this section and treat the unbounded case in Sections 11.2 and 11.3.

Let f be a function with domain a rectangle $R = \{(x, t): a \leqslant x \leqslant b, c \leqslant t \leqslant d\}$ in $\mathbb{R}^2$ and with range in $\mathbb{R}^1$. Let I be the interval $\{x: a \leqslant x \leqslant b\}$ and form the function $\phi: I \to \mathbb{R}^1$ by the formula

$$\phi(x) = \int_c^d f(x, t) \, dt. \tag{11.1}$$

We now seek conditions under which we can obtain the derivative ϕ' by differentiation of the integrand in (11.1). The basic formula is given in the following result.

Theorem 11.1 (Leibniz's rule). *Suppose that f and $f_{,1}$ are continuous on the rectangle R and that ϕ is defined by (11.1). Then*

$$\phi'(x) = \int_c^d f_{,1}(x, t)\, dt, \qquad a < x < b. \tag{11.2}$$

PROOF. Form the difference quotient

$$\frac{\phi(x + h) - \phi(x)}{h} = \frac{1}{h} \int_c^d [f(x + h, t) - f(x, t)]\, dt.$$

Observing that

$$f(x + h, t) - f(x, t) = \int_x^{x+h} f_{,1}(z, t)\, dz,$$

we have

$$\frac{\phi(x + h) - \phi(x)}{h} = \frac{1}{h} \int_c^d \int_x^{x+h} f_{,1}(z, t)\, dz\, dt. \tag{11.3}$$

Since $f_{,1}$ is continuous on the closed, bounded set R, it is uniformly continuous there. Hence, if $\varepsilon > 0$ is given, there is a $\delta > 0$ such that

$$|f_{,1}(z, t) - f_{,1}(x, t)| < \frac{\varepsilon}{d - c}$$

for all t such that $c \leqslant t \leqslant d$ and all z such that $|z - x| < \delta$. We now use the artifice

$$\int_c^d f_{,1}(x, t)\, dt = \frac{1}{h} \int_c^d \int_x^{x+h} f_{,1}(x, t)\, dz\, dt, \tag{11.4}$$

which is valid because z is absent in the integrand on the right. Subtracting (11.4) from (11.3), we find

$$\left| \frac{\phi(x + h) - \phi(x)}{h} - \int_c^d f_{,1}(x, t)\, dt \right|$$

$$= \left| \int_c^d \left\{ \frac{1}{h} \int_x^{x+h} [f_{,1}(z, t) - f_{,1}(x, t)]\, dz \right\} dt \right|. \tag{11.5}$$

Now if $|h|$ is so small that $|z - x| < \delta$ in the integrand on the right side of (11.5), it follows that

$$\left| \frac{\phi(x + h) - \phi(x)}{h} - \int_c^d f_{,1}(x, t)\, dt \right|$$

$$\leqslant \int_c^d \left| \frac{1}{h} \int_x^{x+h} \frac{\varepsilon}{d - c}\, dz \right| dt = \frac{\varepsilon}{d - c} \cdot (d - c) = \varepsilon.$$

Since ε is arbitrary, the left side of the above inequality tends to 0 as $h \to 0$. Formula (11.2) is the result. $\qquad \square$

If the function f in (11.1) can be integrated explicitly with respect to t, then finding the derivative of ϕ is a straightforward computation. However, there are situations in which f cannot be integrated, but $f_{,1}$ can. The next example illustrates this point.

EXAMPLE 1. Define $f: \mathbb{R}^2 \to \mathbb{R}^1$ by

$$f: (x, t) \to \begin{cases} (\sin xt)/t & \text{for } t \neq 0, \\ x & \text{for } t = 0. \end{cases}$$

Find ϕ' where $\phi(x) = \int_0^{\pi/2} f(x, t)\, dt$.

Solution. We have

$$\lim_{t \to 0} \frac{\sin xt}{t} = x \lim_{t \to 0} \frac{\sin(xt)}{(xt)} = x \cdot 1 = x,$$

and so f is continuous on $A \equiv \{(x, t): -\infty < x < \infty, 0 \leq t \leq \pi/2\}$. Also,

$$f_{,1}(x, t) = \begin{cases} \cos xt & \text{for } t \neq 0, \\ 1 & \text{for } t = 0. \end{cases}$$

Hence $f_{,1}$ is continuous on A. We apply Leibniz's rule and find

$$\phi'(x) = \int_0^{\pi/2} \cos xt\, dt = \frac{\sin(\pi/2)x}{x}, \qquad x \neq 0,$$

and so $\phi'(0) = \pi/2$. Observe that the expression for $\phi(x)$ cannot be integrated. □

We now take up an important extension of Leibniz's rule. Suppose that f is defined as before and, setting $I = \{x: a \leq x \leq b\}$ and $J = \{t: c \leq t \leq d\}$, let h_0 and h_1 be two given functions with domain on I and range on J. Suppose that $\phi: I \to \mathbb{R}^1$ is defined by

$$\phi(x) = \int_{h_0(x)}^{h_1(x)} f(x, t)\, dt.$$

We now develop a formula for ϕ'. To do so we consider a function $F: \mathbb{R}^3 \to \mathbb{R}^1$ defined by

$$F(x, y, z) = \int_y^z f(x, t)\, dt. \tag{11.6}$$

Theorem 11.2. Suppose that f and $f_{,1}$ are continuous on $R = \{(x, t): a \leq x \leq b, c \leq t \leq d\}$ and that F is defined by (11.6) with x on I and y, z on J. Then

$$F_{,1} = \int_y^z f_{,1}(x, t)\, dt, \qquad F_{,2} = -f(x, y), \qquad F_{,3} = f(x, z). \tag{11.7}$$

PROOF. The first formula in (11.7) is Theorem 11.1. The second and third formulas hold because of the Fundamental theorem of calculus. □

Theorem 11.3 (General Leibniz rule). *Suppose that f and $f_{,1}$ are continuous on $R = \{(x, t): a \leqslant x \leqslant b, c \leqslant t \leqslant d\}$ and that h_0 and h_1 both have a continuous first derivative on I with range on J. If $\phi: I \to \mathbb{R}^1$ is defined by*

$$\phi(x) = \int_{h_0(x)}^{h_1(x)} f(x, t)\, dt,$$

then

$$\phi'(x) = f[x, h_1(x)]h_1'(x) - f[x, h_0(x)]h_0'(x) + \int_{h_0(x)}^{h_1(x)} f_{,1}(x, t)\, dt. \quad (11.8)$$

PROOF. Referring to F defined in (11.6), we see that $\phi(x) = F(x, h_0(x), h_1(x))$. We apply the Chain rule for finding ϕ' and obtain

$$\phi'(x) = F_{,1} + F_{,2}h_0'(x) + F_{,3}h_1'(x).$$

Now, inserting the values of $F_{,1}$ and $F_{,2}$ and $F_{,3}$ from (11.7) with $y = h_0(x)$ and $z = h_1(x)$, we get the General Leibniz rule (11.8).   □

EXAMPLE 2. Given $\phi: x \to \int_0^{x^2} \arctan\left(\dfrac{t}{x^2}\right) dt$, find ϕ'.

Solution. We have

$$\frac{\partial}{\partial x}\left(\arctan \frac{t}{x^2}\right) = -\frac{2tx}{t^2 + x^4}.$$

Using the General Leibniz rule (11.8), we obtain

$$\phi'(x) = (\arctan 1)(2x) - \int_0^{x^2} \frac{2tx}{t^2 + x^4}\, dt.$$

Setting $t = x^2 u$ in the integral on the right, we get

$$\phi'(x) = \frac{\pi x}{2} - x \int_0^1 \frac{2u\, du}{1 + u^2} = x\left(\frac{\pi}{2} - \log 2\right).   \square$$

EXAMPLE 3. Given

$$F: (x, y) \to \int_y^{\log x} \frac{\sin xt}{t(1 + y)}\, dt,$$

find $F_{,1}$.

Solution. Using the General Leibniz rule (11.8), we find

$$F_{,1} = \frac{\sin(x \log x)}{(1 + y)(\log x)} + \int_y^{\log x} \frac{\cos xt}{1 + y}\, dt$$

$$= \frac{\sin(x \log x)}{(1 + y)(\log x)} + \frac{\sin(x \log x)}{(1 + y)x} - \frac{\sin(xy)}{(1 + y)x}.   \square$$

PROBLEMS

In each of Problems 1 through 12, find ϕ'.

1. $\phi: x \to \displaystyle\int_0^1 \frac{\sin xt}{1+t}\, dt.$

2. $\phi: x \to \displaystyle\int_1^2 \frac{e^{-t}}{1+xt}\, dt.$

3. $\phi: x \to \displaystyle\int_0^1 f(x, t)\, dt$ where $f:(x, t) \to \begin{cases} (t^x - 1)/\log t & \text{for } t \neq 0, 1, \\ 0 & \text{for } t = 0, \\ x & \text{for } t = 1. \end{cases}$

4. $\phi: x \to \displaystyle\int_1^{x^2} \cos(t^2)\,dt.$

5. $\phi: x \to \displaystyle\int_{x^2}^x \sin(xt)\, dt.$

6. $\phi: x \to \displaystyle\int_{x^2}^{e^x} \tan(xt)\, dt.$

7. $\phi: x \to \displaystyle\int_{\cos x}^{1+x^2} \frac{e^{-t}}{1+xt}\, dt.$

8. $\phi: x \to \displaystyle\int_{\pi/2}^{x} \frac{\cos xt}{t}\, dt.$

9. $\phi: x \to \displaystyle\int_{x^2}^{x} \frac{\sin xt}{t}\, dt.$

10. $\phi: x \to \displaystyle\int_{x^m}^{x^n} \frac{dt}{x+t}.$

11. $\phi: x \to \displaystyle\int_0^1 \frac{xt}{\sqrt{1-x^2t^2}}\, dt, \quad |x| < 1.$

12. $\phi: x \to \displaystyle\int_0^\pi \log(1 - 2x \cos t + x^2)\, dt, \quad |x| < 1.$

In each of Problems 13 through 15, compute the indicated partial derivative.

13. $\phi: (x, y) \to \displaystyle\int_y^{x^2} \frac{1}{t} e^{xt}\, dt;$ compute $\phi_{,1}.$

14. $\phi: (x, z) \to \displaystyle\int_{z^3}^{x^2} f(x, t)\, dt$ where $f:(x, t) \to \begin{cases} \dfrac{1}{t} \sin^2(xt), & t \neq 0, \\ 0, & t = 0; \end{cases}$ compute $\phi_{,1}.$

15. $\phi: (x, y) \to \displaystyle\int_{x^2+y^2}^{x^2-y^2} (t^2 + 2x^2 - y^2)\, dt;$ compute $\phi_{,2}.$

16. Show that if m and n are positive integers, then

$$\int_0^1 t^n(\log t)^m\, dt = (-1)^m \frac{m!}{(n+1)^{m+1}}.$$

[Hint: Differentiate $\int_0^1 x^n\, dx$ with respect to n and use induction.]

17. Given

$$F: (x, y) \to \int_{h_0(x,y)}^{h_1(x,y)} f(x, y, t)\, dt,$$

find formulas for $F_{,1}, F_{,2}.$

18. Suppose that the equation

$$\int_{h_0(y)}^{h_1(x)} f(x, y, t)\, dt = 0,$$

which is a relation between x and y, actually defines y as a function of x. If we write $y = \phi(x)$ for this function, find ϕ'.

19. Given

$$\varphi(x, y, z) = \int_{g_0(x,y,z)}^{g_1(x,y,z)} f(x, y, z, t) \, dt,$$

find a formula for $\varphi_{,3}$.

11.2. Convergence and Divergence of Improper Integrals

Suppose that a real-valued function f is defined on the half-open interval $I = \{x : a \leqslant x < b\}$ and that for each $c \in I$, the integral

$$\int_a^c f(x) \, dx$$

exists (see Figure 11.1). We are interested in functions f which are unbounded in a neighborhood of b. For example, the function $f : x \rightarrow (1 - x)^{-1}$ with domain $J = \{x : 0 \leqslant x < 1\}$ is unbounded and has the integral

$$\int_0^c (1 - x)^{-1} \, dx = -\log(1 - c) \quad \text{for } c \in J.$$

As c tends to 1, the value of the integral tends to $+\infty$. On the other hand, for the unbounded function $g : x \rightarrow (1 - x)^{-1/2}$ defined on J, we find

$$\int_0^c (1 - x)^{-1/2} \, dx = 2 - 2\sqrt{1 - c} \quad \text{for } c \in J.$$

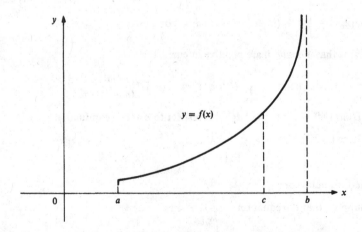

Figure 11.1. Integrating an unbounded function.

Observe that this integral has the finite limiting value 2 as $c \to 1$. With these examples in mind we define an improper integral.

Definitions. Suppose that f is integrable for each number c in the half-open interval $I = \{x: a \leqslant x < b\}$. The integral

$$\int_a^b f(x)\, dx$$

is **convergent** if

$$\lim_{c \to b^-} \int_a^c f(x)\, dx$$

exists. If the limit does not exist the integral is **divergent**. If f is bounded on an interval $J = \{x: a \leqslant x \leqslant b\}$ except in a neighborhood of an interior point $d \in J$, then $\int_a^b f(x)\, dx$ is **convergent** if both limits

$$\lim_{c_1 \to d^-} \int_a^{c_1} f(x)\, dx \quad \text{and} \quad \lim_{c_2 \to d^+} \int_{c_2}^b f(x)\, dx$$

exist. Otherwise the integral is **divergent**.

If a function f is unbounded at several points on an interval I of integration, we decompose the integral into a sum of integrals in which each integral has one endpoint where the function is unbounded. If all the limits of the separate integrals exist, we say the integral is **convergent on** I; otherwise it is **divergent on** I.

EXAMPLE 1. Show that the integrals $\int_a^b (b - x)^{-p}\, dx$ and $\int_a^b (x - a)^{-p}\, dx$ converge for $p < 1$ and diverge for $p \geqslant 1$.

Solution. For $a < c < b$ and $p \neq 1$ we have

$$\int_a^c (b - x)^{-p}\, dx = -\frac{(b - x)^{1-p}}{1 - p}\Bigg]_a^c = \frac{(b - a)^{1-p}}{1 - p} - \frac{(b - c)^{1-p}}{1 - p}.$$

For $p < 1$, the expression on the right tends to $(b - a)^{1-p}/(1 - p)$ as $c \to b^-$. For $p > 1$, there is no limit. The case $p = 1$ yields $\log(b - a) - \log(b - c)$, which has no limit as $c \to b^-$. The analysis for $\int_a^b (x - a)^{-p}\, dx$ is the same. $\square$

When the integral of an unbounded function is convergent we say the integral *exists in an improper sense* or that the **improper integral** *exists*.

In analogy with the convergence of infinite series, it is important to establish criteria which determine when improper integrals exist. The following result, the Comparison test, is a basic tool in determining when integrals converge and diverge.

Theorem 11.4 (Comparison test). *Suppose that f is continuous on the half-open interval $I = \{x: a \leqslant x < b\}$ and that $0 \leqslant |f(x)| \leqslant g(x)$ for all $x \in I$. If $\int_a^b g(x)\, dx$*

converges, then $\int_a^b f(x)\, dx$ converges and

$$\left| \int_a^b f(x)\, dx \right| \leqslant \int_a^b g(x)\, dx.$$

PROOF. First suppose that $f(x) \geqslant 0$ on I and define

$$F: x \to \int_a^x f(t)\, dt, \qquad G: x \to \int_a^x g(t)\, dt.$$

Then F and G are nondecreasing on I and, by hypothesis, $G(x)$ tends to a limit, say M, as $x \to b^-$. Since $F(x) \leqslant G(x) \leqslant M$ on I, we find from the Axiom of continuity that $F(x)$ tends to a limit as $x \to b^-$.

If f is not always nonnegative, define

$$f_1(x) = \frac{|f(x)| + f(x)}{2}, \qquad f_2(x) = \frac{|f(x)| - f(x)}{2}.$$

Then f_1 and f_2 are continuous on I and nonnegative there. Moreover,

$$f_1(x) + f_2(x) = |f(x)| \leqslant g(x), \qquad f_1(x) - f_2(x) = f(x).$$

From the proof for nonnegative functions, the integrals $\int_a^b f_1(x)\, dx$, $\int_a^b f_2(x)\, dx$ exist; using the Theorem on the limit of a sum, we see that $\int_a^b |f(x)|\, dx$ and $\int_a^b f(x)\, dx$ exist. Finally,

$$\left| \int_a^b f(x)\, dx \right| = \left| \int_a^b f_1(x)\, dx - \int_a^b f_2(x)\, dx \right|$$

$$\leqslant \int_a^b f_1(x)\, dx + \int_a^b f_2(x)\, dx \leqslant \int_a^b g(x)\, dx. \qquad \square$$

As a corollary, we have the following comparison test for divergence.

Theorem 11.5. *Suppose that f and g are continuous on the half-open interval $I = \{x: a \leqslant x < b\}$ and that $0 \leqslant g(x) \leqslant f(x)$ for each $x \in I$. If $\int_a^b g(x)\, dx$ diverges, then $\int_a^b f(x)\, dx$ diverges.*

PROOF. If $\int_a^b f(x)\, dx$ were convergent then, by Theorem 11.4, $\int_a^b g(x)\, dx$ would converge, contrary to the hypothesis. $\qquad \square$

EXAMPLE 2. Test for convergence or divergence:

$$\int_0^1 \frac{x^\beta}{\sqrt{1 - x^2}}\, dx, \qquad \beta > 0.$$

Solution. We shall compare $f: x \to x^\beta / \sqrt{1 - x^2}$ with $g: x \to 1/\sqrt{1 - x}$. We have

$$\frac{x^\beta}{\sqrt{1 - x^2}} = \frac{x^\beta}{\sqrt{(1 - x)(1 + x)}} = \frac{x^\beta}{\sqrt{1 + x}} g(x).$$

Since $x^\beta \leqslant \sqrt{1 + x}$ for $0 \leqslant x \leqslant 1$, it follows that $|f(x)| \leqslant g(x)$. By Example 1 we know that $\int_0^1 1/\sqrt{1 - x} \, dx$ is convergent, and so the original integral converges. $\qquad \square$

We now take up the convergence of integrals in which the integrand is bounded but where the interval of integration is unbounded.

Definitions. Let f be defined on $I = \{x: a \leqslant x < +\infty\}$ and suppose that $\int_a^c f(x) \, dx$ exists for each $c \in I$. Define

$$\int_a^{+\infty} f(x) \, dx = \lim_{c \to +\infty} \int_a^c f(x) \, dx$$

whenever the limit on the right exists. In such cases, we say **the integral converges**; when the limit does not exist we say **the integral diverges**. If f is defined on $J = \{x: -\infty < x < +\infty\}$ we may consider expressions of the form

$$\int_{-\infty}^{+\infty} f(x) \, dx$$

which are determined in terms of two limits, one tending to $+\infty$ and the other to $-\infty$. Let d be any point in J. Define

$$\int_{-\infty}^{+\infty} f(x) \, dx = \lim_{c_1 \to -\infty} \int_{c_1}^d f(x) \, dx + \lim_{c_2 \to +\infty} \int_d^{c_2} f(x) \, dx \qquad (11.9)$$

whenever both limits on the right exist. It is a simple matter to see that if f is integrable for every finite interval of J, then the values of the limits in (11.9) do not depend on the choice of the point d.

To illustrate the convergence and divergence properties of integrals when the path of integration is infinite, we show that

$$\int_a^{+\infty} x^{-p} \, dx, \qquad a > 0,$$

converges for $p > 1$ and diverges for $p \leqslant 1$. To see this, observe that $(p \neq 1)$

$$\int_a^c x^{-p} \, dx = \frac{1}{1 - p} [c^{1-p} - a^{1-p}].$$

For $p > 1$, the right side tends to $(1/(p - 1)) a^{1-p}$ as $c \to +\infty$, while for $p < 1$ there is no limit. By the same argument the case $p = 1$ yields divergence. Similarly, the integral

$$\int_{-\infty}^b |x|^{-p} \, dx, \qquad b < 0,$$

converges for $p > 1$ and diverges for $p \leqslant 1$. In analogy with the Comparison test for integrals over a finite path of integration, we state the following result.

Corollary. *For continuous functions f and g defined on the interval* $I = \{x: a \leqslant x < +\infty\}$ *Theorems* 11.4 *and* 11.5 *are valid with respect to integrals* $\int_a^{+\infty} f(x)\, dx$ *and* $\int_a^{+\infty} g(x)\, dx$.

EXAMPLE 3. Test for convergence:

$$\int_1^{+\infty} \frac{\sqrt{x}}{1 + x^{3/2}}\, dx.$$

Solution. For $x \geqslant 1$, observe that

$$\frac{\sqrt{x}}{1 + x^{3/2}} \geqslant \frac{\sqrt{x}}{2x^{3/2}} = \frac{1}{2x} \equiv g(x).$$

However $\int_1^{+\infty} (1/2x)\, dx$ diverges and so the integral is divergent. $\qquad\square$

EXAMPLE 4. Test for convergence:

$$\int_{-\infty}^{+\infty} xe^{-x^2}\, dx.$$

Solution. Consider

$$A \equiv \int_0^c xe^{-x^2}\, dx$$

and set $u = x^2$, $du = 2x\, dx$. Then

$$A = \frac{1}{2} \int_0^{c^2} e^{-u}\, du = \frac{1}{2} - \frac{1}{2} e^{-c^2},$$

and $A \to \frac{1}{2}$ as $c \to +\infty$. Since the integrand is an odd function we see that

$$\int_{-d}^0 xe^{-x^2}\, dx \to -\frac{1}{2} \qquad \text{as } d \to +\infty,$$

and the original integral is convergent. $\qquad\square$

It is clear that the convergence of integrals with unbounded integrands and over an infinite interval may be treated by combining Theorem 11.4 and the Corollary. The next example illustrates the method.

EXAMPLE 5. Test for convergence:

$$\int_0^{+\infty} \frac{e^{-x}}{\sqrt{x}}\, dx.$$

Solution. Because the integrand is unbounded near $x = 0$ we decompose the problem into two parts:

$$\int_0^1 \frac{e^{-x}}{\sqrt{x}}\, dx \qquad \text{and} \qquad \int_1^{+\infty} \frac{e^{-x}}{\sqrt{x}}\, dx.$$

In the first integral we have $e^{-x}/\sqrt{x} \leqslant 1/\sqrt{x}$, and the integral converges by the Comparison test. In the second integral we have $e^{-x}/\sqrt{x} \leqslant e^{-x}$, and once again the Comparison test yields the result since $\int_1^{+\infty} e^{-x}\, dx$ is convergent. Hence, the original integral converges. $\qquad\square$

PROBLEMS

In each of Problems 1 through 12 test for convergence or divergence.

1. $\displaystyle\int_1^{+\infty} \frac{dx}{(x+2)\sqrt{x}}.$

2. $\displaystyle\int_0^1 \frac{dx}{\sqrt{1-x^4}}.$

3. $\displaystyle\int_0^{+\infty} \frac{dx}{\sqrt{1+x^3}}.$

4. $\displaystyle\int_0^{+\infty} \frac{x\, dx}{\sqrt{1+x^4}}.$

5. $\displaystyle\int_{-1}^1 \frac{dx}{\sqrt{1-x^2}}.$

6. $\displaystyle\int_0^{\pi/2} \frac{\sqrt{x}}{\sin x}\, dx.$

7. $\displaystyle\int_1^3 \frac{\sqrt{x}}{\log x}\, dx.$

8. $\displaystyle\int_0^{+\infty} \frac{(\arctan x)^2}{1+x^2}\, dx.$

9. $\displaystyle\int_0^{+\infty} x^2 e^{-x^2}\, dx.$

10. $\displaystyle\int_0^1 \frac{dx}{\sqrt{x-x^2}}.$

11. $\displaystyle\int_0^{+\infty} e^{-x} \sin x\, dx.$

12. $\displaystyle\int_0^\pi \frac{\sin x}{x\sqrt{x}}\, dx.$

13. Show that $\int_2^{+\infty} x^{-1}(\log x)^{-p}\, dx$ converges for $p > 1$ and diverges for $p \leqslant 1$.

14. Assume f is continuous on $I = \{x: 2 \leqslant x < +\infty\}$ and $\lim_{x \to +\infty} x(\log x)^2 f(x) = A$ and $A \neq 0$. Prove that $\int_2^{+\infty} f(x)\, dx$ is convergent.

11.3. The Derivative of Functions Defined by Improper Integrals; the Gamma Function

We consider the integral

$$\int_0^{+\infty} t^{x-1} e^{-t}\, dt$$

which we shall show is convergent for $x > 0$. Since the integrand is unbounded near zero when x is between 0 and 1, the integral may be split into two parts:

$$\int_0^{+\infty} t^{x-1} e^{-t}\, dt = \int_0^1 t^{x-1} e^{-t}\, dt + \int_1^{+\infty} t^{x-1} e^{-t}\, dt \equiv I_1 + I_2.$$

In the first integral on the right, we use the inequality

$$t^{x-1} e^{-t} \leqslant t^{x-1} \quad \text{for} \quad x > 0 \quad \text{and} \quad 0 < t \leqslant 1.$$

The integral $\int_0^1 t^{x-1}\, dt$ converges for $x > 0$ and so I_1 does also. As for I_2, an estimate for the integrand is obtained first by writing

$$t^{x-1}e^{-t} = t^{x+1}e^{-t}t^{-2}$$

and then estimating the function

$$f: t \to t^{x+1}e^{-t}.$$

We find $f'(t) = t^x e^{-t}(x + 1 - t)$, and f has a maximum when $t = x + 1$. This maximum value is $f(x + 1) = (x + 1)^{x+1}e^{-(x+1)}$. Therefore

$$I_2 = \int_1^{+\infty} t^{x-1}e^{-t}\, dt = \int_1^{+\infty} (t^{x+1}e^{-t})t^{-2}\, dt$$

$$\leqslant (x + 1)^{x+1}e^{-(x+1)} \int_1^{+\infty} \frac{1}{t^2}\, dt$$

$$= (x + 1)^{x+1}e^{-(x+1)}.$$

Hence I_2 is convergent for each fixed $x > -1$.

Definition. For $x > 0$ the **Gamma function**, denoted $\Gamma(x)$, is defined by the formula

$$\Gamma(x) = \int_0^{+\infty} t^{x-1}e^{-t}\, dt.$$

The recursion formula

$$\Gamma(x + 1) = x\Gamma(x), \tag{11.10}$$

one of the most important properties of the Gamma function, is derived by means of an integration by parts. To see this, we write

$$\int_0^c t^x e^{-t}\, dt = [t^x(-e^{-t})]_0^c + x \int_0^c t^{x-1}e^{-t}\, dt.$$

Now letting $c \to +\infty$ we obtain (11.10).

It is easy to verify that $\Gamma(1) = 1$, and consequently that $\Gamma(n + 1) = n!$ for positive integers n. Note that the Gamma function is a smooth extension to the positive real numbers of the factorial function which is defined only for the natural numbers.

Leibniz's rule for differentiation of integrals was established for proper integrals. We require a more detailed study to establish similar formulas when the path of integration is infinite or when the integrand is unbounded.

Let $R = \{(x, t): a \leqslant x \leqslant b, c \leqslant t < d\}$ be a rectangle which does not contain its "upper" side. Suppose that $F: R \to \mathbb{R}^1$ is continuous and that

$$\lim_{t \to d-} F(x, t)$$

exists for each $x \in I = \{x: a \leqslant x \leqslant b\}$. We denote the limit above by $f(x)$.

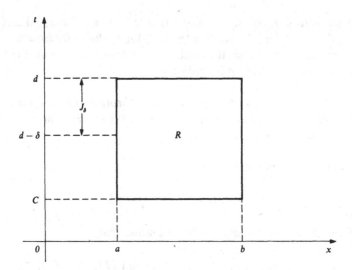

Figure 11.2. $F(x, t) \to f(x)$ as $\delta \to 0$.

Definition. The function $F(x, t)$ **tends to** $f(x)$ **uniformly on** I **as** $t \to d -$ if and only if for every $\varepsilon > 0$ there is a $\delta > 0$ such that

$$|F(x, t) - f(x)| < \varepsilon \qquad (11.11)$$

for all t in the interval $J_\delta = \{t: d - \delta < t < d\}$ and all $x \in I$ (see Figure 11.2). The quantity δ depends on ε but not on x. We also say that F **converges to** f **uniformly as** t **tends to** $d -$. The rectangle R may be replaced by the infinite strip $S = \{(x, t): a \leqslant x \leqslant b, c \leqslant t < \infty\}$. The definition of uniform limit (also called *uniform convergence*) is then modified by stating that for every $\varepsilon > 0$ there is a number T depending on ε such that (11.11) holds for all $t > T$ and all x on I. The number T depends on ε but not on x.

The next two theorems are the basis for Leibniz's rule for improper integrals.

Theorem 11.6. *Suppose that $F: R \to \mathbb{R}^1$ is continuous on I for each $t \in J = \{t: c \leqslant t < d\}$, and that $F(x, t) \to \phi(x)$ uniformly on I as $t \to d -$. Then ϕ is continuous on I. The same result holds if J is the half-infinite interval $\{t: c \leqslant t < +\infty\}$ and $t \to +\infty$.*

PROOF. Let x_1, x_2 be in I. We have

$$\phi(x_1) - \phi(x_2) = \phi(x_1) - F(x_1, t) + F(x_1, t) - F(x_2, t) + F(x_2, t) - \phi(x_2)$$

and

$$|\phi(x_1) - \phi(x_2)| \leqslant |\phi(x_1) - F(x_1, t)|$$
$$+ |F(x_1, t) - F(x_2, t)| + |F(x_2, t) - \phi(x_2)|.$$

Since F tends to ϕ uniformly it follows that for any $\varepsilon > 0$ the first and third terms on the right can be made less than $\varepsilon/3$ for t sufficiently close to d. If x_1 and x_2 are sufficiently close, the continuity of F assures us that the middle term is less than $\varepsilon/3$. Hence ϕ is continuous.   □

Theorem 11.7. *Suppose that the hypotheses of Theorem 11.6 hold and that $F_{,1}$ is continuous on I for each $t \in J$. If $F_{,1}(x, t) \to \psi(x)$ uniformly on I as $t \to d -$ or as $t \to +\infty$, then $\psi(x) = \phi'(x)$ on I.*

PROOF. Let α be any point of I. Then

$$\int_\alpha^x F_{,1}(\xi, t)\, d\xi = F(x, t) - F(\alpha, t).$$

Since $F_{,1}$ converges uniformly to $\psi(x)$, it follows that

$$\int_\alpha^x \psi(\xi)\, d\xi = \lim_{t \to d-} \int_\alpha^x F_{,1}(\xi, t)\, d\xi = \lim_{t \to d-} [F(x, t) - F(\alpha, t)].$$

Hence

$$\int_\alpha^x \psi(\xi)\, d\xi = \phi(x) - \phi(\alpha).$$

We differentiate the left side with respect to x, and the result is $\psi(x) = \phi'(x)$ for any $x \in I$.   □

Leibniz's rule for improper integrals is a direct consequence of Theorems 11.6 and 11.7. We first state a corollary of Theorem 11.6.

Theorem 11.8. *Suppose that $f: R \to \mathbb{R}^1$ is continuous on R. Define*

$$F(x, t) = \int_c^t f(x, \tau)\, d\tau.$$

If the improper integral

$$\phi(x) = \int_c^d f(x, \tau)\, d\tau$$

exists for all $x \in I$, and if $\lim_{t \to d-} F(x, t) = \phi(x)$ exists uniformly for $x \in I$, then ϕ is continuous on I. The same result holds if $J = \{t: c \leqslant t < \delta\}$ is replaced by the interval $\{t: c \leqslant t < +\infty\}$ and $t \to d -$ is replaced by $t \to +\infty$.

Theorem 11.9 (Leibniz's rule for improper integrals). *Suppose that the hypotheses of Theorem 11.8 hold and that $f_{,1}$ is continuous on R. If $F_{,1}$ converges to ψ as $t \to d -$ (or $t \to +\infty$) uniformly in x, then*

$$\psi(x) = \phi'(x) = \int_c^d f_{,1}(x, \tau)\, d\tau$$

or, if d is replaced by $+\infty$,

$$\psi(x) = \phi'(x) = \int_c^{+\infty} f_{,1}(x, \tau) \, d\tau.$$

PROOF. According to Leibniz's rule (Theorem 11.1), for each $t < d$, we have

$$F_{,1}(x, t) = \int_c^t f_{,1}(x, \tau) \, d\tau.$$

Now the result follows from Theorem 11.7. $\qquad\qquad\qquad\qquad\square$

EXAMPLE 1. Let $\phi: x \to \int_0^{+\infty} e^{-xt} \, dt$ be given. Show that ϕ and ϕ' are continuous for $x > 0$ and that

$$\phi'(x) = \int_0^{+\infty} - te^{-xt} \, dt. \qquad\qquad (11.12)$$

Solution. Define

$$F: (x, t) \to \int_0^t e^{-xs} \, ds = \frac{1 - e^{-xt}}{x}.$$

Hence,

$$F_{,1} = -\frac{1 - e^{-xt} - xte^{-xt}}{x^2}.$$

As $t \to +\infty$, we have $F \to 1/x$ and $F_{,1} \to -1/x^2$. To show that the convergence is uniform, we observe that for $h > 0$,

$$\left| F(x, t) - \frac{1}{x} \right| = \frac{e^{-xt}}{x} < \frac{e^{-ht}}{h} \quad \text{for all} \quad x \geqslant h,$$

$$\left| F_{,1}(x, t) - \left(-\frac{1}{x^2} \right) \right| = \frac{e^{-xt}(1 + xt)}{x^2} \leqslant \frac{e^{-ht}(1 + ht)}{h^2} \quad \text{for} \quad x \geqslant h.$$

Therefore the convergence is uniform on any interval $x \geqslant h$ for h positive. Applying Leibniz's rule for improper integrals, we obtain (11.12). $\qquad\square$ ·

The next theorem, a comparison test, shows the utility of Leibniz's rule in cases where the integrals cannot be evaluated directly.

Theorem 11.10

(i) *Suppose that f is continuous on the rectangle*

$$R = \{(x, t): a \leqslant x \leqslant b, c \leqslant t < d\}$$

and that $|f(x, t)| \leqslant g(t)$ *on R. If* $\int_c^d g(t) \, dt$ *converges, then the improper integral*

$$\phi(x) = \int_c^d f(x, t)\, dt$$

is defined for each x on $I = \{x: a \leqslant x \leqslant b\}$ and ϕ is continuous on I.

(ii) *Suppose that $f_{,1}$ is continuous on R and that $|f_{,1}(x, t)| \leqslant h(t)$ on R. If $\int_c^d h(t)\, dt$ converges, then*

$$\phi'(x) = \int_c^d f_{,1}(x, t)\, dt \tag{11.13}$$

for each x in the interior of I. That is, Leibniz's rule holds. The same results hold if d is replaced by $+\infty$ and t tends to $+\infty$.

PROOF. (i) The existence of $\phi(x)$ on I follows from the Comparison test. We define

$$F: (x, t) \to \int_c^t f(x, \tau)\, d\tau.$$

Then

$$|\phi(x) - F(x, t)| = \left| \int_t^d f(x, \tau)\, d\tau \right| \leqslant \int_t^d g(\tau)\, d\tau. \tag{11.14}$$

Since the integral on the right in (11.14) is convergent, it follows that for every $\varepsilon > 0$, the value of $|\int_t^d g(\tau)\, d\tau|$ is less than ε provided t is sufficiently close to d. Therefore

$$|\phi(x) - F(x, t)| \to 0 \quad \text{as} \quad t \to d^-,$$

uniformly for x on I. Hence ϕ is continuous.

(ii) To prove (ii), observe that

$$\left| \int_t^d f_{,1}(x, \tau)\, dt \right| \leqslant \int_t^d h(t)\, dt \to 0 \quad \text{as} \quad t \to d^-.$$

The convergence is uniform and (11.13) is established.   □

Definition. Define $R = \{(x, t): a \leqslant x \leqslant b, c \leqslant t < d\}$ and

$$F(x, t) = \int_c^t f(x, \tau)\, d\tau \quad \text{for} \quad (x, t) \in R,$$

$$\phi(x) = \int_c^d f(x, \tau)\, d\tau \quad \text{for} \quad x \in I = \{x: a \leqslant x \leqslant b\}. \tag{11.15}$$

If $F(x, t) \to \phi(x)$ uniformly for x on I as $t \to d^-$, we say the **improper integral** (11.15) **converges uniformly** *for x on I.* The same definition is used if $d = +\infty$.

EXAMPLE 2. Show that the improper integral

$$\phi(x) = \int_1^{+\infty} \frac{\sin t}{x^2 + t^2} \, dt \tag{11.16}$$

converges uniformly for $-\infty < x < +\infty$.

Solution. For all x and all $t \geq 1$, we have the estimate

$$\left| \frac{\sin t}{x^2 + t^2} \right| \leq \frac{1}{t^2}.$$

Since $\int_1^{+\infty} (1/t^2) \, dt$ converges, we apply Theorem 11.10 to conclude that the integral (11.16) converges uniformly for all x. □

EXAMPLE 3. Given the integral

$$\phi(x) = \int_0^{+\infty} \frac{e^{-xt} - e^{-t}}{t} \, dt, \tag{11.17}$$

and $I = \{x: a \leq x \leq b\}$ with $a > 0$. Show that the integral (11.17) and the integral obtained from (11.17) by differentiating the integrand with respect to x both converge uniformly for x on I. Use these results to evaluate $\phi(x)$.

Solution. We set $f(x, t) = (e^{-xt} - e^{-t})/t$ and obtain

$$f_{,1}(x, t) = -e^{-xt}, \qquad |f_{,1}(x, t)| \leq \begin{cases} e^{-at} & \text{if } a \leq 1, \\ e^{-t} & \text{if } 1 < a. \end{cases}$$

We write the last inequality more compactly:

$$|f_{,1}(x, t)| \leq e^{-ht} \quad \text{where} \quad h = \min(a, 1).$$

Now the Mean-value theorem applied to e^{-xt}/t as a function of x yields

$$\left| \frac{e^{-xt} - e^{-t}}{t} \right| = e^{-\xi t} |1 - x| \quad \text{for} \quad 0 \leq t \leq 1,$$

where ξ is between 1 and x. Also,

$$\left| \frac{e^{-xt} - e^{-t}}{t} \right| \leq e^{-ht} \quad \text{for} \quad t \geq 1.$$

Therefore integral (11.17) and the integral

$$\int_0^{+\infty} f_{,1}(x, t) \, dt$$

converge uniformly for x on I. Using Leibniz's rule, we get

$$\phi'(x) = \int_0^{+\infty} -e^{-xt} \, dt = \lim_{t \to +\infty} \frac{e^{-xt}}{x} \Big]_0^t = -\frac{1}{x}.$$

Integrating $\phi'(x) = -1/x$, we obtain $\phi(x) = C - \log x$. To determine C, we have

$$\phi(1) = \int_0^{+\infty} \frac{e^{-t} - e^{-t}}{t} \, dt = 0,$$

and so

$$\phi(x) = -\log x = \int_0^{+\infty} \frac{e^{-xt} - e^{-t}}{t} \, dt. \qquad \square$$

PROBLEMS

In each of Problems 1 through 8, show that the integrals for ϕ and ϕ' converge uniformly on the given interval. Find ϕ' by Leibniz's rule.

1. $\phi: x \to \displaystyle\int_0^{+\infty} \frac{e^{-xt}}{1+t} \, dt,$ $I = \{x: 0 < a \leqslant x \leqslant b\}.$

2. $\phi: x \to \displaystyle\int_0^1 t^{-1/2} e^{xt} \, dt,$ $I = \{x: -A \leqslant x \leqslant A\}.$

3. $\phi: x \to \displaystyle\int_0^{+\infty} \frac{\cos xt}{1+t^3} \, dt,$ $I = \{x: -A \leqslant x \leqslant A\}.$

4. $\phi: x \to \displaystyle\int_0^{+\infty} \frac{e^{-t}}{1 + xt} \, dt,$ $I = \{x: 0 \leqslant x \leqslant a\}.$

5. $\phi: x \to \displaystyle\int_0^1 t^{-1} \sin(xt)(\log t) \, dt,$ $I = \{x: -a \leqslant x \leqslant a\}.$

6. $\phi: x \to \displaystyle\int_0^1 \frac{(\log t)^2}{1 + xt} \, dt,$ $I = \{x: -1 < a \leqslant x \leqslant b\}.$

7. $\phi: x \to \displaystyle\int_0^1 \frac{dt}{(1 + xt)\sqrt{1 - t}},$ $I = \{x: -1 < a \leqslant x \leqslant b\}.$

8. $\phi: x \to \displaystyle\int_0^{+\infty} \frac{\sin xt}{t(1 + t^2)} \, dt,$ $I = \{x: -a \leqslant x \leqslant a\}.$

9. Use the fact that $\int_0^1 t^x \, dt = (x + 1)^{-1}$, $x > -1$, to deduce that

$$\int_0^1 t^x (-\log t)^m \, dt = \frac{m!}{(x + 1)^{m+1}}, \qquad x > -1.$$

10. Use the fact that

$$\int_0^{+\infty} \frac{t^{x-1}}{1+t} \, dt = \frac{\pi}{\sin \pi x}, \qquad 0 < x < 1,$$

to show that

$$\int_0^{+\infty} \frac{t^{x-1} \log t}{1+t} \, dt = -\frac{\pi^2 \cos \pi x}{\sin^2 \pi x}.$$

11. Verify that

$$\frac{d^n}{dx^n} \Gamma(x) = \int_0^{+\infty} t^{x-1} (\log t)^n e^{-t} \, dt.$$

12. If ϕ is given by $\phi: x \to \int_0^{+\infty} (1 + t)^{-1} e^{-xt} \, dt$, show that $\phi(x) - \phi'(x) = 1/x$.

13. Given

$$\phi: x \to \int_0^{+\infty} \frac{e^{-xt}}{1 + t^2} \, dt,$$

show that $\phi(x) + \phi''(x) = 1/x$.

14. Find ϕ' by Leibniz's rule, given that

$$\phi: x \to \int_0^{+\infty} t^{-1} e^{-t} (1 - \cos xt) \, dt.$$

Find an explicit expression for ϕ.

15. Given

$$\phi: x \to \int_0^1 \frac{t^x - 1}{\log t} \, dt, \qquad x \in I = \{-1 < a \leqslant x < +\infty\}.$$

Find ϕ' by Leibniz's rule and obtain an explicit expression for ϕ.

16. Given

$$\phi: x \to \int_0^{+\infty} t^{-2} e^{-t} (1 - \cos xt) \, dt.$$

Find ϕ' and ϕ'' by Leibniz's rule and then obtain explicit expressions for ϕ' and ϕ. Justify the process.

17. Verify that $\phi(x) = \int_0^{+\infty} e^{-xt} \, dt = 1/x$, and show that

$$\phi^{(n)}(x) = (-1)^n \int_0^{+\infty} t^n e^{-xt} \, dt = (-1)^n n! \, x^{-n-1}, \qquad x > 0.$$

18. Verify the formula

$$\int_0^{+\infty} \frac{dt}{t^2 + x} = \frac{1}{2} \pi x^{-1/2}$$

and show that

$$\int_0^{+\infty} \frac{dt}{(t^2 + x)^{n+1}} = \frac{(2n)!}{2^{2n} (n!)^2} x^{-n-(1/2)}.$$

19. Given the **Legendre polynomial** $P_n(x)$ defined by

$$P_n(x) = \frac{(-1)^n}{n! \sqrt{\pi}} \int_{-\infty}^{+\infty} e^{-t^2(1-x^2)} \frac{d^n}{dx^n} (e^{-x^2 t^2}) \, dt,$$

show that

$$x \frac{d}{dx} P_n(x) - \frac{d}{dx} P_{n-1}(x) = n P_n(x).$$

20. Define $B(x, y) = \int_0^1 t^{x-1} (1 - t)^{y-1} \, dt, 0 < x < \infty, 0 < y < \infty$.
    (a) Show that $B(x, y) = B(y, x)$.
    (b) Find $B(2, 2)$ and $B(4, 3)$.

21. Let $f(x, y) = (e^{-xy} \sin x)/x$ for $x \neq 0$ and $f(0, y) = 1$.

    (a) Show that $f_{,2}(x, y)$ is continuous everywhere. We define

    $$G(y) = \int_0^\infty e^{-xy} \frac{\sin x}{x} dx \quad \text{for } y > 0.$$

    (b) Find $G'(y)$ and show that $G'(y) = -1/(1 + y^2)$. [*Hint*: Integrate by parts twice.]

    (c) Show that $G(y) = \pi/2 - \arctan y$, $y > 0$ and conclude that

    $$\int_0^\infty \frac{\sin x}{x} dx = \frac{\pi}{2}.$$

22. Define

    $$f(y) = \int_0^\infty \frac{\sin xy}{x(x^2 + 1)} dx \quad \text{for } y > 0.$$

    Show that f satisfies the equation $f''(y) = f(y) - \pi/2$. Verify that $f(y) = \frac{1}{2}\pi(1 - e^{-y})$.

CHAPTER 12

The Riemann–Stieltjes Integral and Functions of Bounded Variation

12.1. Functions of Bounded Variation

In earlier chapters we developed the basic properties of functions on $\mathbb{R}^1$ and $\mathbb{R}^N$. Of particular interest are conditions which determine when functions are continuous, differentiable of any order, analytic, and integrable. There are many problems, especially in the applications to physical sciences, in which we require more precise information than we have obtained so far about the behavior of functions. In the simplest case of functions from $\mathbb{R}^1$ to $\mathbb{R}^1$, it is useful, for example, to be able to measure how rapidly a function oscillates. However, the oscillatory character of a function is not easily determined from its continuity or differentiability properties. For this reason, we introduce the notion of the *variation* of a function, defined below. This quantity turns out to be useful for problems in physics, engineering, probability theory, Fourier series, and so forth. In this section we establish the principal theorems concerning the variation of a function on $\mathbb{R}^1$ and in Section 12.2 we show that this concept can be used to define an important extension of the Riemann integral, one which enlarges substantially the class of functions which can be integrated.

Definitions. Let $I = \{x : a \leqslant x \leqslant b\}$ be an interval and $f : I \to \mathbb{R}^1$ a given function. The **variation of f over I**, denoted $V_a^b f$, is the quantity

$$V_a^b f = \sup \sum_{i=1}^n |f(x_i) - f(x_{i-1})|,$$

where the supremum is taken over all possible subdivisions $a = x_0 < x_1 < \cdots < x_n = b$ of I. If the number $V_a^b f$ is finite we say f *is of* **bounded variation on I**. If f is not of bounded variation, we write $V_a^b f = +\infty$.

We begin by establishing several of the fundamental properties of the variation of a function. If the function f is kept fixed, then $V_a^b f$ depends only on the interval $I = \{x: a \leqslant x \leqslant b\}$. The following Theorem shows that the variation is *additive* on intervals. This result (as well as the proof) is similar to the corresponding theorem for integrals, Theorem 5.3(d).

Theorem 12.1. *Let* $f: I \to \mathbb{R}^1$ *be given and suppose that* $c \in I$. *Then*

$$V_a^b f = V_a^c f + V_c^b f. \tag{12.1}$$

PROOF

(a) We first show that $V_a^b f \leqslant V_a^c f + V_c^b f$. We may suppose that both terms on the right are finite. Let $\Delta: a = x_0 < x_1 < \cdots < x_n = b$ be any subdivision. We form the subdivision Δ' by introducing the additional point c which falls between x_{k-1} and x_k, say. Then

$$\sum_{i=1}^{k-1} |f(x_i) - f(x_{i-1})| + |f(c) - f(x_{k-1})| \leqslant V_a^c f,$$

$$\sum_{i=k+1}^{n} |f(x_i) - f(x_{i-1})| + |f(x_k) - f(c)| \leqslant V_c^b f.$$

From the inequality $|f(x_k) - f(x_{k-1})| \leqslant |f(x_k) - f(c)| + |f(c) - f(x_{k-1})|$, it follows that

$$\sum_{i=1}^{n} |f(x_i) - f(x_{i-1})| \leqslant V_a^c f + V_c^b f.$$

Since the subdivision Δ is arbitrary, we obtain $V_a^b f \leqslant V_a^c f + V_c^b f$.

(b) We now show that $V_a^b f \geqslant V_a^c f + V_c^b f$. If $V_a^b f = +\infty$, then the inequality clearly holds. If $V_a^c f = +\infty$, then for every positive number N there is a subdivision $a = x_0 < x_1 < \cdots < x_k = c$ such that $\sum_{i=1}^{k} |f(x_i) - f(x_{i-1})| > N$. Then $a = x_0 < x_1 < \cdots < x_k < x_{k+1} = b$ is a subdivision of I so that

$$V_a^b f \geqslant \sum_{i=1}^{k} |f(x_i) - f(x_{i-1})| + |f(b) - f(x_k)| > N.$$

Since N is arbitrary, it follows that $V_a^b f = +\infty$. Similarly, if $V_c^b f = +\infty$, then $V_a^b f = +\infty$. Hence we may assume that $V_a^c f$ and $V_c^b f$ are both finite. Now let $\varepsilon > 0$ be given. From the definition of supremum, there is a subdivision

$$\Delta_1: a = x_0 < x_1 < \cdots < x_k = c$$

such that

$$\sum_{i=1}^{k} |f(x_i) - f(x_{i-1})| > V_a^c f - \frac{1}{2}\varepsilon.$$

Similarly, there is a subdivision $\Delta_2: c = x_k < x_{k+1} \cdots < x_n = b$ such that

$$\sum_{i=k+1}^{n} |f(x_i) - f(x_{i-1})| > V_c^b f - \frac{1}{2}\varepsilon.$$

Therefore

$$V_a^b f \geqslant \sum_{i=1}^{n} |f(x_i) - f(x_{i-1})| > V_a^c f + V_c^b f - \varepsilon.$$

Since ε is arbitrary, the result follows. ☐

EXAMPLE 1. Suppose that f is a nondecreasing function on $I = \{x : a \leqslant x \leqslant b\}$. Show that $V_a^x f = f(x) - f(a)$ for $x \in I$.

Solution. Let x be in I and let Δ be a subdivision $a = x_0 < x_1 < \cdots < x_{n-1} < x_n = x$. We have

$$\sum_{i=1}^{n} |f(x_i) - f(x_{i-1})| = \sum_{i=1}^{n} [f(x_i) - f(x_{i-1})] = f(x) - f(a),$$

the first equality holding because f is nondecreasing, and the second because all the terms except the first and last cancel. Since the above equalities hold for every subdivision, they hold also for the supremum. Hence $V_a^x f = f(x) - f(a)$. ☐

By the same argument we show easily that if f is nonincreasing on I, then $V_a^x f = f(a) - f(x)$.

Figure 12.1(a) shows a typical function of bounded variation, while Figure 12.1(b) shows the variation of f on the interval $[a, x]$ for x between a and b. We observe that $V_a^x f$ is always an increasing (nondecreasing) positive function of x.

EXAMPLE 2. Given the function

$$f : x \to \begin{cases} \sin(\pi/x) & \text{for } 0 < x \leqslant 1, \\ 0 & \text{for } x = 0. \end{cases}$$

Show that $V_0^1 f = +\infty$ (see Figure 12.2).

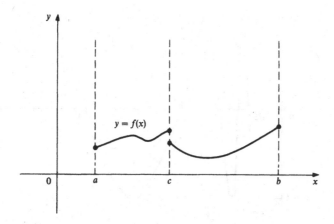

Figure 12.1(a). f, a function of bounded variation.

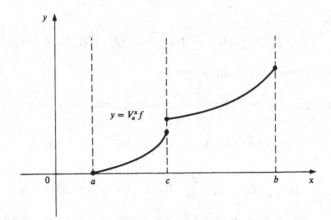

Figure 12.1(b). $V_a^x f$, the variation of f.

Solution. Choose the particular subdivision $a = 0$, $x_1 = 2/(2n + 1)$, $x_2 = 2/(2n - 1)$, ..., $x_{n-1} = 2/3$, $x_n = 1$. Then a computation shows that

$$V_0^1 f \geq \sum_{k=1}^n |f(x_k) - f(x_{k-1})| = 2n.$$

Since n may be arbitrary large, $V_0^1 f = +\infty$.   □

EXAMPLE 3. Suppose that f is continuous on $I = \{x : a \leq x \leq b\}$ and f' is bounded on I. Show that f is of bounded variation.

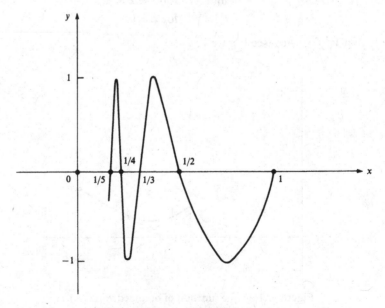

Figure 12.2. $y = \sin(\pi/x)$, $0 < x \leq 1$.

Solution. If $|f'| \leqslant M$ on I, we may apply the Mean-value theorem to obtain $|f(x_k) - f(x_{k-1})| \leqslant M|x_k - x_{k-1}|$ for any two points x_{k-1}, x_k of I. Therefore, for any subdivision Δ,

$$\sum_{k=1}^{n} |f(x_k) - f(x_{k-1})| \leqslant M \sum_{k=1}^{n} |x_k - x_{k-1}| = M(b - a).$$

Hence $V_a^b f \leqslant M(b - a)$. $\square$

The above example gives us an easy sufficient condition for determining when a function is of bounded variation. Since functions may be of bounded variation without being continuous (see Figure 12.1) the set of functions with finite variation is much larger than the set having a bounded first derivative.

We now prove a convergence theorem for sequences of functions whose variations on an interval are uniformly bounded. For this purpose we establish several properties of functions of bounded variation as well as a "diagonal process" for convergence.

Lemma 12.1. *Let f be given on $I = \{x: a \leqslant x \leqslant b\}$, and define $\varphi(x) = V_a^x f$. If x_1, x_2 are any points of I, then $V_{x_1}^{x_2} f = \varphi(x_2) - \varphi(x_1)$ and*

$$V_{x_1}^{x_2} f \geqslant |f(x_2) - f(x_1)|.$$

The function φ is nondecreasing.

The formula $V_{x_1}^{x_2} f = \varphi(x_2) - \varphi(x_1)$ is a restatement of (12.1) in Theorem 12.1. The remainder of the proof is obvious from the definition of variation. $\square$

Theorem 12.2. *Suppose that f is of bounded variation on $I = \{x: a \leqslant x \leqslant b\}$. If f is continuous on the left at b, then $\varphi(x) = V_a^x f$ is continuous on the left at b. Similarly, if f is continuous on the right at a, then φ is also. If f is continuous at any point of I, then φ is continuous at that point.*

PROOF. We prove the first statement, the others being similar. Let $\varepsilon > 0$ be given. Then there is a subdivision $\Delta: a = x_0 < x_1 < \cdots < x_n = b$ such that

$$V_a^b f < \sum_{i=1}^{n} |f(x_i) - f(x_{i-1})| + \frac{1}{2}\varepsilon. \tag{12.2}$$

By inserting an additional point, if necessary, we may suppose (because f is continuous on the left) that $|f(b) - f(x_{n-1})| < \frac{1}{2}\varepsilon$. Then for $x_{n-1} \leqslant x \leqslant b$, it follows from Lemma 12.1 that

$$\varphi(x) \geqslant \varphi(x_{n-1}) = V_a^{x_{n-1}} f$$

$$\geqslant \sum_{i=1}^{n-1} |f(x_i) - f(x_{i-1})| > \sum_{i=1}^{n} |f(x_i) - f(x_{i-1})| - \frac{1}{2}\varepsilon.$$

Therefore, taking (12.2) into account,

$$\varphi(x) > V_a^b f - \varepsilon = \varphi(b) - \varepsilon.$$

Since ε is arbitrary and φ is nondecreasing the result follows.   □

The next theorem exhibits an important relationship between monotone functions and functions of bounded variation. It shows that every function of bounded variation is the difference of two nondecreasing functions.

Theorem 12.3. *Suppose that f is of bounded variation on $I = \{x: a \leqslant x \leqslant b\}$. Then there are nondecreasing functions g and h on I such that*

$$f(x) = g(x) - h(x) \quad for \quad x \in I, \tag{12.3}$$

and $V_a^x f = g(x) + h(x) - f(a)$ for $x \in I$. Moreover, if f is continuous on the left at any point $c \in I$, then g and h are also. Similarly, g and h are continuous on the right wherever f is.

PROOF. Choose

$$g(x) = \tfrac{1}{2}[f(a) + V_a^x f + f(x)], \qquad h(x) = \tfrac{1}{2}[f(a) + V_a^x f - f(x)]. \tag{12.4}$$

To show that g is nondecreasing, observe that $2g(x_2) - 2g(x_1) = V_{x_1}^{x_2} f + f(x_2) - f(x_1)$. From Lemma 12.1 we have $V_{x_1}^{x_2} f \geqslant |f(x_2) - f(x_1)|$ and so $g(x_2) \geqslant g(x_1)$. The proof for h is similar.

Suppose that f is continuous on the left or right at c. Then so is φ where $\varphi(x) = V_a^x f$. The continuity result follows directly from Theorem 12.2 and the explicit formulas for g and h.   □

Remarks. The decomposition of f given by (12.3) and (12.4) is not unique. If ψ is *any* nondecreasing function on I, we also have the decomposition $f = (g + \psi) - (h + \psi)$. In fact, if ψ is strictly increasing, then this decomposition of f is the difference of two strictly increasing functions.

Theorem 12.4. *If f is nondecreasing on $I = \{x: a \leqslant x \leqslant b\}$, then the points of discontinuity of f are at most a countable set. (This result is stated without proof in Section 6.3, which discusses countable sets on the line. See Theorem 6.16.)*

PROOF. For each point c in the interior of I, we know that $\lim_{x \to c+} f(x) = f(c +)$ and $\lim_{x \to c-} f(x) = f(c -)$ both exist. (See Theorem 3.7.) Since f is nondecreasing, it is clear that $f(c +) \geqslant f(c -)$ for each number c. Let k be any positive real number. Define the set

$$E_k = \{x \in I : f(x +) - f(x -) \geqslant k\}.$$

Since f is a bounded monotone function, E_k is a finite set. The set $E = \bigcup_{j=1}^{\infty} E_{1/2^j}$ is countable and contains all the points of discontinuity of f.   □

Corollary. *If f is of bounded variation on $I = \{x: a \leqslant x \leqslant b\}$, then the points of discontinuity form at most a countable set.*

A method for obtaining convergent subsequences of a given sequence, known as the *Cantor diagonal process*, is a widely used and important tool of analysis. Although we shall employ it here only for functions from $\mathbb{R}^1$ to $\mathbb{R}^1$, we prove the result for general mappings from one metric space to another.

Theorem 12.5 (Cantor's diagonal process). *Let X and Y be sets in a metric space with Y compact. For $n = 1, 2, \ldots$, let $f_n: X \to Y$ be given mappings. Suppose that $S = \{x_1, x_2, \ldots, x_n, \ldots\}$ is a countable subset of X. Then there is a subsequence $\{g_n\}$ of $\{f_n\}$ such that $\{g_n\}$ converges at every point of S.*

PROOF. The sequence $\{f_n(x_1)\}$ of Y has a convergent subsequence because Y is compact. Call this convergent subsequence $\{f_{1n}\}$ and suppose $f_{1n}(x_1) \to y_1$ as $n \to \infty$. Because Y is compact the sequence $\{f_{1n}(x_2)\}$ of Y has a convergent subsequence. Call this subsequence $\{f_{2n}\}$ and suppose $f_{2n}(x_2) \to y_2$ as $n \to \infty$. Note that f_{2n}, being a subsequence of f_{1n}, is such that $f_{2n}(x_1) \to y_1$ as $n \to \infty$. We continue taking subsequences so that $\{f_{kn}\}$ converges to $y_1, y_2, \ldots, y_k$ at $x_1, x_2, \ldots, x_k$, respectively.

Now consider the diagonal sequence $\{f_{nn}\}$ and let $x_p \in S$. The sequence $\{f_{pn}\}$ converges at x_p. For $m > p$, the sequence $\{f_{mm}\}$ forms a subsequence of $\{f_{pn}\}$ and so also is convergent at x_p. Hence by choosing $g_n = f_{nn}$ we obtain the desired result. ☐

Let f be a monotone function defined on an interval I. A set of points A is **dense** in an interval I if the closure of A contains I. The next result shows that if a sequence of monotone functions converges to f on a dense set in I, it also converges at every point of I where f is continuous.

Theorem 12.6. *Suppose that f and f_n, $n = 1, 2, \ldots$, are nondecreasing functions from $I = \{x: a \leqslant x \leqslant b\}$ to $\mathbb{R}^1$. Let S be a dense set in I and suppose that $f_n(x) \to f(x)$ for each $x \in S$. If x_0 is a point in the interior of I such that f is continuous at x_0, then $f_n(x_0) \to f(x_0)$.*

PROOF. Let x_0 be a point in the interior of I such that f is continuous at x_0. From the definition of continuity, for every $\varepsilon > 0$ there is a $\delta > 0$ such that

$$|f(x) - f(x_0)| < \tfrac{1}{2}\varepsilon \quad \text{for} \quad |x - x_0| < \delta.$$

Of course, δ is chosen so small that the interval $|x - x_0| < \delta$ is contained in I. Since S is dense in I, there are points x_1 and x_2 in S with $x_0 - \delta < x_1 < x_0$ and $x_0 < x_2 < x_0 + \delta$. (See Figure 12.3.) Because f is nondecreasing we have

$$f(x_0) - \tfrac{1}{2}\varepsilon \leqslant f(x_1) \leqslant f(x_0) \leqslant f(x_2) \leqslant f(x_0) + \tfrac{1}{2}\varepsilon. \tag{12.5}$$

From the fact that $x_1, x_2 \in S$ and $\{f_n\}$ converges for all $x \in S$, there is an integer

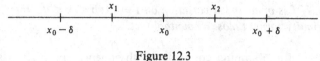

Figure 12.3

N such that

$$|f_n(x_1) - f(x_1)| < \tfrac{1}{2}\varepsilon \quad \text{and} \quad |f_n(x_2) - f(x_2)| < \tfrac{1}{2}\varepsilon \tag{12.6}$$

for all $n > N$. For each n, we have $f_n(x_1) \leqslant f_n(x_0) \leqslant f_n(x_2)$ and this fact is now combined with (12.5) and (12.6) to yield

$$f(x_0) - \varepsilon \leqslant f(x_1) - \tfrac{1}{2}\varepsilon < f_n(x_1) \leqslant f_n(x_0)$$

$$\leqslant f_n(x_2) < f(x_2) + \frac{\varepsilon}{2} \leqslant f(x_0) + \varepsilon.$$

Therefore $|f_n(x_0) - f(x_0)| < \varepsilon$. That is, f_n converges to f at x_0.   □

Remarks. By Theorem 12.4 we know that the points of discontinuity of f in Theorem 12.6 are at most countable. Therefore the points of convergence of a nondecreasing sequence as in the above theorem consists of all of I except for at most a countable set. Because of Theorem 12.3 on decomposition, a theorem analogous to Theorem 12.6 holds for functions of bounded variation.

Since there are countable dense sets in every interval (for example, the rational numbers), the Cantor diagonal process may be combined with Theorem 12.6 to yield subsequences which converge throughout the interval. The principal result in this direction, known as Helly's theorem, merely requires that the functions f_n and their **total variation**, i.e., the quantity $V_a^b f_n$, remain uniformly bounded.

Theorem 12.7 (Helly's theorem). *Let f_n, $n = 1, 2, \ldots$, be functions of bounded variation on $I = \{x : a \leqslant x \leqslant b\}$. Suppose that the functions and their total variations are uniformly bounded. That is, there are constants L and M such that*

$$|f_n(x)| \leqslant L \quad \text{and} \quad V_a^b f_n \leqslant M \quad \text{for all } x, n.$$

Then there is a subsequence $\{g_n\}$ of $\{f_n\}$ which converges at every point of I to a function f. Furthermore, $V_a^b f \leqslant M$.

PROOF. Define $g_n = \tfrac{1}{2}[f_n(a) + V_a^x f_n + f_n(x)]$, $h_n = \tfrac{1}{2}[f_n(a) + V_a^x f_n - f_n(x)]$ as in (12.4); then g_n, h_n are nondecreasing and uniformly bounded. Therefore it is sufficient to prove the theorem for nondecreasing functions f_n. Let $S = \{x_1, x_2, \ldots\}$ be a countable dense subset of I which contains a and b. Since the interval $J = \{x : -L \leqslant x \leqslant L\}$ is a compact subset of $\mathbb{R}^1$, we may employ the Cantor diagonal process. Hence there is a subsequence $\{H_n\}$ of $\{f_n\}$ such that $H_n(x_i) \to y_i$ as $n \to \infty$ for each $x_i \in S$. The functions H_n are nondecreasing and therefore for every n we have $H_n(x_i) \leqslant H_n(x_j)$ if $x_i < x_j$. It follows that $y_p \leqslant y_q$ whenever $x_p < x_q$ for $x_p, x_q \in S$. Define

$$f_0(x) = \sup_{x_p \leqslant x} y_p \quad \text{for} \quad x \in I, x_p \in S.$$

The function f_0 is nondecreasing since, for $z < w$, the points in S less than z form a subset of the points in S which are less than w. Also, for $x_p \in S$, we have $H_n(x_p) \to f_0(x_p) = y_p$. From Theorem 12.6 it follows that $H_n(x) \to f_0(x)$ at any point x where f_0 is continuous. Since f_0 is continuous except at possibly a countable set T in I, there is a further subsequence $\{G_n\}$ of $\{H_n\}$ which converges at each point of T and continues to converge at the remaining points of I. Let f denote the limit function of $\{G_n\}$.

Let $a = x_0' < x_1' < x_2' < \cdots < x_k' = b$ be any subdivision of I. Then

$$\sum_{i=1}^{k} |f(x_i') - f(x_{i-1}')| = \lim_{n \to \infty} \sum_{i=1}^{k} |G_n(x_i') - G_n(x_{i-1}')| \leqslant M,$$

since for each n, we have $V_a^b g_n \leqslant M$. Hence $V_a^b f \leqslant M$. $\qquad\square$

The next result gives a sufficient condition for determining when a function is of bounded variation, and also a method for computing its variation (see Example 3).

Theorem 12.8. *Suppose that f and f' are continuous on an interval $I = \{x: a \leqslant x \leqslant b\}$. Then f is of bounded variation on I and*

$$V_a^b f = \int_a^b |f'(x)| \, dx.$$

PROOF. The first part of the Theorem was established in Example 3 earlier. Let $\Delta: a = x_0 < x_1 < \cdots < x_n = b$ be any subdivision. Then by the Mean-value theorem, there are numbers ξ_i such that $x_{i-1} \leqslant \xi_i \leqslant x_i$ with

$$\sum_{i=1}^{n} |f(x_i) - f(x_{i-1})| = \sum_{i=1}^{n} |f'(\xi_i)| \cdot |x_i - x_{i-1}|.$$

Let $\varepsilon > 0$ be given. From the definition of integral, there is a $\delta > 0$ such that

$$\left| \sum_{i=1}^{n} |f'(\xi_i)||x_i - x_{i-1}| - \int_a^b |f'(x)| \, dx \right| < \frac{1}{2}\varepsilon$$

for *every* subdivision with mesh less than δ. Now we use the definition of bounded variation to assert that for the above ε, there is a subdivision $\Delta_0: a = z_0 < z_1 < \cdots < z_m = b$ such that

$$V_a^b f \geqslant \sum_{i=1}^{m} |f(z_i) - f(z_{i-1})| > V_a^b f - \frac{1}{2}\varepsilon.$$

Let $\Delta_1: a = x_0' < x_1' < \cdots < x_p' = b$, be the common refinement of Δ and Δ_0.

Then Δ_1 has mesh less than δ. Hence

$$V_a^b f \geqslant \sum_{i=1}^{p} |f(x_i') - f(x_{i-1}')| \geqslant \sum_{i=1}^{m} |f(z_i) - f(z_{i-1})| > V_a^b f - \frac{1}{2}\varepsilon$$

and

$$\left| \sum_{i=1}^{p} |f(x_i') - f(x_{i-1}')| - \int_a^b |f'(x)| \, dx \right| < \frac{1}{2}\varepsilon.$$

Combining these inequalities, we obtain

$$\left| V_a^b f - \int_a^b |f'(x)| \, dx \right| < \varepsilon.$$

Since ε is arbitrary, the result follows.   □

Remark. A function f may be of bounded variation without having a bounded derivative. For example, the function $f: x \to x^{2/3}$ on $I = \{x: 0 \leqslant x \leqslant 1\}$ is continuous and increasing on I. Hence it is of bounded variation. However, f' is unbounded at the origin.

PROBLEMS

1. Suppose that f is of bounded variation on $I = \{x: a \leqslant x \leqslant b\}$. Show that f is bounded on I. In fact, show that $|f(x)| \leqslant |f(a)| + V_a^b f$.

2. Suppose that f and g are of bounded variation on $I = \{x: a \leqslant x \leqslant b\}$. Show that $f - g$ and fg are functions of bounded variation.

3. Given $f(x) = \sin^2 x$ for $x \in I = \{x: 0 \leqslant x \leqslant \pi\}$. Find $V_0^\pi f$.

4. Given $f(x) = x^3 - 3x + 4$ for $x \in I = \{x: 0 \leqslant x \leqslant 2\}$. Find $V_0^2 f$.

5. Given

$$f(x) = \begin{cases} 1 & \text{for } 0 \leqslant x < 1, \\ \frac{1}{2} & \text{for } 1 \leqslant x < 2, \\ 2 & \text{for } 2 \leqslant x \leqslant 3. \end{cases}$$

Find $V_0^3 f$.

6. Let $I_i = \{x: i - 1 \leqslant x < i\}$, $i = 1, 2, \ldots, n$. Let $f(x) = c_i$ for $x \in I_i$ and $f(n) = c_n$ where each c_i is a constant. Find $V_0^n f$.

7. Show that the function

$$f: x \to \begin{cases} x \sin(1/x) & \text{for } x \neq 0, \\ 0 & \text{for } x = 0, \end{cases}$$

is not of bounded variation on $I = \{x: 0 \leqslant x \leqslant 1\}$. However, prove that

$$g: x \to \begin{cases} x^2 \sin(1/x) & \text{for } x \neq 0, \\ 0 & \text{for } x = 0, \end{cases}$$

is of bounded variation on I.

8. Find those values of α and β for which the function

$$f: x \to \begin{cases} x^\alpha \sin(x^{-\beta}) & \text{for } x \neq 0, \\ 0 & \text{for } x = 0, \end{cases}$$

is of bounded variation on $I = \{x: 0 \leqslant x \leqslant 1\}$.

9. Let f and g be functions of bounded variation on $I = \{x: a \leqslant x \leqslant b\}$. Show that

$$V_a^b(f + g) \leqslant V_a^b f + V_a^b g,$$

$$V_a^b(kf) = |k| V_a^b f, \quad k \text{ a constant}.$$

10. Let $\Delta: a = x_0 < x_1 < \cdots < x_n = b$ be a subdivision of $I = \{x: a \leqslant x \leqslant b\}$ and suppose f is defined on I. Then $\sum_{i=1}^n [(x_i - x_{i-1})^2 + (f(x_i) - f(x_{i-1}))^2]^{1/2}$ is the length of the inscribed polygonal arc of f. We define the **length of f on I**, denoted $L_a^b f$, by the formula

$$L_a^b f = \sup \sum_{i=1}^n [(x_i - x_{i-1})^2 + (f(x_i) - f(x_{i-1}))^2]^{1/2},$$

where the supremum is taken over all possible subdivisions.
(a) Show that for any function f, the inequalities

$$V_a^b f + (b - a) \geqslant L_a^b f \geqslant [(V_a^b f)^2 + (b - a)^2]^{1/2}$$

hold. Hence conclude that a function is of bounded variation if and only if it has finite length.
(b) Show that if $a < c < b$ then

$$L_a^b f = L_a^c f + L_c^b f.$$

[*Hint*: Follow the proof of Theorem 12.1.]

11. A function f defined on $I = \{x: a \leqslant x \leqslant b\}$ is said to satisfy a **uniform Hölder condition with exponent** α if there is a number M such that

$$|f(x_1) - f(x_2)| \leqslant M \cdot |x_1 - x_2|^\alpha \quad \text{for all} \quad x_1, x_2 \in I.$$

(a) If f satisfies a uniform Hölder condition with $\alpha = 1$ show that f is of bounded variation.
(b) Give an example of a function which satisfies a uniform Hölder condition with $0 < \alpha < 1$ and which is not of bounded variation.

12. Compute $V_0^a f$ for $f(x) = x^n e^{-x}$ on $I = \{x: 0 \leqslant x \leqslant a\}$, where $a > n > 0$.

13. Complete the proof of Lemma 12.1.

14. Complete the proof of Theorem 12.2.

15. Show that if f is continuous and has a finite number of maxima and minima on an interval $I = \{x: a \leqslant x \leqslant b\}$, then f is of bounded variation on I. Conclude that every polynomial function is of bounded variation on every finite interval.

16. Suppose that f is of bounded variation on $I = \{x: a \leqslant x \leqslant b\}$. If $|f(x)| \geqslant c > 0$ for all $x \in I$ where c is a constant, show that $g(x) = 1/f(x)$ is of bounded variation on I.

12.2. The Riemann–Stieltjes Integral

We introduce a generalization of the Riemann integral, one in which a function f is integrated with respect to a second function g. If $g(x) = x$ then the generalized integral reduces to the Riemann integral. This new integral, called the Riemann–Stieltjes integral, has many applications not only in various branches of mathematics, but in physics and engineering as well. By choosing the function g appropriately we shall see that the Riemann–Stieltjes integral allows us to represent discrete as well as continuous processes in terms of integrals. This possibility yields applications to probability theory and statistics.

Definitions. Let f and g be functions from $I = \{x: a \leqslant x \leqslant b\}$ into $\mathbb{R}^1$. Suppose that there is a number A such that for each $\varepsilon > 0$ there is a $\delta > 0$ for which

$$\left| \sum_{i=1}^{n} f(\zeta_i)[g(x_i) - g(x_{i-1})] - A \right| < \varepsilon \qquad (12.7)$$

for every subdivision Δ of mesh size less than δ and for every sequence $\{\zeta_i\}$ with $x_{i-1} \leqslant \zeta_i \leqslant x_i, i = 1, 2, \ldots, n$. Then we say that f **is integrable with respect to** g **on** I. We also say that the integral exists in the **Riemann–Stieltjes** sense. The number A is called the R–S **integral of** f **with respect to** g and we write

$$A = \int_a^b f \, dg = \int_a^b f(x) \, dg(x).$$

As in the case of a Riemann integral, it is a simple matter to show that when the number A exists it is unique. Furthermore, when $g(x) = x$, the sum in (12.7) is a Riemann sum and the R–S integral reduces to the Riemann integral as described in Chapter 5.

It is important to observe that the R–S integral may exist when g is not continuous. For example, with $I = \{x: 0 \leqslant x \leqslant 1\}$ let $f(x) \equiv 1$ on I,

$$g(x) = \begin{cases} 0 & \text{for } 0 \leqslant x < \tfrac{1}{2}, \\ 1 & \text{for } \tfrac{1}{2} \leqslant x \leqslant 1. \end{cases}$$

The quantity $\sum_{i=1}^{n} f(\zeta_i)[g(x_i) - g(x_{i-1})]$ reduces to $\sum_{i=1}^{n} [g(x_i) - g(x_{i-1})]$. However, all these terms are zero except for the subinterval which contains $x = \tfrac{1}{2}$. In any case, the terms of the sum "telescope" so that its value is $g(1) - g(0) = 1$. Therefore, for every subdivision the Riemann–Stieltjes sum has the value 1, and this is the value of the R–S integral.

The R–S integral may not exist if f has a single point of discontinuity provided that the function g is also discontinuous at the same point. For example, with $I = \{x: 0 \leqslant x \leqslant 1\}$ define

$$f(x) = \begin{cases} 1 & \text{for } 0 \leqslant x < \tfrac{1}{2}, \\ 2 & \text{for } \tfrac{1}{2} \leqslant x \leqslant 1, \end{cases} \qquad g(x) = \begin{cases} 0 & \text{for } 0 \leqslant x < \tfrac{1}{2}, \\ 1 & \text{for } \tfrac{1}{2} \leqslant x \leqslant 1. \end{cases}$$

The quantity $\sum_{i=1}^{n} f(\zeta_i)[g(x_i) - g(x_{i-1})]$ reduces to the single term which corresponds to the subinterval containing $x = \frac{1}{2}$. That is, we have one term of the form

$$f(\zeta_k)[g(x_k) - g(x_{k-1})]$$

where $x_{k-1} \leqslant \frac{1}{2} \leqslant x_k$ and $x_{k-1} \leqslant \zeta_k \leqslant x_k$. Then $g(x_k) - g(x_{k-1}) = 1$, but the value of $f(\zeta_k)$ will be 1 or 2 depending on whether ζ_k is chosen less than $\frac{1}{2}$ or greater than or equal to $\frac{1}{2}$. Since these two choices may be made regardless of the mesh of the subdivision, the R–S integral does not exist.

The R–S integral has properties of additivity and homogeneity similar to those of the Riemann integral. These are stated in the next theorem, and we leave the proof of the reader.

Theorem 12.9

(a) *Suppose that $\int_a^b f \, dg_1$ and $\int_a^b f \, dg_2$ both exist. Define $g = g_1 + g_2$. Then f is integrable with respect to g and*

$$\int_a^b f \, dg = \int_a^b f \, dg_1 + \int_a^b f \, dg_2.$$

(b) *Suppose that $\int_a^b f_1 \, dg$ and $\int_a^b f_2 \, dg$ both exist. Define $f = f_1 + f_2$. Then f is integrable with respect to g and*

$$\int_a^b f \, dg = \int_a^b f_1 \, dg + \int_a^b f_2 \, dg.$$

(c) *Suppose that $\int_a^b f \, dg$ exists and that c is a constant. Then*

$$\int_a^b (cf) \, dg = c \int_a^b f \, dg.$$

(d) *Suppose $a < c < b$. Assume that not both f and g are discontinuous at c. If $\int_a^c f \, dg$ and $\int_c^b f \, dg$ exist, then $\int_a^b f \, dg$ exists and*

$$\int_a^b f \, dg = \int_a^c f \, dg + \int_c^b f \, dg.$$

The next theorem shows that if g is smooth, then the R–S integral is reducible to an ordinary Riemann integral. This reduction is useful for the calculation of Riemann–Stieltjes integrals. We show later (Theorem 12.16) that the R–S integral exists for much larger classes of functions.

Theorem 12.10. *Suppose that f, g, and g' are continuous on the interval $I = \{x: a \leqslant x \leqslant b\}$. Then $\int_a^b f \, dg$ exists and*

$$\int_a^b f \, dg = \int_a^b f(x) g'(x) \, dx. \tag{12.8}$$

PROOF. Let $\varepsilon > 0$ be given. We wish to show that

$$\left| \sum_{i=1}^{n} f(\zeta_i)[g(x_i) - g(x_{i-1})] - \int_{a}^{b} f(x)g'(x)\, dx \right| < \varepsilon \qquad (12.9)$$

provided that the mesh of the subdivision is sufficiently small. We apply the Mean-value theorem to the Riemann–Stieltjes sum in (12.9) getting

$$\sum_{i=1}^{n} f(\zeta_i)[g(x_i) - g(x_{i-1})] = \sum_{i=1}^{n} f(\zeta_i)g'(\eta_i)(x_i - x_{i-1}) \qquad (12.10)$$

where $x_{i-1} \leqslant \eta_i \leqslant x_i$. The sum on the right would be a Riemann sum if η_i where equal to ζ_i. We show that for subdivisions with sufficiently small mesh this sum is close to a Riemann sum. Let M denote the maximum of $|f(x)|$ on I. Since g' is continuous on I, it is uniformly continuous there. Hence there is a $\delta > 0$ such that for $|\zeta_i - \eta_i| < \delta$ it follows that

$$|g'(\zeta_i) - g'(\eta_i)| < \frac{\varepsilon}{2M(b-a)}. \qquad (12.11)$$

From the definition of Riemann integral there is a subdivision with mesh so small (and less than δ) that

$$\left| \sum_{i=1}^{N} f(\zeta_i)g'(\zeta_i)(x_i - x_{i-1}) - \int_{a}^{b} f(x)g'(x)\, dx \right| < \frac{1}{2}\varepsilon. \qquad (12.12)$$

By means of (12.11), we have

$$\left| \sum_{i=1}^{N} f(\zeta_i)[g'(\eta_i) - g'(\zeta_i)](x_i - x_{i-1}) \right|$$

$$< \sum_{i=1}^{N} M \left| \frac{\varepsilon}{2M(b-a)}(x_i - x_{i-1}) \right| = \frac{1}{2}\varepsilon. \qquad (12.13)$$

From (12.12) and (12.13), for any ζ_i, η_i such that $x_{i-1} \leqslant \zeta_i \leqslant x_i$ and $x_{i-1} \leqslant \eta_i \leqslant x_i$, we get the inequality

$$\left| \sum_{i=1}^{N} f(\zeta_i)g'(\eta_i)(x_i - x_{i-1}) - \int_{a}^{b} f(x)g'(x)\, dx \right| < \varepsilon.$$

Now taking (12.9) and (12.10) into account, we get the desired result. $\square$

EXAMPLE 1. Find the value of

$$\int_{-1}^{2} x^5\, d(|x|^3).$$

Solution. Consider the integrals $\int_{-1}^{0} x^5\, d(-x^3)$ and $\int_{0}^{2} x^5\, d(x^3)$. According to

Theorem 12.10, we have

$$\int_{-1}^{0} x^5 \, d(-x^3) = -3 \int_{-1}^{0} x^7 \, dx = -\frac{3}{8} x^8 \Big]_{-1}^{0} = \frac{3}{8}$$

$$\int_{0}^{2} x^5 \, d(x^3) = 3 \int_{0}^{2} x^7 \, dx = \frac{3}{8} x^8 \Big]_{0}^{2} = 96.$$

Now Property (d) of Theorem 12.9 yields

$$\int_{-1}^{2} x^5 (d|x|^3) = \frac{3}{8} + 96. \qquad \square$$

The next theorem shows how to change variables in R–S integrals, a result which is useful when actually performing integrations. This is the analogue of Theorem 5.13, the change of variables formula for Riemann integrals.

Theorem 12.11. *Suppose that f is integrable with respect to g on $I = \{x: a \leqslant x \leqslant b\}$. Let $x = x(u)$ be a continuous, increasing function on $J = \{u: c \leqslant u \leqslant d\}$ with $x(c) = a$ and $x(d) = b$. Define*

$$F(u) = f[x(u)] \text{ and } G(u) = g[x(u)].$$

Then F is integrable with respect to G on J, and

$$\int_{c}^{d} F(u) \, dG(u) = \int_{a}^{b} f(x) \, dg(x). \tag{12.14}$$

If $x(u)$ is continuous and decreasing on J with $x(c) = b$ and $x(d) = a$, then

$$\int_{c}^{d} F(u) \, dG(u) = -\int_{a}^{b} f(x) \, dg(x). \tag{12.15}$$

PROOF. Let $\varepsilon > 0$ be given. From the definition of R–S integral, there is a $\delta > 0$ such that

$$\left| \sum_{i=1}^{n} f(\zeta_i) [g(x_i) - g(x_{i-1})] - \int_{a}^{b} f(x) \, dg(x) \right| < \varepsilon \tag{12.16}$$

for all subdivisions $\Delta: a = x_0 < x_1 < \cdots < x_n = b$ with mesh less than δ and any $\{\zeta_i\}$ in which $x_{i-1} \leqslant \zeta_i \leqslant x_i$. Since x is (uniformly) continuous on J, there is a δ_1 such that $|x(u') - x(u'')| < \delta$ whenever $|u' - u''| < \delta_1$. Consider the subdivision $\Delta_1 : c = u_0 < u_1 < \cdots < u_n = d$ of J with mesh less than δ_1. Let η_i be such that $u_{i-1} \leqslant \eta_i \leqslant u_i$ for $i = 1, 2, \ldots, n$. Since $x(u)$ is increasing on J, denote $x(u_i) = x_i$ and $x(\eta_i) = \zeta_i$. Then the subdivision Δ of I has mesh less than δ and $x_{i-1} \leqslant \zeta_i \leqslant x_i$ for $i = 1, 2, \ldots, n$. Therefore

$$\sum_{i=1}^{n} F(\eta_i) [G(u_i) - G(u_{i-1})] = \sum_{i=1}^{n} f(\zeta_i) [g(x_i) - g(x_{i-1})].$$

Taking into account the definition of the R–S integral, (12.7), we see that the integrals in (12.14) are equal.

If $x(u)$ is decreasing on J, let $x(u_i) = x_{n-i}$, $x(\eta_i) = \zeta_{n+1-i}$. Then

$$\sum_{i=1}^{n} F(\eta_i)[G(u_i) - G(u_{i-1})] = \sum_{i=1}^{n} f(\zeta_{n+1-i})[g(x_{n-i}) - g(x_{n+1-i})].$$

Setting $k = n + 1 - i$, $i = 1, 2, \ldots, n$, we find the sum on the right is

$$-\sum_{k=1}^{n} f(\zeta_k)[g(x_k) - g(x_{k-1})].$$

Then (12.16) shows that the integrals in (12.15) are equal. $\square$

The next result is a generalization of the customary integration by parts formula studied in calculus. Formula (12.17) below is most useful for the actual computation of R–S integrals. It also shows that if either of the integrals $\int_a^b f\, dg$ or $\int_a^b g\, df$ exists then the other one does.

Theorem 12.12 (Integration by parts). *If $\int_a^b f\, dg$ exists, then so does $\int_a^b g\, df$ and*

$$\int_a^b g\, df = g(b)f(b) - g(a)f(a) - \int_a^b f\, dg. \tag{12.17}$$

PROOF. Let $\varepsilon > 0$ be given. From the definition of R–S integral, there is a $\delta' > 0$ such that

$$\left| \sum_{i=1}^{m} f(\zeta_i')[g(x_i') - g(x_{i-1}')] - \int_a^b f\, dg \right| < \varepsilon \tag{12.18}$$

for any subdivision Δ': $a = x_0' < x_1' < \cdots < x_m' = b$ of mesh less than δ' and any ζ_i' with $x_{i-1}' \leqslant \zeta_i' \leqslant x_i'$. Let $\delta = \frac{1}{2}\delta'$ and choose a subdivision Δ: $a = x_0 < x_1 < \cdots < x_n = b$ of mesh less than δ and points ζ_i such that $x_{i-1} \leqslant \zeta_i \leqslant x_i$, $i = 1, 2, \ldots, n$. We further select $\zeta_0 = a$ and $\zeta_{n+1} = b$. Then we observe that $a = \zeta_0 \leqslant \zeta_1 \leqslant \cdots \leqslant \zeta_n \leqslant \zeta_{n+1} = b$ is a subdivision with mesh size less than δ', and furthermore $\zeta_{i-1} \leqslant x_{i-1} \leqslant \zeta_i$, $i = 1, 2, \ldots, n + 1$.

Therefore

$$\sum_{i=1}^{n} g(\zeta_i)[f(x_i) - f(x_{i-1})]$$

$$= \sum_{i=1}^{n} g(\zeta_i)f(x_i) - \sum_{i=1}^{n} g(\zeta_i)f(x_{i-1})$$

$$= \sum_{i=2}^{n+1} g(\zeta_{i-1})f(x_{i-1}) + g(a)f(a) - g(a)f(a)$$

$$\quad - \sum_{i=1}^{n} g(\zeta_i)f(x_{i-1}) - g(b)f(b) + g(b)f(b)$$

$$= \sum_{i=1}^{n+1} g(\zeta_{i-1})f(x_{i-1}) - g(a)f(a) - \sum_{i=1}^{n+1} g(\zeta_i)f(x_{i-1}) + g(b)f(b)$$

or

$$\sum_{i=1}^{n} g(\zeta_i)[f(x_i) - f(x_{i-1})] = g(b)f(b) - g(a)f(a)$$
$$- \sum_{i=1}^{n+1} f(x_{i-1})[g(\zeta_i) - g(\zeta_{i-1})]. \quad (12.19)$$

The sum on the right is an R–S sum which satisfies (12.18). Hence the right side of (12.19) differs from the right side of (12.17) by less than ε. But the left side of (12.19) is the R–S sum for the left side of (12.17). The result follows since ε is arbitrary. $\qquad\square$

We shall show below that if f is continuous and g of bounded variation on an interval $I = \{x: a \leqslant x \leqslant b\}$, then $\int_a^b f \, dg$ exists. Theorem 12.12 shows that if the hypotheses on f and g are reversed, the R–S integral will still exist.

EXAMPLE 2. Let f be any function with a continuous derivative on $I = \{x: a \leqslant x \leqslant b\}$. Let $a = a_0 < a_1 < \cdots < a_N = b$ be any finite sequence of numbers, and define

$$g(x) = \begin{cases} c_i & \text{for} \quad a_{i-1} < x \leqslant a_i, \quad i = 1, 2, \ldots, N, \\ c_0 & \text{for} \quad x = a, \end{cases}$$

where $c_0, c_1, \ldots, c_N$ are any constants. Show that

$$\int_a^b g(x)f'(x) \, dx = c_N f(b) - c_0 f(a) - \sum_{i=0}^{N-1} f(a_i)(c_{i+1} - c_i). \quad (12.20)$$

Solution. For the subinterval $I_i = \{x: a_{i-1} \leqslant x \leqslant a_i\}$, we have

$$\int_{a_{i-1}}^{a_i} f(x) \, dg(x) = f(a_{i-1})(c_i - c_{i-1}),$$

which can be seen by considering the R–S sum and proceeding to the limit. Property (d) of Theorem 12.9 yields

$$\int_a^b f \, dg = \sum_{i=0}^{N-1} f(a_i)(c_{i+1} - c_i).$$

Now we employ integration by parts to obtain $\int_a^b g \, df$. Observe that Theorem 12.10 can be proved when the integrand f in Theorem 12.10 is a step function and, applying the result to $\int_a^b g \, df$, we obtain (12.20). $\qquad\square$

We recall that any function of bounded variation may be represented as the difference of two monotone functions. Therefore if the existence of the R–S integral $\int_a^b f \, dg$ when f is continuous and g is nondecreasing is established, it follows that the R–S integral exists when f is continuous and g is of bounded variation. To establish the existence of the R–S integral for as large a class as possible, we shall employ the method of Darboux described in Chapter 5. We begin by defining upper and lower Darboux–Stieltjes sums and integrals.

Definitions. Suppose that f and g are defined on $I = \{x: a \leqslant x \leqslant b\}$ and g is nondecreasing on I. Let $\Delta: a = x_0 < x_1 < \cdots < x_n = b$ be a subdivision of I. Define $\Delta_i g = g(x_i) - g(x_{i-1}), i = 1, 2, \ldots, n$. Note that $\Delta_i g \geqslant 0$ for all i; also set

$$M_i = \sup_{x_{i-1} \leqslant x \leqslant x_i} f(x), \qquad m_i = \inf_{x_{i-1} \leqslant x \leqslant x_i} f(x).$$

Define the **upper** and **lower Darboux–Stieltjes sums**

$$S^+(f, g, \Delta) = \sum_{i=1}^{n} M_i \Delta_i g, \qquad S_-(f, g, \Delta) = \sum_{i=1}^{n} m_i \Delta_i g.$$

Since g is nondecreasing, observe that the numbers $I_i = \Delta_i g = g(x_i) - g(x_{i-1})$ are nonnegative and that $\sum_{i=1}^{n} \Delta_i g = g(b) - g(a)$. Thus $\{I_i\}$ form a subdivision $\bar{\Delta}$ of the interval $g(b) - g(a)$, and if Δ' is a refinement of Δ, then there is induced a corresponding refinement $\bar{\Delta}'$ of $\bar{\Delta}$.

The next result is completely analogous to Theorem 5.1 concerning Darboux sums, and we leave the proof to the reader.

Lemma 12.2. *Suppose that f and g are defined on $I = \{x: a \leqslant x \leqslant b\}$ and g is nondecreasing on I.*

(a) *If $m \leqslant f(x) \leqslant M$ on I and Δ is any subdivision, then*

$$m[g(b) - g(a)] \leqslant S_-(f, g, \Delta) \leqslant S^+(f, g, \Delta) \leqslant M[g(b) - g(a)].$$

(b) *If Δ' is a refinement of Δ, then*

$$S_-(f, g, \Delta') \geqslant S_-(f, g, \Delta) \quad and \quad S^+(f, g, \Delta') \leqslant S^+(f, g, \Delta).$$

(c) *If Δ_1 and Δ_2 are any subdivisions of I, then*

$$S_-(f, g, \Delta_1) \leqslant S^+(f, g, \Delta_2).$$

Definitions. *The* **upper** *and* **lower Darboux–Stieltjes integrals** *are given by*

$$\overline{\int_a^b} f \, dg = \inf S^+(f, g, \Delta) \quad \text{for all subdivisions } \Delta \text{ of } I.$$

$$\underline{\int_a^b} f \, dg = \sup S_-(f, g, \Delta) \quad \text{for all subdivisions } \Delta \text{ of } I.$$

These integrals are defined for all functions f on I and all functions g which are nondecreasing on I. If

$$\overline{\int_a^b} f \, dg = \underline{\int_a^b} f \, dg \tag{12.21}$$

we say f is **integrable with respect to g on** I (in the Darboux–Stieltjes sense). The *integral* is the common value (12.21).

The following elementary properties of upper and lower Darboux–Stieltjes integrals are similar to the corresponding ones for upper and lower Darboux integrals given in Chapter 5.

Theorem 12.13. *Suppose that* f, f_1, *and* f_2 *are bounded on the interval* $I = \{x: a \leqslant x \leqslant b\}$ *and that* g *is nondecreasing on* I.

(a) *If* $m \leqslant f(x) \leqslant M$ *on* I, *then*

$$m[g(b) - g(a)] \leqslant \underline{\int_a^b} f\, dg \leqslant \overline{\int_a^b} f\, dg \leqslant M[g(b) - g(a)].$$

(b) *If* $f_1(x) \leqslant f_2(x)$ *for* $x \in I$, *then*

$$\underline{\int_a^b} f_1\, dg \leqslant \underline{\int_a^b} f_2\, dg \qquad and \qquad \overline{\int_a^b} f_1\, dg \leqslant \overline{\int_a^b} f_2\, dg.$$

(c) *Suppose that* $a < c < b$. *Then*

$$\underline{\int_a^c} f\, dg + \underline{\int_c^b} f\, dg = \underline{\int_a^b} f\, dg, \qquad \overline{\int_a^c} f\, dg + \overline{\int_c^b} f\, dg = \overline{\int_a^b} f\, dg.$$

(d) $\int_a^c f\, dg$ *and* $\int_c^b f\, dg$ *both exist if and only if* $\int_a^b f\, dg$ *exists. In this case, we have*

$$\int_a^c f\, dg + \int_c^b f\, dg = \int_a^b f\, dg.$$

The proof of Theorem 12.13 is similar to the proof of the corresponding results for Riemann integrals as described in Theorems 5.2 and 5.3. We leave the details to the reader.

Theorem 12.14. *Suppose that* f *is continuous on* $I = \{x: a \leqslant x \leqslant b\}$ *and that* g *is nondecreasing on* I. *Then the Riemann–Stieltjes integral* $\int_a^b f\, dg$ *exists.*

PROOF. Since f is uniformly continuous on I, for every $\varepsilon > 0$ there is a $\delta > 0$ such that

$$|f(x) - f(y)| < \frac{\varepsilon}{2[1 + g(b) - g(a)]}$$

whenever $|x - y| < \delta$. Let $\Delta: a = x_0 < x_1 < \cdots < x_n = b$ be a subdivision of I with mesh less than δ. Define $I_i = \{x: x_{i-1} \leqslant x \leqslant x_i\}$. Since f is continuous on each I_i, there are numbers η_i and ζ_i on I_i such that $m_i = f(\eta_i)$ and $M_i = f(\zeta_i)$. Then we have

$$S^+(f, g, \Delta) - S_-(f, g, \Delta) = \sum_{i=1}^{n} [f(\zeta_i) - f(\eta_i)]\Delta_i g$$

$$\leqslant \sum_{i=1}^{n} \frac{\varepsilon}{2[1 + g(b) - g(a)]}\Delta_i g < \frac{1}{2}\varepsilon.$$

Since ε is arbitrary it follows that the Darboux–Stieltjes integral denoted

(D–S) $\int_a^b f\, dg$ exists. Now let ξ_i be any point in I_i. From the fact that the Darboux–Stieltjes integral is between S^+ and S_-, we find

$$(\text{D–S}) \int_a^b f\, dg - \varepsilon < S_-(f, g, \Delta)$$

$$= \sum_{i=1}^n f(\eta_i)\Delta_i g \leqslant \sum_{i=1}^n f(\xi_i)\Delta_i g$$

$$\leqslant \sum_{i=1}^n f(\zeta_i)\Delta_i g = S^+(f, g, \Delta)$$

$$< (\text{D–S}) \int_a^b f\, dg + \varepsilon.$$

Since $\sum_{i=1}^n f(\xi_i)\Delta_i g$ tends to (D–S) $\int_a^b f\, dg$ as $\varepsilon \to 0$, the (R–S) integral exists and, in fact, is equal to the (D–S) integral. $\qquad\square$

Corollary. *Suppose that f is continuous on $I = \{x: a \leqslant x \leqslant b\}$ and that g is of bounded variation on I. Then the Riemann–Stieltjes integral $\int_a^b f\, dg$ exists.*

We give an example of functions f and g such that f is not Riemann integrable on an interval I but $\int f\, dg$ exists on that interval. To see this, we define

$$f(x) = \begin{cases} 0 & \text{for} \quad x = 0, \\ 1/x & \text{for} \quad 0 < x < 1, \\ 1 & \text{for} \quad 1 \leqslant x \leqslant 2, \end{cases} \qquad g(x) = \begin{cases} 1 & \text{for} \quad 0 \leqslant x \leqslant 1, \\ x & \text{for} \quad 1 < x \leqslant 2. \end{cases}$$

Then clearly $\int_0^2 f(x)\, dx$ does not exist since $f(x) = 1/x$ is not integrable for $0 < x < 1$. However, by considering Riemann–Stieltjes sums, we find that all such sums vanish for $0 \leqslant x \leqslant 1$. Hence

$$\int_0^2 f\, dg = \int_1^2 f\, dg = \int_1^2 dx = 1.$$

The next result gives an important basic upper bound for the value of any R–S integral.

Theorem 12.15. *Suppose that f is continuous on $I = \{x: a \leqslant x \leqslant b\}$ and that g is of bounded variation on I. Let $M = max_{x \in I}|f(x)|$. Then*

$$\left| \int_a^b f\, dg \right| \leqslant M V_a^b g.$$

The proof is a direct consequence of the definition of the (R–S) integral and the Corollary to Theorem 12.14. We leave the details to the reader.

Suppose that $\{f_n\}$ and $\{g_n\}$ are sequences of functions which converge to functions f and g for all x on an interval $I = \{x: a \leqslant x \leqslant b\}$. We wish to establish conditions which guarantee that $\int_a^b f_n \, dg_n$ converges to $\int_a^b f \, dg$. For this purpose we prove the next technical result.

Lemma 12.3. *Suppose that f is continuous and g is of bounded variation on $I = \{x: a \leqslant x \leqslant b\}$. Let $\Delta: a = x_0 < x_1 < \cdots < x_n = b$ be a subdivision of mesh less than δ. Define*

$$\omega(f, \delta) = \sup |f(x) - f(y)| \quad \text{for all } x, y \text{ with } |x - y| < \delta.$$

Suppose that ζ_i is in the interval $I_i = \{x: x_{i-1} \leqslant x \leqslant x_i\}$ for $i = 1, 2, \ldots, n$. Then

$$\left| \sum_{i=1}^{n} f(\zeta_i) \Delta_i g - \int_a^b f \, dg \right| \leqslant \omega(f, \delta) V_a^b g.$$

PROOF. We may write

$$\int_a^b f \, dg = \sum_{i=1}^{n} \int_{x_{i-1}}^{x_i} f \, dg$$

and

$$\int_{x_{i-1}}^{x_i} f(\zeta_i) \, dg = f(\zeta_i) \Delta_i g.$$

Therefore

$$\sum_{i=1}^{n} f(\zeta_i) \Delta_i g - \int_a^b f \, dg = \sum_{i=1}^{n} \int_{x_{i-1}}^{x_i} [f(\zeta_i) - f(x)] \, dg.$$

If the mesh of Δ is less than δ, then for ζ_i and x in I_i, it follows that

$$|f(\zeta_i) - f(x)| \leqslant \omega(f, \delta),$$

and so

$$\left| \sum_{i=1}^{n} f(\zeta_i) \Delta_i g - \int_a^b f \, dg \right| \leqslant \sum_{i=1}^{n} \left| \int_{x_{i-1}}^{x_i} [f(\zeta_i) - f(x)] \, dg \right|$$

$$\leqslant \omega(f, \delta) \sum_{i=1}^{n} V_{x_{i-1}}^{x_i} g = \omega(f, \delta) V_a^b g. \qquad \square$$

The next theorem shows that for uniformly convergent sequences $\{f_n\}$ and for functions $\{g_n\}$ of uniformly bounded variation which tend to a limit on a dense set of points, the passage to the limit extends through Riemann–Stieltjes integrals.

Theorem 12.16. *Suppose that f_n, $n = 1, 2, \ldots$, are continuous and $f_n \to f$ as $n \to \infty$ uniformly on $I = \{x: a \leqslant x \leqslant b\}$. Suppose that g_n and g are of bounded variation with $V_a^b g \leqslant L$ and $V_a^b g_n \leqslant L$ for all n. Let S be any dense set of points on I which contains a and b. If $g_n \to g$ on S, then*

$$\lim_{n \to \infty} \int_a^b f_n \, dg_n = \int_a^b f \, dg.$$

PROOF. Let $\varepsilon_n = \max_{x \in I} |f_n(x) - f(x)|$. Then from Theorem 12.15, we have

$$\left| \int_a^b (f_n - f) \, dg_n \right| \leq \varepsilon_n V_a^b g_n \leq \varepsilon_n L.$$

Observe that $\varepsilon_n \to 0$ as $n \to \infty$. Therefore

$$\left| \int_a^b f_n \, dg_n - \int_a^b f \, dg \right| = \left| \int_a^b (f_n - f) \, dg_n + \int_a^b f(dg_n - dg) \right|$$

$$\leq \varepsilon_n L + \left| \int_a^b f \, dg_n - \int_a^b f \, dg \right|.$$

We shall show that the second term on the right tends to zero as $n \to \infty$. Let $\varepsilon > 0$ be given. Choose $\delta > 0$ so small that $\omega(f, \delta) < \varepsilon/3L$. Next select a subdivision $\Delta : a = x_0 < x_1 < \cdots < x_k = b$ of mesh less than δ and with all subdivision points in S. From Lemma 12.3, we find

$$\left| \sum_{i=1}^k f(\zeta_i)[g_n(x_i) - g_n(x_{i-1})] - \int_a^b f \, dg_n \right| < \frac{\varepsilon}{3L} \cdot L = \frac{\varepsilon}{3}. \qquad (12.22)$$

Also, by the same reasoning

$$\left| \sum_{i=1}^k f(\zeta_i)[g(x_i) - g(x_{i-1})] - \int_a^b f \, dg \right| < \frac{1}{3}\varepsilon. \qquad (12.23)$$

We now use the fact that $g_n \to g$ on the dense set S. For each i there is an integer N such that

$$\left| \sum_{i=1}^k f(\zeta_i)\{[g_n(x_i) - g_n(x_{i-1})] - [g(x_i) - g(x_{i-1})]\} \right| < \frac{\varepsilon}{3} \qquad (12.24)$$

for all $n > N$. The result follows by inserting the three inequalities (12.22), (12.23), and (12.24) in the "triangle" inequality

$$\left| \int_a^b f \, dg_n - \int_a^b f \, dg \right|$$

$$\leq \left| \int_a^b f \, dg_n - \sum_{i=1}^k f(\zeta_i)[g_n(x_i) - g_n(x_{i-1})] \right|$$

$$+ \left| \sum_{i=1}^k f(\zeta_i)[g_n(x_i) - g_n(x_{i-1})] - \sum_{i=1}^k f(\zeta_i)[g(x_i) - g(x_{i-1})] \right|$$

$$+ \left| \sum_{i=1}^k f(\zeta_i)[g(x_i) - g(x_{i-1})] - \int_a^b f \, dg \right| < \frac{\varepsilon}{3} + \frac{\varepsilon}{3} + \frac{\varepsilon}{3} = \varepsilon. \qquad \square$$

EXAMPLE 3. Let $f_n = 1 - (1/n)\sin nx$, $g_n = 1 + x^n$, $n = 1, 2, \ldots$ Show that $\int_0^1 f_n \, dg_n \to 1$ as $n \to \infty$.

Solution. We observe that $f_n \to 1$ uniformly on $I = \{x: 0 \leqslant x \leqslant 1\}$ and that g_n tends to the function

$$g(x) = \begin{cases} 1 & \text{for} \quad 0 \leqslant x < 1, \\ 2 & \text{for} \qquad x = 1. \end{cases}$$

Therefore the hypotheses of Theorem 12.16 are satisfied. Since $\int_0^1 1 \, dg = 1$, the result follows. $\qquad\qquad\square$

PROBLEMS

1. (a) Let $a < c_1 < c_2 < \cdots < c_n < b$ be any points of $I = \{x: a \leqslant x \leqslant b\}$. Suppose that

$$g(x) = \begin{cases} 1 & \text{for } x \neq c_i, \quad i = 1, 2, \ldots, n, \\ d_i & \text{for } x = c_i, \quad i = 1, 2, \ldots, n, \end{cases}$$

where d_i is a constant for each i.
Suppose that f is continuous on I. Find an expression for $\int_a^b f \, dg$.
   (b) Work Part (a) if $a = c_1 < c_2 < \cdots < c_n = b$. [*Hint:* First do Part (a) with $a = c_1 < c_2 = b$.]

2. Let $g(x) = \sin x$ for $0 \leqslant x \leqslant \pi$. Find the value of $\int_0^\pi x \, dg$.

3. Let $g(x) = e^{|x|}$ for $-1 \leqslant x \leqslant 1$. Find the value of $\int_{-1}^1 x \, dg$.

4. Let $g(x) = k$ for $k - 1 < x \leqslant k$, $k = 1, 2, 3, \ldots$. Find the value of $\int_1^4 x \, dg$.

5. Show that with g as in Problem 4, $\int_1^5 g \, dg$ does not exist.

6. Prove Theorem 12.9.

7. Use Theorem 12.11 to evaluate $\int_0^{\sqrt{\pi}} \cos(u^2) d(\cos(u^2))$.

8. Show that if f and g have a common point of discontinuity on an interval $I = \{x: a \leqslant x \leqslant b\}$, then $\int_a^b f \, dg$ cannot exist.

9. Suppose that $f \equiv c$ on $I = \{x: a \leqslant x \leqslant b\}$ where c is any constant. If g is of bounded variation on I, use integration by parts (Theorem 12.12) to show that $\int_a^b f \, dg = c[g(b) - g(a)]$.

10. Let $[x]$ denote the largest integer less than or equal to the number x. Find the value of $\int_0^3 (x^2 + 1)d([x])$.

11. With $[x]$ defined as in Problem 10, show that

$$\int_1^n \frac{[x]}{x^{t+1}} \, dx = \frac{-1}{t(n^{t-1})} + \sum_{i=1}^n \frac{1}{i^t}, \qquad t \neq 1.$$

Use the R–S integral.

12. Same as Problem 11 for

$$\int_1^{2n} \frac{2[\frac{1}{2}x] - [x]}{x^{t+1}} \, dx = \frac{1}{t} \sum_{i=1}^{2n} \frac{(-1)^i}{i^t}, \qquad t \neq 1.$$

13. Suppose that f_n, f, g_n, g satisfy the hypotheses of Theorem 12.16. Under what circumstances is it true that

$$\lim_{n \to \infty} \int_a^b g_n \, df_n \to \int_a^b g \, df?$$

14. Prove Lemma 12.2.

15. Prove Theorem 12.13.

16. Let f be continuous and g of bounded variation on $I = \{x: a \leqslant x < \infty\}$. Define $\int_a^\infty f \, dg = \lim_{x \to \infty} \int_a^x f \, dg$ when the limit exists. Show that if f is bounded on I and $g(x) = 1/k^2$ for $k - 1 \leqslant x < k$, $k = 1, 2, \ldots$, then

$$\int_1^\infty f \, dg = - \sum_{k=1}^\infty f(k + 1) \left[\frac{1}{(k + 1)^2} - \frac{1}{(k + 2)^2} \right].$$

17. Use the function $g(x) = b_k$ for $k - 1 < x \leqslant k$, $k = 1, 2, \ldots$, b_k constant, and the definition of integral over a half-infinite interval, as in Problem 16, to show that *any infinite series $\sum_{k=1}^\infty a_k$ may be represented as a Riemann–Stieltjes integral.*

18. Prove Theorem 12.15.

19. Let $f_n(x) = nx/(1 + n^3 x^2)$ and $g_n(x) = x^n$ on $I = \{x: 0 \leqslant x \leqslant 1\}$. Show that $\lim_{n \to \infty} \int_0^1 f_n(x) \, dg_n(x)$ exists and find its value.

20. Let

$$f(x) = \begin{cases} 1 & \text{if } x \text{ is rational,} \\ 0 & \text{if } x \text{ is irrational,} \end{cases}$$

and let $g(x)$ be any nonconstant, nondecreasing function on $I = \{x: 0 \leqslant x \leqslant 1\}$. Show that $\int_0^1 f \, dg$ does not exist.

21. Let f be continuous and g nondecreasing on $I = \{x: 0 \leqslant x \leqslant b\}$. Prove the **Mean-value theorem:**

$$\int_a^b f(x) \, dg(x) = f(\zeta) \int_a^b dg(x),$$

where ζ is some point of I. [*Hint*: Use Property (a) of Theorem 12.13, and the Intermediate-value theorem applied to the function f.]

22. Suppose that f and g are continuous on $I = \{x: a \leqslant x \leqslant b\}$. Define $h(x) = \int_a^x g(t) \, dt$. Show that

$$\int_a^b f \, dh = \int_a^b f(x) g(x) \, dx.$$

CHAPTER 13

Contraction Mappings, Newton's Method, and Differential Equations

13.1. A Fixed Point Theorem and Newton's Method

The main result of this section is a simple theorem which proves to be useful in the solution of algebraic and differential equations. It will also be used to prove the important Implicit function theorem in Chapter 14.

We first recall the definition of a Cauchy sequence in a metric space S (see Chapter 6). A sequence of points $\{p_n\}$ in S is called a **Cauchy sequence** if and only if for every $\varepsilon > 0$ there is an integer N such that $d(p_m, p_n) < \varepsilon$ whenever $m, n > N$. We also recall Theorem 6.22 in Chapter 6 which states that *every convergent sequence in a metric space is a Cauchy sequence*. Since not every Cauchy sequence in a metric space is convergent, we introduce the following class of metric spaces.

Definition. A metric space S is said to be **complete** if and only if every Cauchy sequence in S converges to a point in S.

We observe that a compact metric space is always complete (Theorem 6.23). Although the space $\mathbb{R}^1$ is not compact it is complete, as was established in Theorem 3.14. With the aid of this result we now show that all the Euclidean spaces are complete.

Theorem 13.1. *For every positive integer N, the space $\mathbb{R}^N$ is complete.*

PROOF. We already showed that $\mathbb{R}^1$ is complete. Let $N \geqslant 2$. Let $x_1, x_2, \ldots, x_n, \ldots$ be a Cauchy sequence in $\mathbb{R}^N$. With the notation $x_n = (x_n^1, x_n^2, \ldots, x_n^N)$, the distance between the points x_m and x_n is

$$d(x_m, x_n) = \left[\sum_{i=1}^{N} (x_m^i - x_n^i)^2 \right]^{1/2}.$$

Fix j and observe that for each $j = 1, 2, \ldots, N$,

$$\sum_{i=1}^{N} (x_m^i - x_n^i)^2 \geqslant (x_m^j - x_n^j)^2.$$

Hence for each fixed j with $1 \leqslant j \leqslant N$ the sequence $\{x_n^j\}$ is a Cauchy sequence in $\mathbb{R}^1$ and thus converges to a limit which is denoted x_0^j. We conclude that $x_n \to x_0$ where $x_0 = (x_0^1, x_0^2, \ldots, x_0^N)$. The space $\mathbb{R}^N$ is complete.   □

Definition. Let f be a mapping from a metric space S into S. A point x_0 in S is called a **fixed point** of the mapping f if $f(x_0) = x_0$.

We now show that certain classes of mappings from a complete metric space into itself always have a fixed point.

Theorem 13.2 (A fixed point theorem). *Suppose that f is a mapping on a complete metric space S into S such that*

$$d(f(x), f(y)) \leqslant k \, d(x, y) \tag{13.1}$$

for all x, y in S with $0 < k < 1$. Then the mapping has a unique fixed point. That is, there is a unique point $x_0 \in S$ such that $f(x_0) = x_0$. Moreover, if x_1 is any element of S, and the sequence $\{x_n\}$ is defined by setting $x_{n+1} = f(x_n)$, $n = 1, 2, \ldots$, then x_n converges to x_0.

PROOF. Start with any x_1 in S and set $x_{n+1} = f(x_n)$, $n = 1, 2, \ldots$. Then $d(x_2, x_3) = d(f(x_1), f(x_2)) \leqslant kd(x_1, x_2)$. Continuing,

$$d(x_3, x_4) = d(f(x_2), f(x_3)) \leqslant kd(x_2, x_3) \leqslant k^2 d(x_1, x_2).$$

In general,

$$d(x_n, x_{n+1}) \leqslant k^{n-1} d(x_1, x_2) \quad \text{for} \quad n = 1, 2, \ldots. \tag{13.2}$$

Let m, n be any positive integers with $m > n$. Then by the triangle inequality in a metric space,

$$d(x_n, x_m) \leqslant d(x_n, x_{n+1}) + d(x_{n+1}, x_{n+2}) + \cdots + d(x_{m-1}, x_m).$$

Applying inequality (13.2) to each term on the right, we obtain

$$d(x_n, x_m) \leqslant d(x_1, x_2)[k^{n-1} + k^n + \cdots + k^{m-2}]$$

$$= d(x_1, x_2)k^{n-1}[1 + k + \cdots + k^{m-n-1}].$$

Since $\sum_{j=0}^{\infty} k^j = 1/(1 - k)$, we find

$$d(x_n, x_m) < d(x_1, x_2)k^{n-1}\frac{1}{1 - k}.$$

Now let $\varepsilon > 0$ be given. Since $0 < k < 1$, the right side tends to 0 as $n \to \infty$. Therefore there is an N such that the right side is less than ε for all $n > N$. Thus $\{x_n\}$ is a Cauchy sequence. Let x_0 be the limit of $\{x_n\}$. By (13.1), we know that f is continuous on S and so $f(x_n) \to f(x_0)$ as $n \to \infty$. Since $x_{n+1} \to x_0$ and $x_{n+1} = f(x_n)$, it follows that $f(x_0) = x_0$.

To establish the uniqueness of the point x_0, suppose there is a point x' such that $x' = f(x')$. Using (13.1), we find that

$$d(x_0, x') = d[f(x_0), f(x')] \leqslant kd(x_0, x').$$

Since $0 < k < 1$, it is clear that $d(x_0, x') = 0$, and so $x_0 = x'$. $\square$

A mapping f which satisfies (13.1) with $0 < k < 1$ is called a **contraction mapping**.

Corollary. *Let $I = \{x : a \leqslant x \leqslant b\}$ be an interval in $\mathbb{R}^1$ and suppose that $f : I \to I$ is a differentiable function at each interior point of I with $|f'(x)| \leqslant k, 0 < k < 1$ for each x. Then there is a unique point x_0 of I such that $f(x_0) = x_0$.*

We leave the proof to the reader.

EXAMPLE 1. Let $I = \{x : a \leqslant x \leqslant b\}$ be an interval and suppose that f is differentiable on I with $f(a)$ and $-f(b)$ contained in the interval $J = \{x : 0 \leqslant x \leqslant b - a\}$. Furthermore, suppose that there are numbers k and k' such that $-1 < k' \leqslant f'(x) \leqslant k < 0$ for all $x \in I$. Show that there is an $x_0 \in I$ such that $f(x_0) = 0$ and prove that for any $x_1 \in I$, the iteration $x_{n+1} = f(x_n), n = 1, 2, \ldots$ converges to x_0.

Solution. Set $g(x) = f(x) + x$. Then, by hypothesis, $g(a) \in I, g(b) \in I$, and, since $g'(x) = 1 + f'(x)$, it follows that g is an increasing function on I. Hence g maps I into I. Also, $g(x) - g(y) = x - y + f(x) - f(y)$ and, applying the Mean-value Theorem, we find

$$|g(x) - g(y)| = |(x - y)(1 + f'(\zeta))|$$

with $a < \zeta < b$. Thus

$$|g(x) - g(y)| \leqslant (1 + k)|x - y|,$$

and from Theorem 13.2 it follows that there is an x_0 such that $g(x_0) = x_0$. Therefore $x_0 + f(x_0) = x_0$ and $f(x_0) = 0$. $\square$

Remarks. The above example shows that for any differentiable function f, if $f(a)$ and $f(b)$ are of *opposite* sign and if the derivative of f is small (but bounded away from zero) then the zero of f which must exist between a and b may always be found by a simple iteration method. There are many iteration methods for finding the zeros of functions on an interval. We now describe a method of wide applicability, especially when the first derivative of the function is large in the neighborhood of the zero.

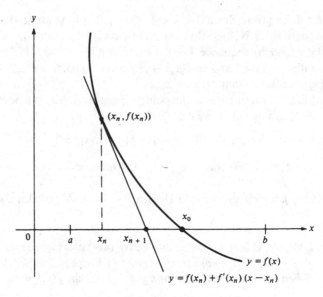

Figure 13.1. Newton's method.

Newton's method

Let f be a twice differentiable function on $I = \{x : a \leqslant x \leqslant b\}$ with $f(a) > 0$ and $f(b) < 0$. Then there is an $x_0 \in I$ such that $f(x_0) = 0$. Newton's method consists of an iteration process which converges to x_0, one of the zeros of f in I. We assume that there is a positive number M such that $|f'(x)| \geqslant 1/M$ and $|f''(x)| \leqslant 2M$ for all $x \in I$. Let x_1 be any point in I. We define a sequence $\{x_n\}$ by the formula, known as **Newton's method**,

$$x_{n+1} = x_n - \frac{f(x_n)}{f'(x_n)}, \qquad n = 1, 2, \ldots. \tag{13.3}$$

The determination of x_{n+1} from x_n is shown geometrically in Figure 13.1. The tangent line to the graph of $y = f(x)$ is constructed at the point $(x_n, f(x_n))$. The point where this line intersects the x axis is x_{n+1}. To show that the sequence $\{x_n\}$ converges to x_0, first observe that

$$|x_{n+1} - x_n| \leqslant M|f(x_n)|. \tag{13.4}$$

Next, apply Taylor's theorem with remainder to f at x_{n+1}, getting

$$f(x_{n+1}) = f(x_n) + f'(x_n)(x_{n+1} - x_n) + \tfrac{1}{2}f''(\zeta)(x_{n+1} - x_n)^2.$$

Because of (13.3) the first two terms on the right cancel, and so

$$|f(x_{n+1})| \leqslant M|x_{n+1} - x_n|^2. \tag{13.5}$$

By means of (13.4) and (13.5), it is not difficult to show by induction that $\{x_n\}$ is a convergent sequence provided $|f(x_1)|$ and M are less than 1. Hence it follows from (13.5) that $f(x_n) \to 0$ as $n \to \infty$. In terms of the Fixed point theorem, we define the function

$$F(x) = x - \frac{1}{f'(x)} f(x);$$

then Newton's method consists of forming $x_{n+1} = F(x_n)$, $n = 1, 2, \ldots$, and concluding from the convergence of the sequence that there is a fixed point x_0 such that $F(x_0) = x_0$. The existence of such a point in general does not follow from Theorem 13.2 since the range of F is not necessarily contained in its domain and so we may not have a mapping of a complete metric space I into itself.

EXAMPLE 2. Find a positive root of the equation $2x^4 + 2x^3 - 3x^2 - 5x - 5 = 0$ with an accuracy of three decimal places.

Solution. We construct a table of values of $f(x) = 2x^4 + 2x^3 - 3x^2 - 5x - 5$ and obtain

x	0	1	2
$f(x)$	-5	-9	21

Thus there is a positive root x_0 between $x = 1$ and $x = 2$. We choose $x_1 = 1.6$ as a first approximation and apply Newton's method. We calculate $f'(x) = 8x^3 + 6x^2 - 6x - 5$. Then, according to Formula (13.3) we have

$$x_2 = 1.6 - \frac{f(1.6)}{f'(1.6)} = 1.6 - \frac{0.6193}{33.528} = 1.5815.$$

The next approximation yields

$$x_3 = x_2 - \frac{f(x_2)}{f'(x_2)} = 1.5815 - \frac{0.114}{32.1623} = 1.5780.$$

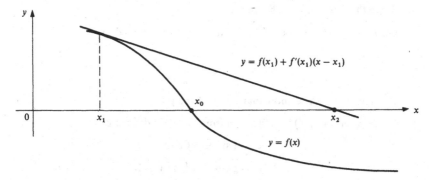

Figure 13.2. Newton's method fails because x_1 is not close to x_0.

Continuing, we find

$$x_4 = x_3 - \frac{f(x_3)}{f'(x_3)} = 1.5780 + 0.0031 = 1.5811.$$

The root, accurate to three decimal places is 1.581.   □

Newton's method may fail if the first approximation x_1 is not sufficiently close to the root x_0 or if the slope of f at x_1 is not sufficiently large. Figure 13.2 illustrates how the second approximation, x_2, may be further from the root than the first approximation, x_1.

PROBLEMS

1. Find the positive root of $2x^4 - 3x^2 - 5 = 0$ with an accuracy of three decimal places.

2. Find the positive root of $3x^3 + 6x^2 - 7x - 14 = 0$ with an accuracy of three decimal places.

3. Find the negative root of $x^4 - 2x^3 - 3x^2 - 2x - 4 = 0$ with an accuracy of two decimal places.

4. Find the root of $x - \cos x = 0$, $0 \leqslant x \leqslant \pi/2$ with an accuracy of two decimal places.

5. Find all the roots of $3x^3 + x^2 - 11x + 6 = 0$ with an accuracy of two decimal places.

6. Find the positive root of $3x^3 + 16x^2 - 8x - 16 = 0$ with an accuracy of four decimal places.

7. Given $f: x \to (1/4)(1 - x - (1/10)x^5)$ defined on $I = \{x: 0 \leqslant x \leqslant 1\}$. By applying the Fixed point theorem to $g(x) = f(x) + x$, show that f has a zero in I. Set up an iteration process and find the first three terms.

8. Consider the function $f: x \to \sqrt{1 + x^2}$ defined on $I = \{x: 0 \leqslant x < \infty\}$. Show that $|f(x) - f(y)| < |x - y|$ for all $x \neq y$ and that f does not have a fixed point. Conclude that Theorem 13.2 is false if $k = 1$.

9. Prove the Corollary to Theorem 13.2.

10. Let $f: (x, y) \to (x', y')$ be a function from $\mathbb{R}^2$ to $\mathbb{R}^2$ defined by

$$x' = \tfrac{1}{4}x + \tfrac{1}{3}y - 2,$$
$$y' = \tfrac{1}{3}x - \tfrac{1}{3}y + 3.$$

Use Theorem 13.2 to show that f has a fixed point.

11. Let $f: (x, y) \to (x', y')$ be a function from $\mathbb{R}^2$ to $\mathbb{R}^2$ defined by

$$x' = \tfrac{1}{3}\sin x - \tfrac{1}{3}\cos y + 2,$$
$$y' = \tfrac{1}{6}\cos x + \tfrac{1}{2}\sin y - 1.$$

Use Theorem 13.2 to show that f has a fixed point.

12. Let f be a mapping of a complete metric space S into S, and suppose that $f^2 = f \circ f$ satisfies the conditions of Theorem 13.2. Show that f has a unique fixed point.

13. Use Newton's method to find to three decimal places the root of $e^{-x} - 5x = 0$ on the interval $I = \{x: 0 \leqslant x \leqslant 1\}$.

14. Complete the details of the induction in the proof of convergence of Newton's method.

15. Suppose that an iteration, similar to Newton's method, is used according to the formula

$$x_{n+1} = x_n - \frac{f(x_n)}{f'(x_1)}, \qquad n = 1, 2, \dots .$$

State and prove a theorem which establishes convergence of this method.

13.2. Application of the Fixed Point Theorem to Differential Equations

Let D be an open set in $\mathbb{R}^2$ and suppose that $f: D \to \mathbb{R}^1$ is continuous. We now investigate the problem of determining solutions in D of the first order differential equation

$$\frac{dy}{dx} = f(x, y) \tag{13.6}$$

by means of the fixed point theorem of Section 13.1.

Definitions. Let $P(x_0, y_0)$ be a point of D. A function $y = \varphi(x)$ is a solution of the **initial value problem** of (13.6) if and only if $\varphi'(x) = f(x, \varphi(x))$ for x in some interval $I = \{x: x_0 - h < x < x_0 + h\}$ and if φ satisfies the **initial condition** $y_0 = \varphi(x_0)$.

We shall show that when f satisfies certain smoothness conditions, a solution to the initial value problem exists and is unique. Since the method employs the Fixed point theorem of Section 1, the solution may always be found by an iteration technique.

Definitions. Let S be a metric space and suppose that $f: S \to \mathbb{R}^1$ and $g: S \to \mathbb{R}^1$ are bounded continuous functions on S. We define the **distance between f and g** by

$$d(f, g) = \sup_{x \in S} |f(x) - g(x)|. \tag{13.7}$$

It is easy to verify that the collection of all bounded continuous functions from S to $\mathbb{R}^1$ forms a metric space with the distance function given by (13.7). We denote this space by $C(S)$.

Theorem 13.3. *Let S be a metric space and $C(S)$ the metric space of bounded continuous functions on S. Let $\{f_n\}$ be a sequence of elements of $C(S)$. Then $f_n \to f$ in $C(S)$ as $n \to \infty$ if and only if $f_n \to f$ uniformly on S as $n \to \infty$.*

PROOF. Suppose f_n converges to f uniformly on S. Thus, suppose that for every $\varepsilon > 0$ there is an integer n_0 such that

$$|f_n(x) - f(x)| < \varepsilon \quad \text{for all} \quad n > n_0,$$

and the number n_0 is independent of $x \in S$. Consequently,

$$\sup_{x \in S} |f_n(x) - f(x)| \leqslant \varepsilon \quad \text{for} \quad n > n_0,$$

and we conclude that $d(f_n, f) \to 0$ as $n \to \infty$. That is, f_n converges to f in $C(S)$. The converse is obtained by reversing the steps in the argument. $\square$

We leave to the reader the proof of the next two theorems.

Theorem 13.4. *If S is any metric space, the space $C(S)$ is complete.*

Theorem 13.5. *Let S be a complete metric space and A a closed subset of S. Then A, considered as a metric space, is complete.*

Remarks. Let S be a metric space and suppose that $f \colon S \to \mathbb{R}^N$ and $g \colon S \to \mathbb{R}^N$ are bounded continuous functions on S. That is, f and g have the form $f = (f_1, f_2, \ldots, f_N)$, $g = (g_1, g_2, \ldots, g_N)$ where f_i, g_i, $i = 1, 2, \ldots, N$, are bounded continuous functions from S into $\mathbb{R}^1$. If we define $|f(x) - g(x)|^2 = \sum_{i=1}^{N} |f_i(x) - g_i(x)|^2$, then the distance formula (13.7) determines a metric space denoted $C^N(S)$. It is now a straightforward matter to establish the analog of Theorem 13.3 for the space $C^N(S)$.

The next lemma enables us to reduce the problem of solving a differential equation to that of solving an integral equation. This reduction is used extensively in problems involving ordinary and partial differential equations.

Lemma 13.1. *Suppose that $f \colon D \to \mathbb{R}^1$ is continuous, that φ is defined and continuous on $I = \{x \colon x_0 - h < x < x_0 + h\}$ to $\mathbb{R}^1$, and that $(x_0, y_0) \in D$ with $\varphi(x_0) = y_0$. Then a necessary and sufficient condition that φ be a solution of*

$$\frac{d\varphi}{dx} = f[x, \varphi(x)] \tag{13.8}$$

on I is that φ satisfy the integral equation

$$\varphi(x) = y_0 + \int_{x_0}^{x} f[t, \varphi(t)] \, dt \quad \text{for} \quad x \in I. \tag{13.9}$$

PROOF. We integrate (13.8) between x_0 and x obtaining (13.9). Then the Fundamental theorem of calculus establishes the required equivalence. $\square$

Definitions. Let F be defined on a set S in $\mathbb{R}^1$ with values in $\mathbb{R}^1$. We say that F satisfies a **Lipschitz condition** on S if and only if there is a constant M such that

$$|F(x_1) - F(x_2)| \leqslant M|x_1 - x_2| \tag{13.10}$$

for all values of x_1, x_2 in S. The smallest number M for which (13.10) holds is called the **Lipschitz constant**.

If f has a bounded derivative on an interval I, then it satisfies a Lipschitz condition there. To see this, observe that by the Mean-value theorem there is a value ζ such that $f(x_1) - f(x_2) = f'(\zeta)(x_1 - x_2)$ for any $x_1, x_2 \in I$. If $|f'(\zeta)| \leqslant M$ for all $\zeta \in I$, then (13.10) clearly holds for the function f. On the other hand, a function may satisfy a Lipschitz condition and not be differentiable at certain points of I. A function whose graph consists of several connected straight line segments illustrates this property.

Theorem 13.6. *Suppose that $f: R \to \mathbb{R}^1$ is continuous where R is the rectangle $R = \{(x, y): |x - x_0| < h, |y - y_0| \leqslant k\}$, that $|f(x, y)| \leqslant M_0$ for $(x, y) \in R$, and that f satisfies a Lipschitz condition with respect to y,*

$$|f(x, y_1) - f(x, y_2)| \leqslant M_1 |y_1 - y_2|$$

whenever (x, y_1) and (x, y_2) are in R. Then if

$$M_0 h \leqslant k \quad and \quad M_1 h < 1,$$

there is a unique continuously differentiable function φ defined on $I = \{x: x_0 - h < x < x_0 + h\}$ for which

$$\varphi(x_0) = y_0, \quad |\varphi(x) - y_0| \leqslant k \quad and \quad \frac{d\varphi}{dx} = f[x, \varphi(x)].$$

PROOF. According to Lemma 13.1 it suffices to find a solution of the integral equation

$$\varphi(x) = y_0 + \int_{x_0}^x f[t, \varphi(t)]\, dt \tag{13.11}$$

with $|\varphi(x) - y_0| \leqslant k$ for $x \in I$. Let $C(I)$ be the metric space of bounded continuous functions on I, and let E be the subset of $C(I)$ for which $|\varphi(x) - y_0| \leqslant k$. If $\{\varphi_n\}$ is a sequence of functions in E which converges in $C(I)$ to a function φ_0, then the inequality

$$|\varphi_0 - y_0| \leqslant |\varphi_0 - \varphi_n| + |\varphi_n - y_0|$$

shows that φ_0 must also belong to E. Thus E is a closed subset of $C(I)$ and, by Theorem 13.5, a complete metric space. We define a mapping T on E by the formula

$$T(\varphi) = \psi \quad where \quad \psi(x) = y_0 + \int_{x_0}^x f[t, \varphi(t)]\, dt, \qquad \varphi \in E.$$

We have

$$|\psi(x) - y_0| = \left| \int_{x_0}^{x} f[t, \varphi(t)] \, dt \right| \leqslant M_0 |x - x_0| < M_0 h \leqslant k.$$

Hence ψ is in E and the mapping T takes the metric space E into itself. Now we show that T is a contraction map. Let φ_1, φ_2 be in E and denote $\psi_1 = T(\varphi_1)$, $\psi_2 = T(\varphi_2)$. Then

$$|\psi_2(x) - \psi_1(x)| = \left| \int_{x_0}^{x} \{f[t, \varphi_2(t)] - f[t, \varphi_1(t)]\} \, dt \right|$$

$$\leqslant \left| \int_{x_0}^{x} M_1 |\varphi_2(t) - \varphi_1(t)| \, dt \right|$$

$$\leqslant M_1 h \sup_{t \in I} |\varphi_2(t) - \varphi_1(t)| \leqslant M_1 h d(\varphi_2, \varphi_1).$$

Therefore $d(\psi_2, \psi_1) \leqslant M_1 h d(\varphi_2, \varphi_1)$. Since $M_1 h < 1$ by hypothesis, Theorem 13.2 shows that the mapping T has a unique fixed point. That is, Equation (13.11) has a unique solution and the result is established.   □

Remarks. If f is defined in an open set D in $\mathbb{R}^2$ and M_0, M_1 of Theorem 13.6 hold for all of D, then the solution φ, valid in a rectangle, can be extended throughout D. We observe that in the proof of Theorem 13.6, the size of h depends only on M_0, M_1 and k and not on φ. Once a solution φ is found in a rectangle centered at (x_0, y_0), we may take any other point on this solution curve, say (x_1, y_1) and solve the initial value problem in a new rectangle centered at (x_1, y_1) (see Figure 13.3). It is not difficult to show that in general this new solution will extend beyond the original one. By the uniqueness result, the solutions must coincide in the overlapping portions of the rectangles. Proceeding step by step we obtain a solution of the initial value problem throughout D.

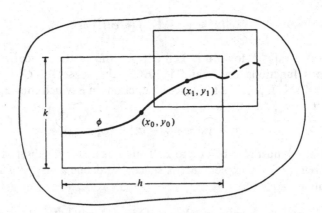

Figure 13.3. Extending a solution.

Theorem 13.6 has an immediate extension to solutions of the system of equations

$$\frac{dy_i}{dx} = f_i(x, y_1, y_2, \ldots, y_N), \qquad i = 1, 2, \ldots, N.$$

We consider mappings from I into $\mathbb{R}^N$ and the corresponding space of functions $C^N(I)$ as described in the remarks following Theorem 13.5. We obtain an existence and uniqueness theorem for the initial value problem provided the f_i are continuous and satisfy a Lipschitz condition with respect to each of the variables $y_1, y_2, \ldots, y_N$.

The uniqueness conclusion of Theorem 13.6 fails if we drop the hypothesis of a Lipschitz condition and merely assume that f is continuous. To see this, consider the equation

$$\frac{dy}{dx} = y^{1/3} \tag{13.12}$$

in any open set containing $(0, 0)$. Then it is clear that $y = \varphi(x) \equiv 0$ is a solution to (13.12) which satisfies the initial condition $\varphi(0) = 0$. A second solution is given by $y = \varphi(x) = (2x/3)^{3/2}$. It is not difficult to see that the function $f(x, y) = y^{1/3}$ does not satisfy a Lipschitz condition in any open set which contains the point $(0, 0)$.

PROBLEMS

1. Let S be a metric space. Show that the space $C(S)$ is complete (Theorem 13.4).

2. Let S be a complete metric space and A a closed subset of S. Show that A, considered as a metric space, is complete (Theorem 13.5).

3. Prove Theorem 13.3 with the metric space $C^N(S)$ in place of the space $C(S)$.

4. Consider the integral equation

$$\varphi(x) = g(x) + \lambda \int_a^b K(x, y)\varphi(y)\, dy,$$

where λ is a constant, g is continuous on $I = \{x: a \leqslant x \leqslant b\}$, and K is continuous on the square $S = \{(x, y): a \leqslant x \leqslant b, a \leqslant y \leqslant b\}$. Define the mapping $\psi = T\varphi$ by

$$\psi = T\varphi \equiv g(x) + \lambda \int_a^b K(x, y)\varphi(y)\, dy.$$

Use the fixed point theorem to show that for sufficiently small values of λ there is a unique solution to the integral equation.

5. Given the differential equation $dy/dx = x^2 + y^2$ and the initial condition $\varphi(0) = 1$, use the method of reduction to an integral equation and successive approximation to find the first six terms in the Taylor expansion solution $y = \varphi(x)$.

6. In the proof of Theorem 13.6, write a complete proof of the statement that E is a closed subset of $C(I)$.

7. Given the linear differential equation $dy/dx = A(x)y + B(x)$, show that if $A(x)$ and $B(x)$ are bounded and integrable on $I = \{x: a \leqslant x \leqslant b\}$, then the fixed point theorem yields a solution to the initial value problem on I.

8. Given the second order linear equation

$$\frac{d^2y}{dx^2} + A(x)\frac{dy}{dx} + B(x)y + C(x) = 0.$$

Let $y_1 = y$ and $y_2 = dy/dx$. Then the second order equation reduces to the pair of first order equations:

$$\frac{dy_1}{dx} = y_2, \qquad \frac{dy_2}{dx} = -A(x)y_2 - B(x)y_1 - C(x).$$

If A, B, and C are continuous on $I = \{x: a \leqslant x \leqslant b\}$, show that a theorem similar to Theorem 13.6 holds.

9. Consider the system of differential equations

$$\frac{dy_i}{dx} = f_i(x, y_1, y_2, \ldots, y_N), \qquad i = 1, 2, \ldots, N. \tag{13.13}$$

State and prove the analog of Lemma 13.1 for such a system.

10. State and prove the analog of Theorem 13.6 for systems of the form (13.13).

CHAPTER 14

Implicit Function Theorems and Lagrange Multipliers

14.1. The Implicit Function Theorem for a Single Equation

Suppose we are given a relation in $\mathbb{R}^2$ of the form

$$F(x, y) = 0. \tag{14.1}$$

Then to each value of x there may correspond one or more values of y which satisfy (14.1)—or there may be no values of y which do so. If $I = \{x: x_0 - h < x < x_0 + h\}$ is an interval such that for each $x \in I$ there is exactly one value of y satisfying (14.1), then we say that $F(x, y) = 0$ defines y as a function of x **implicitly** on I. Denoting this function by f, we have $F[x, f(x)] = 0$ for x on I.

An Implicit function theorem is one which determines conditions under which a relation such as (14.1) defines y as a function of x or x as a function of y. The solution is a local one in the sense that the size of the interval I may be much smaller than the domain of the relation F. Figure 14.1 shows the graph of a relation such as (14.1). We see that F defines y as a function of x in a region about P, but not beyond the point Q. Furthermore, the relation does not yield y as a function of x in any region containing the point Q in its interior.

The simplest example of an Implicit function theorem states that if F is smooth and if P is a point at which $F_{,2}$ (that is, $\partial F/\partial y$) does not vanish, then it is possible to express y as a function of x in a region containing this point. More precisely we have the following result.

Theorem 14.1. *Suppose that F, $F_{,1}$ and $F_{,2}$ are continuous on an open set A in $\mathbb{R}^2$ containing the point $P(x_0, y_0)$, and suppose that*

$$F(x_0, y_0) = 0, \qquad F_{,2}(x_0, y_0) \neq 0.$$

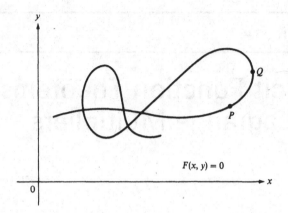

Figure 14.1

(a) *Then there are positive numbers h and k which determine a rectangle R contained in A (see Figure 14.2) given by*

$$R = \{(x, y): |x - x_0| < h, |y - y_0| < k\},$$

such that for each x in $I = \{x: |x - x_0| < h\}$ there is a unique number y in $J = \{y: |y - y_0| < k\}$ which satisfies the equation $F(x, y) = 0$. The totality of the points (x, y) forms a function f whose domain contains I and whose range is in J.

(b) *The function f and its derivative f′ are continuous on I.*

We shall give two proofs of Part (a), one which uses the elementary proper-

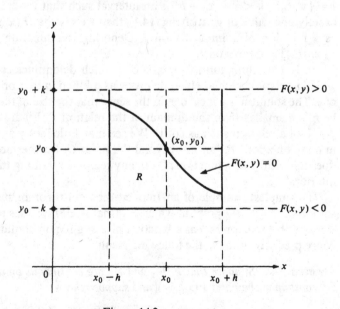

Figure 14.2

ties of continuous functions and the Intermediate-value theorem (Theorem 3.3), and a second which employs the Fixed point theorem in Chapter 13 (Theorem 13.2).

First Proof of Part (a). We assume $F_{,2}(x_0, y_0) > 0$; otherwise we replace F by $-F$ and repeat the argument. Since $F_{,2}$ is continuous there is a (sufficiently small) square $S = \{(x, y): |x - x_0| \leqslant k, |y - y_0| \leqslant k\}$ which is contained in A and on which $F_{,2}$ is positive. For each fixed value of x such that $|x - x_0| < k$ we see that $F(x, y)$, considered as a function of y, is an increasing function. Since $F(x_0, y_0) = 0$, it is clear that

$$F(x_0, y_0 + k) > 0 \quad \text{and} \quad F(x_0, y_0 - k) < 0.$$

Because F is continuous on S, there is a (sufficiently small) number h such that $F(x, y_0 + k) > 0$ on $I = \{x: |x - x_0| < h\}$ and $F(x, y_0 - k) < 0$ on I. We fix a value of x in I and examine solutions of $F(x, y) = 0$ in the rectangle R (see Figure 14.2). Since $F(x, y_0 - k)$ is negative and $F(x, y_0 + k)$ is positive, there is a value $\bar{y}$ in R such that $F(x, \bar{y}) = 0$. Also, because $F_{,2} > 0$, there is precisely one such value. The correspondence $x \to \bar{y}$ is the function we seek, and we denote it by f.   □

(b) To show that f is continuous at x_0 let $\varepsilon > 0$ be given and suppose that ε is smaller than k. Then we may construct a square S_ε with side 2ε and center at (x_0, y_0) as in the proof of Part (a). There is a value $h' < h$ such that f is a function on $I' = \{x: |x - x_0| < h'\}$. Therefore

$$|f(x) - f(x_0)| < \varepsilon \quad \text{whenever} \quad |x - x_0| < h',$$

and f is continuous at x_0. At any other point $x_1 \in I$, we construct a square S_1 with center at $(x_1, f(x_1))$ and repeat the above argument.

To show that f' exists and is continuous we use the Fundamental lemma on differentiation (Theorem 7.2). Let $x \in I$ and choose a number ρ such that $x + \rho \in I$. Then

$$F(x + \rho, f(x + \rho)) = 0 \quad \text{and} \quad F(x, f(x)) = 0.$$

Writing $f(x + \rho) = f + \Delta f$ and using Theorem 7.2, we obtain

$$[F_{,1}(x, f) + \varepsilon_1(\rho, \Delta f)]\rho + [F_{,2}(x, f) + \varepsilon_2(\rho, \Delta f)]\Delta f = 0 \qquad (14.2)$$

where ε_1 and ε_2 tend to zero as $\rho, \Delta f \to 0$. From the continuity of f, which we established, it follows that $\Delta f \to 0$ as $\rho \to 0$. From (14.2) it is clear that

$$\frac{\Delta f}{\rho} = \frac{f(x + \rho) - f(x)}{\rho} = -\frac{F_{,1}(x, f) + \varepsilon_1(\rho, \Delta f)}{F_{,2}(x, f) + \varepsilon_2(\rho, \Delta f)}.$$

Since the right side tends to a limit as $\rho \to 0$, we see that

$$f'(x) = -\frac{F_{,1}(x, f)}{F_{,2}(x, f)}. \qquad (14.3)$$

By hypothesis the right side of (14.3) is continuous, and so f' is also.   □

SECOND PROOF OF PART (a). For fixed x in the rectangle R we consider the mapping

$$T_x y = y - \frac{F(x, y)}{F_{,2}(x_0, y_0)},$$

which takes a point y in J into $\mathbb{R}^1$. We shall show that for h and k sufficiently small, the mapping takes J into J and has a fixed point. That is, there is a y such that $T_x y = y$ or, in other words, there is a y such that $F(x, y) = 0$. To accomplish this, we first write the mapping T_x in the more complicated form:

$$T_x y = y_0 - \frac{F_{,1}(x_0, y_0)}{F_{,2}(x_0, y_0)}(x - x_0)$$

$$- \frac{1}{F_{,2}(x_0, y_0)}[F(x, y) - F_{,1}(x_0, y_0)(x - x_0) - F_{,2}(x_0, y_0)(y - y_0)].$$

We define

$$c = \frac{F_{,1}(x_0, y_0)}{F_{,2}(x_0, y_0)}, \qquad \psi(x, y) = \frac{1}{F_{,2}(x_0, y_0)}[F(x, y) - F_{,1}(x_0, y_0)(x - x_0)$$

$$- F_{,2}(x_0, y_0)(y - y_0)].$$

Then the mapping $T_x y$ can be written

$$T_x y = y_0 - c(x - x_0) - \psi(x, y).$$

Since $F(x_0, y_0) = 0$, we see that

$$\psi(x_0, y_0) = 0, \qquad \psi_{,1}(x_0, y_0) = 0, \qquad \psi_{,2}(x_0, y_0) = 0.$$

Because $\psi_{,1}$ and $\psi_{,2}$ are continuous we can take k so small that

$$|\psi_{,1}(x, y)| \leqslant \tfrac{1}{2}, \qquad |\psi_{,2}(x, y)| \leqslant \tfrac{1}{2},$$

for (x, y) in the square $S = \{(x, y): |x - x_0| \leqslant k, |y - y_0| \leqslant k\}$. We now expand $\psi(x, y)$ in a Taylor series in S about the point (x_0, y_0) getting

$$\psi(x, y) = \psi_{,1}(\xi, \eta)(x - x_0) + \psi_{,2}(\xi, \eta)(y - y_0), \qquad (\xi, \eta) \in S.$$

Hence for $h \leqslant k$, we have the estimate in the rectangle R:

$$|\psi(x, y)| \leqslant \tfrac{1}{2}h + \tfrac{1}{2}k.$$

Next we show that if we reduce h sufficiently, the mapping T_x takes the interval (space) J into J. We have

$$|T_x y - y_0| \leqslant |c(x - x_0)| + |\psi(x, y)|$$

$$\leqslant |c|h + \tfrac{1}{2}h + \tfrac{1}{2}k = (\tfrac{1}{2} + |c|)h + \tfrac{1}{2}k.$$

We choose h so small that $(\tfrac{1}{2} + |c|)h \leqslant k$. Then $T_x y$ maps J into J for each x in $I = \{x: |x - x_0| \leqslant h\}$. The mapping T_x is a contraction map; in fact, by the Mean-value theorem

$$|T_x y_1 - T_x y_2| = |-\psi(x, y_1) + \psi(x, y_2)| \leqslant \tfrac{1}{2}|y_1 - y_2|.$$

We apply Theorem 13.2 and for each fixed x in I, there is a unique y in J such that $F(x, y) = 0$. That is, y is a function of x for $(x, y) \in R$. ▢

The Implicit function theorem has a number of generalizations and applications. If F is a function from $\mathbb{R}^{N+1}$ to $\mathbb{R}^1$, we may consider whether or not the relation $F(x_1, x_2, \ldots, x_N, y) = 0$ defines y as a function from $\mathbb{R}^N$ into $\mathbb{R}^1$. That is, we state conditions which show that $y = f(x_1, x_2, \ldots, x_N)$. The proof of the following theorem is a straightforward extension of the proof of Theorem 14.1 and we leave the details to the reader.

Theorem 14.2. *Suppose that $F, F_{,1}, F_{,2}, \ldots, F_{,N+1}$ are continuous on an open set A in $\mathbb{R}^{N+1}$ containing the point $P(x_1^0, x_2^0, \ldots, x_N^0, y^0)$. We use the notation $x = (x_1, x_2, \ldots, x_N)$, $x^0 = (x_1^0, x_2^0, \ldots, x_N^0)$ and suppose that*

$$F(x^0, y^0) = 0, \qquad F_{,N+1}(x^0, y^0) \neq 0.$$

(a) *Then there are positive numbers h and k which determine a cell R contained in A given by*

$$R = \{(x, y): |x_i - x_i^0| < h, \quad i = 1, 2, \ldots, N, |y - y^0| < k\},$$

such that for each x in the N-dimensional hypercube

$$I_N = \{x: |x_i - x_i^0| < h, \quad i = 1, 2, \ldots, N\}$$

there is a unique number y in the interval

$$J = \{y: |y - y^0| < k\}$$

which satisfies the equation $F(x, y) = 0$. That is, y is a function of x which may be written $y = f(x)$. The domain of f contains I_N and its range is in J.
(b) *The function f and its partial derivatives $f_{,1}, f_{,2}, \ldots, f_N$ are continuous on I_N.*

A special case of Theorem 14.1 is the Inverse function theorem which was established in Chapter 4 (Theorems 4.17 and 4.18). If f is a function from $\mathbb{R}^1$ to $\mathbb{R}^1$, denoted $y = f(x)$, we wish to decide when it is true that x may be expressed as a function of y. Set

$$F(x, y) = y - f(x) = 0$$

and, in order to apply Theorem 14.1, f' must be continuous and $F_{,1} = -f'(x) \neq 0$. We state the result in the following Corollary to Theorem 14.1.

Corollary (Inverse function theorem). *Suppose that f is defined on an open set A in $\mathbb{R}^1$ with values in $\mathbb{R}^1$. Also assume that f' is continuous on A and that $f(x_0) = y_0$, $f'(x_0) \neq 0$. Then there is an interval I containing y_0 such that the inverse function of f, denoted f^{-1}, exists on I and has a continuous derivative there. Furthermore, the derivative $(f^{-1})'$ is given by the formula*

$$(f^{-1}(y))' = \frac{1}{f'(x)}, \tag{14.4}$$

where $y = f(x)$.

Since $f^{-1}(f(x)) = x$, we can use the Chain rule to obtain (14.4). However, (14.4) is also a consequence of Formula (14.3), with $F(x, y) = y - f(x)$, and we find

$$(f^{-1}(y))' = -\frac{F_{,2}}{F_{,1}} = -\frac{1}{-f'(x)}.$$

Observe that in Theorems 4.17 and 4.18 the inverse mapping is one-to-one over the entire interval in which f' does not vanish.

EXAMPLE 1. Given the relation

$$F(x, y) = y^3 + 2x^2y - x^4 + 2x + 4y = 0, \tag{14.5}$$

show that this relation defines y as a function of x for all values of x in $\mathbb{R}^1$.

Solution. We have

$$F_{,2} = 3y^2 + 2x^2 + 4,$$

and so $F_{,2} > 0$ for all x, y. Hence for each fixed x, the function F is an increasing function of y. Furthermore, from (14.5) it follows that $F(x, y) \to -\infty$ as $y \to -\infty$ and $F(x, y) \to +\infty$ as $y \to +\infty$. Since F is continuous, for each fixed x there is exactly one value of y such that $F(x, y) = 0$. Applying Theorem 14.1, we conclude that there is a function f on $\mathbb{R}^1$ which is continuous and differentiable such that $F[x, f(x)] = 0$ for all x.   □

EXAMPLE 2. Given the relation

$$F(x, y) = x^3 + y^3 - 6xy = 0, \tag{14.6}$$

find the values of x for which the relation defines y as a function of x (on some interval) and find the values of y for which the relation defines x as a function of y (on some interval).

Solution. The graph of the relation is shown in Figure 14.3. We see that

$$F_{,1} = 3x^2 - 6y, \qquad F_{,2} = 3y^2 - 6x,$$

and both partial derivatives vanish at $(0, 0)$. We also observe that $F_{,2} = 0$ when $x = \frac{1}{2}y^2$, and substituting this value into the relation (14.6) we get $x = 2\sqrt[3]{4}$, $y = 2\sqrt[3]{2}$. The curve has a vertical tangent at this point, denoted P in Figure 14.3. Hence y is expressible as a function of x in a neighborhood of all points on the curve except P and the origin 0. Similarly $F_{,1} = 0$ yields the point Q with coordinates $(2\sqrt[3]{2}, 2\sqrt[3]{4})$. Then x is expressible as a function of y in a neighborhood of all points except Q and the origin 0.   □

PROBLEMS

In each of Problems 1 through 4 show that the relation $F(x, y) = 0$ yields y as a function of x in an interval I about x_0 where $F(x_0, y_0) = 0$. Denote the function by f and compute f'.

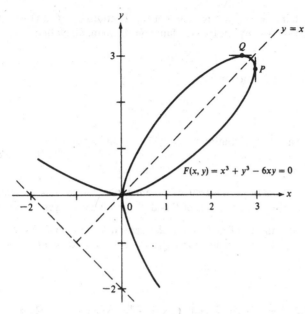

Figure 14.3

1. $F(x, y) \equiv y^3 + y - x^2 = 0$; $(x_0, y_0) = (0, 0)$.

2. $F(x, y) \equiv x^{2/3} + y^{2/3} - 4 = 0$; $(x_0, y_0) = (1, 3\sqrt{3})$.

3. $F(x, y) \equiv xy + 2 \ln x + 3 \ln y - 1 = 0$; $(x_0, y_0) = (1, 1)$.

4. $F(x, y) \equiv \sin x + 2 \cos y - \frac{1}{2} = 0$; $(x_0, y_0) = (\pi/6, 3\pi/2)$.

5. Give an example of a relation $F(x, y) = 0$ such that $F(x_0, y_0) = 0$ and $F_{,2}(x_0, y_0) = 0$ at a point $O = (x_0, y_0)$, and yet y is expressible as a function of x in an interval about x_0.

In each of Problems 6 through 9 show that the relation $F(x_1, x_2, y) = 0$ yields y as a function of (x_1, x_2) in a neighborhood of the given point $P(x_1^0, x_2^0, y^0)$. Denoting this function by f, compute $f_{,1}$ and $f_{,2}$ at P.

6. $F(x_1, x_2, y) \equiv x_1^3 + x_2^3 + y^3 - 3x_1 x_2 y - 4 = 0$; $P(x_1^0, x_2^0, y^0) = (1, 1, 2)$.

7. $F(x_1, x_2, y) \equiv e^y - y^2 - x_1^2 - x_2^2 = 0$; $P(x_1^0, x_2^0, y^0) = (1, 0, 0)$.

8. $F(x_1, x_2, y) \equiv x_1 + x_2 - y - \cos(x_1 x_2 y) = 0$; $P(x_1^0, x_2^0, y^0) = (0, 0, -1)$.

9. $F(x_1, x_2, y) \equiv x_1 + x_2 + y - e^{x_1 x_2 y} = 0$; $P(x_1^0, x_2^0, y^0) = (0, \frac{1}{2}, \frac{1}{2})$.

10. Prove Theorem 14.2.

11. Suppose that F is a function from $\mathbb{R}^2$ to $\mathbb{R}^1$ which we write $y = F(x_1, x_2)$. State hypotheses on F which imply that x_2 may be expressed as a function of x_1 and y (extension of the Inverse function theorem). Use Theorem 14.2.

12. Suppose that $F(x, y, z) = 0$ is such that the functions $z = f(x, y)$, $x = g(y, z)$, and $y = h(z, x)$ all exist by the Implicit function theorem. Show that

$$f_{,1} \cdot g_{,1} \cdot h_{,1} = -1.$$

This formula is frequently written

$$\frac{\partial z}{\partial x} \cdot \frac{\partial x}{\partial y} \cdot \frac{\partial y}{\partial z} = -1.$$

13. Find an example of a relation $F(x_1, x_2, y) = 0$ and a point $P(x_1^0, x_2^0, y^0)$ such that P satisfies the relation, and $F_{,1}(x_1^0, x_2^0, y^0) = F_{,2}(x_1^0, x_2^0, y^0) = F_{,3}(x_1^0, x_2^0, y^0) = 0$, yet y is a function of (x_1, x_2) in a neighborhood of P.

14. Suppose that the Implicit function theorem applies to $F(x, y) = 0$ so that $y = f(x)$. Find a formula for f'' in terms of F and its partial derivatives.

15. Suppose that the Implicit function theorem applies to $F(x_1, x_2, y) = 0$ so that $y = f(x_1, x_2)$. Find formulas for $f_{,1,1}$; $f_{,1,2}$; $f_{,2,2}$ in terms of F and its partial derivatives.

14.2. The Implicit Function Theorem for Systems

We shall establish an extension of the Implicit function theorem of Section 14.1 to systems of equations which define functions implicitly. A vector x in $\mathbb{R}^m$ has components denoted $(x_1, x_2, \ldots, x_m)$ and a vector y in $\mathbb{R}^n$ will have its components denoted by $(y_1, y_2, \ldots, y_n)$. An element in $\mathbb{R}^{m+n}$ will be written (x, y). We consider vector functions from $\mathbb{R}^{m+n}$ to $\mathbb{R}^n$ and write $F(x, y)$ for such a function. That is, F will have components

$$F^1(x, y), \qquad F^2(x, y), \ldots, F^n(x, y)$$

with each F^i a function from $\mathbb{R}^{m+n}$ to $\mathbb{R}^1$.

In order to establish the Implicit function theorem for systems we need several facts from linear algebra and a number of useful inequalities. We suppose the reader is familiar with the elements of linear algebra and in the next three lemmas we establish the needed inequalities.

Definition. Let A be an $m \times n$ matrix with elements

$$\begin{pmatrix} a_1^1 & a_2^1 & \cdots & a_n^1 \\ \vdots & \vdots & & \vdots \\ a_1^m & a_2^m & \cdots & a_n^m \end{pmatrix}.$$

The **norm** *of A*, written $|A|$, is defined by

$$|A| = \left[\sum_{i=1}^m \sum_{j=1}^n (a_j^i)^2 \right]^{1/2}.$$

Observe that for a vector, i.e., a $1 \times n$ matrix, the norm is the Euclidean length of the vector.

Lemma 14.1. *Let A be an $m \times n$ matrix, and suppose that $\zeta = (\zeta^1, \zeta^2, \ldots, \zeta^n)$ is a column vector (that is, an $n \times 1$ matrix) with n components and that $\eta = (\eta^1, \eta^2, \ldots, \eta^m)$ is a column vector with m components such that*

$$\eta = A\zeta,$$

or equivalently

$$\eta^i = \sum_{j=1}^n a_j^i \zeta^j, \qquad i = 1, 2, \ldots, m. \tag{14.7}$$

Then

$$|\eta| \leqslant |A| |\zeta|. \tag{14.8}$$

PROOF. For fixed i in (14.7) we square both sides and apply the Schwarz inequality (Section 6.1), getting

$$(\eta^i)^2 \leqslant \sum_{j=1}^n (a_j^i)^2 \sum_{j=1}^n (\zeta^j)^2.$$

Then (14.8) follows by summing on i and taking the square root. $\qquad\square$

The next lemma shows that with the above norm for matrices (and vectors) we can obtain an inequality for the estimation of integrals which resembles the customary one for absolute values.

Lemma 14.2. *Let $b: \mathbb{R}^m \to \mathbb{R}^n$ be a continuous vector function on a bounded, closed figure H in $\mathbb{R}^m$. Suppose that ζ is the $n \times 1$ column vector defined by*

$$\zeta = \int_H b \, dV_m.$$

That is,

$$\zeta^i = \int_H b^i \, dV_m, \qquad i = 1, 2, \ldots, n,$$

where $b^i: \mathbb{R}^m \to \mathbb{R}^1$, $i = 1, 2, \ldots, n$ are the components of b. Then

$$|\zeta| \leqslant \int_H |b| \, dV_m.$$

PROOF. Define $\lambda_i = \zeta^i / |\zeta|$. Then multiplying by ζ^i and summing on i, we have $\sum_{i=1}^n \lambda_i \zeta^i = |\zeta|$. Therefore

$$|\zeta| = \sum_{i=1}^n \lambda_i \zeta^i = \sum_{i=1}^n \lambda_i \int_H b^i \, dV_m = \int_H \sum_{i=1}^n \lambda_i b^i \, dV_m.$$

We apply Lemma 14.1 and note that since $|\lambda| = 1$ we obtain

$$|\zeta| \leqslant \int_H |\lambda| |b| \, dV_m = \int_H |b| \, dV_m. \qquad\square$$

Definitions. Let G be an open set in $\mathbb{R}^{m+n}$ and suppose that $F: G \to \mathbb{R}^n$ is a vector function $F(x, y)$ with continuous first partial derivatives. We *define the $n \times m$ and the $n \times n$ matrices* $\nabla_x F$ and $\nabla_y F$ by the formulas

$$\nabla_x F = \begin{pmatrix} \dfrac{\partial}{\partial x_1} F^1 & \cdots & \dfrac{\partial}{\partial x_m} F^1 \\ \vdots & \vdots & \\ \dfrac{\partial}{\partial x_1} F^n & \cdots & \dfrac{\partial}{\partial x_m} F^n \end{pmatrix}, \qquad \nabla_y F = \begin{pmatrix} \dfrac{\partial}{\partial y_1} F^1 & \cdots & \dfrac{\partial}{\partial y_n} F^1 \\ \vdots & \vdots & \\ \dfrac{\partial}{\partial y_1} F^n & \cdots & \dfrac{\partial}{\partial y_n} F^n \end{pmatrix}.$$

The Fixed point theorem of Chapter 13 will be used to establish the Implicit function theorem for systems. We note that in proving this theorem for a single equation we made essential use of the Mean-value theorem. The next lemma provides an appropriate generalization to systems of the Mean-value theorem.

Lemma 14.3. *Let G be an open set in $\mathbb{R}^{m+n}$ and $F: G \to \mathbb{R}^n$ a vector function with continuous first partial derivatives. Suppose that the straight line segment L joining $(\bar{x}, \bar{y})$ and $(\bar{\bar{x}}, \bar{\bar{y}})$ is in G and that there are two positive constants M_1, M_2 such that*

$$|\nabla_x F| \leqslant M_1 \quad and \quad |\nabla_y F| \leqslant M_2$$

for all points (x, y) on the segment L. Then

$$|F(\bar{\bar{x}}, \bar{\bar{y}}) - F(\bar{x}, \bar{y})| \leqslant M_1 \cdot |\bar{\bar{x}} - \bar{x}| + M_2 \cdot |\bar{\bar{y}} - \bar{y}|.$$

PROOF. Any point on the segment joining $(\bar{x}, \bar{y})$ to $(\bar{\bar{x}}, \bar{\bar{y}})$ has coordinates $(\bar{x} + t(\bar{\bar{x}} - \bar{x}), \bar{y} + t(\bar{\bar{y}} - \bar{y}))$ for $0 \leqslant t \leqslant 1$. We define the vector function

$$f(t) = F(\bar{x} + t(\bar{\bar{x}} - \bar{x}), \bar{y} + t(\bar{\bar{y}} - \bar{y}))$$

and use the simple fact that

$$f(1) - f(0) = \int_0^1 f'(t) \, dt.$$

Since $f(1) = F(\bar{\bar{x}}, \bar{\bar{y}})$, $f(0) = F(\bar{x}, \bar{y})$, it follows that

$$F(\bar{\bar{x}}, \bar{\bar{y}}) - F(\bar{x}, \bar{y}) = \int_0^1 \frac{d}{dt} F(\bar{x} + t(\bar{\bar{x}} - \bar{x}), \bar{y} + t(\bar{\bar{y}} - \bar{y})) \, dt.$$

Carrying out the differentiation with respect to t, and using the Chain rule, we find for each component F^i,

$$F^i(\bar{\bar{x}}, \bar{\bar{y}}) - F^i(\bar{x}, \bar{y})$$

$$= \int_0^1 \left\{ \sum_{j=1}^m \frac{\partial}{\partial x_j} (F^i)(\bar{\bar{x}}_j - \bar{x}_j) + \sum_{k=1}^n \frac{\partial}{\partial y_k} (F^i)(\bar{\bar{y}}_k - \bar{y}_k) \right\} dt.$$

In matrix notation we write

$$F(\bar{\bar{x}}, \bar{\bar{y}}) - F(\bar{x}, \bar{y}) = \int_0^1 [\nabla_x F \cdot (\bar{\bar{x}} - \bar{x}) + \nabla_y F \cdot (\bar{\bar{y}} - \bar{y})] \, dt.$$

From Lemma 14.2, it is clear that

$$|F(\bar{\bar{x}}, \bar{\bar{y}}) - F(\bar{x}, \bar{y})| \leqslant \int_0^1 [|\nabla_x F| \cdot |\bar{\bar{x}} - \bar{x}| + |\nabla_y F| \cdot |\bar{\bar{y}} - \bar{y}|] \, dt$$

$$\leqslant M_1 \cdot |\bar{\bar{x}} - \bar{x}| + M_2 \cdot |\bar{\bar{y}} - \bar{y}|. \qquad \square$$

For later use we next prove a simple proposition on nonsingular linear transformations.

Lemma 14.4. *Let B be an $n \times n$ matrix and suppose that $|B| < 1$. Define $A = I - B$ where I is the $n \times n$ identity matrix. Then A is nonsingular.*

PROOF. Consider the mapping from $\mathbb{R}^n$ to $\mathbb{R}^n$ given by $y = Ax$, where $x \in \mathbb{R}^n$, $y \in \mathbb{R}^n$. We show that the mapping is 1-1 thereby implying that A is non-singular. Let $x_1, x_2 \in \mathbb{R}^n$; we have

$$Ax_1 - Ax_2 = (x_1 - x_2) - (Bx_1 - Bx_2)$$

and

$$|Bx_1 - Bx_2| = |B \cdot (x_1 - x_2)| \leqslant |B| \cdot |x_1 - x_2|.$$

Therefore

$$|Ax_1 - Ax_2| \geqslant |x_1 - x_2| - |Bx_1 - Bx_2| \geqslant |x_1 - x_2| - |B| \cdot |x_1 - x_2|$$

$$\geqslant |x_1 - x_2|(1 - |B|).$$

We conclude that if $x_1 \neq x_2$ then $Ax_1 \neq Ax_2$ and so the mapping is one-to-one. $\qquad \square$

The next lemma, a special case of the Implicit function theorem for systems, contains the principal ingredients for the proof of the main theorem. We establish the result for functions $F: \mathbb{R}^{m+n} \to \mathbb{R}^n$ which have the form

$$F(x, y) = y - Cx - \psi(x, y)$$

where C is a constant $n \times m$ matrix and ψ is such that it and its first partial derivatives vanish at the origin. Note the relation of this form of F with the second proof of the Implicit Function theorem for a single equation given in Theorem 14.1. Although the proof is lengthy, the reader will see that with the aid of the fixed point theorem of Chapter 13 and Lemma 14.3 the arguments proceed in a straightforward manner.

Lemma 14.5. *Let G be an open set in $\mathbb{R}^{m+n}$ which contains the origin. Suppose that $\psi: G \to \mathbb{R}^n$ is a continuous function with continuous first partial derivatives* *in G and that*

$$\psi(0, 0) = 0, \qquad \nabla_x\psi(0, 0) = 0, \qquad \nabla_y\psi(0, 0) = 0. \tag{14.9}$$

Suppose that C is a constant $n \times m$ matrix, and define the function $F: G \to \mathbb{R}^n$ by the formula

$$F(x, y) = y - Cx - \psi(x, y).$$

For any positive numbers r and s, denoted by $B_m(0, r)$ and $B_n(0, s)$ the balls in $\mathbb{R}^m$ and $\mathbb{R}^n$ with center at the origin and radii r and s, respectively. Then

(a) *There are (sufficiently small) positive numbers h and k with $B_m(0, h) \times B_n(0, k)$ in G and such that for each $x \in B_m(0, h)$ there is a unique element $y \in B_n(0, k)$ whereby*

$$F(x, y) = 0 \quad \text{or, equivalently,} \quad y = Cx + \psi(x, y).$$

(b) *If g denotes the function from $B_m(0, h)$ to $B_n(0, k)$ given by these ordered pairs (x, y), then g is continuous on $B_m(0, h)$ and all first partial derivatives of g are continuous on $B_m(0, h)$.*

PROOF

(a) Since G is open and ψ is continuous on G, there is a positive number k such that the closed set $B = \overline{B_m(0, k)} \times \overline{B_n(0, k)}$ is contained in G with ψ continuous on B. Also, because of (14.9) and the fact that the partial derivatives of ψ are continuous, k can be chosen so small that

$$|\nabla_x\psi| \leqslant \tfrac{1}{2}, \qquad |\nabla_y\psi| \leqslant \tfrac{1}{2} \quad \text{on } B.$$

We *fix* x in $B_m(0, k)$ and define the mapping T from $B_n(0, k)$ into $\mathbb{R}^n$ by the formula[1]

$$T(y) = Cx + \psi(x, y). \tag{14.10}$$

We apply Lemma 14.3 to $\psi(x, y)$, getting for $x \in B_m(0, k)$, $y \in B_n(0, k)$

$$|\psi(x, y)| = |\psi(x, y) - \psi(0, 0)| \leqslant \max|\nabla_x\psi| \cdot |x - 0| + \max|\nabla_y\psi| \cdot |y - 0|$$

$$\leqslant \tfrac{1}{2}|x| + \tfrac{1}{2}|y|.$$

Therefore, for $x \in B_m(0, k)$, $y \in B_n(0, k)$ it follows that

$$|T(y)| \leqslant |C| \cdot |x| + \tfrac{1}{2}|x| + \tfrac{1}{2}|y|. \tag{14.11}$$

Since C is a constant matrix there is a positive number M such that $|C| \leqslant M$. Now choose a positive number h which satisfies the inequality $h < k/(2M + 1)$. The mapping (14.10) will be restricted to those values of x in the ball $B_m(0, h)$. Then, from (14.10), for each fixed $x \in B_m(0, h)$ and $y \in B_n(0, k)$ we have

$$|T(y)| \leqslant (M + \tfrac{1}{2})h + \tfrac{1}{2}k < \tfrac{1}{2}k + \tfrac{1}{2}k = k;$$

---

[1] In the second proof of Theorem 14.1 we denoted this mapping by $T_x y$.

hence T maps $B_n(0, k)$ into itself. Furthermore, for $y_1, y_2 \in B_n(0, k)$ we find

$$|T(y_1) - T(y_2)| = |\psi(x, y_1) - \psi(x, y_2)| \leqslant \tfrac{1}{2}|y_1 - y_2|,$$

where Lemma 14.3 is used for the last inequality. Thus the mapping T is a contraction and the Fixed point theorem of Chapter 13 (Theorem 13.2) can be applied. For each *fixed* $x \in B_m(0, k)$ there is a unique $y \in B_n(0, k)$ such that

$$y = T(y) \quad \text{or} \quad y = Cx + \psi(x, y).$$

That is, y is a function of x which we denote by g. Writing $y = g(x)$, we observe that the equation $F(x, g(x)) = 0$ holds for all $x \in B_m(0, h)$.

   (b) We show that g is continuous. Let $x_1, x_2 \in B_m(0, h)$ and $y_1, y_2 \in B_n(0, k)$ be such that $y_1 = g(x_1)$, $y_2 = g(x_2)$ or

$$y_1 = Cx_1 + \psi(x_1, y_1) \quad \text{and} \quad y_2 = Cx_2 + \psi(x_2, y_2).$$

Then

$$|y_2 - y_1| \leqslant |C(x_2 - x_1)| + |\psi(x_2, y_2) - \psi(x_1, y_1)|.$$

We use Lemma 14.3 for the last term on the right, getting

$$|y_2 - y_1| \leqslant M \cdot |x_2 - x_1| + \tfrac{1}{2}|x_2 - x_1| + \tfrac{1}{2}|y_2 - y_1|$$

or

$$|y_2 - y_1| \leqslant (2M + 1)|x_2 - x_1|.$$

Hence

$$|g(x_2) - g(x_1)| \leqslant (2M + 1)|x_2 - x_1|,$$

and g is continuous on $B_m(0, h)$.

   We now show that the first partial derivatives of g exist and are continuous. Let the components of g be denoted by $g^1, g^2, \ldots, g^n$. We shall prove the result for a typical partial derivative $(\partial/\partial x_p)g^i$ where $1 \leqslant p \leqslant m$. In $\mathbb{R}^m$ let e^p denote the unit vector in the p-direction. That is, e^p has components $(e_1^p, e_2^p, \ldots, e_m^p)$ where $e_p^p = 1$ and $e_j^p = 0$ for $j \neq p$. Fix $\bar{x}$ in $B_m(0, h)$ and choose a positive number t_0 so small that the points $\bar{x} + te^p$ lie in $B_m(0, h)$ for all t such that $|t| \leqslant t_0$. Now set $x = \bar{x} + te^p$ and write

$$g(x) = Cx + \psi(x, g(x)).$$

The ith component of this equation reads

$$g^i(x) = \sum_{j=1}^{m} c_j^i x_j + \psi^i(x, g(x))$$

where the c_j^i are the components of the matrix C. Let Δg^i be defined by

$$\Delta g^i = g^i(\bar{x} + te^p) - g^i(\bar{x}).$$

Then from the Fundamental lemma on differentiation (Theorem 7.2), it follows that

$$\frac{\Delta g^i}{t} = c_p^i + \frac{\partial}{\partial x_p}\psi^i(x, g(x)) + \bar{\varepsilon}_p^i(te^p, \Delta g)$$

$$+ \sum_{k=1}^n \left[\frac{\partial}{\partial y_k}\psi^i(x, g(x)) + \bar{\bar{\varepsilon}}_k^i(te^p, \Delta g)\right]\frac{\Delta g^k}{t},$$

where $\bar{\varepsilon}_p^i(\rho, \sigma)$ and $\bar{\bar{\varepsilon}}_k^i(\rho, \sigma)$ are continuous at $(0, 0)$ and vanish there. Taking the definition of the vector e^p into account, we can write the above expression in the form

$$\frac{\Delta g^i}{t} = \sum_{j=1}^m c_j^i e_j^p + \sum_{j=1}^m \frac{\partial}{\partial x_j}\psi^i(x, g(x))e_j^p + \sum_{j=1}^m \bar{\varepsilon}_j^i(te^p, \Delta g)e_j^p$$

$$+ \sum_{k=1}^n \left[\frac{\partial}{\partial y_k}\psi^i(x, g(x)) + \bar{\bar{\varepsilon}}_k^i(te^p, \Delta g)\right]\frac{\Delta g^k}{t} \qquad (14.12)$$

where $\bar{\varepsilon}_j^i(\rho, \sigma)$, $j = 1, 2, \ldots, m$, are continuous at $(0, 0)$ and vanish there. We define the matrices

$$A_1(t) = C + \nabla_x\psi(x, g(x)) + \bar{\varepsilon}(te^p, \Delta g),$$

$$A_2(t) = \nabla_y\psi(x, g(x)) + \bar{\bar{\varepsilon}}(te^p, \Delta g)$$

where the components of $\bar{\varepsilon}$ are $\bar{\varepsilon}_j^i$ and the components of $\bar{\bar{\varepsilon}}$ are $\bar{\bar{\varepsilon}}_k^i$, i, $k = 1, 2, \ldots, n$, $j = 1, 2, \ldots, m$. Then (14.12) can be written as the single vector equation

$$\frac{\Delta g}{t} = A_1 e^p + A_2\frac{\Delta g}{t}. \qquad (14.13)$$

Define $B = I - A_2$ where I is the $n \times n$ unit matrix. Then (14.13) becomes

$$B\frac{\Delta g}{t} = A_1 e^p. \qquad (14.14)$$

According to (14.9) we have $|A_2(0)| \leq \frac{1}{2}$. Therefore, by Lemma 14.4 the matrix $B(0)$ is nonsingular. Since g is continuous on $B_m(0, h)$, we know that $\Delta g \to 0$ as $t \to 0$. Therefore the matrices $A_1(t)$, $A_2(t)$, and $B(t)$ are continuous at $t = 0$. Consequently $B(t)$ is nonsingular for t sufficiently close to zero. We allow t to tend to zero in (14.14) and conclude that the limit of $\Delta g/t$ exists; that is, $(\partial/\partial x_p)g^i$ exists for every i and every p. The formula

$$\lim_{t\to 0} \frac{\Delta g}{t} = B^{-1}(0)A_1(0)e^p$$

shows that the partial derivatives are continuous functions of x.   □

Theorem 14.3 (Implicit function theorem for systems). *Let G be an open set in $\mathbb{R}^{m+n}$ containing the point $(\bar{x}, \bar{y})$. Suppose that $F: G \to \mathbb{R}^n$ is continuous and has continuous first partial derivatives in G. Suppose that*

$$F(\bar{x}, \bar{y}) = 0 \quad and \quad \det \nabla_y F(\bar{x}, \bar{y}) \neq 0.$$

Then positive numbers h and k can be chosen so that: (a) *the direct product of the closed balls* $\overline{B_m(\bar{x}, h)}$ *and* $\overline{B_n(\bar{y}, k)}$ *with centers at* $\bar{x}$, $\bar{y}$ *and radii h and k, respectively, is in G; and* (b) *h and k are such that for each* $x \in B_m(\bar{x}, h)$ *there is a unique* $y \in B_n(\bar{y}, k)$ *satisfying* $F(x, y) = 0$. *If f is the function from* $B_m(\bar{x}, h)$ *to* $B_n(\bar{y}, k)$ *defined by these ordered pairs* (x, y), *then* $F(x, f(x)) = 0$; *furthermore, f and all its first partial derivatives are continuous on* $B_m(\bar{x}, h)$.

PROOF. We define the matrices

$$A = \nabla_x F(\bar{x}, \bar{y}), \qquad B = \nabla_y F(\bar{x}, \bar{y}),$$

and write F in the form

$$F(x, y) = A \cdot (x - \bar{x}) + B \cdot (y - \bar{y}) + \phi(x, y), \qquad (14.15)$$

where ϕ is *defined*[2] by Equation (14.15). It is clear that ϕ has the properties

$$\phi(\bar{x}, \bar{y}) = 0, \qquad \nabla_x \phi(\bar{x}, \bar{y}) = 0, \qquad \nabla_y \phi(\bar{x}, \bar{y}) = 0.$$

By hypothesis, B is a nonsingular matrix. We multiply (14.15) by B^{-1}, getting

$$B^{-1}F = B^{-1}A \cdot (x - \bar{x}) + (y - \bar{y}) + B^{-1}\phi(x, y).$$

Now we may apply Lemma 14.5 with $B^{-1}F$ in place of F in that lemma, $x - \bar{x}$ in place of x; also, $y - \bar{y}$ in place of y, $-B^{-1}A$ in place of C, and $B^{-1}\phi$ in place of ψ. It is simple to verify that all the hypotheses of the lemma are fulfilled. The theorem follows for $B^{-1}F$. Since B^{-1} is a constant nonsingular matrix, the result holds for F. $\qquad\square$

Remarks. The first partial derivatives of the implicitly defined function f may be found by a direct computation in terms of partial derivatives of F. To see this suppose that F has components $F^1, F^2, \ldots, F^n$ and that f has components $f^1, f^2, \ldots, f^n$. We write

$$F^i(x_1, x_2, \ldots, x_m, y_1, y_2, \ldots, y_n) = 0, \qquad (14.16)$$

where $y_i = f^i(x_1, x_2, \ldots, x_m)$. To find the partial derivatives of f^i, we take the derivative of F^i with respect to x_p in (14.16), getting (by the Chain rule)

$$\frac{\partial F^i}{\partial x_p} + \sum_{k=1}^{n} \frac{\partial F^i}{\partial y_k} \frac{\partial f^k}{\partial x_p} = 0, \qquad i = 1, 2, \ldots, n, \; p = 1, 2, \ldots, m. \qquad (14.17)$$

Treating $\partial f^k/\partial x_p$ (for fixed p) as a set of n unknowns, we see that the above equations form an algebraic system of n equations in n unknowns in which, by hypothesis, the determinant of the coefficients does not vanish at $(\bar{x}, \bar{y})$. Therefore by Cramer's rule the system can always be solved uniquely.

---

[2] In the second proof of Theorem 14.1, the function φ is defined by: $F_{,2}(x_0, y_0)\psi(x, y)$.

EXAMPLE 1. Let $F(x, y)$ be a function from $\mathbb{R}^4$ to $\mathbb{R}^2$ given by

$$F^1(x_1, x_2, y_1, y_2) = x_1^2 - x_2^2 - y_1^3 + y_2^2 + 4,$$

$$F^2(x_1, x_2, y_1, y_2) = 2x_1 x_2 + x_2^2 - 2y_1^2 + 3y_2^4 + 8.$$

Let $P(\bar{x}, \bar{y}) = (2, -1, 2, 1)$. It is clear that $F(\bar{x}, \bar{y}) = 0$. Verify that $\det \nabla_y F(\bar{x}, \bar{y}) \neq 0$ and find the first partial derivatives of the function $y = f(x)$ defined implicitly by F at the point P.

Solution. We have

$$\frac{\partial F^1}{\partial y_1} = -3y_1^2, \qquad \frac{\partial F^1}{\partial y_2} = 2y_2, \qquad \frac{\partial F^2}{\partial y_1} = -4y_1, \qquad \frac{\partial F^2}{\partial y_2} = 12y_2^3.$$

At P, we find

$$\det(\nabla_y F) = \frac{\partial F^1}{\partial y_1} \frac{\partial F^2}{\partial y_2} - \frac{\partial F^1}{\partial y_2} \frac{\partial F^2}{\partial y_1} = -128.$$

Also,

$$\frac{\partial F^1}{\partial x_1} = 2x_1, \qquad \frac{\partial F^1}{\partial x_2} = -2x_2, \qquad \frac{\partial F^2}{\partial x_1} = 2x_2, \qquad \frac{\partial F^2}{\partial x_2} = 2x_1 + 2x_2.$$

Substituting the partial derivatives evaluated at P in (14.17) and solving the resulting systems of two equations in two unknowns first with $p = 1$ and then with $p = 2$, we get

$$\frac{\partial f^1}{\partial x_1} = \frac{13}{32}, \qquad \frac{\partial f^2}{\partial x_1} = \frac{7}{16}, \qquad \frac{\partial f^1}{\partial x_2} = \frac{5}{32}, \qquad \frac{\partial f^2}{\partial x_2} = -\frac{1}{16}. \qquad \square$$

EXAMPLE 2. Given $F: \mathbb{R}^5 \to \mathbb{R}^3$ defined according to the formulas

$$F^1(x_1, x_2, y_1, y_2, y_3) = x_1^2 + 2x_2^2 - 3y_1^2 + 4y_1 y_2 - y_2^2 + y_3^3,$$

$$F^2(x_1, x_2, y_1, y_2, y_3) = x_1 + 3x_2 - 4x_1 x_2 + 4y_1^2 - 2y_2^2 + y_3^2,$$

$$F^3(x_1, x_2, y_1, y_2, y_3) = x_1^3 - x_2^3 + 4y_1^2 + 2y_2 - 3y_3^2.$$

Assume that $P(x, y)$ is a point where $F(x, y) = 0$ and $\nabla_y F$ is nonsingular. Denoting the implicit function by f, determine $\partial f^i / \partial x_j$ at P.

Solution. According to (14.17) a straightforward computation yields

$$(-6y_1 + 4y_2)\frac{\partial f^1}{\partial x_1} + (4y_1 - 2y_2)\frac{\partial f^2}{\partial x_1} + 3y_3^2 \frac{\partial f^3}{\partial x_1} = -2x_1,$$

$$8y_1 \frac{\partial f^1}{\partial x_1} - 4y_2 \qquad \frac{\partial f^2}{\partial x_1} + 2y_3 \frac{\partial f^3}{\partial x_1} = 4x_2 - 1,$$

$$8y_1 \frac{\partial f^1}{\partial x_1} + 2 \qquad \frac{\partial f^2}{\partial x_1} - 6y_3 \frac{\partial f^3}{\partial x_1} = -3x_1^2.$$

We solve this linear system of three equations in three unknowns by Cramer's rule and obtain expressions for $\partial f^1/\partial x_1$, $\partial f^2/\partial x_1$, $\partial f^3/\partial x_1$. To find the partial derivatives of f with respect to x_2 we repeat the entire procedure, obtaining a similar linear system which can be solved by Cramer's rule. We leave the details to the reader. ☐

Definition. Let G be an open set in $\mathbb{R}^m$ and suppose that $F: G \to \mathbb{R}^n$ is a given vector function. The function F *is of* **class C^k** *on* G, where k is a nonnegative integer if and only if F and all its partial derivatives up to and including those of order k are continuous on G.

The Inverse function theorem which is a Corollary to Theorem 14.1 has a natural generalization for vector functions.

Theorem 14.4 (Inverse function theorem). *Let G be an open set in $\mathbb{R}^m$ containing the point $\bar{x}$. Suppose that $f: G \to \mathbb{R}^m$ is a function of class C^1 and that*

$$\bar{y} = f(\bar{x}), \qquad \det \nabla_x f(\bar{x}) \neq 0.$$

Then there are positive numbers h and k such that the ball $B_m(\bar{x}, k)$ is in G and for each $y \in B_m(\bar{y}, h)$ there is a unique point $x \in B_m(\bar{x}, k)$ with $f(x) = y$. If g is defined to be the inverse function determined by the ordered pairs (y, x) with the domain of g consisting of $B_m(\bar{y}, h)$ and range of g in $B_m(\bar{x}, k)$, then g is a function of class C^1. Furthermore, $f[g(y)] = y$ for $y \in B_m(\bar{y}, h)$.

PROOF. This theorem is a corollary of Theorem 14.3 in which

$$F(y, x) = y - f(x).$$ ☐

Remarks. The Inverse function theorem for functions of one variable (Corollary to Theorem 14.1) has the property that the function is one-to-one over the entire domain in which the derivative does not vanish. In Theorem 14.4, the condition $\det \nabla_x f \neq 0$ does not guarantee that the inverse (vector) function will be one-to-one over its domain. To see this consider the function $f: \mathbb{R}^2 \to \mathbb{R}^2$ given by

$$f^1 = x_1^2 - x_2^2, \qquad f^2 = 2x_1 x_2, \qquad (14.18)$$

with domain the annular ring $G \equiv \{(x_1, x_2): r_1 < (x_1^2 + x_2^2)^{1/2} < r_2\}$ where r_1, r_2 are positive numbers. A computation shows that

$$\nabla_x f = \begin{pmatrix} 2x_1 & -2x_2 \\ 2x_2 & 2x_1 \end{pmatrix},$$

and so $\det \nabla_x f = 4(x_1^2 + x_2^2)$, which is positive in G. However, setting $y = f(x)$, we see from (14.18) that there are two distinct values of x for each value of y. The inverse relation is a function in a sufficiently small ball of G, but if one considers the entire ring G there are two distinct values of $x = (x_1, x_2)$ in G which correspond to a given value of $y = (y_1, y_2)$.

PROBLEMS

In each of Problems 1 through 4 a function F and a point P are given. Verify that the Implicit function theorem is applicable. Denoting the implicitly defined function by f, find the values of all the first partial derivatives of f at P.

1. $F = (F^1, F^2)$, $P = (0, 0, 0, 0)$ where $F^1 = 2x_1 - 3x_2 + y_1 - y_2$, and $F^2 = x_1 + 2x_2 + y_1 + 2y_2$.

2. $F = (F^1, F^2)$, $P = (0, 0, 0, 0)$ where $F^1 = 2x_1 - x_2 + 2y_1 - y_2$, and $F^2 = 3x_1 + 2x_2 + y_1 + y_2$.

3. $F = (F^1, F^2)$, $P = (3, -1, 2, 1)$ where $F^1 = x_1 - 2x_2 + y_1 + y_2 - 8$, and $F^2 = x_1^2 - 2x_2^2 - y_1^2 + y_2^2 - 4$.

4. $F = (F^1, F^2)$, $P = (2, 1, -1, 2)$ where $F^1 = x_1^2 - x_2^2 + y_1 y_2 - y_2^2 + 3$, and $F^2 = x_1 + x_2^2 + y_1^2 + y_1 y_2 - 2$.

5. Suppose that $x = (x_1, x_2)$, $y = (y_1, y_2)$ and $F: \mathbb{R}^4 \to \mathbb{R}^2$ are such that $F(x, y) = 0$ and the Implicit function theorem is applicable for all (x, y). Denoting the implicitly defined function by f, find a formula in terms of the first partial derivatives of F for $\partial f^1/\partial x_1$, $\partial f^1/\partial x_2$, $\partial f^2/\partial x_1$, $\partial f^2/\partial x_2$.

6. Suppose that $F(x, y) = 0$ where $x = (x_1, \ldots, x_m)$ and $y = (y_1, y_2)$ and that the Implicit function theorem is applicable. Denoting the implicitly defined function by f, find $\partial f^i/\partial x_j$, $i = 1, 2$, $j = 1, 2, \ldots, m$, in terms of the partial derivatives of F.

7. Complete Example 2.

8. Given $F = (F^1, F^2)$ where $F: \mathbb{R}^2 \to \mathbb{R}^2$ and $F^1 = e^{2x+y}$, $F^2 = (4x^2 + 4xy + y^2 + 6x + 3y)^{2/3}$. Show that there is no value of x for which the Implicit function theorem is applicable. Find a relation between F^1 and F^2.

In each of Problems 9 through 12 a vector function $f: \mathbb{R}^2 \to \mathbb{R}^2$ is given. Verify that the Inverse function theorem is applicable and find the inverse function g.

9. $y_1 = x_1$, $y_2 = x_1^2 + x_2$.

10. $y_1 = 2x_1 - 3x_2$, $y_2 = x_1 + 2x_2$.

11. $y_1 = x_1/(1 + x_1 + x_2)$, $y_2 = x_2/(1 + x_1 + x_2)$, $x_1 + x_2 > -1$.

12. $y_1 = x_1 \cos(\pi x_2/2)$, $y_2 = x_1 \sin(\pi x_2/2)$, $x_1 > 0$, $-1 < x_2 < 1$.

13. Given the function $f: \mathbb{R}^3 \to \mathbb{R}^3$ where $f^1 = e^{x_2} \cos x_1$, $f^2 = e^{x_2} \sin x_1$, and $f^3 = 2 - \cos x_3$. Find the points $P(x_1, x_2, x_3)$ where the Inverse function theorem holds.

14. Given the function $f: \mathbb{R}^2 \to \mathbb{R}^2$ and suppose that the Inverse function theorem applies. We write $x = g(y)$ for the inverse function. Find formulas for $\partial g^i/\partial y_j$, $i, j = 1, 2$ in terms of partial derivatives of f^1 and f^2. Also find a formula for $\partial^2 g^1/\partial y_2^2$.

15. Given $F: \mathbb{R}^4 \to \mathbb{R}^2$ and suppose that $F(x, y) = 0$ for all $x = (x_1, x_2)$ and $y =$

(y_1, y_2). State conditions which guarantee that the equation

$$\frac{\partial y_1}{\partial x_1}\frac{\partial x_2}{\partial y_1} + \frac{\partial y_2}{\partial x_1}\frac{\partial x_2}{\partial y_2} = 0$$

holds.

14.3. Change of Variables in a Multiple Integral

For functions of one variable, an integral of the form

$$\int f(x)\, dx$$

can be transformed into

$$\int f[g(u)]g'(u)\, du$$

by the "change of variable" $x = g(u)$, $dx = g'(u)\, du$. Such transformations are useful in the actual evaluation of many integrals. The corresponding result for multiple integrals is more complicated and, in order to establish the appropriate formula for such a change of variables, we employ several results in linear algebra. In this section we assume the reader is familiar with the basic facts concerning matrices and linear transformations.

Definition. Let G be an open set in $\mathbb{R}^m$ and let $f: G \to \mathbb{R}^m$ be a C^1 function. That is, f has components $f^1, f^2, \ldots, f^m$ and $f^i: G \to \mathbb{R}^1$ are C^1 functions for $i = 1, 2, \ldots, m$. The **Jacobian** of f is the $m \times m$ matrix having the first partial derivative f^i_j as the entry in the ith row and jth column, $i, j = 1, 2, \ldots, m$. We also use the terms **Jacobian matrix** and **gradient**, and we denote this matrix by ∇f.

In the next theorem we restate for vector functions the Fundamental lemma of differentiation (Part (a)) and the Chain rule (Part (b)). In Part (c) we give an extension to vector functions of Equation (14.4), the formula for the derivative of the inverse of a function.

Theorem 14.5. *Let G and G_1 be open sets in $\mathbb{R}^m$ with $\bar{x}$ a point in G. Let $f: G \to G_1$ be a C^1 function and denote $f = (f^1, f^2, \ldots, f^m)$.*

(a) *We have the formula (Fundamental Lemma of Differentiation)*

$$f^i(\bar{x} + h) - f^i(\bar{x}) = \nabla f^i(\bar{x})h + \varepsilon^i(h)$$

$$= \sum_{j=1}^{m} f^i_{,j}h_j + \varepsilon^i(h), \qquad i = 1, 2, \ldots, m, \quad (14.19)$$

where $h = (h_1, h_2, \ldots, h_m)$ *and* $\varepsilon(h) = (\varepsilon^1(h), \ldots, \varepsilon^m(h))$ *are vectors, and* $\lim_{|h| \to 0} \varepsilon^i(h)/|h| = 0$.

(b) *Let* $g: G_1 \to \mathbb{R}^m$ *be of class* C^1 *and define* $F(x) = g[f(x)]$ *for* $x \in G$. *Then we have the Chain Rule*:

$$\nabla F(x) = \nabla g[f(x)] \cdot \nabla f(x). \tag{14.20}$$

(c) *Suppose that* f *is one-to-one with* $\det \nabla f(x) \neq 0$ *on* G. *Then the image* $f(G) = G_0$ *is open and the inverse function* $g_1 = f^{-1}$ *is one-to-one on* G_0 *and of class* C^1. *Furthermore,*

$$\nabla g_1[f(x)] = [\nabla f(x)]^{-1} \quad \text{with} \quad \det([\nabla f(x)]^{-1}) \neq 0 \quad \text{for} \quad x \in G \tag{14.21}$$

or

$$\nabla g_1(u) = \{\nabla f[f^{-1}(u)]\}^{-1} \neq 0, \qquad u \in G_0.$$

PROOF

(a) Formula (14.19) follows directly from the Fundamental lemma of differentiation for functions in $\mathbb{R}^m$ as given in Theorem 7.2.

(b) Formula (14.20) is a consequence of the Chain rule for partial derivatives as stated in Theorem 7.3. Each component of ∇F may be written (according to Theorem 7.3)

$$F^i_{,j}(x) = \sum_{j=1}^{m} g^i_{,k}[f(x)] \cdot f^k_{,j}(x),$$

which is (14.20) precisely.

(c) Since f is one-to-one, it is clear that f^{-1} is a function. Let $\bar{y} \in G_0$ where G_0 is the image of G and suppose $f(\bar{x}) = \bar{y}$. From the Inverse function theorem, which is applicable since $\nabla f(\bar{x}) \neq 0$, there are positive numbers h and k such that the ball $B(\bar{x}, k)$ is in G and also such that for each $y \in B(\bar{y}, h)$ there is a unique $x \in B(\bar{x}, k)$ with the property that $f(x) = y$. We define $g_1(y)$ to be the function given by the pairs (y, x). Then g_1 is of class C^1 on $B(\bar{y}, h)$ and the domain of g_1 contains $B(\bar{y}, h)$. Hence for each $\bar{y}$ in G_0, there is a ball with $\bar{y}$ as center which is also in G_0. We conclude that G_0 is open. Formula (14.21) follows from (14.20) and the Inverse function theorem. $\qquad\square$

In establishing the change of variables formula we shall see that an essential step in the proof is the reduction of any C^1 function f into the composition of a sequence of functions which have a somewhat simpler character. This process can be carried out whenever the Jacobian of f does not vanish.

Definition. Let $(i_1, i_2, \ldots, i_m)$ be a permutation of the numbers $(1, 2, \ldots, m)$. A linear transformation τ from $\mathbb{R}^m$ into $\mathbb{R}^m$ is **simple** if τ has the form

$$\tau(x_1, x_2, \ldots, x_m) = (\pm x_{i_1}, \pm x_{i_2}, \ldots, \pm x_{i_m}).$$

The next lemma is an immediate consequence of the above definition.

Lemma 14.6. *The product of simple transformations is simple and the inverse of a simple transformation is simple.*

If f_1 and f_2 are functions on $\mathbb{R}^m$ to $\mathbb{R}^m$ such that the range of f_2 is in the domain of f_1, we use the notation $f_1 \circ f_2$ for the composition $f_1[f_2(x)]$ of the two functions.

The next lemma gives the precise reduction of a function on $\mathbb{R}^m$ as the composition of functions each of which has an essentially simpler character.

Lemma 14.7. *Let G be an open set in $\mathbb{R}^m$, $\bar{x} \in G$, and let $f: G \to \mathbb{R}^m$ be a C^1 function with $\det \nabla f(\bar{x}) \neq 0$. Then there is an open subset G_1 of G containing $\bar{x}$ such that f can be written on G_1 as the composition of $m + 1$ functions*

$$f = g_{m+1} \circ g_m \circ \cdots \circ g_1. \tag{14.22}$$

The first m functions $g_1, g_2, \ldots, g_m$ are each defined on an open set G_i in $\mathbb{R}^m$ with range on an open set in $\mathbb{R}^m$ such that $g_i: G_i \to G_{i+1}, i = 1, 2, \ldots, m$. Moreover, the components $(g_i^1, g_i^2, \ldots, g_i^m)$ of g_i have the form

$$g_i^j(x_1, x_2, \ldots, x_m) = x_j \quad for \quad j \neq i \quad and \quad g_i^i = \varphi^i(x_1, x_2, \ldots, x_m). \tag{14.23}$$

The functions φ^i are determined in terms of f and have the property that $\varphi_{,i}^i > 0$ on G_i. The function g_{m+1} is simple.

PROOF. Since all the components except one in the definition of g_i given by (14.23) are coordinate functions, a straightforward computation shows that the determinant of the matrix ∇g_i, denoted $\det \nabla g_i$, is equal to $\varphi_{,i}^i$. We shall establish that $\varphi_{,i}^i$ is positive and so these determinants will all be positive.

Since the Jacobian $\nabla f(\bar{x})$ is nonsingular, there is a linear transformation τ_1 such that $\tau_1 \circ f$ has the property that all the principal minors of the Jacobian $\nabla(\tau_1 \circ f)$ are positive at $\bar{x}$. Define $f_0 = \tau_1 \circ f$ and denote the components of f_0 by $(f_0^1, f_0^2, \ldots, f_0^m)$. Next define m functions $h_1, h_2, \ldots, h_m$ as follows:

$$h_i \text{ has components } (f_0^1, f_0^2, \ldots, f_0^i, x_{i+1}, x_{i+2}, \ldots, x_m)$$

for $i = 1, 2, \ldots, m - 1$. We set $h_m = f_0$. Since all the principal minors of ∇f_0 are positive, it is not difficult to see that each $\nabla h_i(\bar{x})$ is nonsingular and, in fact, $\det \nabla h_i(\bar{x}) > 0$ for each i. According to Part (c) of Theorem 14.5 and the manner in which the h_i are defined, for each i there is an open set H_i on which $\det \nabla h_i(x) > 0$. Also, h_i is one-to-one from H_i onto an open set. Define

$$G_1 = H_1 \cap H_2 \cap \cdots \cap H_m.$$

Now, define sets $G_2, G_3, \ldots, G_{m+1}$ as follows:

$$G_{i+1} = h_i(G_1), \qquad i = 1, 2, \ldots, m.$$

Henceforth we restrict the domain of $h_1, \ldots, h_m$ to be G_1 without relabeling the functions. Define

$$g_1 = h_1, \qquad g_i = h_i \circ h_{i-1}^{-1}, \qquad i = 2, 3, \ldots, m. \tag{14.24}$$

To define g_{m+1} consider the function inverse to τ_1, denoted τ^{-1} which, like τ_1 is a linear function. Define

$$g_{m+1} = \tau^{-1} \quad \text{restricted to} \quad G_{m+1}.$$

We observe that each function g_i is a one-to-one mapping from G_i onto G_{i+1}, and that det $\nabla g_i(x) \neq 0$ on G_i. Also,

$$g_{m+1} \circ g_m \circ \cdots \circ g_2 \circ g_1 = \tau^{-1} \circ h_m \circ h_{m-1}^{-1} \circ h_{m-1} \circ h_{m-2}^{-1} \circ \cdots \circ h_2 \circ h_1^{-1} \circ h_1$$

$$= \tau^{-1} \circ h_m$$

$$= \tau^{-1} \circ \tau_1 \circ f = f,$$

and so (14.22) holds.

Once we show that each function g_i has the form given by (14.23) the proof will be complete. Since

$$g_1 = h_1 = (f_0^1, x_2, \ldots, x_m)$$

it is clear that g_1 has the proper form. Now $g_2 = h_2 \circ h_1^{-1}$ and since $h_1^{-1} = (f_0^{-1}, x_2, \ldots, x_m)$, $h_2 = (f_0^1, f_0^2, x_3, \ldots, x_m)$, we see that

$$g_2 = (x_1, f_0^2, x_3, \ldots, x_m).$$

The argument for each g_i is similar. Since all the principal minors of ∇f_0 are positive, we know that $f_{0,i}^i > 0$ and, from the way we selected the vectors h_i and g_i we conclude that $\varphi_{,i}^i = f_{0,i}^i > 0$. $\qquad\square$

In Lemma 14.7 we express an arbitrary C^1 function f as the composition of functions each of which is the identity in all components except one (plus a simple function). The one component which is not the identity, for example the ith, has the property that its partial derivative with respect to x_i is positive on G_i.

The next step (Lemma 14.9) establishes the change of variables formula for a typical function which appears in such a decomposition.

Lemma 14.8. *Let G be a set in $\mathbb{R}^m$ and $\varphi: G \to \mathbb{R}^1$ a bounded function such that $|\varphi(x) - \varphi(y)| \leqslant \varepsilon$ for all $x, y \in G$. Define $m = \inf\{\varphi(x): x \in G\}$ and $M = \sup\{\varphi(x): x \in G\}$. Then*

$$M - m \leqslant \varepsilon.$$

The proof is left to the reader.

Lemma 14.9. *Let G be an open set in $\mathbb{R}^m$ and suppose that $f: G \to \mathbb{R}^m$ is a one-to-one function of class C^1. We denote the components of f by $(u^1, u^2, \ldots, u^m)$ and let k be a fixed integer between 1 and m. Suppose the u^i have the form*

$$u^i(x_1, x_2, \ldots, x_m) = x_i, \quad \text{the ith coordinate in } \mathbb{R}^m \text{ for } i \neq k,$$

$$u^k = f^k(x) \quad \text{with} \quad f_{,k}^k(x) > 0 \quad \text{on} \quad G.$$

(a) *If F is a figure with $\bar{F} \subset G$, then the set $f(F)$ is a figure in $\mathbb{R}^m$.*

(b) *Denote $f(G)$ by G_1 and let $K\colon G_1 \to \mathbb{R}^1$ be uniformly continuous on G_1. Then the change of variables formula holds:*

$$\int_F K[f(x)]|J(x)|\, dV_m = \int_{f(F)} K(u)\, dV_m, \qquad \text{where} \quad J(x) = \det \nabla f(x). \quad (14.25)$$

PROOF

(a) Let R be a closed cell in G with $a_i \leqslant x_i \leqslant b_i$, $i = 1, 2, \ldots, m$ (see Section 8.1). Then the set $S = f(R)$ is given by

$$a_i \leqslant u^i \leqslant b_i, \qquad i \neq k,$$

and

$$f^k(x_1, \ldots, x_{k-1}, a_k, x_{k+1}, \ldots, x_m) \leqslant u^k \leqslant f^k(x_1, \ldots, x_{k-1}, b_k, x_{k+1}, \ldots, x_m).$$

We now employ the Corollary to Theorem 8.13 to conclude not only that S is a figure in $\mathbb{R}^m$ but that its volume, denoted $V(S) = V[f(R)]$, is given by

$$V(S) = \int_{a_k'}^{b_k'} [f^k(x_1, \ldots, x_{k-1}, b_k, x_{k+1}, \ldots, x_m)$$
$$- f^k(x_1, \ldots, x_{k-1}, a_k, x_{k+1}, \ldots, x_m)]\, dx_1 \ldots dx_{k-1} dx_{k+1} \ldots dx_m.$$

The symbols a_k', b_k' mean that the integration with respect to each variable x_i is between the limits a_i and b_i. Since the integrand above can be written as

$$\int_{a_k}^{b_k} f_{,k}^k\, dx_k,$$

we find

$$V(S) = \int_R f_{,k}^k(x)\, dV_m,$$

where dV_m is the usual element of volume in $\mathbb{R}^m$. From the way we defined f, a simple computation shows that $|J(x)| = f_{,k}^k(x)$, and we conclude that

$$V[f(R)] = \int_R |J(x)|\, dV_m.$$

Part (a) is now established when F is a cell. Next, let F be any figure such that $\bar{F} \subset G$. For any positive integer n we may cover F with hypercubes of side 2^{-n}, denoting by F_n^- the collection of inner hypercubes and by F_n^+ the collection of inner and boundary hypercubes. From the Lebesgue lemma (Theorem 3.16 and Theorem 6.27) which is valid in $\mathbb{R}^m$ it follows that there is a positive number ρ such that all members of F_n^+ are entirely in G and, in fact are at least at distance ρ from the boundary of G. Since no two hypercubes of F_n^+ have interior points in common, it follows that

$$V[f(F_n^+)] = \int_{F_n^+} |J(x)|\, dV,$$

and a similar formula holds for F_n^-. Denoting the inner and outer volume of $f(F)$ by $V^-[f(F)]$ and $V^+[f(F)]$, respectively, we find

$$\int_{F_n^-} |J(x)|\, dV = V[f(F_n^-)] \leqslant V^-[f(F)]$$

$$\leqslant V^+[f(F)] \leqslant V[f(F_n^+)]$$

$$= \int_{F_n^+} |J(x)|\, dV. \tag{14.26}$$

Since f is of class C^1, the function $|J(x)|$ is uniformly continuous on F_n^+ for all sufficiently large n and hence bounded by a constant which we denote by M. Therefore,

$$\int_{F-F_n^-} |J(x)|\, dV \leqslant MV(F - F_n^-), \qquad \int_{F_n^+-F} |J(x)|\, dV \leqslant MV(F_n^+ - F).$$

Since F is a figure, we let $n \to \infty$ and these integrals tend to zero. Employing this fact in (14.26) we conclude that $V^-[f(F)] = V^+[f(F)]$, and so $f(F)$ is a figure. Moreover,

$$V[f(F)] = \int_F |J(x)|\, dV,$$

that is, in addition to Part (a) we showed that Formula (14.25) holds in the special case $K(x) \equiv 1$.

(b) Let F be a figure such that $\bar{F} \subset G$. Since f and $|J|$ are continuous on $\bar{F}$, a closed bounded set, they are uniformly continuous on $\bar{F}$. Since, by hypothesis, $K[f(x)]$ is uniformly continuous on G_1, we see that the function $K[f(x)] \cdot |J(x)|$ is uniformly continuous on $\bar{F}$ and hence integrable on F. We shall establish Formula (14.25) by approximating each of the integrals in (14.25) by a Riemann sum and then by showing that the two Riemann sums are arbitrarily close if the subdivision is sufficiently fine. Let $\varepsilon > 0$ be given, and let $\Delta: \{F_1, F_2, \ldots, F_n\}$ be a subdivision of F. Choose $\xi_i \in F_i$, $i = 1, 2, \ldots, n$. Then

$$\left| \sum_{i=1}^n K[f(\xi_i)]|J(\xi_i)| V_m(F_i) - \int_F K[f(x)]|J(x)|\, dV_m \right| < \frac{\varepsilon}{3}, \tag{14.27}$$

if the mesh $\|\Delta\|$ is sufficiently small, say less than some number δ. Similarly, if $\Delta_1: \{F_1', F_2', \ldots, F_n'\}$ is a subdivision of $f(F)$ with $\|\Delta_1\| < \eta$ and with $\xi_i' \in F_i'$, $i = 1, 2, \ldots, n$, then for sufficiently small η, it follows that

$$\left| \sum_{i=1}^n K(\xi_i') V_m(F_i') - \int_{f(F)} K(u)\, dV_m \right| < \frac{\varepsilon}{3}. \tag{14.28}$$

Let $M = \sup_{u \in f(F)} |K(u)|$. Because of the uniform continuity of f and $|J|$, we may choose δ so small that for all $x', x'' \in F$ with $|x' - x''| < \delta$, we have

$$|f(x') - f(x'')| < \eta \quad \text{and} \quad \big||J(x')| - |J(x'')|\big| < \frac{\varepsilon}{3MV_m(F)}. \tag{14.29}$$

We now assume δ is chosen in this way and that δ is made smaller, if necessary, so that (14.27) holds. Select $\xi_i' = f(\xi_i)$ and $F_i' = f(F_i)$, $i = 1, 2, \ldots, n$. Then $\Delta_1: \{F_1', \ldots, F_n'\}$ is a subdivision of $f(F)$, and from the first inequality in (14.29), we have $\|\Delta_1\| < \eta$. Thus (14.28) holds. Next denote by m_i and M_i the infimum and supremum of $|J(x)|$ on F_i, respectively. Then from the proof of Part (a) and the Mean-value theorem for integrals, it follows that

$$V_m(F_i') = \int_{F_i} |J(x)| \, dV_m = |J_i| V_m(F_i), \tag{14.30}$$

where $|J_i|$ is a number such that $m_i \leqslant |J_i| \leqslant M_i$. We also have $m_i \leqslant |J(\xi_i)| \leqslant M_i$ and so, by Lemma 14.8 and the second inequality in (14.29), we find

$$\||J_i| - |J(\xi_i)|\| \leqslant M_i - m_i \leqslant \frac{\varepsilon}{3M V_m(F)}. \tag{14.31}$$

We wish to estimate the difference of the Riemann sums

$$\left| \sum_{i=1}^n K(\xi_i') V_m(F_i') - \sum_{i=1}^n K[f(\xi_i)]|J(\xi_i)| V_m(F_i) \right|. \tag{14.32}$$

Using (14.30) and the fact that $\xi_i' = f(\xi_i)$, we obtain for (14.32)

$$\left| \sum_{i=1}^n K(\xi_i')[|J_i| - |J(\xi_i)|] V_m(F_i) \right|.$$

Inserting (14.31) into this expression, we find that

$$\left| \sum_{i=1}^n K(\xi_i') V_m(F_i') - \sum_{i=1}^n K[f(\xi_i)]|J(\xi_i)| V_m(F_i) \right| < \frac{\varepsilon}{3}. \tag{14.33}$$

Combining (14.27), (14.28), and (14.33), we conclude that

$$\left| \int_F K[f(x)]|J(x)| \, dV_m - \int_{f(F)} K(u) \, dV_m \right| < \varepsilon.$$

Since ε is arbitrary Formula (14.25) holds. $\qquad \square$

Lemma 14.10. *Suppose that $f: G \to G_1$ is simple. Then the conclusions of Lemma 14.9 hold.*

PROOF. If f is simple the image of any cell in G is a cell in G_1 (perhaps with the sides arranged in a different order). Also, for f simple, we have $|J(x)| = 1$, and $|f(x') - f(x'')| = |x' - x''|$ for any two points $x', x'' \in G$. The remaining details may be filled in by the reader. $\qquad \square$

In Lemma 14.7 we showed how to express a function f as the composition of essentially simpler functions $g_1, g_2, \ldots, g_m$. Then, in Lemmas 14.9 and 14.10 we established the change of variables formula for these simpler functions. Now we show that the change of variables formula, (14.25), holds in general.

Theorem 14.6 (Change of variables formula). *Let G be an open set in $\mathbb{R}^m$ and suppose that $f: G \to \mathbb{R}^m$ is one-to-one and of class C^1 with $\det \nabla f(x) \neq 0$ on G. Let F be a closed bounded figure contained in G. Suppose that $K: f(F) \to \mathbb{R}^1$ is continuous on $f(F)$. Then $f(F)$ is a figure, $K[f(x)]$ is continuous on F, and*

$$\int_{f(F)} K(u)\, dV_m = \int_F K[f(x)] \cdot |J(x)|\, dV_m, \quad \text{where} \quad J(x) = \det \nabla f(x). \quad (14.34)$$

PROOF. Let x_0 be any point of G. Then, according to Lemma 14.7, there is an open set G_1 with $x_0 \in G_1$, $G_1 \subset G$, and such that on G_1 we have $f = g_{m+1} \circ g_m \circ \cdots \circ g_1$ with the g_i satisfying all the conditions of Lemma 14.7. In Lemmas 14.9 and 14.10, we established the change of variables formula for each g_i.

Let F be a closed bounded figure in G_1 and suppose K is continuous on $f(F)$. The set $f(F)$ is given by

$$f(F) = g_{m+1} \circ g_m \circ \cdots \circ g_1(F).$$

Applying Lemma 14.10 to the simple mapping g_{m+1}, we see that the set $g_m \circ g_{m-1} \circ \cdots \circ g_1(F)$ is a figure. Define the function

$$K_1(u) = K[g_{m+1}(u)] \cdot |\det \nabla g_{m+1}(u)|.$$

Then K_1 is continuous on $g_m \circ g_{m-1} \circ \cdots \circ g_1(F)$ and

$$\int_{f(F)} K(u)\, dV_m = \int_{g_{m+1} \circ \cdots \circ g_1(F)} K(u)\, dV_m$$

$$= \int_{g_m \circ \cdots \circ g_1(F)} K_1(u)\, dV_m. \quad (14.35)$$

Next apply Lemma 14.9 to the mapping g_m. We define

$$K_2(u) = K_1[g_m(u)] \cdot |\det \nabla g_m(u)|,$$

and we observe that $g_{m-1} \circ g_{m-2} \circ \cdots \circ g_1(F)$ is a figure with K_2 continuous on this set. Therefore from Lemma 14.9, we have

$$\int_{f(F)} K(u)\, dV_m = \int_{g_m \circ \cdots \circ g_1(F)} K_1(u)\, dV_m = \int_{g_{m-1} \circ \cdots \circ g_1(F)} K_2(u)\, dV_m.$$

By substitution, we find

$$K_2(u) = K_1[g_m(u)] \cdot |\det \nabla g_m(u)|$$

$$= K[g_{m+1}(g_m(u))] \cdot |\det \nabla g_{m+1}[g_m(u)]| \cdot |\det \nabla g_m(u)|.$$

Set $h_2(u) = g_{m+1} \circ g_m(u)$ and then the above formula becomes

$$K_2(u) = K[h_2(u)] \cdot |\det \nabla h_2(u)|,$$

where the Chain rule and the formula for the product of determinants have been used. We continue this process by defining $h_p(u) = g_{m+1} \circ g_m \circ \cdots \circ g_{m-p+2}$, $p = 2, 3, \ldots, m + 1$, and $K_p(u) = K[h_p(u)] \cdot |\det \nabla h_p(u)|$. We arrive at the formula

$$\int_{g_{m+1} \circ \cdots \circ g_1(F)} K(u)\, dV_m = \int_{g_1(F)} K[h_m(u)] \cdot |\det \nabla h_m(u)|\, dV_m$$

$$= \int_F K[h_{m+1}(u)] \cdot |\det \nabla h_{m+1}(u)|\, dV_m$$

$$= \int_F K[f(u)] \cdot |\det \nabla f(u)|\, dV_m,$$

which is the desired result for a figure F in G_1.

To complete the proof, let F be any closed bounded figure in G and suppose that K is continuous on F. From the Lebesgue lemma (Theorems 3.16 and 6.27), there is a number ρ such that any ball $B(x, \rho)$ with center at a point of F lies in some open set G_1. We subdivide F into a finite number of figures $F_1, F_2, \ldots, F_s$ such that each F_i is contained in a single ball $B(x, \rho)$. For $i = 1, 2, \ldots, s$, we have

$$\int_{f(F_i)} K(u)\, dV_m = \int_{F_i} K[f(v)] |\det \nabla f(v)|\, dV_m.$$

The formula (14.34) follows by addition on i. $\square$

EXAMPLE. Evaluate $\int_F x_1\, dV_2(x)$ where F is the region bounded by the curves $x_1 = -x_2^2$, $x_1 = 2x_2 - x_2^2$, and $x_1 = 2 - 2x_2 - x_2^2$ (see Figure 14.4(a)). Introduce new variables (u_1, u_2) by

$$f: x_1 = u_1 - \tfrac{1}{4}(u_1 + u_2)^2, \qquad x_2 = \tfrac{1}{2}(u_1 + u_2), \qquad (14.36)$$

and use Theorem 14.6.

Solution. Figure 14.4 shows G, the image of F in the (u_1, u_2)-plane. Solving (14.36) for u_1, u_2 in terms of x_1, x_2, we get

$$u_1 = x_1 + x_2^2, \qquad u_2 = -x_1 + 2x_2 - x_2^2,$$

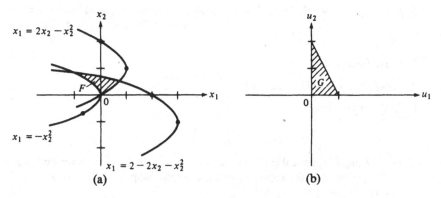

(a)   (b)

Figure 14.4. Changing variables from (x_1, x_2) to (u_1, u_2).

and so f is a one-to-one transformation of $\mathbb{R}^2$ onto itself. The equations of the bounding curves of G are

$$u_1 = 0, \qquad u_2 = 0, \qquad 2u_1 + u_2 = 2.$$

The Jacobian of f is

$$\det \nabla f = \begin{vmatrix} 1 - \frac{1}{2}(u_1 + u_2) & -\frac{1}{2}(u_1 + u_2) \\ \frac{1}{2} & \frac{1}{2} \end{vmatrix} = \frac{1}{2}.$$

Therefore

$$\int_F x_1 \, dV_2(x) = \int_G \left[u_1 - \frac{1}{4}(u_1 + u_2)^2 \right] \cdot \frac{1}{2} \, dV_2(u)$$

$$= \frac{1}{2} \int_0^1 \int_0^{2-2u_1} \left[u_1 - \frac{1}{4}(u_1 + u_2)^2 \right] du_2 \, du_1 = \frac{1}{48}. \qquad \square$$

PROBLEMS

In each of Problems 1 through 6 evaluate $\int_F K(x_1, x_2) \, dV_2(x)$, where F is bounded by the curves whose equations are given. Perform the integration by introducing variables u_1, u_2 as indicated. Draw a graph of F and the corresponding region in the u_1, u_2-plane. Find the inverse of each transformation.

1. $K(x_1, x_2) = x_1 x_2$. F is bounded by $x_2 = 3x_1$, $x_1 = 3x_2$, and $x_1 + x_2 = 4$. Mapping: $x_1 = 3u_1 + u_2$, $x_2 = u_1 + 3u_2$.

2. $K(x_1, x_2) = x_1 - x_2^2$. F is bounded by $x_2 = 2$, $x_1 = x_2^2 - x_2$, $x_1 = 2x_2 + x_2^2$. Mapping: $x_1 = 2u_1 - u_2 + (u_1 + u_2)^2$, $x_2 = u_1 + u_2$.

3. $K(x_1, x_2) = x_2$. F is bounded by $x_1 + x_2 - x_2^2 = 0$, $2x_1 + x_2 - 2x_2^2 = 1$, $x_1 - x_2^2 = 0$. Mapping: $x_1 = u_1 + (u_2 - u_1)^2$, $x_2 = u_2$.

4. $K(x_1, x_2) = (x_1^2 + x_2^2)^{-3}$. F is bounded by $x_1^2 + x_2^2 = 2x_1$, $x_1^2 + x_2^2 = 4x_1$, $x_1^2 + x_2^2 = 2x_2$, $x_1^2 + x_2^2 = 6x_2$. Mapping: $x_1 = u_1/(u_1^2 + u_2^2)$, $x_2 = u_2/(u_1^2 + u_2^2)$.

5. $K(x_1, x_2) = 4x_1 x_2$. F is bounded by $x_1 = x_2$, $x_1 = -x_2$, $(x_1 + x_2)^2 + x_1 - x_2 - 1 = 0$. Mapping: $x_1 = \frac{1}{2}(u_1 + u_2)$, $x_2 = \frac{1}{2}(-u_1 + u_2)$. Assume $x_1 + x_2 > 0$.

6. $K(x_1, x_2) = x_1^2 + x_2^2$. F is the region in the first quadrant bounded by $x_1^2 - x_2^2 = 1$, $x_1^2 - x_2^2 = 2$, $x_1 x_2 = 1$, $x_1 x_2 = 2$. The inverse mapping is: $u_1 = x_1^2 - x_2^2$, $u_2 = 2x_1 x_2$.

7. Prove Lemma 14.8.

8. Complete the proof of Lemma 14.10.

9. Evaluate the integral

$$\int_F x_3 \, dV_3(x)$$

by changing to spherical coordinates: $x_1 = \rho \cos \varphi \sin \theta$, $x_2 = \rho \sin \varphi \sin \theta$, $x_3 = \rho \cos \theta$, where F is the region determined by the inequalities $0 \leqslant x_1^2 + x_2^2 \leqslant x_3^2$, $0 \leqslant x_1^2 + x_2^2 + x_3^2 \leqslant 1$, $x_3 \geqslant 0$.

10. Write a proof of the Fundamental Lemma of Differentiation for vector functions (Theorem 14.5, Part (a)).

11. Show that the product of simple transformations is simple and that the inverse of a simple transformation is simple (Lemma 14.6).

12. Let $g = (g^1, g^2, \ldots, g^m)$ where $g^j = x_j$ for $j \neq i$; $j = 1, 2, \ldots, m$, and $g^i = \varphi(x)$. Show that $\nabla g = \varphi_{,i}$ (see Lemma 14.7).

13. If f is of class C^1 on a closed bounded region $\bar{G}$ in $\mathbb{R}^m$, show that det ∇f is uniformly continuous on $\bar{G}$.

14.4. The Lagrange Multiplier Rule

Let D be a region in $\mathbb{R}^m$ and suppose that $f: D \to \mathbb{R}^1$ is a C^1 function. At any local maximum or minimum of $f(x) = f(x_1, \ldots, x_m)$, we know that $f_{,i} = 0$, $i = 1, 2, \ldots, m$. In many applications we wish to find the local maxima and minima of such a function f subject to certain constraints. These constraints are usually given by a set of equations such as

$$\varphi^1(x_1, x_2, \ldots, x_m) = 0, \qquad \varphi^2(x_1, x_2, \ldots, x_m) = 0, \ldots,$$

$$\varphi^k(x_1, x_2, \ldots, x_m) = 0. \tag{14.37}$$

Equations (14.37) are called **side conditions**. Throughout we shall suppose that k is less than m. Otherwise, if there were say m side conditions, Equations (14.37) when solved simultaneously might yield a unique solution $\bar{x} = (\bar{x}_1, \ldots, \bar{x}_m)$. Then this value when inserted in f would give a solution to the problem without further calculation. We reject the case $k > m$ since there may be no solution to the system given by (14.37). We shall suppose that the functions $\varphi^i: D \to \mathbb{R}^1$, $i = 1, 2, \ldots, k$, are C^1 functions, and furthermore that the $k \times m$ matrix

$$\begin{pmatrix} \varphi^1_{,1} & \varphi^1_{,2} \cdots \varphi^1_{,m} \\ \vdots & \vdots \\ \varphi^k_{,1} & \varphi^k_{,2} \cdots \varphi^k_{,m} \end{pmatrix}$$

is of rank k. That is, we suppose that at least one of the $k \times k$ minors of the above matrix has determinant different from zero in D. Without loss of generality, we assume that the square matrix consisting of the first k columns has nonvanishing determinant in D. This may always be achieved by relabeling the variables. Then according to the Implicit function theorem, in the neighborhood of any point $\bar{x} \in D$ we may solve for $x_1, x_2, \ldots, x_k$ in terms of $x_{k+1}, \ldots, x_m$. That is, there are functions $g^1, \ldots, g^k$ of class C^1 such that Equations (14.37) can be written

$$x_1 = g^1(x_{k+1}, \ldots, x_m), \quad x_2 = g^2(x_{k+1}, \ldots, x_m), \quad \ldots, \quad x_k = g^k(x_{k+1}, \ldots, x_m).$$

$$\tag{14.38}$$

A customary way of finding a local maximum or minimum of f subject to the side Conditions (14.38) consists of the following procedure. First, solve the system given by (14.37) for $x_1, \ldots, x_k$ and obtain Equations (14.38). We assume that this is valid throughout D. Second, insert the functions in (14.38) in f, obtaining a function of the variables $x_{k+1}, \ldots, x_m$ given by

$$H(x_{k+1}, \ldots, x_m) \equiv f[g^1(x_{k+1}, \ldots, x_m),$$
$$g^2(x_{k+1}, \ldots, x_m), \ldots, g^k(x_{k+1}, \ldots, x_m), x_{k+1}, \ldots, x_m].$$

Finally, find the local maxima and minima of H as a function of $x_{k+1}, \ldots, x_m$ in the ordinary way. That is, compute

$$H_{,i} = \sum_{j=1}^{k} f_{,j} g^j_{,i} + f_{,i}, \qquad i = k+1, \ldots, m, \tag{14.39}$$

and then find the solutions of the system of $m - k$ equations

$$H_{,i} = 0. \tag{14.40}$$

The values of $x_{k+1}, \ldots, x_m$ obtained in this way are inserted in (14.38) to yield values for $x_1, \ldots, x_k$. In this way we obtain the critical points of f which also satisfy (14.37). Various second derivative tests may then be used to decide whether the critical points are local maxima, local minima, or neither.

The method of **Lagrange multipliers** employs a simpler technique for achieving the same purpose. The method is especially useful when it is difficult or not possible to solve the system given by (14.37) in order to obtain the functions $g^1, \ldots, g^k$ given by (14.38).

The Lagrange multiplier rule is frequently explained but seldom proved. In Theorem 14.7 below we establish the validity of this rule which we now describe. We introduce k new variables (or parameters), denoted by $\lambda = (\lambda_1, \lambda_2, \ldots, \lambda_k)$, and we form the function of $m + k$ variables

$$F(x, \lambda) = F(x_1, \ldots, x_m, \lambda_1, \ldots, \lambda_k) \equiv f(x) + \sum_{j=1}^{k} \lambda_j \varphi^j(x).$$

For this function F we compute the critical points when x is in D and λ in $\mathbb{R}^k$ *without* side conditions. That is, we find solutions to the $m + k$ equations formed by all the first derivatives of $F(x, \lambda)$:

$$F_{,i} = 0, \qquad i = 1, 2, \ldots, m,$$
$$F_{,j} = 0, \qquad j = 1, 2, \ldots, k. \tag{14.41}$$

We shall show that the critical points given by solutions of (14.40) are among the solutions of the system given by (14.41).

Suppose that f takes on its minimum at x^0, a point in the set D_0 consisting of all points x in D where the side conditions (14.37) hold. Suppose there is a function $g = (g^1, g^2, \ldots, g^m)$ from $I = \{t: -t_0 < t < t_0\}$ into $\mathbb{R}^m$ which is of class C^1 and has the properties

$$g(0) = x^0 \quad \text{and} \quad \phi^j[g(t)] = 0 \quad \text{for} \quad j = 1, 2, \ldots, k; t \in I. \tag{14.42}$$

Then the function $\Phi\colon I \to \mathbb{R}^m$ defined by

$$\Phi(t) = f[g(t)] \tag{14.43}$$

takes on its minimum at $t = 0$. Differentiating (14.42) and (14.43) with respect to t and setting $t = 0$, we get

$$\sum_{i=1}^{m} \phi_{,i}^{j}(x^0)\frac{dg^i(0)}{dt} = 0 \quad \text{and} \quad \sum_{i=1}^{m} f_{,i}(x_0)\frac{dg^i(0)}{dt} = 0. \tag{14.44}$$

Now let $h = (h_1, h_2, \ldots, h_m)$ be *any* vector[3] in V_m which is orthogonal to the k vectors $(\phi_{,1}^{j}(x^0), \phi_{,2}^{j}(x^0), \ldots, \phi_{,m}^{j}(x^0))$, $j = 1, 2, \ldots, k$. That is, suppose that

$$\sum_{i=1}^{m} \phi_{,i}^{j}(x^0)h_i = 0 \quad \text{or} \quad \nabla\phi^j(x^0)\cdot h = 0, \qquad j = 1, 2, \ldots, k.$$

From the Implicit function theorem, it follows that we may solve (14.37) for $x_1, \ldots, x_k$ in terms of $x_{k+1}, \ldots, x_m$, getting

$$x_i = \mu^i(x_{k+1}, \ldots, x_m), \qquad i = 1, 2, \ldots, k.$$

If we denote $x^0 = (x_1^0, \ldots, x_m^0)$ and define

$$g^i(t) = \begin{cases} \mu^i(x_{k+1}^0 + th_{k+1}, \ldots, x_m^0 + th_m), & i = 1, 2, \ldots, k, \\ x_i^0 + th_i, & i = k+1, \ldots, m, \end{cases}$$

then $g = (g^1(t), \ldots, g^m(t))$ satisfies Conditions (14.42) and (14.44). We have thereby proved the following lemma.

Lemma 14.11. *Suppose that $f, \phi^1, \phi^2, \ldots, \phi^k$ are C^1 functions on an open set D in $\mathbb{R}^m$ containing a point x^0, that the vectors $\nabla\phi^1(x^0), \ldots, \nabla\phi^k(x^0)$ are linearly independent, and that f takes on its minimum among all points of D_0 at x^0, where D_0 is the subset of D on which the side conditions (14.37) hold. If h is any vector in V_m orthogonal to $\nabla\phi^1(x^0), \ldots, \nabla\phi^k(x^0)$, then*

$$\nabla f(x^0)\cdot h = 0.$$

The next lemma, concerning a simple fact about vectors in V_m, is needed in the proof of the Lagrange multiplier rule.

Lemma 14.12. *Let $b^1, b^2, \ldots, b^k$ be linearly independent vectors in the vector space V_m. Suppose that a is a vector in V_m with the property that a is orthogonal to any vector h which is orthogonal to all the b^i. Then there are numbers $\lambda_1, \lambda_2, \ldots, \lambda_k$ such that*

$$a = \sum_{i=1}^{k} \lambda_i b^i.$$

That is, a is in the subspace spanned by $b^1, b^2, \ldots, b^k$.

---

[3] In this argument we assume the reader is familiar with the customary m-dimensional vector space, denoted V_m. See Appendix 4.

PROOF. Let B be the subspace of V_m spanned by $b^1, b^2, \ldots, b^k$. Then there are vectors $c^{k+1}, c^{k+2}, \ldots, c^m$, such that the set $b^1, \ldots, b^k, c^{k+1}, \ldots, c^m$ form a linearly independent set (basis) of vectors in V_m. Let h be any vector orthogonal to all the b^i; then h will have components $h_1, \ldots, h_m$ in terms of the above basis with $h_1 = h_2 = \cdots = h_k = 0$. The vector a with components $(a_1, \ldots, a_m)$ and with the property $a \cdot h = 0$ for *all* such h must have $a_{i+1} = a_{i+2} = \cdots = a_m = 0$. Therefore, $a = \sum_{i=1}^k a_i b^i$. We set $a_i = \lambda_i$ to obtain the result. □

Theorem 14.7 (Lagrange multiplier rule). *Suppose that $f, \phi^1, \phi^2, \ldots, \phi^k$ and x^0 satisfy the hypotheses of Lemma 14.11. Define*

$$F(x, \lambda) = f(x) - \sum_{i=1}^k \lambda_i \phi^i(x).$$

Then there are numbers $\lambda_1^0, \lambda_2^0, \ldots, \lambda_k^0$ such that

$$F_{x_i}(x^0, \lambda^0) = 0, \qquad i = 1, 2, \ldots, m,$$

and

$$F_{\lambda_j}(x^0, \lambda^0) = 0, \qquad j = 1, 2, \ldots, k. \tag{14.45}$$

PROOF. The Equations (14.45) are

$$\nabla f(x_0) = \sum_{i=1}^k \lambda_i^0 \nabla \phi^i(x^0) \qquad \text{and} \qquad \phi^j(x^0) = 0, \qquad j = 1, 2, \ldots, k.$$

We set $a = \nabla f(x^0)$ and $b^j = \nabla \phi^j(x^0)$. Then Lemma 14.11 and 14.12 combine to yield the result. □

Remark. This theorem shows that the minimum (or maximum) of f subject to the side conditions $\phi^1 = \phi^2 = \cdots = \phi^k = 0$ is among the minima (or maxima) of the function F without any constraints.

EXAMPLE. Find the maximum of the function $x_1 + 3x_2 - 2x_3$ on the sphere $x_1^2 + x_2^2 + x_3^2 = 14$.

Solution. Let $F(x_1, x_2, x_3, \lambda) \equiv x_1 + 3x_2 - 2x_3 + \lambda(x_1^2 + x_2^2 + x_3^2 - 14)$. Then $F_{,1} = 1 + 2\lambda x_1$, $F_{,2} = 3 + 2\lambda x_2$, $F_{,3} = -2 + 2\lambda x_3$, $F_{,4} = x_1^2 + x_2^2 + x_3^2 - 14$. Setting $F_{,i} = 0, i = 1, \ldots, 4$, we obtain

$$x_1 = -\frac{1}{2\lambda}, \qquad x_2 = -\frac{3}{2\lambda}, \qquad x_3 = \frac{1}{\lambda}, \qquad 14 = \frac{14}{4\lambda^2}.$$

The solutions are $(x_1, x_2, x_3, \lambda) = (1, 3, -2, -\frac{1}{2})$ or $(-1, -3, 2, \frac{1}{2})$. The first solution gives the maximum value of 14. □

PROBLEMS

In each of Problems 1 through 10 find the solution by the Lagrange multiplier rule.

1. Find the minimum value of $x_1^2 + 3x_2^2 + 2x_3^2$ subject to the condition $2x_1 + 3x_2 + 4x_3 - 15 = 0$.

2. Find the minimum value of $2x_1^2 + x_2^2 + 2x_3^2$, subject to the condition $2x_1 + 3x_2 - 2x_3 - 13 = 0$.

3. Find the minimum value of $x_1^2 + x_2^2 + x_3^2$ subject to the conditions $2x_1 + 2x_2 + x_3 + 9 = 0$ and $2x_1 - x_2 - 2x_3 - 18 = 0$.

4. Find the minimum value of $4x_1^2 + 2x_2^2 + 3x_3^2$ subject to the conditions $x_1 + 2x_2 + 3x_3 - 9 = 0$ and $4x_1 - 2x_2 + x_3 + 19 = 0$.

5. Find the minimum value of $x_1^2 + x_2^2 + x_3^2 + x_4^2$ subject to the condition $2x_1 + x_2 - x_3 - 2x_4 - 5 = 0$.

6. Find the minimum value of $x_1^2 + x_2^2 + x_3^2 + x_4^2$ subject to the conditions $x_1 - x_2 + x_3 + x_4 - 4 = 0$ and $x_1 + x_2 - x_3 + x_4 + 6 = 0$.

7. Find the points on the curve $4x_1^2 + 4x_1x_2 + x_2^2 = 25$ which are nearest to the origin.

8. Find the points on the curve $7x_1^2 + 6x_1x_2 + 2x_2^2 = 25$ which are nearest to the origin.

9. Find the points on the curve $x_1^4 + y_1^4 + 3x_1y_1 = 2$ which are farthest from the origin.

10. Let $b_1, b_2, \ldots, b_k$ be positive numbers. Find the maximum value of $\sum_{i=1}^{k} b_i x_i$ subject to the side condition $\sum_{i=1}^{k} x_i^2 = 1$.

11. (a) Find the maximum of the function $x_1^2 \cdot x_2^2 \cdots x_n^2$ subject to the side condition $\sum_{i=1}^{n} x_i^2 = 1$.
    (b) If $\sum_{i=1}^{n} x_i^2 = 1$, show that $(x_1^2 x_2^2 \cdots x_n^2)^{1/n} \leqslant 1/n$.
    (c) If $a_1, a_2, \cdots, a_n$ are positive numbers, prove that

    $$(a_1 \cdot a_2 \cdots a_n)^{1/n} \leqslant \frac{a_1 + a_2 + \cdots + a_n}{n}.$$

[The geometric mean of n numbers is always less than or equal to the arithmetic mean.]

CHAPTER 15

Functions on Metric Spaces; Approximation

15.1. Complete Metric Spaces

We developed many of the basic properties of metric spaces in Chapter 6. A complete metric space was defined in Chapter 13 and we saw the importance of such spaces in the proof of the fundamental fixed point theorem (Theorem 13.2). This theorem, which has many applications, was used to prove the existence of solutions of ordinary differential equations and the Implicit function theorem. We now discuss in more detail functions whose domain is a metric space, and we prove convergence and approximation theorems which are useful throughout analysis.

We recall that a sequence of points $\{p_n\}$ in a metric space S is a **Cauchy sequence** if and only if for every $\varepsilon > 0$ there is an integer N such that $d(p_m, p_n) < \varepsilon$ whenever $m, n > N$. A **complete metric space** S is one with the property that every Cauchy sequence in S converges to a point in S. We remind the reader that Theorem 3.14 shows that $\mathbb{R}^1$ is a complete metric space and that Theorem 13.1 establishes the completeness of $\mathbb{R}^N$ for every positive integer N.

The notion of compactness, defined in Section 6.4, plays an essential part in the study of metric spaces. Repeating the definition, we say that a set A in a metric space S is **compact** if and only if each sequence of points $\{p_n\}$ in A contains a subsequence which converges to a point in A. Taking into account the definitions of completeness and compactness, we see that Theorem 6.23 implies that *every compact metric space S is complete*. Also, since a closed subset of a compact metric space is compact (Theorem 6.21), then such a subset, considered as a metric space in its own right, is complete.

Let S be any metric space and f a bounded, continuous function on S into $\mathbb{R}^1$. The totality of all such functions f forms a metric space, denoted $C(S)$, when we define the distance d between two such functions f and g by the formula

$$d(f, g) = \sup_{x \in S} |f(x) - g(x)|. \tag{15.1}$$

It is a simple matter to verify that d has all the properties of a metric on $C(S)$. We conclude directly from Theorem 13.3 on the uniform convergence of sequences of continuous functions that for *any* metric space S, *the space $C(S)$ is complete.*

EXAMPLES. Let $I = \{x: a \leqslant x \leqslant b\}$ be a closed interval in $\mathbb{R}^1$. Then I, considered as a metric space (with the metric of $\mathbb{R}^1$), is complete. On the other hand an open interval $J = \{x: a < x < b\}$ is not a complete metric space since there are Cauchy sequences in J converging to a and b, two points which are not in the space. The collection of all rational numbers in $\mathbb{R}^1$ is an example of a metric space which is not complete, with respect to the metric of $\mathbb{R}^1$. Let 0 be the function in $C(S)$ which is identically zero, and define $B(0, r)$ as the set of all functions f is $C(S)$ such that $d(f, 0) < r$. That is, $B(0, r)$ is the ball with center 0 and radius r. Then $B(0, r)$ is a metric space which is not complete. However, the set of all functions f which satisfy $d(f, 0) \leqslant r$ with d given by (15.1) does form a complete metric space. □

Definitions. Let S be a metric space. We say that a set A in S is **dense in** S if and only if $\bar{A} = S$. If A and B are sets in S, we say that A **is dense in** B if and only if $\bar{A} \supset B$. We do not require that A be a subset of B.

For example, let S be $\mathbb{R}^1$, let A be all the rational numbers, and let B be all the irrational numbers. Since $\bar{A} = \mathbb{R}^1$ we see that $\bar{A} \supset B$ and so A is dense in B. Note that A and B are disjoint.

Definition. Let S be a metric space and A a subset of S. We say that A is **nowhere dense in** S if and only if $\bar{A}$ contains no ball of S. For example, if S is $\mathbb{R}^1$, then any finite set of points is nowhere dense. A convergent sequence of points and the set of integer points in $\mathbb{R}^1$ are other examples of nowhere dense sets. It is important to observe that being dense and nowhere dense are *not* complementary properties. It is possible for a set to be neither dense nor nowhere dense. For example, any bounded open or closed interval in $\mathbb{R}^1$ is neither dense nor nowhere dense in $\mathbb{R}^1$.

The next result, an equivalent formulation of the notion of a nowhere dense set, is useful in many applications.

Theorem 15.1. *A set A is nowhere dense in a metric space S if and only if every open ball B of S contains an open ball B_1 which is disjoint with A.*

PROOF

(a) Suppose A is nowhere dense in S. Then $\bar{A}$ contains no ball of S. We shall show that every open ball B of S contains a ball B_1 which is disjoint from A. Suppose this is not true. Then there must be a ball B_0 of S such that every ball B_1 in B_0 contains points of A. Let $p \in B_0$. Then every ball with p as center and radius sufficiently small is in B_0. All these balls have points of A and so p is a limit point of A. Since p is an arbitrary point of B_0, we have $\bar{A} \supset B_0$. We have contradicted the fact that $\bar{A}$ contains no ball of S, and (a) is proved.

(b) If A is not nowhere dense, then $\bar{A}$ must contain a ball B. If there were a ball B_1 in B disjoint with A, then no point of B_1 could belong to $\bar{A}$. Hence every ball in B contains of A, and (b) is proved. $\qquad\square$

Definitions. A set A is a metric space S is of the **first category** if and only if A is the union of a countable number of nowhere dense sets. A set C is of the **second category** if and only if C is not of the first category.

EXAMPLES. Since the set in $\mathbb{R}^N$ consisting of a single point is nowhere dense, we see that every countable set in $\mathbb{R}^N$ is of the first category. Also, because the set of rational points in $\mathbb{R}^N$ is countable and dense, it is possible to have dense sets of first category. In fact, if the entire space consists of the rational points, say of $\mathbb{R}^1$, then every set is of the first category, and there are no sets of the second category. On the other hand, an isolated point in a metric space is an open set and in such cases a single point is of the second category. $\qquad\square$

No complete metric space can be of the first category as we show in the next theorem and its Corollary.

Theorem 15.2. *Let S be a complete metric space. Let C be the complement in S of a set of the first category. Then C is dense in S.*

PROOF. We shall show that every ball B of S must contain a point of C, implying that C is dense in S. Define $A = S - C$. Then A is of the first category and consequently $A = \bigcup_{n=1}^{\infty} T_n$ where T_n is nowhere dense for every n. Let $B_0 = B(p_0, r_0)$ be any ball in S with center p_0 and radius r_0. We shall show that B_0 contains a point of C. Since T_1 is nowhere dense, there is a ball $B(p_1, r_1)$ contained in B_0 which is disjoint with T_1. We may take r_1 so small that $r_1 < r_0/2$, and also so that $\overline{B(p_1, r_1)} \subset B_0$. Since T_2 is nowhere dense the ball $B(p_1, r_1)$ contains a ball $B(p_2, r_2)$ with $r_2 < r_1/2$, so that $B(p_2, r_2)$ contains no point of T_2 and $\overline{B(p_2, r_2)} \subset B(p_1, r_1)$. We continue in this way, obtaining a decreasing sequence of balls $B(p_n, r_n) \equiv B_n$ such that $r_n < r_{n-1}/2$ and $\bar{B}_n \subset B_{n-1}$. Also, T_n and B_n have no points in common. For $n \geqslant m$, we see that $d(p_m, p_n) \leqslant r_m < r_0/2^m$. Hence $\{p_n\}$ is a Cauchy sequence. Since S is complete, there is a point $q \in S$ such that $p_n \to q$ as $n \to \infty$. For each m, all the p_n are in B_m whenever $n > m$. We conclude that $q \in \bar{B}_m$ for every m and so $q \notin T_m$ for every m. Thus q

belongs to C and because $q \in B_0(p_0, r_0)$, which is an arbitrary ball, the result is established.   □

Corollary. *A complete metric space S is of the second category.*

PROOF. If S were of the first category its complement would be empty, contradictng Theorem 15.2.   □

By an argument similar to that used in Theorem 15.2, the next result known as the **Baire Category Theorem** is easily established. We omit the details.

Theorem 15.3 (Baire category theorem). *A nonempty open set in a complete metric space is of the second category.*

We recall that a mapping f from a metric space S_1 with metric d_1 to a metric space S_2 with metric d_2 is **uniformly continuous on S_1** if and only if for every $\varepsilon > 0$, there is a $\delta > 0$ such that $d_2(f(p), f(q)) < \varepsilon$ whenever $d_1(p, q) < \delta$, and the number δ depends only on ε and not on the particular points p and q. With the aid of Lemma 15.1 which follows it is not difficult to show that a function which is uniformly continuous on a set A in a metric space can be extended to $\bar{A}$, the closure of A, in such a way that f remains uniformly continuous. Moreover, the extension is unique.

Lemma 15.1. *Suppose that A is a subset of a metric space S.*

(a) *Then any point $p \in \bar{A} - A$ is a limit point of A.*
(b) *If $p \in \bar{A}$, then there exists a sequence $\{p_n\} \subset A$ such that $p_n \to p$ as $n \to \infty$.*
(c) *Suppose that $f: A \to S_2$ is uniformly continuous with S_2 a metric space. If $\{p_n\}$ is a Cauchy sequence in A, then $\{f(p_n)\}$ is a Cauchy sequence in S_2.*

PROOF

(a) Let A' be the set of limit points of A. Then by definition, $\bar{A} - A = \{p: p \in (A \cup A') - A\}$, and (a) is proved.

(b) If $p \in \bar{A} - A$, then p is a limit point of A and the result follows from Theorem 6.4. If $p \in A$, we choose $p_n = p$ for all n.

(c) Let $\varepsilon > 0$ be given. From the uniform continuity, there is a $\delta > 0$ such that $d_2(f(p'), f(p'')) < \varepsilon$ whenever $d_1(p', p'') < \delta$. Since $\{p_n\}$ is a Cauchy sequence, there is an integer N such that $d_1(p_n, p_m) < \delta$ for $m, n > N$. Hence $d_2(f(p_n), f(p_m)) < \varepsilon$ for $n, m > N$. Thus $\{f(p_n)\}$ is a Cauchy sequence.   □

Theorem 15.4. *Let S_1, S_2 be metric spaces and suppose that S_2 is complete. Let A be a subset of S_1 and $f: A \to S_2$ be a mapping which is uniformly continuous on A. Then there is a unique mapping $f^*: \bar{A} \to S_2$ which is uniformly continuous on $\bar{A}$ and such that $f^*(p) = f(p)$ for all $p \in A$.*

PROOF. For $p \in A$, we define $f^*(p) = f(p)$. If $p \in \bar{A} - A$, let $\{p_n\}$ be a *particular sequence* of elements of A such that $p_n \to p$ as $n \to \infty$. Such a sequence exists according to Lemma 15.1. Since $\{p_n\}$ is a Cauchy sequence, it follows from Lemma 15.1, Part (c), that $\{f(p_n)\}$ is a Cauchy sequence. Since S_2 is complete, we define $f^*(p)$ to be the limit approached by the Cauchy sequence $\{f(p_n)\}$. We show f^* is uniformly continuous. Let $\varepsilon > 0$ be given. Since f is uniformly continuous on A, there is a $\delta > 0$ such that if $p', p'' \in A$, then

$$d_2(f^*(p'), f^*(p'')) < \varepsilon/3 \quad \text{whenever} \quad d_1(p', p'') < 2\delta.$$

Now let $q', q'' \in \bar{A}$ with $d_1(q', q'') < \delta$. Let $\{p_n'\}, \{p_n''\}$ be the particular Cauchy sequences described above such that $p_n' \to q', p_n'' \to q''$ as $n \to \infty$ and $f^*(p_n') \to f^*(q'), f^*(p_n'') \to f^*(q'')$. Then there is an integer $N > 0$ such that for $n > N$, we have

$$d_1(p_n', p_n'') \leqslant d_1(p_n', q') + d_1(q', q'') + d_1(q'', p_n'') < 2\delta,$$

and

$$d_2[f^*(p_n'), f^*(p_n'')] < \varepsilon/3.$$

Therefore, for n sufficiently large, it follows that

$$d_2(f^*(q'), f^*(q'')) < d_2(f^*(q'), f^*(p_n')) + d_2(f^*(p_n'), f^*(p_n''))$$
$$+ d_2(f^*(p_n''), f^*(q'')) < \varepsilon.$$

Since q' and q'' are arbitrary elements of $\bar{A}$ which are closer together than δ, the uniform continuity is established.

To show that f^* is unique, let f^{**} be uniformly continuous on $\bar{A}$ with the property that $f^{**}(p) = f(p)$ for $p \in A$. Let $p \in \bar{A} - A$ and suppose that $\{p_n\}$ is *any* sequence in A such that $p_n \to p$ as $n \to \infty$. Then

$$f^{**}(p) = \lim_{n \to \infty} f^{**}(p_n) = \lim_{n \to \infty} f(p_n) = \lim_{n \to \infty} f^*(p_n) = f^*(p). \qquad \square$$

Among the mappings of one metric space into another, those which leave distances unchanged are of particular importance. Let $f: S_1 \to S_2$ be a mapping from the metric space S_1 into the metric space S_2. Then f is an **isometry** if and only if f is one-to-one and $d_2(f(p), f(q)) = d_1(p, q)$ for all $p, q \in S_1$. Translations and rotations in Euclidean spaces are the simplest examples of isometries. More specifically, if $f(x) = x + c$ with $S_1 = S_2 = \mathbb{R}^1$ and c a constant, then f is clearly an isometry. Also, with $S_1 = S_2 = \mathbb{R}^2$, the rotation given by

$$y_1 = x_1 \cos \theta + x_2 \sin \theta, \qquad y_2 = -x_1 \sin \theta + x_2 \cos \theta,$$

for any fixed θ, is an isometry. Geometrically, any mapping which is a "rigid motion" is an isometry.

Theorem 15.5. *Let S_1, S_2 be metric spaces and suppose that S_1 is complete. Let $f: S_1 \to S_2$ be an isometry. Then $f(S_1)$ considered as a metric space is complete.*

PROOF. Suppose $\{f(p_n)\}$ is a Cauchy sequence in S_2. Then because of the isometry, $\{p_n\}$ is a Cauchy sequence in S_1. Let p be the limit of $\{p_n\}$. Since f is defined on all of S_1, there is a point $q \in S_2$ such that $f(p) = q$. Then $f(p_n) \to q$ as $n \to \infty$ and so every Cauchy sequence in $f(S_1)$ has a limit in $f(S_1)$. That is, $f(S_1)$ is complete. $\qquad\square$

For many purposes it is desirable to deal with complete metric spaces. In the next theorems we show that *any* metric space S can be embedded in a complete metric space $\bar{S}$ (Theorem 15.7). That is, we can adjoin certain points to S so that every Cauchy sequence in the new space will have a limit. Also the distance functions in $\bar{S}$ will coincide with that of S for all those points of $\bar{S}$ which are in S.

Definition. Let S be a metric space with distance function d. Two Cauchy sequences $\{p_n\}$ and $\{q_n\}$ are **equivalent** if and only if $d(p_n, q_n) \to 0$ as $n \to \infty$. We write $\{p_n\} \approx \{q_n\}$ for such equivalent sequences.

The next theorem consists of elementary statements about equivalent Cauchy sequences.

Theorem 15.6. *Let $\{p_n\}$ and $\{q_n\}$ be Cauchy sequences in a metric space S.*

(a) *The numbers $d(p_n, q_n)$ tend to a limit as $n \to \infty$.*
(b) *If $\{p_n\} \approx \{q_n\}$ and $p_n \to p$ as $n \to \infty$, then $q_n \to p$ as $n \to \infty$.*
(c) *If $\{p_n'\}$, $\{q_n'\}$ are Cauchy sequences and $\{p_n\} \approx \{p_n'\}$, $\{q_n\} \approx \{q_n'\}$, then $\lim_{n \to \infty} d(p_n, q_n) = \lim_{n \to \infty} d(p_n', q_n')$.*

PROOF OF (a). From the triangle inequality, we have

$$d(p_n, q_n) \leqslant d(p_n, p_m) + d(p_m, q_m) + d(q_m, q_n)$$

and, by interchanging m and n, we find

$$|d(p_n, q_n) - d(p_m, q_m)| \leqslant d(p_n, p_m) + d(q_m, q_n).$$

That is, $\{d(p_n, q_n)\}$ is a Cauchy sequence of real numbers. Therefore, it has a limit. $\qquad\square$

We leave the proofs of (b) and (c) to the reader.

Definition. Let S be a metric space and suppose that $\{p_n\}$ is a Cauchy sequence which does not converge to an element in S. The class of all Cauchy sequences in S which are equivalent to $\{p_n\}$ is called an **ideal element** of S.

For example, if Q is the space of all rational numbers in $\mathbb{R}^1$ with the usual distance function, then all Cauchy sequences which tend to a specific irrational number would form an ideal element of Q. Actually, we identify the particular irrational number with this ideal class and, in fact, we may *define* an irrational number in this way. Let S' denote the set of all ideal elements of a metric space

S. We define the space $\bar{S}$ as the union of S and S', and we shall show that $\bar{S}$ is a complete metric space. For this purpose the distance function $\bar{d}$ on $\bar{S}$ is defined in the following way. For points p and q in S' there are Cauchy sequences $\{p_n\}$, $\{q_n\}$ in S which are in the equivalence classes determining p and q, respectively. These choices may be made arbitrarily in the definition of $\bar{d}$. Let

$$
\bar{d}(p, q) = \begin{cases}
d(p, q) & \text{if } p, q \in S, \\
\lim_{n \to \infty} d(p, q_n) & \text{if } p \in S, q \in S', \\
\lim_{n \to \infty} d(p_n, q) & \text{if } p \in S', q \in S, \\
\lim_{n \to \infty} d(p_n, q_n) & \text{if } p, q \in S'.
\end{cases}
$$

It is a simple matter to verify that the space $\bar{S}$ with the distance function $\bar{d}$ is a metric space. We leave this verification to the reader. Also, we see at once that the mapping of S into $\bar{S}$ which associates to each point of S the same point, considered as a point of $\bar{S}$, is an isometry. Therefore an equivalence class of Cauchy sequences in S which converges to a point in S may be considered as an ideal element in $\bar{S}$ without loss of consistency.

Theorem 15.7. *The space $\bar{S}$ with distance $\bar{d}$ is a complete metric space.*

PROOF. Let $\{p_n\}$ be a Cauchy sequence in $\bar{S}$. We wish to show that there is a point $\bar{p} \in \bar{S}$ such that $\bar{d}(\bar{p}, p_n) \to 0$ as $n \to \infty$. For each p_n, there is, by definition, a Cauchy sequence in S, denoted $\{q_{kn}\}$, such that $\bar{d}(p_n, q_{kn}) \to 0$ as $k \to \infty$. Let $\varepsilon_n > 0$ be a sequence tending to 0 as $n \to \infty$. From the double sequence $\{q_{kn}\}$ we can extract a subsequence q'_n such that $\bar{d}(p_n, q'_n) < \varepsilon_n$ for all n. From the triangle inequality, it follows that

$$
\bar{d}(q'_n, q'_m) \leqslant \bar{d}(q'_n, p_n) + \bar{d}(p_n, p_m) + \bar{d}(p_m, q'_m). \tag{15.2}
$$

Since $\{p_n\}$ is a Cauchy sequence, given $\varepsilon > 0$, there is an $N > 0$ such that $\bar{d}(p_n, p_m) < \varepsilon$ for $m, n > N$. We choose m and n so large that $\varepsilon_n, \varepsilon_m < \varepsilon$. Thus (15.2) shows that $\{q'_n\}$ is a Cauchy sequence in S. Let $\bar{p}$ be the corresponding ideal element in $\bar{S}$. Since

$$
\bar{d}(\bar{p}, p_n) \leqslant \bar{d}(\bar{p}, q'_n) + \bar{d}(q'_n, p_n) < 2\varepsilon \quad \text{for} \quad n > N,
$$

we conclude that $p_n \to \bar{p}$ as $n \to \infty$. That is, $\bar{S}$ is complete. □

PROBLEMS

1. Let S be a metric space and $C(S)$ the collection of all bounded, continuous functions on S. Show that

$$
d(f, g) = \sup_{x \in S} |f(x) - g(x)|
$$

is a metric. Let $M(S)$ be the totality of bounded functions on S. Is $M(S)$ with the above metric a metric space?

2. Let $x = (a_1, a_2, \ldots, a_n, \ldots)$ and $y = (b_1, b_2, \ldots, b_n, \ldots)$ where $\{a_n\}$ and $\{b_n\}$ are bounded sequences of real numbers. Define

$$d(x, y) = \sup_n |a_n - b_n|.$$

Show that the space S consisting of the totality of such bounded sequences with the distance defined above is a metric space. Is the metric space complete?

3. Let A be the set of irrational numbers on the interval $I = \{x: 0 \leqslant x \leqslant 1\}$. Show that A is of the second category with respect to $\mathbb{R}^1$.

4. Let $P(x) = a_n x^n + a_{n-1} x^{n-1} + \cdots + a_1 x + a_0$ be a polynomial with integer coefficients. Let A be the set of all roots of all such polynomials for $n = 1, 2, \ldots$ and with all possible integer coefficients. Show that A is of category 1 in the complex plane. (The metric in $\mathbb{C}$, the complex plane, is $d(z_1, z_2) = |z_1 - z_2| = [(x_1 - x_2)^2 + (y_1 - y_2)^2]^{1/2}$ where $z_1 = x_1 + iy_1, z_2 = x_2 + iy_2$.)

5. Let $I = \{x: 0 \leqslant x \leqslant 1\}$ be the unit interval in $\mathbb{R}^1$ and let every $x \in I$ be represented in a ternary expansion:

$$x = 0 \cdot a_1 a_2 a_3, \ldots,$$

where each a_i has the value 0, 1, or 2. Let A be the subset of I such that every point of A has only zeros or twos in its expansion.
(a) Show that A is an uncountable set.
(b) Show that A is nowhere dense in I.

6. Write a detailed proof of the Baire category theorem (Theorem 15.3).

7. Let $I = \{x: 0 \leqslant x \leqslant 1\}$ be the unit interval with the metric of $\mathbb{R}^1$. Find a set A in I and a continuous mapping $f: A \to \mathbb{R}^1$ such that the mapping cannot be extended to be continuous from $\bar{A}$ into $\mathbb{R}^1$. That is, show that in Theorem 15.4 the hypothesis that f is uniformly continuous cannot be dropped.

8. Let S_1, S_2 be metric spaces and suppose that $f: S_1 \to S_2$ is an isometry. Suppose that S_2 is complete. Is the space $f(S_1)$ complete? Justify your statement.

9. Prove Parts (b) and (c) of Theorem 15.6.

10. Verify that the function $\bar{d}$ defined before Theorem 15.7 is a metric and hence that $\bar{S}$ is a metric space.

11. Prove that the mapping of $\mathbb{R}^2 \to \mathbb{R}^2$ given by

$$y_1 = x_1 \cos \theta + x_2 \sin \theta,$$

$$y_2 = -x_1 \sin \theta + x_2 \cos \theta,$$

is an isometry.

15.2. Convex Sets and Convex Functions

The geometric developments in this section will be established for sets in an N-dimensional space. Unless otherwise noted, we shall suppose throughout

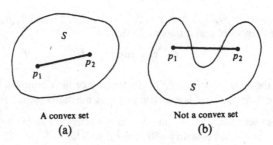

A convex set   Not a convex set
(a)   (b)

Figure 15.1

that all sets are contained in $\mathbb{R}^N$. However, many of the results given here are valid if $\mathbb{R}^N$ is replaced by an arbitrary linear space.

Definition. A set S in $\mathbb{R}^N$ is **convex** if and only if every point on the straight line segment $\overline{p_1 p_2}$ is in S whenever p_1 and p_2 are in S (see Figure 15.1).

Theorem 15.8. *The intersection of any family of convex sets is a convex set.*

We leave to the reader the proof of this simple fact.

Theorem 15.9. *If S is a convex set, then $\bar{S}$ is convex.*

PROOF. Let $p, q \in \bar{S}$. Then there is a sequence $\{p_n\}$ such that $p_n \in S$ and $p_n \to p$. Similarly, there is a sequence $\{q_n\}$ in S with $q_n \to q$. Since S is convex, all the segments $\overline{p_n q_n}$ are in S. Any point r_n on the line segment $\overline{p_n q_n}$ has the form

$$r_n = \lambda p_n + (1 - \lambda)q_n, \qquad 0 \leqslant \lambda \leqslant 1.$$

That is, if p_n, q_n have coordinates $x_i^n, y_i^n, i = 1, 2, \ldots, N$, respectively, then r_n has coordinates $\lambda x_i^n + (1 - \lambda)y_i^n, i = 1, 2, \ldots, N$. As $n \to \infty$, the sequence $\{r_n\}$ tends to $\lambda p + (1 - \lambda)q$ which is a point on $\overline{pq}$. Consequently, $\overline{pq}$ must be in $\bar{S}$. □

Theorem 15.10. *Suppose that S is a convex set and $p_1, p_2, \ldots, p_k \in S$. Let $\lambda_1, \lambda_2, \ldots, \lambda_k$ be nonnegative real numbers with $\lambda_1 + \lambda_2 + \cdots + \lambda_k = 1$. Then the point $\lambda_1 p_1 + \lambda_2 p_2 + \cdots + \lambda_k p_k$ is in S.*

PROOF. For two points, p_1 and p_2, the theorem is merely the definition of convexity. We proceed by induction. Assume the result holds for $k - 1$ points. We wish to show that it holds for k points. Hence for $\mu_i \geqslant 0, i = 1, 2, \ldots, k - 1$ and $\mu_1 + \cdots + \mu_{k-1} = 1$, we have $\sum_{i=1}^{k-1} \mu_i p_i \in S$. Therefore, by definition,

$$(1 - \lambda_k)(\mu_1 p_1 + \cdots + \mu_{k-1} p_{k-1}) + \lambda_k p_k \in S.$$

Choosing $\mu_i = \lambda_i/(1 - \lambda_k), i = 1, 2, \ldots, k - 1$, we obtain the result for k points. □

Definitions. Let a point x in $\mathbb{R}^N$ have coordinates $x_1, x_2, \ldots, x_N$. Then a **hyperplane** in $\mathbb{R}^N$ is the graph of an equation of the form

$$\sum_{i=1}^{N} a_i x_i + b = 0$$

where the a_i, b are real numbers and not all the a_i are zero. By dividing the above equation by $(\sum_{i=1}^{N} a_i^2)^{1/2}$ we may assume without loss of generality that $a_1^2 + \cdots + a_N^2 = 1$. Note that a hyperplane is a line if $N = 2$ and an ordinary plane if $N = 3$. A **line** l in $\mathbb{R}^N$ is the graph of the system of N equations

$$l = \{x: x_i = x_i^0 + \lambda_i t, i = 1, 2, \ldots, N, -\infty < t < \infty\} \tag{15.3}$$

in which $x^0 = (x_1^0, \ldots, x_N^0)$ is a fixed point and the quantities $\lambda_1, \ldots, \lambda_N$ (not all zero) are called the **direction numbers** of l. The real parameter t varies throughout $\mathbb{R}^1$. If we replace t by $t/(\lambda_1^2 + \cdots + \lambda_N^2)^{1/2}$ we may, without loss of generality assume that $\lambda_1^2 + \cdots + \lambda_N^2 = 1$, in which case we call the λ_i the **direction cosines** of l. The equation of the *hyperplane perpendicular to the line* l given by (15.3) and passing through the point $x^1 = (x_1^1, x_2^1, \ldots, x_N^1)$ is

$$\sum_{i=1}^{N} \lambda_i (x_i - x_i^1) = 0. \tag{15.4}$$

Let A be any set in a metric space S. Then we recall that p is an **interior point** *of* A if there is a ball B in S with p as center which is contained in A.

Theorem 15.11. *Let S be a convex set in $\mathbb{R}^N$ which has no interior points. Then there is a hyperplane H which contains S.*

PROOF. Suppose that S does not lie in a hyperplane. Then there are $N + 1$ points of S which are contained in no hyperplane. Let these points be denoted $O, x^1, x^2, \ldots, x^N$ where O is the origin of the coordinate system. Then the vectors from O to $x^i, i = 1, 2, \ldots, N$ are linearly independent. Since S is convex, every point $\lambda_1 x^1 + \cdots + \lambda_N x^N$ is in S where $0 \leqslant \lambda_i \leqslant 1$ and $\lambda_1 + \lambda_2 + \cdots + \lambda_N \leqslant 1$. Choose

$$\bar{x} = \bar{\lambda}_1 x^1 + \bar{\lambda}_2 x^2 + \cdots + \bar{\lambda}_N x^N. \tag{15.5}$$

with $0 < \bar{\lambda}_i < 1$ for every i and $\bar{\lambda}_1 + \bar{\lambda}_2 + \cdots + \bar{\lambda}_N < 1$. Then $\bar{x}$ is in S and every point near $\bar{x}$ has the same representation with λ_i near $\bar{\lambda}_i$ for every i. That is, S contains a ball with $\bar{x}$ as center and sufficiently small radius. Thus $\bar{x}$ is an interior point. We conclude that if S has no interior points, it must lie in a hyperplane. $\square$

Theorem 15.12. *Let S be a convex set in $\mathbb{R}^N$ which is not the entire space. Let $S^{(0)}$ denote the set of interior points of S. Then for each point $x^0 \in \bar{S} - S^{(0)}$ there is a hyperplane H through x_0 such that S is in one of the two half-spaces bounded by H (see Figure 15.2).*

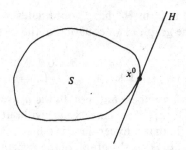

Figure 15.2. S is on one side of the hyperplane H.

PROOF. Consider $x^0 \in \bar{S} - S^{(0)}$. Since S is not the entire space, there is a sequence of points $\{x^m\}$ such that $x^m \in \mathbb{R}^N - \bar{S}$ and $x^m \to x^0$ as $m \to \infty$. Since $\bar{S}$ is a closed set, for each x^m there is a point $y^m \in \bar{S}$ such that x^m is at least as close to y^m as it is to any other point of $\bar{S}$. In fact, $y^m = \inf_{y \in S} d(y, x^m)$, where d is distance in $\mathbb{R}^N$ (see Figure 15.3). Denote by H_m the hyperplane perpendicular to the line segment $\overline{x^m y^m}$ and which passes through x^m. If $\lambda_{m1}, \lambda_{m2}, \ldots, \lambda_{mN}$ are the direction cosines of the line through x^m and y^m, then the equation of H_m is given by

$$H_m = \left\{ x: \sum_{i=1}^{N} \lambda_{mi}(x_i - x_i^m) = 0 \right\}, \tag{15.6}$$

where x^m has coordinates $x_1^m, x_2^m, \ldots, x_N^m$. Suppose there is a point $z^m \in S$ on the opposite side of H_m from y^m. By convexity, the entire segment $\overline{y^m z^m}$ is in S and in such a case y^m is not the closest point of $\bar{S}$ to x^m. Hence there is no such point z^m, and $\bar{S}$ is contained in the half space

$$P_m = \left\{ x: \sum_{i=1}^{N} \lambda_{mi}(x_i - x_i^m) \geq 0 \right\}. \tag{15.7}$$

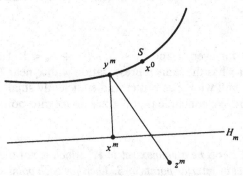

Figure 15.3. H_m is perpendicular to the line segment $\overline{x^m y^m}$.

Since $x^m \to x^0$, there is a subsequence $x^{m_k} \to x^0$ such that $\lambda_{m_k i} \to \lambda_{0i}$, $i = 1, 2,$ $\ldots, N$, as $m_k \to \infty$. Of course $\lambda_{01}^2 + \cdots + \lambda_{0N}^2 = 1$, and we form the hyperplane through x^0:

$$H = \left\{ x \colon \sum_{i=1}^{N} \lambda_{0i}(x_i - x_i^0) = 0 \right\}.$$

The hyperplanes H_{k_m} tend to H and since $\bar{S} \subset P_{k_m}$, we have in the limit

$$\bar{S} \subset P_0 = \left\{ x \colon \sum_{i=1}^{N} \lambda_{0i}(x_i - x_i^0) \geqslant 0 \right\}. \qquad \square$$

We now study functions from $\mathbb{R}^N \to \mathbb{R}^1$ and, in particular, those functions which have certain convexity properties. We shall show (Theorem 15.17 below) that a convex function is necessarily continuous at interior points of its domain.

Definition. Let S be a convex set in $\mathbb{R}^N$ and let $f \colon S \to \mathbb{R}^1$ be a real-valued function. We say that f *is a* **convex function on** S if and only if

$$f[\lambda x^1 + (1 - \lambda)x^2] \leqslant \lambda f(x^1) + (1 - \lambda)f(x^2)$$

for all $x^1, x^2 \in S$ and for all λ such that $0 \leqslant \lambda \leqslant 1$. Note that *convex functions are not defined if the domain is not a convex set.*

Theorem 15.13. *Let S be a convex set in $\mathbb{R}^N$ and suppose $f \colon S \to \mathbb{R}^1$ is convex. Then the set $G = \{(x, y) \colon x \in S, y \in \mathbb{R}^1, y \geqslant f(x)\}$ is a convex set in $\mathbb{R}^{N+1}$. Conversely, if S and G are convex, then f is a convex function.*

The proof is a direct consequence of the definition of convex set and convex function, and we leave the details to the reader (see Figure 15.4). The set G is usually called the **epigraph** of the function f.

The next three theorems describe simple properties of convex functions, ones which are useful not only in the proof of the continuity properties of convex functions but also in the applications given in Chapter 16.

Theorem 15.14. *Let $\mathscr{F} = \{f\}$ be a family of convex real-valued functions defined on a convex set S in $\mathbb{R}^N$. Suppose there is a function $g \colon S \to \mathbb{R}^1$ such that $f(x) \leqslant g(x)$ for all $x \in S$ and all $f \in \mathscr{F}$. Define*

$$F(x) = \sup_{f \in \mathscr{F}} f(x).$$

Then F is a convex function on S.

PROOF. Let $\varepsilon > 0$ be given and suppose that x^1, x^2 are any points of S. From the definition of supremum there is a function $f \in \mathscr{F}$ such that

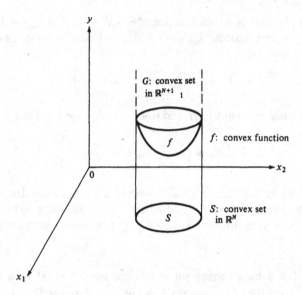

Figure 15.4

$$F[\lambda x^1 + (1 - \lambda)x^2] < f[\lambda x^1 + (1 - \lambda)x^2] + \varepsilon. \text{ Since } f \text{ is convex, we have}$$

$$F[\lambda x^1 + (1 - \lambda)x^2] < f[\lambda x^1 + (1 - \lambda)x^2] + \varepsilon$$

$$\leqslant \lambda f(x^1) + (1 - \lambda)f(x^2) + \varepsilon$$

$$\leqslant \lambda F(x^1) + (1 - \lambda)F(x^2) + \varepsilon, \qquad 0 \leqslant \lambda \leqslant 1.$$

The result follows because ε is arbitrary.   □

Theorem 15.15. *Let S be a convex set in $\mathbb{R}^N$. Suppose that $x^1, x^2, \ldots, x^m$ are points of S and $\lambda_1, \lambda_2, \ldots, \lambda_m$ are nonnegative numbers with $\lambda_1 + \lambda_2 + \cdots + \lambda_m = 1$. If $f: S \to \mathbb{R}^1$ is convex on S, then*

$$f(\lambda_1 x^1 + \cdots + \lambda_m x^m) \leqslant \lambda_1 f(x^1) + \cdots + \lambda_m f(x^m).$$

PROOF. We proceed by induction (see the proof of Theorem 15.10). For $m = 2$, the result is the statement of convexity of f. Suppose the result holds for $k = m - 1$. We show that it holds for $k = m$. By the induction hypothesis it follows that

$$f(\mu_1 x^1 + \cdots + \mu_{m-1} x^{m-1}) \leqslant \mu_1 f(x^1) + \cdots + \mu_{m-1} f(x^{m-1})$$

where $\mu_i \geqslant 0$, $\mu_1 + \cdots + \mu_{m-1} = 1$. Choose $\mu_i = \lambda_i/(1 - \lambda_m)$, $i = 1, 2, \ldots, m - 1$. Then

$$f(\lambda_1 x^1 + \cdots + \lambda_m x^m)$$

$$= f[(1 - \lambda_m)(\mu_1 x^1 + \cdots + \mu_{m-1} x^{m-1}) + \lambda_m x^m]$$

$$\leqslant (1 - \lambda_m) f(\mu_1 x^1 + \cdots + \mu_{m-1} x^{m-1}) + \lambda_m f(x^m)$$
$$\leqslant (1 - \lambda_m)[\mu_1 f(x^1) + \cdots + \mu_{m-1} f(x^{m-1})] + \lambda_m f(x^m)$$
$$\leqslant \lambda_1 f(x^1) + \cdots + \lambda_{m-1} f(x^{m-1}) + \lambda_m f(x^m),$$

and the induction is established. $\qquad\square$

In some sense linear functions are the simplest convex functions. The next theorem establishes this fact.

Theorem 15.16. *Let S be a convex set in $\mathbb{R}^N$. Let $A = (a_{ij})$ be an $N \times N$ matrix and $b = (b_1, b_2, \ldots, b_N)$ a vector. Define the linear transformation $T \colon \mathbb{R}^N \to \mathbb{R}^N$ by $y = Ax + b$. Then*

(i) *The set $S' = T(S)$ is a convex set in $\mathbb{R}^N$.*
(ii) *If $f \colon S \to \mathbb{R}^1$ is a convex function and if A is nonsingular, then the function $g \colon S' \to \mathbb{R}^1$ defined by $g(y) = f(x) = f[A^{-1}(y - b)]$ is a convex function.*
(iii) *Any linear function $y = A_1 x + b$ is convex on any convex domain.*

PROOF

(i) Let $x^1, x^2 \in S$. Then $\lambda x^1 + (1 - \lambda) x^2 \in S$ for $0 \leqslant \lambda \leqslant 1$. Because T is linear, we have

$$T(\lambda x^1 + (1 - \lambda) x^2) = \lambda T(x^1) + (1 - \lambda) T(x^2).$$

These equalities also establish (iii).
(ii) Let $y^1, y^2 \in S'$. Then, by definition,

$$g(\lambda y^1 + (1 - \lambda) y^2) = f[\lambda A^{-1}(y^1 - b) + (1 - \lambda) A^{-1}(y^2 - b)].$$

Since f is convex, we find

$$g(\lambda y^1 + (1 - \lambda) y^2) \leqslant \lambda f[A^{-1}(y^1 - b)] + (1 - \lambda) f[A^{-1}(y^2 - b)]$$
$$= \lambda g(y^1) + (1 - \lambda) g(y^2). \qquad\square$$

Lemma 15.2. *Let $I = \{x \colon a \leqslant x \leqslant b\}$ be any interval and $f \colon I \to \mathbb{R}^1$ a convex function on I. Let x^0 be given with $a < x^0 < b$ and define*

$$m = \frac{f(x^0) - f(a)}{x^0 - a}, \qquad l(x) = f(a) + m(x - a), \qquad x \in I.$$

Then $f(x) \geqslant l(x)$ for $x^0 \leqslant x \leqslant b$.

PROOF. The result is obvious geometrically. The linear function $l(x)$ is below f for $x^0 \leqslant x \leqslant b$ (Figure 15.5). For the actual proof, observe first that $f(x^0) = l(x^0)$. Let x be such that $x^0 \leqslant x \leqslant b$. Then from the convexity of f, we find

$$f(x^0) \leqslant (1 - \lambda) f(a) + \lambda f(x).$$

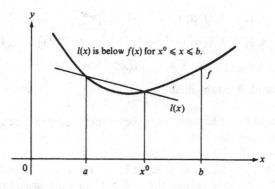

Figure 15.5

Choose $\lambda = (x^0 - a)/(x - a)$. Then $1 - \lambda = (x - x^0)/(x - a)$ and

$$f(x) \geqslant \frac{1}{\lambda}[f(x^0) - (1 - \lambda)f(a)]$$

$$= f(a) + \frac{f(x^0) - f(a)}{x^0 - a}(x - a) = l(x). \qquad \square$$

The next technical lemma is essential for the proof that a convex function is smooth at interior points of its domain.

Lemma 15.3. *Let $I = \{x: x^1 \leqslant x \leqslant x^4\}$ be an interval and suppose that $f: I \to \mathbb{R}^1$ is a convex function on I. Let x^2, x^3 be such that $x^1 < x^2 < x^3 < x^4$ and define*

$$m_{ij} = \frac{f(x^i) - f(x^j)}{x^i - x^j}, \qquad i, j = 1, 2, 3, 4, \quad i \neq j.$$

Suppose that $|f(x)| \leqslant M$ on I and let $\delta > 0$ be given such that $x^1 + \delta \leqslant x^2 < x^3 \leqslant x^4 - \delta$. Then

(i) $m_{12} \leqslant m_{23} \leqslant m_{34}$. $\hspace{4cm}$ (15.8)

(ii) $|f(x^3) - f(x^2)| \leqslant \dfrac{2M}{\delta}|x^3 - x^2|$. $\hspace{3cm}$ (15.9)

PROOF. The geometric interpretation of (i) is shown in Figure 15.6, where m_{ij} is the slope of the line through $(x^i, f(x^i))$ and $(x^j, f(x^j))$. Because of the convexity of f, the line segment through x^2 and x^3 lies above the graph of f between x^2 and x^3. That is,

$$f(x^2) + m_{23}(x - x^2) \geqslant f(x) \quad \text{for } x^2 \leqslant x \leqslant x^3. \qquad (15.10)$$

Also, by Lemma 15.2 with $x^1 = a$, $x^2 = x^0$, we have

$$f(x) \geqslant f(x^1) + m_{12}(x - x^1) \quad \text{for } x \geqslant x^2. \qquad (15.11)$$

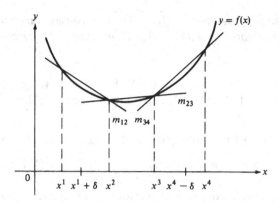

Figure 15.6

Combining (15.10) and (15.11) we find

$$f(x^2) + m_{23}(x - x^2) \geqslant f(x^1) + m_{12}(x - x^1), \qquad x^2 \leqslant x \leqslant x^3,$$

or

$$m_{23}(x - x^2) \geqslant m_{12}(x - x^1) - m_{12}(x^2 - x^1) = m_{12}(x - x^2).$$

That is, $m_{23} \geqslant m_{12}$. The proof that $m_{34} \geqslant m_{23}$ is the same. To establish (ii), observe that

$$m_{34} \leqslant \frac{|f(x^4) - f(x^3)|}{|x^4 - x^3|} \leqslant \frac{|f(x^4)| + |f(x^3)|}{|x^4 - x^3|} \leqslant \frac{2M}{\delta}.$$

Similarly, $m_{12} \geqslant -2M/\delta$. Hence we find

$$-\frac{2M}{\delta} < m_{12} \leqslant m_{23} \leqslant m_{34} \leqslant \frac{2M}{\delta} \qquad \text{and} \qquad |m_{23}| \leqslant \frac{2M}{\delta},$$

or

$$|f(x^3) - f(x^2)| \leqslant \frac{2M}{\delta}|x^3 - x^2|. \qquad \square$$

Remark. Part (ii) of Lemma 15.3 shows that a convex function of one variable is continuous at every interior point of its interval of definition. Moreover, it satisfies a Lipschitz condition on every closed subinterval of the interior of its domain. In Theorem 15.17 we extend this result to functions defined on convex sets in $\mathbb{R}^N$.

Definitions. We denote by $Q_N = \{x: -1 \leqslant x_i \leqslant 1, i = 1, 2, \dots, N\}$ a **hypercube** of side length 2 in $\mathbb{R}^N$. A point ξ in $\mathbb{R}^N$ with coordinates $\xi_1, \xi_2, \dots, \xi_N$ is called a **lattice point** of Q_N if every ξ_i has one of the values 0, 1, or -1. Thus the origin is a lattice point as are, for example, the points $(1, 0, 0, \dots, 0)$ and $(0, -1, 0, 1, 0, 0, \dots, 0)$. A hypercube is clearly a convex set.

Lemma 15.4. *Suppose that* $f: Q_N \to \mathbb{R}^1$ *is a convex function. Let* $M_0 = \max |f(\xi)|$ *for all lattice points* $\xi \in Q_N$. *Then* $|f(x)| \leqslant M$ *for all* $x \in Q_N$ *where* $M = 3^N M_0$.

PROOF. We proceed by induction on N. For $N = 1$, we note that Q_1 is the interval $I = \{x: -1 \leqslant x \leqslant 1\}$. We first apply Lemma 15.2 with $a = -1$ and $x^0 = 0$. Then

$$f(-1) + [f(0) - f(-1)](x + 1) \leqslant f(x) \qquad \text{for} \quad 0 \leqslant x \leqslant 1.$$

Also, since f is convex we have $f(x) \leqslant (1 - x)f(0) + xf(1)$ for $0 \leqslant x \leqslant 1$. Set $M_0 = \max[|f(-1)|, |f(0)|, |f(1)|]$. Then for $0 \leqslant x \leqslant 1$, it follows that $f(x)$ is between the maximum of $f(0)$ and $f(1)$ and the minimum of $f(0)$ and $2f(0) - f(-1)$. That is, $|f(x)| \leqslant 3M_0$. The function $g(y) = f(-x)$ yields to the same treatment for $0 \leqslant y \leqslant 1$, and we conclude that $|f(x)| \leqslant 3M_0$ for $-1 \leqslant x \leqslant 1$.

Suppose now that the theorem is true for $N = k$; we wish to establish the result for $k + 1$. Let f be convex on Q_{k+1}. Define $\varphi_i(x_1, x_2, \ldots, x_k) = f(x_1, x_2, \ldots, x_k, i)$ for $i = -1, 0, 1$. Since φ_i is convex on Q_k it follows from the induction hypothesis that

$$|f(x_1, x_2, \ldots, x_k, i)| \leqslant 3^k \max_{\xi} |f(\xi_1, \xi_2, \ldots, \xi_k, i)|, \qquad (15.12)$$

where the maximum is taken over all lattice points $\xi \in Q_k$. Now the function $f(x_1, x_2, \ldots, x_{k+1})$ is a convex function of x_{k+1} for each *fixed* $x_1, x_2, \ldots, x_k \in Q_k$. We use the result for $N = 1$ and Inequality (15.12) to obtain

$$|f(x_1, x_2, \ldots, x_{k+1})| \leqslant 3^k \cdot 3M_0. \qquad \square$$

The next theorem establishes the fact that convex functions are continuous at interior points of their domain of definition. Inequality (15.13) shows that such functions satisfy a Lipschitz condition.

Theorem 15.17. *Let* S *be a convex set in* $\mathbb{R}^N$ *and suppose that* $f: S \to \mathbb{R}^1$ *is a convex function. Then* f *is continuous on* $S^{(0)}$, *the interior of* S. *Furthermore, if* $|f(x)| \leqslant M$ *on* S, *then*

$$|f(x^1) - f(x^2)| \leqslant \frac{2M}{\delta} |x^1 - x^2| \qquad (15.13)$$

where x^1, x^2 *are such that the balls* $B(x^1, \delta)$ *and* $B(x^2, \delta)$ *are in* S.

PROOF. Suppose that the interior of S is not empty. Let $x^0 \in S^{(0)}$. Then x^0 is the center of a hypercube contained in $S^{(0)}$ (Theorem 15.11). Consequently, according to Lemma 15.4, there is a smaller hypercube with x^0 as center on which f is bounded. We conclude that f is bounded on any compact subset K of $S^{(0)}$.

We first establish Inequality (15.13). Suppose that $|f(x)| \leqslant M$ on S and let x^1, x^2 be interior points with $B(x^1, \delta)$, $B(x^2, \delta)$ in S. Let L be the line passing through x^1 and x^2. Then the intersection $l = L \cap S$ is a line segment (convex set). With $\bar{x}$ any point on l, and using the notation $\lambda = (\lambda_1, \lambda_2, \ldots, \lambda_N)$, we may write the parametric equations of l in the form

$$x = \bar{x} + \lambda t, \qquad \lambda_1^2 + \lambda_2^2 + \cdots + \lambda_N^2 = 1.$$

If t_1 corresponds to x^1 and t_2 to x^2, then all t such that $t_1 - \delta < t < t_2 + \delta$ correspond to points of l, that is, to interior points of S. We define

$$\varphi(t) = f(\bar{x} + \lambda t), \qquad t_1 - \delta < t < t_2 + \delta.$$

Then $|\varphi(t)| \leqslant M$ on its domain. We apply Inequality (15.9) of Lemma 15.3 to obtain

$$|f(x^2) - f(x^1)| = |\varphi(t_2) - \varphi(t_1)| \leqslant \frac{2M}{\delta}|t_2 - t_1| = \frac{2M}{\delta}|x^2 - x^1|.$$

Since f is bounded on any compact subset of $S^{(0)}$, we now employ (15.13) and conclude that f is continuous at each interior point of S. In fact, f satisfies a Lipschitz condition at each interior point. $\square$

Remark. Convex functions may be discontinuous at boundary points. To see this, with $I = \{x: 0 \leqslant x \leqslant 1\}$, define $f: I \to \mathbb{R}^1$ so that $f(x) = x, 0 < x < 1$, $f(0) = 1, f(1) = 2$. Then f is convex on I and discontinuous for $x = 0, 1$.

If a convex function is differentiable at interior points, then we can obtain more specific information about hyperplanes of support of such a function. The next theorem gives the precise results.

Theorem 15.18. *Let S be a convex set in $\mathbb{R}^N$ and suppose that $f: S \to \mathbb{R}^1$ is convex. Let x^0 be an interior point of S.*

(i) *Then there are real numbers $a_1, a_2, \ldots, a_N$ such that*

$$f(x) \geqslant f(x^0) + \sum_{i=1}^{N} a_i(x_i - x_i^0), \; x \in S.$$

(ii) *If $f \in C^1$ on $S^{(0)}$, then*

$$a_i = \frac{\partial f}{\partial x_i}\bigg|_{x=x^0}.$$

(iii) *If $f \in C^2$ on $S^{(0)}$, then the convexity of f on $S^{(0)}$ is equivalent to the inequality*

$$\sum_{i,j=1}^{N} \left(\frac{\partial^2 f}{\partial x_i \partial x_j}\bigg|_{x=x^0}\right)\lambda_i \lambda_j \geqslant 0 \quad \text{for all } \lambda \in \mathbb{R}^N \text{ and all } x^0 \in S^{(0)}.$$

PROOF

(i) Let G be the set in $\mathbb{R}^{N+1}$ defined by $G = \{(x, y): x \in S, y \in \mathbb{R}^1, y \geqslant f(x)\}$. Then G is a convex set and, setting $y^0 = f(x^0)$, we see that $(x^0, y^0) \in \partial G$. Hence there is a hyperplane H_0 in $\mathbb{R}^{N+1}$ passing through (x^0, y^0) with the property that G lies entirely on one side of H_0. Since f is continuous, the point $(x^0, y^0 + 1)$ is an interior point of G. Thus H_0 is not a "vertical" hyperplane and therefore is of the form

$$y = f(x^0) + \sum_{i=1}^{N} a_i(x_i - x_i^0).$$

The region G is above H_0. $\square$

The proofs of (ii) and (iii) are obtained by expanding f in a Taylor series with remainder and using the result obtained in Part (i). The details are left to the reader.

PROBLEMS

1. Let A and B be convex, disjoint, closed, nonempty sets in $\mathbb{R}^N$. Show that $A \cup B$ can never be convex.

2. Show that every convex polygon in $\mathbb{R}^2$ is the intersection of a finite number of half-planes.

3. Show directly (without using Theorem 15.11) that if a convex set S in $\mathbb{R}^2$ has three points which do not lie on a line, then S must have interior points.

4. Let $f_1, f_2, \ldots, f_p$ be convex functions defined on a convex set S in $\mathbb{R}^N$. Let $\alpha_i \geq 0$, $i = 1, 2, \ldots, p$ be given numbers. Show that $\sum_{i=1}^{p} \alpha_i f_i$ is a convex function. Give an example to show that the conclusion may be false if any α_i is negative.

*5. Let S be a convex region in $\mathbb{R}^2$ with interior points and such that ∂S has a unique tangent line at each point. Show that S can be inscribed in a square.

6. Consider the family of conical regions in 3-space which satisfy inequalities of the form $(\alpha x^2 + \beta y^2)^{1/2} \leq z, \alpha > 0, \beta > 0$, constants. Show that the intersection of any number of such regions is a convex set.

7. Let f, g be real-valued, nonnegative convex functions defined on an interval $I \subset \mathbb{R}^1$. Find conditions on f and g such that $f \cdot g$ is convex.

8. Prove Theorem 15.13.

9. Given $f(x) = \sum_{n=0}^{\infty} a_n x^n$ is a convergent series on $I = \{x: 0 \leq x \leq 1\}$. If $a_n \geq 0$ for all n, show that f is convex on I.

10. Let S be any set in $\mathbb{R}^N$. The **convex hull** of S is the intersection of all convex sets which contain S. Show that the convex hull of S consists of all points of the form $x = \lambda_0 x_0 + \cdots + \lambda_k x_k$ where $x_i \in S$ and $\lambda_i \geq 0$ for $i = 0, 1, \ldots, k$ and $\lambda_1 + \cdots + \lambda_k = 1$.

11. Show that if a set S in $\mathbb{R}^N$ is open, its convex hull is open (see Problem 10).

12. A subset K of $\mathbb{R}^N$ is called a **cone** if, whenever $x \in K$, then $\lambda x \in K$ for every $\lambda \geq 0$. Show that a cone K is convex if and only if $x + y \in K$ whenever $x, y \in K$. Give an example of a nonconvex cone. Show that if K_1 and K_2 are cones in $\mathbb{R}^N$, then $K_1 \cap K_2$ and $K_1 \cup K_2$ are cones. If K_1 and K_2 are convex, under what conditions, does it follows that $K_1 \cap K_2$ and $K_1 \cup K_2$ are convex cones?

13. Let K be a convex cone in $\mathbb{R}^N$ which contains an entire line l through the origin. If K also contains an interior point, show that K is either all of $\mathbb{R}^N$ or a half-space. (See Problem 12.)

14. In the notation of Lemma 15.3 show that if $f(x^1) \leqslant f(x^4)$, then $m_{13} \leqslant m_{24}$.

15. Prove Parts (ii) and (iii) of Theorem 15.18.

16. Show that the intersection of any family (finite or infinite) of convex sets is a convex set (Theorem 15.8).

17. Let S be a convex set in $\mathbb{R}^N$ which contains three points P, Q, R, not on a straight line. Show that the interior of the triangle with vertices P, Q, R lies in S.

18. Let $B(p_0, r)$ be a ball in $\mathbb{R}^N$ with center at p_0 and radius r. Show that B is a convex set.

19. In $\mathbb{R}^N$, define the cylinder $C = \{(x_1, x_2, \ldots, x_N): x_1^2 + x_2^2 + \cdots + x_{N-1}^2 \leqslant 1, 0 \leqslant x_N \leqslant 1\}$. Show that C is a convex set.

20. In $\mathbb{R}^N$, define $A = \{(x_1, x_2, \ldots, x_N): x_N^2 \leqslant x_1^2 + \cdots + x_{N-1}^2, -1 \leqslant x_N \leqslant 1\}$. Show that A is not a convex set.

21. Define in $\mathbb{R}^2$ the set $B = \{(x_1, x_2): x_1^2 + x_2^2 \leqslant 1\}$. Show that the function $y = f(x) = \alpha x_1^2 + \beta x_2^2, \alpha > 0, \beta > 0$ defined in B is a convex function.

22. In $\mathbb{R}^2$ consider the square $S = \{(x_1, x_2): 0 \leqslant x_1 \leqslant 1, 0 \leqslant x_2 \leqslant 1\}$. Show that the function $y = g(x) = x_1 x_2$ is not convex in S.

15.3. Arzela's Theorem; the Tietze Extension Theorem

Let S be a metric space and $\mathscr{F}$ a family of functions from S into $\mathbb{R}^1$. In many applications we are given such a family and wish to extract from it a sequence of functions $f_n: S \to \mathbb{R}^1$ which converges at every point p in S to some function f. If S is an arbitrary metric space and $\mathscr{F}$ is, say, any collection of continuous functions, then not much can be said about convergent sequences in $\mathscr{F}$. However, if the members of $\mathscr{F}$ have certain uniformity properties which we describe below and if S is restricted properly, then it is always possible to extract such convergent subsequences (Theorem 15.20 below).

Definitions. A metric space S is **separable** if and only if S contains a countable dense set. It is easy to see that $\mathbb{R}^N$ is separable for every N. In fact, the **rational points**, that is, the points $x: (x_1, x_2, \ldots, x_N)$ such that all x_i are rational numbers form a countable dense set in $\mathbb{R}^N$.

We now show that large classes of metric spaces are separable.

Theorem 15.19. *A compact metric space S is separable.*

PROOF. We recall Theorem 6.25 which states that for every $\delta > 0$, there is a finite number of points $p_1, p_2, \ldots, p_k$ (the number k depending on δ) such that S is contained in $\bigcup_{i=1}^{k} B(p_i, \delta)$. Consider the sequence $\delta = 1, 1/2, 1/3, \ldots,$ $1/n, \ldots$, and the totality of points which are centers of the finite number of balls of radii $1/n$ which cover S. This set is countable and dense, as is readily verified. $\square$

Remark. We give an example of a space which is not separable. A sequence of real numbers $x_1, x_2, \ldots, x_n, \ldots$ is **bounded** if there is a number M such that $|x_i| \leqslant M$ for all i. Denote such a sequence by x and consider the space S of all such elements x. It is not difficult to verify that the function

$$d(x, y) = \sup_i |x_i - y_i| \tag{15.14}$$

where $x = (x_1, \ldots, x_n, \ldots)$ and $y = (y_1, \ldots, y_n, \ldots)$ is a metric on S. Now suppose S were separable. Then there would be a countable set $\{x^m\}$, $m = 1,$ $2, \ldots$, which is dense in S. We write $x^m = (x_1^m, x_2^m, \ldots, x_n^m, \ldots)$, and we construct the element $y = (y_1, y_2, \ldots, y_n, \ldots)$ as follows:

$$y_i = \begin{cases} 1 & \text{if } x_i^i \leqslant 0, \\ -1 & \text{if } x_i^i > 0, \end{cases} \quad i = 1, 2, \ldots.$$

According to (15.14) it is clear that $d(x^m, y) \geqslant 1$ for every m. Thus $\{x^m\}$ is not dense and S cannot be separable.

Definitions. Let S be a metric space and A a set in S. We denote by $\mathscr{F}$ a family of functions $f: A \to \mathbb{R}^1$. For $p_0 \in A$, we say that $\mathscr{F}$ is **equicontinuous at** p_0 if and only if for every $\varepsilon > 0$ there is a $\delta > 0$ such that

$$|f(p) - f(p_0)| < \varepsilon \tag{15.15}$$

for all $p \in A \cap \overline{B(p_0, \delta)}$ and for *all* f in $\mathscr{F}$. The family is **equicontinuous on a set** A if it is equicontinuous at each point of A. Clearly, each f in $\mathscr{F}$ is continuous if $\mathscr{F}$ is an equicontinuous family. It is important to note that given the number ε the same value δ yields (15.15) for all functions of the family.

EXAMPLES. Let $I = \{x: 0 \leqslant x \leqslant 1\}$ and consider the family $\mathscr{F} = \{f_n\}$ where $f_n(x) = x^n$ for $x \in I$. Then it is not difficult to verify that this family is equicontinuous at every point except $x = 1$. In any neighborhood of $x = 1$, the δ required for a given ε shrinks to zero as n tends to infinity (see Figure 15.7). Thus there is no single value of δ valid for all x^n. On the other hand, the family $\mathscr{F} = \{f_n\}$ given by $f_n(x) = n \sin(x/n)$ defined on $J = \{x: 0 \leqslant x < \infty\}$ is equicontinuous. To see this, we apply the Mean-value theorem and find that for any n there is an $\bar{x}_n$ between x_1 and x_2 such that

$$\left| n \sin\left(\frac{x_1}{n}\right) - n \sin\left(\frac{x_2}{n}\right) \right| \leqslant \left| \cos\left(\frac{\bar{x}_n}{n}\right) \right| |x_1 - x_2| \leqslant |x_1 - x_2|.$$

Hence for any $\varepsilon > 0$ we may choose $\delta = \varepsilon$ to obtain (15.15).

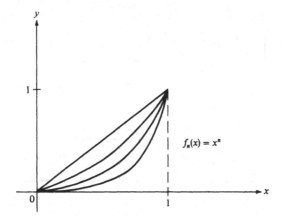

Figure 15.7. An equicontinuous family except at $x = 1$.

It is important to have simple criteria for determining when a family of functions is equicontinuous. Let $I = \{x: a \leqslant x \leqslant b\}$ and suppose $\mathscr{F}$ is a family of functions $f: I \to \mathbb{R}^1$. If all the functions are differentiable and the first derivative is bounded for all $x \in I$ and $f \in \mathscr{F}$, then the family is equicontinuous. To see this let $|f'(x)| \leqslant M$ for all $x \in I$ and all $f \in \mathscr{F}$. From the Mean-value theorem it follows at once that

$$|f(x_1) - f(x_2)| \leqslant |f'(\zeta)||x_1 - x_2| \leqslant M|x_1 - x_2|.$$

Choosing $\delta = \varepsilon/M$, we see that the definition of equicontinuity is satisfied. For a more general criterion see Problem 3 at the end of this section.

Definition. Let $\mathscr{F}$ be a family of functions from a metric space S into $\mathbb{R}^1$. The family $\mathscr{F}$ is **uniformly bounded** if there is a number $M > 0$ such that $|f(p)| \leqslant M$ for all $p \in S$ and all $f \in \mathscr{F}$.

We observe that it is possible for a family to be equicontinuous and not uniformly bounded or even bounded. Perhaps the simplest example is the sequence of functions $f_n(p) = n$, for $n = 1, 2, \ldots$, and for all p in a metric space S. The functions $f_n(x) = (1/n)x$ for $x \in \mathbb{R}^1$ provide another example of an unbounded equicontinuous family as does the sequence $f_n(x) = n \sin(x/n)$ given above.

The next technical lemma is needed for the proof of the existence of convergent subsequences given in Theorem 15.20.

Lemma 15.5. *Let A be a compact set in a metric space S. Let $\mathscr{F}$ be a family of functions from A into $\mathbb{R}^1$ which is equicontinuous and uniformly bounded on A. We set $D = \sup_{p,q \in A} d(p, q)$. ($D$ is called the **diameter** of A.) We define*

$$K = \sup_{p,q \in A} |f(p) - f(q)|, \qquad f \in \mathscr{F}.$$

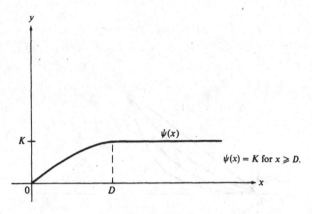

Figure 15.8

Then there is a function $\psi: I \to \mathbb{R}^1$ *where* $I = \{x: 0 < x < \infty\}$ *such that:* (i) ψ *is nondecreasing;* (ii) $\psi(x)$ *has the constant value* K *for all* $x \geq D$; (iii) $|f(p) - f(q)| \leq \psi(x)$ *whenever* $d(p, q) \leq x$, *for all* $f \in \mathscr{F}$; (iv) $\lim_{x \to 0^+} \psi(x) = 0$. *(See Figure 15.8.)*

PROOF. We use (iii) to define $\psi(x) = \sup |f(p) - f(q)|$ where the supremum is taken for all $p, q \in A$ with $d(p, q) \leq x$, and for all $f \in \mathscr{F}$. Since $\mathscr{F}$ is uniformly bounded, ψ is well defined. Clearly ψ is a nondecreasing function of x and so (i) holds. Since $d(p, q) \leq D$ for all $p, q \in A$, the function $\psi(x)$ has exactly the value K for $x \geq D$. Thus (ii) is established. Property (iii) follows from the definition of ψ. To show (iv) holds, let $\varepsilon > 0$ be given. Since $\mathscr{F}$ is equicontinuous, each point $p \in A$ is the center of a ball $B(p, r)$ of radius r such that for any $q \in A$ is this ball, we have

$$|f(q) - f(p)| \leq \frac{\varepsilon}{2} \qquad \text{for all} \quad f \in \mathscr{F}.$$

That fact that A is compact now allows us to use the Lebesgue lemma (Theorem 6.27). We conclude that there is a positive number δ such that every ball $B(q, \delta)$ with $q \in A$ must lie in a single one of the balls $B(p, r)$ defined above. Suppose p_0, q_0 are any points of A with $d(p_0, q_0) < \delta$. Then $p_0, q_0 \in B(p_0, \delta)$, and so $p_0, q_0 \in B(\bar{p}, \bar{r})$ for some $\bar{p}$ and $\bar{r}$. Therefore

$$|f(p_0) - f(q_0)| \leq |f(p_0) - f(\bar{p})| + |f(\bar{p}) - f(q_0)| \leq \frac{\varepsilon}{2} + \frac{\varepsilon}{2} = \varepsilon$$

for all $f \in \mathscr{F}$.

Hence $\psi(\delta) < \varepsilon$, and since ε is arbitrary, $\lim_{x \to 0^+} \psi(x) = 0$. $\qquad \square$

One method of obtaining solutions of differential and integral equations consists of the following procedure. First, a family of functions is found which approximate the solution. Next, it is shown that this family is equicontinuous and bounded. Then, with the aid of the next theorem (Arzela), a convergent subsequence is extracted from the family. Finally, the limit of the subsequence is shown to be the desired solution of the differential or integral equation.

Theorem 15.20 (Arzela's theorem). *Let S be a separable metric space, and suppose that $f_n\colon S \to \mathbb{R}^1$, $n = 1, 2, \ldots$, form on S an equicontinuous, uniformly bounded family $\mathcal{F}$.*

(i) *Then there is a subsequence $f_{k_1}, f_{k_2}, \ldots, f_{k_n}, \ldots$ of $\{f_n\}$ which converges at each point of S to a continuous function f.*

(ii) *if the space S is compact, then the subsequence converges uniformly to f.*

PROOF

(i) Let $\{p_n\} \equiv A$ be a countable, dense subset of S. Since the $\{f_n\}$ are uniformly bounded, $\{f_n(p_1)\}$ is a bounded set of real numbers and so has a convergent subsequence. Denote the subsequence by $g_{11}(p_1), g_{12}(p_1), \ldots,$ $g_{1k}(p_1), \ldots$. Consider $\{g_{1k}(p_2)\}$. This sequence of real numbers has a convergent subsequence which we denote by $g_{21}(p_2), g_{22}(p_2), \ldots, g_{2k}(p_2), \ldots$. We continue the process. We now observe that the "diagonal sequence" of functions of $g_{11}, g_{22}, \ldots, g_{kk}, \ldots$ converges for every p_n. We denote the limit function (defined on A) by f, and we simplify the notation by setting $g_{kk} = g_k$.

Let p be any point of S and let $\varepsilon > 0$ be given. From the equicontinuity of $\mathcal{F}$, it follows that

$$|g_k(p) - g_k(q)| < \frac{\varepsilon}{3} \qquad \text{whenever} \quad d(p, q) < \delta \tag{15.16}$$

for all k. Because A is dense in S, there is a $q_0 \in A$ such that $d(p, q_0) < \delta$. We now show that the sequence $\{g_k(p)\}$ is a Cauchy sequence. We have

$$|g_k(p) - g_l(p)| \leqslant |g_k(p) - g_k(q_0)| + |g_k(q_0) - g_l(q_0)| + |g_l(q_0) - g_l(p)|.$$

Because of (15.16), the first and third terms on the right are less than $\varepsilon/3$. Also, there is an $N > 0$ such that $|g_k(q_0) - g_l(q_0)| < \varepsilon/3$ if $k, l > N$ (because g_k converges at all points of A). Therefore $\{g_k(p)\}$ is a Cauchy sequence for all $p \in S$. The limit function f is now defined on S and we show that it is continuous. Let $p_0 \in S$ and $\varepsilon > 0$ be given. From the equicontinuity, it follows that $|g_k(p) - g_k(p_0)| \leqslant \varepsilon$ whenever $d(p, p_0) < \delta$, and this holds for all k. However, $g_k(p) \to f(p)$ and $g_k(p_0) \to f(p_0)$ as $k \to \infty$. The inequality holds in the limit and f is continuous at p_0.

(ii) We establish the uniform convergence when S is compact. Let $\varepsilon > 0$ be given. From Lemma 15.5 there is an $x > 0$ so small that

$$|g_n(p) - g_n(q)| \leqslant \psi(x) = \frac{\varepsilon}{3} \qquad \text{whenever} \quad d(p, q) \leqslant x,$$

and this inequality holds for all n. Consider all balls of radius x in S. Since S

is compact, there is a finite number of such balls with centers say at $\bar{p}_1, \bar{p}_2,$ $\ldots, \bar{p}_m$ such that $\bigcup_{i=1}^{m} B(\bar{p}_i, x) \supset S$. We know that $\lim_{n \to \infty} g_n(\bar{p}_i) = f(\bar{p}_i), i = 1,$ $2, \ldots, m$. Choose an N so large that

$$|f(\bar{p}_i) - g_n(\bar{p}_i)| \leqslant \frac{\varepsilon}{3} \qquad \text{for all} \quad n \geqslant N, i = 1, 2, \ldots, m.$$

Let p be any point of S. Then p is in one of the covering balls, say $B(\bar{p}_i, x)$. Consequently,

$$|g_n(p) - f(p)| \leqslant |g_n(p) - g_n(\bar{p}_i)| + |g_n(\bar{p}_i) - f(\bar{p}_i)| + |f(\bar{p}_i) - f(p)|$$
$$\leqslant \tfrac{1}{3}\varepsilon + \tfrac{1}{3}\varepsilon + \tfrac{1}{3}\varepsilon = \varepsilon,$$

provided that $n > N$. Since the choice of N does not depend on the point p chosen, the convergence is uniform.

Since the space S is compact, the limit function f is uniformly continuous. $\qquad \square$

Remarks

(i) If $\mathscr{F}$ is a family of functions defined on a metric space S with range in $\mathbb{R}^N$, equicontinuity may be defined in a way completely analogous to that given when $N = 1$. We simply interpret the quantity $|f(p) - f(q)|$ to be the distance in $\mathbb{R}^N$. The proofs of Lemma 15.5 and Arzela's theorem hold with the modifications required when replacing distance in $\mathbb{R}^1$ by distance in $\mathbb{R}^N$.

(ii) Let S be any compact, separable metric space and let $C(S)$ be the space of continuous real-valued bounded functions on S. We define

$$d(f, g) = \sup_{p \in S} |f(p) - g(p)|$$

and it is easily verified that $C(S)$ is a metric space. Then Arzela's theorem states that an equicontinuous, closed bounded set in the metric space $C(S)$ is compact. If $C(S, \mathbb{R}^N)$ denotes the metric space of bounded continuous functions $f: S \to \mathbb{R}^N$, a similar statement holds on the compactness of bounded, equicontinuous sets.

Let $f: I \to \mathbb{R}^1$ be a continuous function with $I = \{x: a \leqslant x \leqslant b\}$. We may define

$$g(x) = \begin{cases} f(a) & \text{for } x < a, \\ f(x) & \text{for } x \in I, \\ f(b) & \text{for } x > b. \end{cases}$$

Clearly, g is continuous on all of $\mathbb{R}^1$, coincides with f for $x \in I$, and $\max_{x \in \mathbb{R}^1} |g(x)| = \max_{x \in I} |f(x)|$. This example is typical of the fact that continuous functions defined on a closed subset of a metric space may be extended as a continuous function on the entire space in such a way that the supremum of the extended function does not exceed that of the original function.

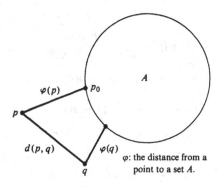

Figure 15.9

In the above example it is essential that the interval I is closed. The function $f(x) = 1/(x - a)(x - b)$ is continuous on $J = \{x: a < x < b\}$, and there is no way of extending f in a continuous manner beyond this open interval.

Definition. Let A be a set in a metric space S. The **distance of a point p from A**, denoted $d(p, A)$, is defined by the formula

$$d(p, A) = \inf_{q \in A} d(p, q).$$

The following useful lemma shows that the distance function we just defined is continuous on the entire metric space.

Lemma 15.6. *Let A be a set in a metric space S. Let $\varphi(p)$ denote the distance from p to A. Then φ is continuous on S. In fact, $|\varphi(p) - \varphi(q)| \leq d(p, q)$ for p, $q \in S$. (See Figure 15.9.)*

PROOF. For any $\varepsilon > 0$ there is a point $p_0 \in A$ such that $\varphi(p) > d(p, p_0) - \varepsilon$. From the triangle inequality we find

$$d(q, p_0) \leq d(q, p) + d(p, p_0)$$
$$\leq d(p, q) + \varphi(p) + \varepsilon.$$

From the definition of $\varphi(q)$, we get

$$\varphi(q) \leq d(q, p_0) \leq d(p, q) + \varphi(p) + \varepsilon,$$

and therefore

$$\varphi(q) - \varphi(p) \leq d(p, q) + \varepsilon.$$

Since ε is arbitrary, it follows that $\varphi(q) - \varphi(p) \leq d(p, q)$. Interchanging p and q, we get the result. $\square$

Lemma 15.7. *Let A and B be disjoint closed sets in a metric space S. Then there is a continuous function $\psi: S \to \mathbb{R}^1$ such that $\psi(p) = 0$ for $p \in A$, $\psi(p) = 1$ for $p \in B$, and ψ is between 0 and 1 for all $p \in S$.*

PROOF. Define

$$\psi(p) = \frac{d(p, A)}{d(p, A) + d(p, B)},$$

and it is immediate that ψ has the desired properties. $\square$

Remark. Let $a > 0$ be any number. Then the function

$$\psi_1(p) = -a + 2a\psi(p)$$

has the property that $\psi_1(p) = -a$ for $p \in A$, $\psi_1(p) = a$ for $p \in B$ and ψ_1 is between $-a$ and a for all $p \in S$. Also, ψ_1 is continuous on S.

In Section 15.4 we establish several theorems which show that continuous functions may be approximated uniformly by sequences of smooth functions. In order to make such approximations it is important to be able to enlarge the domain of the continuous function. The next result gives the basic extension theorem for continuous functions.

Theorem 15.21 (Tietze extension theorem). *Let A be a closed set in a metric space S and $f: A \to \mathbb{R}^1$ a continuous, bounded function. Define $M = \sup_{p \in A} |f(p)|$. Then there is a continuous function $g: S \to \mathbb{R}^1$ such that $g(p) = f(p)$ for $p \in A$ and $|g(p)| \leqslant M$ for all $p \in S$.*

PROOF. We shall obtain g as the limit of a sequence $g_1, g_2, \ldots, g_n, \ldots$ which will be determined by sequences $f_1, f_2, \ldots, f_n, \ldots$ and $\psi_1, \psi_2, \ldots, \psi_n, \ldots$ which we now define. We begin by setting $f_1(p) = f(p)$ for $p \in A$, and we define the sets

$$A_1 = \{p: p \in A \quad \text{and} \quad f_1(p) \leqslant -\tfrac{1}{3}M\},$$

$$B_1 = \{p: p \in A \quad \text{and} \quad f_1(p) \geqslant \tfrac{1}{3}M\}.$$

The sets A_1, B_1 are disjoint and closed. Now according to the Remark following Lemma 15.7, there is a function $\psi_1(p)$, continuous on S and such that

$$\psi_1(p) = -\tfrac{1}{3}M \quad \text{on } A_1,$$

$$\psi_1(p) = \tfrac{1}{3}M \quad \text{on } B_1,$$

$$|\psi_1(p)| \leqslant \tfrac{1}{3}M \quad \text{on } S.$$

We define

$$f_2(p) = f_1(p) - \psi_1(p) \quad \text{for } p \in A.$$

We show that $|f_2(p)| \leqslant (2/3)M$ for $p \in A$. To see this, let $p \in A_1$. Then $f_2(p) = (1/3)M + f_1(p)$. But $-M \leqslant f_1(p) \leqslant -(1/3)M$ on this set. Hence $|(f_2(p)| \leqslant (2/3)M$. Now let $p \in B_1$. Then $f_2(p) = f_1(p) - (1/3)M$. From the definition of B_1, we get $|f_2(p)| \leqslant M - (1/3)M = (2/3)M$. Similarly, if $p \in A - A_1 - B_1$, then both f_1 and ψ_1 are bounded in absolute value by $(1/3)M$. Now we define

$$A_2 = \{p: p \in A \quad \text{and} \quad f_2(p) \leqslant -\tfrac{1}{3} \cdot \tfrac{2}{3}M\},$$
$$B_2 = \{p: p \in A \quad \text{and} \quad f_2(p) \geqslant \tfrac{1}{3} \cdot \tfrac{2}{3}M\}.$$

The sets A_2, B_2 are disjoint and closed. In the same way as before, there is a function $\psi_2(p)$ such that

$$\psi_2(p) = -\tfrac{1}{3} \cdot \tfrac{2}{3}M \quad \text{on } A_2,$$
$$\psi_2(p) = \tfrac{1}{3} \cdot \tfrac{2}{3}M \quad \text{on } B_2,$$
$$|\psi_2(p)| \leqslant \tfrac{1}{3} \cdot \tfrac{2}{3}M \quad \text{on } S.$$

We define

$$f_3(p) = f_2(p) - \psi_2(p)$$

and find

$$|f_3(p)| \leqslant (\tfrac{2}{3})^2 M.$$

Continuing in this manner, we define

$$A_n = \{p: p \in A \quad \text{and} \quad f_n(p) \leqslant -\tfrac{1}{3}(\tfrac{2}{3})^{n-1}M\},$$
$$B_n = \{p: p \in A \quad \text{and} \quad f_n(p) \geqslant \tfrac{1}{3}(\tfrac{2}{3})^{n-1}M\}.$$

The function ψ_n is defined in the analogous manner and furthermore

$$|\psi_n(p)| \leqslant \tfrac{1}{3}(\tfrac{2}{3})^{n-1}M, \qquad p \in S.$$

We define

$$f_{n+1}(p) = f_n(p) - \psi_n(p).$$

The condition $|f_{n+1}(p)| \leqslant (2/3)^n M$ holds. Next we define

$$g_n(p) = \psi_1(p) + \cdots + \psi_n(p), \qquad p \in S,$$

and we show that g_n is a uniformly convergent sequence. In fact, for $m > n$, it follows that

$$|g_m(p) - g_n(p)| = |\psi_{n+1}(p) + \cdots \psi_m(p)|$$
$$\leqslant \tfrac{1}{3}[(\tfrac{2}{3})^n + \cdots + (\tfrac{2}{3})^m]M$$
$$\leqslant \tfrac{1}{3}(\tfrac{2}{3})^n[1 + \tfrac{2}{3} + \cdots + (\tfrac{2}{3})^{m-n}]M \leqslant (\tfrac{2}{3})^n M.$$

The $\{g_n(p)\}$ form a Cauchy sequence and so converge uniformly to a continuous function which we denote by g. We now show that $g = f$ for $p \in A$. To see this, observe that

$$g_n = \psi_1 + \cdots + \psi_n = \sum_{i=1}^{n} (f_i - f_{i+1}) = f_1 - f_{n+1}.$$

Since $|f_{n+1}(p)| \leqslant (2/3)^n M$, we know that $f_{n+1}(p) \to 0$ as $n \to \infty$. Therefore $g_n \to$

$g = f_1 = f$ as $n \to \infty$. Also,

$$|g(p)| \leqslant \tfrac{1}{3}M[1 + \tfrac{2}{3} + (\tfrac{2}{3})^2 + \cdots] = M \quad \text{for all } p \in S. \qquad \square$$

PROBLEMS

1. Let $I = \{x: 0 \leqslant x \leqslant 1\}$ and consider the totality of functions $f: I \to \mathbb{R}^1$ such that f is bounded. Denote this space by $\mathscr{B}$ and for $f, g \in \mathscr{B}$, let

$$d(f, g) = \sup_{x \in I} |f(x) - g(x)|.$$

Show that $d(f,g)$ is a metric and that $\mathscr{B}$ is a metric space which is not separable.

2. Let $x = (c_1, c_2, \ldots, c_n, \ldots)$ where the $\{c_n\}$ are real numbers such that $c_n \to 0$ as $n \to \infty$. Let $\mathscr{C}_0$ be the totality of all such sequences. Let $y = (d_1, d_2, \ldots, d_m, \ldots)$ and for $x, y \in \mathscr{C}_0$ define

$$d(x, y) = \sup_{1 \leqslant i < \infty} |c_i - d_i|.$$

Show that $\mathscr{C}_0$ is a metric space which is separable. [*Hint*: Consider the elements of $\mathscr{C}_0$ of the form $(r_1, 0, 0, \ldots), (r_1, r_2, 0, 0, \ldots)$, etc., where r_i are rational numbers.]

3. Let S be a metric space and suppose that $f: S \to \mathbb{R}^1$ has the following property: there are positive numbers M, α with $0 < \alpha \leqslant 1$ such that

$$|f(p) - f(q)| \leqslant M[d(p, q)]^\alpha$$

for any two points p, q in S. Then we say f satisfies a **Hölder condition** on S with constant M and exponent α. Show that a family $\mathscr{F}$ of functions satisfying a Hölder condition with constant M and exponent α is equicontinuous.

4. Let A be an open convex set in $\mathbb{R}^N$ and $\{f_n\}$ a sequence of functions $f_n: A \to \mathbb{R}^1$ which are convex. Show that if $\{f_n\}$ converges at each point of A, then $\{f_n\}$ converges uniformly on each compact subset of A.

5. Let A be an open convex set in $\mathbb{R}^N$ and $\{f_n\}$ a sequence of functions $f_n: A \to \mathbb{R}^1$ which are convex. Suppose that $\{f_n\}$ are uniformly bounded. Show that $\{f_n\}$ contains a convergent subsequence and that the subsequence converges uniformly on each compact subset of A.

6. Let $I = \{x: a \leqslant x < \infty\}$ and let $\mathscr{F}$ be a family of functions $f: I \to \mathbb{R}^1$ which are differentiable on I. Let M_0, M_1 be constants such that $|f(x)| \leqslant M_0$, $|f'(x)| \leqslant M_1$ for all $f \in \mathscr{F}$. Is there always a subsequence of $\mathscr{F}$ which converges uniformly on I?

7. Let A be a bounded convex set in $\mathbb{R}^N$ and suppose $A^{(0)}$ is not empty. Let $\{f_n\}$ be a sequence of functions $f_n: A \to \mathbb{R}^1$ such that $f_n \in C^1$ for each n and $|f_n(x)| \leqslant M_0$ for all $x \in A$ and $|\nabla f_n(x)| \leqslant M_1$ for all $x \in A$. Then a subsequence of $\{f_n\}$ converges uniformly on A.

8. Let $I = \{x: a \leqslant x \leqslant b\}$ and $S = \{(x, y): a \leqslant x \leqslant b, a \leqslant y \leqslant b\}$, and suppose that $K: S \to \mathbb{R}^1$ is continuous on S. We set

$$f(x) = \int_a^b K(x, y)g(y)\, dy \qquad (15.17)$$

where f and g are real-valued functions defined on I. For any family of functions $\mathscr{G}$, Equation (15.17) defines a family of functions $\mathscr{F}$. Show that if for all $g \in \mathscr{G}$, we have $|g(y)| \leqslant M$ for all $y \in I$, then the family $\mathscr{F}$ contains a uniformly convergent subsequence.

9. Prove Arzela's theorem (Theorem 15.20) for functions defined on a separable metric space with range in $\mathbb{R}^N$, $N > 1$.

10. Give an example of a function defined on $I = \{x: a < x < b\}$ with range in $\mathbb{R}^1$ which is continuous on I, bounded, and which cannot be extended as a continuous function to any larger set. i.e., show that the conclusion of Tietze's theorem does not hold.

11. In Lemma 15.7, suppose that the sets A and B are disjoint but not closed. Show that it may not be possible to find a function ψ with the stated properties.

12. Let $I = \{x: a \leqslant x < \infty\}$ and let $f: I \to \mathbb{R}^1$ be continuous but not necessarily bounded. Show that f can be extended to all $\mathbb{R}^1$ as a continuous function. Prove the same result if the domain of f is any closed set $A \subset \mathbb{R}^1$.

13. In $\mathbb{R}^2$ let $S = \{(x_1, x_2): 0 \leqslant x_1 \leqslant 1, 0 \leqslant x_2 \leqslant 1\}$, and define $f(x_1, x_2) = x_1 x_2$ for $(x_1, x_2) \in S$. Show explicitly how f can be extended to all of $\mathbb{R}^2$ as a continuous function with no increase in the maximum of its absolute value.

14. In $\mathbb{R}^3$ let $C = \{(x_1, x_2, x_3): 0 \leqslant x_1 \leqslant 1, 0 \leqslant x_2 \leqslant 1, 0 \leqslant x_3 \leqslant 1\}$, and define $f(x) = x_1 x_2 x_3$ for $x \in C$. Show explicitly how f can be extended to all of $\mathbb{R}^3$ as a continuous function with no increase in the maximum of its absolute value.

15.4. Approximations and the Stone–Weierstrass Theorem

Let f, defined on an interval $I = \{x: a < x < b\}$ with values in $\mathbb{R}^1$, have derivatives of all orders at a point c in I. We recall that f is **analytic** *at* c if and only if f can be expanded in a power series of the form

$$f(x) = \sum_{n=0}^{\infty} \frac{f^{(n)}(c)}{n!}(x - c)^n, \tag{15.18}$$

and this series has a positive radius of convergence. A function is **analytic on a set** A if it is analytic at each point of A. As we saw in Chapter 9, a function may possess derivatives of all orders (we say it is *infinitely differentiable*, and write $f \in C^\infty$) at a point and yet not be analytic at that point. To illustrate this fact consider the function

$$f(x) = \begin{cases} e^{-1/x^2}, & x \neq 0, \\ 0, & x = 0. \end{cases}$$

It is not difficult to verify that $f^{(n)}(0) = 0$ for $n = 1, 2, \ldots$, and therefore the power series expansion (15.18) does not have a positive radius of convergence

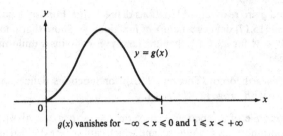

Figure 15.10

when $c = 0$. However, clearly f is analytic for all $x \neq 0$. We now define the functions g_1 and g:

$$g_1(x) = \begin{cases} e^{-1/x^2}, & x > 0, \\ 0, & x \leqslant 0, \end{cases}$$

and $g(x) = g_1(x) \cdot g_1(1 - x)$. Then $g(x)$ vanishes outside the unit interval $I = \{x : 0 < x < 1\}$ and is positive on I (see Figure 15.10). We set

$$M = \int_0^1 g(x)\, dx$$

and define

$$h(x) = \frac{1}{M} \int_0^x g(t)\, dt.$$

Then $h \in C^\infty$ and h has the properties: $h(x) = 0$ for $x \leqslant 0$, $h(x) = 1$ for $x \geqslant 1$ and $h(x)$ is nondecreasing (see Figure 15.11). Finally, we introduce the function $k: \mathbb{R}^1 \to \mathbb{R}^1$ by the formula

$$k(x) = \begin{cases} 1 - h(2x - 1) & \text{for } x \geqslant 0, \\ k(-x) & \text{for } x \leqslant 0. \end{cases}$$

Then we verify easily that $k(x)$ is nonnegative everywhere, and

$$k(x) = \begin{cases} 1 & \text{for } -\tfrac{1}{2} \leqslant x \leqslant \tfrac{1}{2}, \\ 0 & \text{for } x \geqslant 1, \text{ and for } x \leqslant -1. \end{cases}$$

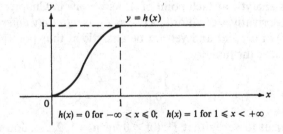

Figure 15.11

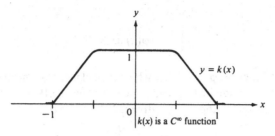

Figure 15.12

In fact, k is nondecreasing for $x \leqslant 0$ and nonincreasing for $k \geqslant 0$ (see Figure 15.12). The function k is analytic for all values of x except $x = \pm 1, \pm \frac{1}{2}$, at which points k is a C^∞ function. Smooth functions which are equal to 1 in a neighborhood of some point and which vanish outside a larger neighborhood of the same point form a useful tool in the problem of approximation of continuous functions by classes of smooth functions.

The above functions have counterparts in any number of dimensions. We define

$$k_2(x, y) = k(x) \cdot k(y)$$

and obtain a function which has the value 1 on the square

$$S_0 = \{(x, y): -\tfrac{1}{2} \leqslant x \leqslant \tfrac{1}{2}, \quad -\tfrac{1}{2} \leqslant y \leqslant \tfrac{1}{2}\},$$

which vanishes outside the square

$$S_1 = \{(x, y): -1 \leqslant x \leqslant 1, \quad -1 \leqslant y \leqslant 1\},$$

and is such that $0 \leqslant k_2(x, y) \leqslant 1$ for all (x, y). Furthermore, k_2 has partial derivatives of all orders with respect to both x and y. In $\mathbb{R}^N$, we set $k_N(x_1, \ldots, x_N) = k(x_1)k(x_2)\ldots k(x_N)$ and obtain a C^∞ function which vanishes outside the hypercube of side 2, center at 0, and which has the value 1 on the hypercube of side 1, center at 0.

For many purposes it is convenient to replace k_N by a function of $r = (\sum_{i=1}^N x_i^2)^{1/2}$ with the same essential properties. To do this we define $\varphi_1(x_1, x_2, \ldots, x_N) = k(r)$ for $r > 0$. Then with $x = (x_1, x_2, \ldots, x_N)$, the function $\varphi_1(x)$ vanishes outside the unit ball $B(0, 1)$ in $\mathbb{R}^N$ and has the value 1 inside the ball $B(0, 1/2)$. We set

$$M_0 = \int_{B(0, 1)} \varphi_1(x)\, dx$$

and define

$$\varphi(x) = \frac{\varphi_1(x)}{M_0}.$$

Then φ is a C^∞ function which vanishes outside $B(0, 1)$ and has the property that

$$\int_{B(0,1)} \varphi(x)\, dx = 1. \tag{15.19}$$

Definitions. A function $\varphi \colon \mathbb{R}^N \to \mathbb{R}^1$ which is of class C^∞, is nonnegative everywhere, vanishes outside $B(0, 1)$, and which satisfies (15.19) is called a **mollifier**. Let x be a point in $\mathbb{R}^N$ and ρ a positive number. We define

$$\varphi_\rho^*(x) = \frac{\varphi(x/\rho)}{\rho^N}.$$

We note that φ_ρ^* vanishes outside the ball $B(0, \rho)$ and has the property that

$$\int_{B(0,\rho)} \varphi_\rho^*(x)\, dx = 1.$$

We also call φ_ρ^* a mollifier or sometimes a **modified mollifier**. Suppose that $f \colon \mathbb{R}^N \to \mathbb{R}^1$ is a continuous function. We define the **mollified function**, denoted $f_{\varphi,\rho}$, or simply f_ρ, by the formula

$$f_\rho \equiv f_{\varphi,\rho}(x) = \int_{\mathbb{R}^N} \varphi_\rho^*(x - \xi) f(\xi)\, d\xi. \tag{15.20}$$

The mollified functions f_ρ are smooth and tend to f as $\rho \to 0$, as we show in the next theorem.

Theorem 15.22. *Suppose that* $f \colon \mathbb{R}^N \to \mathbb{R}^1$ *is continuous and* φ *is a mollifier on* $\mathbb{R}^N$. *Then for every* $\rho > 0$, *the function* $f_{\varphi,\rho}$ *is of class* C^∞ *on* $\mathbb{R}^N$ *and* $f_{\varphi,\rho}$ *converges uniformly to* f *on any compact subset of* $\mathbb{R}^N$ *as* $\rho \to 0^+$.

PROOF. The formula for f_ρ given by (15.20) shows that since $\varphi_\rho^*(x - \xi)$ vanishes outside a ball of radius ρ with center at x, the integration is over a bounded region rather than over all of $\mathbb{R}^N$. Therefore we may use the rule for differentiating under the integral sign given in Section 11.1, to deduce that f_ρ has partial derivatives of all orders.

Now let $R > 0$ be given. Since f is uniformly continuous on the closed ball $B(0, R + 1)$, it follows that there is a nondecreasing function $\psi(\rho)$ defined on $I = \{\rho \colon 0 < \rho \leqslant 1\}$ (see Lemma 15.5) such that

$$\psi(\rho) \to 0 \quad \text{as } \rho \to 0^+ \quad \text{and} \quad |f(x) - f(y)| \leqslant \psi(|x - y|)$$

for all $x, y \in B(0, R + 1)$ whenever $|x - y| \leqslant \rho \leqslant 1$. Accordingly, if $\rho < 1$ and $x \in \overline{B(0, R)}$, then

$$|f_\rho(x) - f(x)| = \left| \int_{B(x,\rho)} \varphi_\rho^*(x - \xi)[f(\xi) - f(x)]\, d\xi \right|$$

$$\leqslant \int_{B(x,\rho)} \varphi_\rho^*(x - \xi)|f(\xi) - f(x)|\, d\xi$$

$$\leqslant \psi(\rho) \int_{B(x,\rho)} \varphi_\rho^*(x - \xi)\, d\xi.$$

The last inequality holds because $|\xi - x| \leqslant \rho$. From the definition of φ_ρ^*, the last integral on the right has the value 1, and so

$$|f_\rho(x) - f(x)| \leqslant \psi(\rho)$$

for all $x \in \overline{B(0, R)}$. Thus f_ρ tends to f uniformly on this ball, and since R is arbitrary, the result follows. □

The next corollaries show that any continuous function on a compact set may be approximated uniformly by a sequence of C^∞ functions. Moreover, if a function possesses derivatives, then the function and all its derivatives up to any finite order may be approximated uniformly by the same sequence.

Corollary 1. *Let A be a compact subset of $\mathbb{R}^N$ and suppose $f: A \to \mathbb{R}^1$ is continuous on A. Let G be an open set in $\mathbb{R}^N$ containing A. Then there is a sequence $\{f_n\}$ such that f_n is C^∞ on G for each n and f_n converges uniformly to f on A.*

PROOF. By the Tietze extension theorem (Theorem 15.21) there is a continuous function $F: \mathbb{R}^N \to \mathbb{R}^1$ such that $F = f$ on A. Let $\rho_n = 1/n$ and denote by F_{φ,ρ_n} the mollifier of F. Then setting $f_n(x) = F_{\varphi,\rho_n}(x)$ for $x \in G$, we use Theorem 15.22 to conclude that $f_n \to f$ uniformly on A. □

Corollary 2. *Suppose that in Corollary 1 the interior $A^{(0)}$ of A is not empty and that f is of class C^k on $A^{(0)}$. Let $D^\alpha f$ denote any partial derivative of f not exceeding order k. Then the sequence f_n in Corollary 1 may be chosen so that $D^\alpha f_n$ converges uniformly to $D^\alpha f$ on each compact subset of $A^{(0)}$.*

PROOF. The function F in Corollary 1 has partial derivatives on $A^{(0)}$, and we set $G_0 = D^\alpha F$. We form the mollified function G_{φ,ρ_n} and it is clear that $D^\alpha F_{\varphi,\rho_n} = G_{\varphi,\rho_n}$ for $x \in A^{(0)}$. Then $G_{\varphi,\rho_n} \to G_0$ as before. □

In Corollaries 1 and 2 above we obtained sequences of C^∞ functions in $\mathbb{R}^N$ which can be used to approximate an arbitrary continuous function to any desired degree of accuracy. If S is any compact metric space and if $C(S)$ is the space of continuous functions from S into $\mathbb{R}^1$ with the usual metric, then we seek those subsets of $C(S)$ which yield approximations to all elements of $C(S)$. The basic property required of any approximating subset is that it "separate points" in S.

Definitions. Let $\mathscr{L}$ be a subset of $C(S)$ with the property that for each pair of distinct points p, q in S there is a function f in $\mathscr{L}$ such that $f(p) \neq f(q)$. Then we say that the set $\mathscr{L}$ **separates points in** S. If $\mathscr{L}$ has the further property that for every pair of distinct points p, $q \in S$ and every pair of real numbers a, b there is a function $f \in \mathscr{L}$ such that $f(p) = a$ and $f(q) = b$, we say that $\mathscr{L}$ **separates points in S and $\mathbb{R}^1$.**

Lemma 15.8. *Let f, g be in $C(S)$ where S is any metric space. Define $h(p) =$ $\max(f(p), g(p))$ and $k(p) = \min(f(p), g(p))$ for all $p \in S$. Then h, k are members of $C(S)$.*

The proof of Lemma 15.8 depends on the fact that for any two real numbers a, b, the relations

$$\max(a, b) = \tfrac{1}{2}(a + b + |a - b|), \qquad \min(a, b) = \tfrac{1}{2}(a + b - |a - b|)$$

hold. We leave the details of the remainder of the proof to the reader.

Theorem 15.23 (Stone approximation theorem). *Let S be a compact metric space. Let $\mathscr{L}$ be a subset of $C(S)$, the space of continuous functions from S into $\mathbb{R}^1$, which separates points in S and $\mathbb{R}^1$. In addition, suppose that $\mathscr{L}$ has the property that $\max(f, g)$ and $\min(f, g)$ belong to $\mathscr{L}$ whenever f and g do. Then any $F \in C(S)$ can be approximated uniformly on S by functions in $\mathscr{L}$.*

PROOF. Let $F \in C(S)$ be given. Suppose p, q are in S and let $a = F(p)$, $b = F(q)$. Then, since $\mathscr{L}$ separates points in S and $\mathbb{R}^1$, there is a function $g = g_{pq}$ in $\mathscr{L}$ such that

$$g_{pq}(p) = a, \qquad g_{pq}(q) = b.$$

Since g and F are continuous at q, for any $\varepsilon > 0$ there is a neighborhood $N(q)$ such that $g(s) > b - \tfrac{1}{2}\varepsilon$ and $F(s) < b + \tfrac{1}{2}\varepsilon$ for $s \in N(q)$. Hence

$$g(s) > F(s) - \varepsilon \quad \text{for } s \in N(q). \tag{15.21}$$

We fix p and obtain a function g_{pq} and a neighborhood $N(q)$ for each $q \in S$. Since S is compact, there is a finite subset of such neighborhoods which covers S. We denote them

$$N(q_1), N(q_2), \ldots, N(q_m).$$

With these neighborhoods we associated the functions

$$g_{pq_1}, g_{pq_2}, \ldots, g_{pq_m}$$

each of which satisfies an inequality similar to (15.21). We now define

$$h_p = \max(g_{pq_1}, g_{pq_2}, \ldots, g_{pq_m}).$$

Then h_p is in $\mathscr{L}$ and it follows from (15.21) that

$$h_p(s) > F(s) - \varepsilon \quad \text{for all } s \in S. \tag{15.22}$$

From the way we defined h_p, it follows that $h_p(p) = a$ since $g_{pq_1}(p) = g_{pq_2}(p) = \cdots = g_{pq_m}(p) = a$. Because h_p and F are continuous, there is a neighborhood $N(p)$ such that

$$h_p(s) < F(s) + \varepsilon \quad \text{for } s \in N(p). \tag{15.23}$$

We can find a function h_p and a neighborhood $N(p)$ for each point p in S and, because S is compact, there is a finite subset $N(p_1), N(p_2), \ldots, N(p_n)$ of such neighborhoods which covers S. We set

$$h = \min(h_{p_1}, h_{p_2}, \ldots, h_{p_n})$$

and, by virtue of (15.22) and (15.23), it follows that

$$h(s) > F(s) - \varepsilon \text{ for all } s \in S,$$
$$h(s) < F(s) + \varepsilon \text{ for all } s \in S. \tag{15.24}$$

That is, $|F(s) - h(s)| < \varepsilon$ for all $s \in S$. Since ε is arbitrary the result is established. $\qquad \square$

Definition. Let $\mathscr{A}$ be a subset of $C(S)$ such that for every two functions f, g in $\mathscr{A}$: (i) $\alpha f + \beta g \in \mathscr{A}$ for all real numbers α, β; and (ii) $f \cdot g \in \mathscr{A}$. We then say that the set $\mathscr{A}$ forms an **algebra of functions** in $C(S)$.

If S is a compact metric space and $\mathscr{A}$ is any algebra of functions in $C(S)$ which separates points of S and which contains the constant functions, then we shall show that $\mathscr{A}$ is dense in $C(S)$. More precisely, we shall prove that every function f in $C(S)$ can be approximated uniformly by polynomials f_n where each f_n is a polynomial in functions of $\mathscr{A}$. To establish this result, known as the Stone–Weierstrass theorem, we employ two elementary facts concerning the functions $|x|$ and $\sqrt{1 + x}$. The most important special case of this approximation theorem, due to Weierstrass, states that any real-valued continuous function defined on a closed interval $I = \{x : a \leqslant x \leqslant b\}$ can be approximated uniformly by polynomials. Equivalently, we say that the collection of polynomials is dense in the space $C(I)$.

Lemma 15.9. *Let $I = \{x : -1 \leqslant x \leqslant 1\}$, and let $\varphi : I \to \mathbb{R}^1$ be the function $\varphi(x) = \sqrt{1 + x}$. Then the series expansion*

$$\varphi(x) = 1 + \frac{1}{2} \sum_{n=0}^{\infty} \frac{(-1)^n (n + 1)(2n)!}{2^{2n}[(n + 1)!]^2} x^{n+1} \tag{15.25}$$

converges uniformly on I. Hence φ can be approximated uniformly on I by the polynomials consisting of the partial sums in (15.25).

PROOF. The ratio test shows that the series converges at all interior points of I. To establish the uniform convergence on I, we use Stirling's formula[1] (valid for large n)

$$n! \sim e^{-(n+1)}(n + 1)^{n+(1/2)} \sqrt{2\pi}(1 + o(1/n))$$

---

[1] A proof of Stirling's formula may be found, for example, in *Calculus* by T.M. Apostol, Blaisdell, New York, p. 450.

and we find that for some constant K

$$\frac{(n+1)(2n)!}{2^{2n}[(n+1)!]^2} \leqslant \frac{K}{\sqrt{2\pi}} \frac{(n+1)e^{-2n+1}(2n+1)^{2n+(1/2)}}{2^{2n}e^{-2n-4}(n+2)^{2n+2}}$$

$$\leqslant \frac{Ke^5}{\sqrt{2\pi}(n+2)^{3/2}}.$$

Hence the series for φ is dominated by the series $\sum_{n=1}^{\infty} K_1/n^{3/2}$ for all x on I, and so the convergence is uniform. $\square$

Remark. For an elementary proof of Lemma 15.9 (without use of Stirling's formula) see Problem 11 at the end of this section and the hint given there.

Lemma 15.10. *Let* $I_M = \{x: -M \leqslant x \leqslant M\}$ *and let* $\psi: I_M \to \mathbb{R}^1$ *be the function* $\psi(x) = |x|$. *Then* ψ *can be approximated uniformly on* I_M *by polynomials in* x.

PROOF. Let $M = 1$. Then we define $\psi_1(x) = (1 + u)^{1/2}$ where $u = x^2 - 1$. By Lemma 15.9 the function $(1 + u)^{1/2}$ may be approximated uniformly on I by polynomials. Since u is in I whenever x is in I, we can approximate ψ_1 by polynomials in $x^2 - 1$, that is, polynomials in x. For the general case, we write $\psi(x) = M\psi_1(x/M)$, and the result follows. $\square$

Theorem 15.24 (Stone–Weierstrass theorem). *Let* S *be a compact metric space and let* $\mathscr{A}$ *be an algebra in* $C(S)$ *which separates points of* S. *Suppose that the function* f_0 *defined by* $f_0(p) = 1$ *for all* $p \in S$ *is in* $\mathscr{A}$. *Then any function* f *in* $C(S)$ *can be approximated uniformly on* S *by functions in* $\mathscr{A}$.

PROOF. Let $\mathscr{L}$ be the subset of functions of $C(S)$ which can be approximated uniformly by functions of $\mathscr{A}$. We wish to show that $\mathscr{L} = C(S)$. We shall show that $\mathscr{L}$ satisfies the hypothesis of Theorem 15.23. For this purpose we prove that $\mathscr{A}$ separates points in S and $\mathbb{R}^1$. Let $a, b \in \mathbb{R}^1$, and $p, q \in S$ with $p \neq q$. There is an $f \in \mathscr{A}$ such that $f(p) \neq f(q)$. Since $f_0(p) = f_0(q) = 1$, there are real numbers α, β such that

$$\alpha f(p) + \beta f_0(p) = a, \qquad \alpha f(q) + \beta f_0(q) = b.$$

Hence the function $h = \alpha f + \beta f_0$ is in $\mathscr{A}$ and satisfies the conditions $h(p) = a$, $h(q) = b$. That is, $\mathscr{A}$ separate points in S and $\mathbb{R}^1$. We now show that for f and g in $\mathscr{A}$, the functions $\max(f, g)$ and $\min(f, g)$ are in $\mathscr{L}$. Let $f \in \mathscr{A}$ be given. Then there is an $M > 0$ such that $|f(p)| \leqslant M$ for all $p \in S$. According to Lemma 15.10 the function $\psi[f(p)] = |f(p)|$ can be approximated uniformly by polynomials in f. Since $\mathscr{A}$ is an algebra, every polynomial in f is also in $\mathscr{A}$ and so ψ can be approximated by polynomials in $\mathscr{A}$. Let f and g be arbitrary elements of $\mathscr{A}$. Then $f + g$ and $f - g$ are in $\mathscr{A}$ and $|f - g|$ can be approximated uniformly by elements of $\mathscr{A}$. We know that

$$\max(f, g) = \tfrac{1}{2}(f + g + |f - g|) \quad \text{and} \quad \min(f, g) = \tfrac{1}{2}(f + g - |f - g|).$$

Hence for any f and g in $\mathscr{A}$, we conclude that $\max(f, g)$ and $\min(f, g)$ are in $\mathscr{L}$. Next, suppose that f and g are in $\mathscr{L}$. We show that $\max(f, g)$ and $\min(f, g)$ are in $\mathscr{L}$. To see this, choose sequences $f_n, g_n, n = 1, 2, \ldots$, of functions in $\mathscr{A}$ which approximate f and g uniformly on S. Then $F_n \equiv \max(f_n, g_n)$ and $G_n \equiv \min(f_n, g_n)$ are functions in $\mathscr{L}$. It is clear from the formula

$$F_n = \max(f_n, g_n) = \tfrac{1}{2}(f_n + g_n + |f_n - g_n|)$$

and the similar formula for G_n that $F_n \to \max(f, g)$ and $G_n \to \min(f, g)$ as $n \to \infty$. Thus $\max(f, g)$ and $\min(f, g)$ are in $\mathscr{L}$ whenever f and g are.

Finally, let f be any element in $C(S)$. According to Theorem 15.23, the function f can be approximated uniformly by elements of $\mathscr{L}$. Therefore given $\varepsilon > 0$, there is a function f_1 in $\mathscr{L}$ such that $|f(p) - f_1(p)| < \tfrac{1}{2}\varepsilon$ for all $p \in S$. Moreover, there is a function f_2 in $\mathscr{A}$ such that $|f_1(p) - f_2(p)| < \tfrac{1}{2}\varepsilon$ for all $p \in S$. We conclude that $|f(p) - f_2(p)| < \varepsilon$ for all p in S, and so $\mathscr{L} = C(S)$. $\square$

Corollary (Weierstrass approximation theorem). *Let S be a closed bounded set in $\mathbb{R}^N$. Then any continuous function on S can be approximated uniformly on S by a polynomial in the coordinates $x_1, x_2, \ldots, x_N$.*

To establish the Corollary it is only necessary to show that the set of polynomials separates points of S. We leave the details to the reader.

PROBLEMS

1. Given the function $f(x) = \exp(-1/(x^2(1 - x)^2))$ for $x \neq 0, 1$ and $f(0) = f(1) = 0$, show that $f^{(n)}(0) = f^{(n)}(1) = 0$ for $n = 1, 2, \ldots$.

2. Let G be a bounded region in $\mathbb{R}^N$ with smooth boundary ∂G. To each x_0 of G we can associate an infinitely differentiable mollifier $\varphi(x)$ which vanishes outside some neighborhood of x_0 contained in G. Show that it is possible to obtain a **partition of unity**. That is, find k functions $\psi_1, \psi_2, \ldots, \psi_k$, each infinitely differentiable, each vanishing outside some neighborhood in G, and such that $\psi_1(x) + \cdots + \psi_k(x) = 1$ for all $x \in G$. [*Hint*: Use the Heine–Borel theorem to obtain $\varphi_1, \varphi_2, \ldots, \varphi_k$ covering $\bar{G}$ and then set $\psi_i(x) = \varphi_i(x)/\sum_{i=1}^{k} \varphi_i(x)$.]

3. Prove Lemma 15.8.

4. Show that the set P of all polynomials in x defined on $I = \{x: -1 \leqslant x \leqslant 1\}$ forms an algebra in $C(I)$ which separates points in I and $\mathbb{R}^1$. However, show that P is not a subset of the type $\mathscr{L}$ described in the Stone approximation theorem (Theorem 15.23).

5. Let $I = \{x: 0 \leqslant x \leqslant 1\}$ be given and let $\mathscr{L}$ be the subset of $C(I)$ consisting of all piecewise linear (continuous) functions. Show that $\mathscr{L}$ satisfies the hypotheses of Theorem 15.23.

6. Show that Theorem 15.24 fails to hold if S is not compact. [*Hint*: Let $I = \{x: 0 \leqslant x < \infty\}$, let $\mathscr{A}$ be the collection of all polynomials, and consider the function $f(x) = e^x$.]

7. Let A be a hypercube in $\mathbb{R}^N$. Show that the space $C(A)$ is separable. [*Hint*: Consider the family of all polynomials in N variables which have rational coefficients.]

8. Let $I = \{x: 0 \leqslant x \leqslant 1\}$ be given and denote by $\mathscr{L}$ the collection of all members f in $C(I)$ such that $f(0) = 0$. Is $\mathscr{L}$ dense in $C(I)$?

9. Let $I = \{x: 0 < x \leqslant 1\}$ and $C(I)$ be given. Show that the conclusion of the Stone–Weierstrass theorem fails for the algebra $\mathscr{A}$ of all polynomials on I in spite of the fact that $\mathscr{A}$ satisfies the hypotheses of Theorem 15.24.

10. Prove that it is not possible to have a mollifier in $\mathbb{R}^N$ which is analytic everywhere.

11. Employ the following technique to provide an elementary proof of Lemma 15.9: Define

$$b_n = \left(1 - \frac{1}{2}\right)\left(1 - \frac{1}{4}\right)\cdots\left(1 - \frac{1}{2n-2}\right), \qquad n = 2, 3, \ldots,$$

and let $a_n = (-1)^{n+1}b_n/2n$. Show that the series (15.25) has the form $1 + \frac{1}{2}x + \sum_{n=2}^{\infty} a_n x^n$. Use the fact that $\log(1 - x) \leqslant -x$ for $0 < x < 1$ to obtain the inequality

$$\log b_n \leqslant -\frac{1}{2}\left(1 + \frac{1}{2} + \cdots + \frac{1}{n-1}\right) \leqslant -\frac{1}{2}\log n.$$

Finally show that $b_n \leqslant 1/\sqrt{n}$, and consequently that $|a_n| \leqslant 1/2n^{3/2}$.

12. Let $k_3(x) = k(x_1)k(x_2)k(x_3)$ where k is the function defined on page 404 and $x = (x_1, x_2, x_3) \in \mathbb{R}^3$. Describe the set in $\mathbb{R}^3$ where k is analytic and the set where k is C^∞ but not analytic.

13. Let $\varphi(x, y)$ be a C^∞ function for $0 \leqslant x \leqslant 1, 0 \leqslant y \leqslant 1$ and suppose $f(y)$ is defined for $0 \leqslant y \leqslant 1$. What conditions must f satisfy in order that the function

$$f_\varphi(x) = \int_0^1 \varphi(x, y)f(y)\,dy$$

is C^∞ for $0 \leqslant x \leqslant 1$?

14. Prove the Corollary to Theorem 15.24.

CHAPTER 16

Vector Field Theory; the Theorems of Green and Stokes

16.1. Vector Functions on $\mathbb{R}^1$

Let $\overrightarrow{OP}$ be the directed line segment in $\mathbb{R}^N$ having its base at the origin and its head at the point $P = (1, 0, \ldots, 0)$. We define the unit vector e_1 as the equivalence class of all directed line segments of length 1 which are parallel to $\overrightarrow{OP}$ and directed similarly. By considering directed line segments from the origin to the points $(0, 1, 0, \ldots, 0), (0, 0, 1, \ldots, 0), \ldots, (0, 0, \ldots, 0, 1)$, we obtain the set of unit vectors $e_1, e_2, \ldots, e_N$. We denote by $V_N(\mathbb{R}^N)$ or simply V_N the linear space formed by taking all linear combinations of these unit vectors with real scalars. That is, any vector v in V_N is of the form

$$v = a_1 e_1 + a_2 e_2 + \cdots + a_N e_N$$

where the a_i are real numbers. Addition of vectors and multiplication of vectors by scalars follow the usual rules for a linear space and are a direct generalization of the rules for vectors in two and three dimensions which the reader has encountered earlier.[1] The **length** of a vector, denoted $|v|$, is

$$|v| = (a_1^2 + a_2^2 + \cdots + a_N^2)^{1/2}.$$

Let D be a subset of $\mathbb{R}^1$ and suppose that $f_1, f_2, \ldots, f_N$ are functions each of which is a mapping from D into $\mathbb{R}^1$. The mapping $f: D \to V_N$ defined by

$$f(t) = f_1(t)e_1 + f_2(t)e_2 + \cdots + f_N(t)e_N, \qquad t \in D, \tag{16.1}$$

defines f as a **vector function from D into V_N**.

Let $f: D \to V_N$ be given and suppose that c is in V_N. Then $f(t)$ **tends to c as**

---

[1] Appendix 4 gives a brief introduction to vectors in N-dimensional Euclidean space.

t **tends to** a if and only if $|f(t) - c| \to 0$ as $t \to a$. We write

$$\lim_{t \to a} f(t) = c,$$

and observe that if $c = c_1 e_1 + \cdots + c_N e_N$, then $f(t) \to c$ if and only if $f_i(t) \to c_i$, $i = 1, 2, \ldots, N$, as $t \to a$.

Given $f: D \to V_N$, we define the **derivative of f at a** by the formula

$$f'(a) = \lim_{h \to 0} \frac{f(a + h) - f(a)}{h}. \tag{16.2}$$

If f is given by (16.1), the $f'(a)$ exists if and only if the ordinary derivatives $f_i'(a)$ exist for $i = 1, 2, \ldots, N$. We have the formula

$$f'(a) = \sum_{i=1}^{N} f_i'(a) e_i.$$

If P is in $\mathbb{R}^N$, we denote by $v(\overrightarrow{OP})$ the vector in V_N which has the directed line segment $\overrightarrow{OP}$ as one of its representatives. In terms of such directed line segments, the derivative given by (16.2) has the geometric interpretation shown in Figure 16.1.

In terms of a Cartesian coordinate system $(x_1, x_2, \ldots, x_N)$ in $\mathbb{R}^N$, the function f is a mapping from D into $\mathbb{R}^N$ given by

$$x_i = f_i(t), \qquad i = 1, 2, \ldots, N, t \in D. \tag{16.3}$$

If all the f_i and f_i' are continuous in a neighborhood of some point t_0 in D, it may be possible to eliminate the parameter t in (16.3). If one of the numbers $f_i'(t_0), i = 1, 2, \ldots, N$, is different from zero, say $f_k'(t_0)$, then there is an interval $t_0 - h \leqslant t \leqslant t_0 + h$ on which f_k' does not vanish (assuming this interval is contained in D). In such a case, we may employ the Implicit function theorem to write

$$t = g_k(x_k) \quad \text{with } f_k[g_k(t)] = t$$

in some neighborhood of t_0. Therefore, by substitution in (16.3), we find

$$x_1 = f_1[g_k(x_k)], \qquad x_2 = f_2[g_k(x_k)], \ldots, \qquad x_N = f_N[g_k(x_k)],$$

and the parameter t has been eliminated.

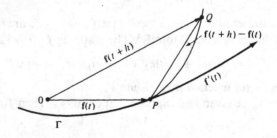

Figure 16.1. Geometric interpretation of the derivative.

an arc   not an arc
(a)   (b)

Figure 16.2

Let f be a continuous function from a closed interval I in $\mathbb{R}^1$ into V_N. If the mapping $f: I \to V_N$ is one-to-one, we say that the range of f is an **arc in** V_N (see Figure 16.2). In general, many functions with range in V_N will have the same arc as an image. Each such function determines a parametric representation of the arc in terms of a parameter t which takes on values in a domain I. The following lemma is needed to show the relationship between any two parameters which represent the same arc.

Lemma 16.1. *Let I be an interval in $\mathbb{R}^1$, and suppose that $S: I \to \mathbb{R}^1$ is a continuous function such that $S(t_1) \neq S(t_2)$ for $t_1 \neq t_2$. That is, S is one-to-one. Then S is strictly monotone on I.*

PROOF. Suppose that S is not strictly monotone. Then there are two points t_1, t_2 in I such that

$$t_2 > t_1 \quad \text{and} \quad S(t_2) > S(t_1)$$

and, in addition, there are two points t_3, t_4 in I such that

$$t_4 > t_3 \quad \text{and} \quad S(t_4) < S(t_3).$$

We define

$$\varphi(s) = S[t_2 + s(t_4 - t_2)] - S[t_1 + s(t_3 - t_1)], \qquad 0 \leqslant s \leqslant 1,$$

and clearly φ is continuous since S is. Moreover, $\varphi(0) > 0$, $\varphi(1) < 0$. Hence there is a value $\bar{s}$ such that $\varphi(\bar{s}) = 0$. That is,

$$S[t_2 + \bar{s}(t_4 - t_2)] = S[t_1 + \bar{s}(t_3 - t_1)].$$

On the other hand, $t_2 + \bar{s}(t_4 - t_2) > t_1 + \bar{s}(t_3 - t_1)$ for all $\bar{s}$ between 0 and 1 and so we contradict the fact that S is one-to-one. We conclude that S is strictly monotone. $\qquad \square$

Theorem 16.1. *Let $I = \{x: a \leqslant x \leqslant b\}$ and $J = \{x: c \leqslant x \leqslant d\}$ be intervals of $\mathbb{R}^1$, and suppose that $f: I \to V_N$, $g: J \to V_N$ are continuous functions which have the same arc C as image. Then there is a continuous monotone function S from I onto J such that*

$$f(t) = g[S(t)] \quad \text{for } t \in I.$$

PROOF. Since g is one-to-one and continuous, we may apply Part (d) of Theorem 6.42 to conclude that g^{-1} is a continuous map of the arc C onto the interval J. Since the composition of continuous functions is continuous, the function $S = g^{-1}f$ is a one-to-one continuous map of I onto J. Lemma 16.1 yields the final result.   □

Definitions. A **path** is a class of continuous functions from intervals in $\mathbb{R}^1$ into V_N, any two of which are related as in Theorem 16.1. Any function in this class is called a **parametric representation** (or simply a **representation**) of the path. If every two representations in the class are related by an increasing $S(t)$, as given in Theorem 16.1, then the path is said to be a **directed path**. A representation f of a path will sometimes be identified with the path itself. That is, we write "let f be a path..." when there is no danger of confusion.

Remarks. The definitions of path and arc remain unchanged for functions with domain an interval of $\mathbb{R}^1$ and range in an arbitrary metric space. Since Theorem 6.42 holds for continuous functions with range in any metric space, the proof of Theorem 16.1 is unchanged in this more general setting. Similarly, the following results on the length of paths in V_N could be extended to paths in any metric space.

Definitions. Let $f: I \to V_N$ be a continuous function. We define the **length of the path** f by the formula

$$l(f) = \sup \sum_{i=1}^{n} |f(t_i) - f(t_{i-1})| \tag{16.4}$$

where the supremum is taken over all subdivisions

$$\Delta: a = t_0 < t_1 < \cdots < t_n = b \text{ of the interval } I = \{x: a \leqslant x \leqslant b\}.$$

If $l(f)$ is finite, we say the path is **rectifiable**.

Remark. The length of a path $f: I \to V_1$ is given by the supremum of $\sum_{i=1}^{n} |f(t_i) - f(t_{i-1})|$ where $f: I \to \mathbb{R}^1$ is a continuous function. Recalling the definition of the total variation $V_a^b f$, of a function defined on $I = \{x: a \leqslant x \leqslant b\}$, we see at once that $l(f) = V_a^b f$ in this case.

The next result is similar to Theorems 12.1 and 12.2.

Theorem 16.2

(a) *Let $f: I \to V_N$ be a path, and suppose that*

$$\Delta: a = T_0 < T_1 < \cdots < T_n = b$$

is a subdivision of I. Define g_k as the restriction of f to the interval $I_k = \{t: T_{k-1} \leqslant t \leqslant T_k\}$. Then

$$l(f) = l(g_1) + l(g_2) + \cdots + l(g_n).$$

(b) *Let $I_T = \{t: a \leqslant t \leqslant T\}$ and define f_T as the restriction of f to I_T. Define $s(T) = l(f_T)$. Then s is a continuous function for T on I.*

The proof of Part (a) for $N = 2$ is similar to the proof of Theorem 12.1. Then the general result can be established by induction. We leave the details to the reader. The proof of Part (b) follows the procedure given in the proof of Theorem 12.2, and we again leave the details to the reader.

Part (a) of Theorem 16.2 simply states that if a path is the union of disjoint subpaths, the total length is the sum of the lengths of the subpaths.

The next result establishes the important fact that the length of an arc is independent of the parameter chosen to describe it.

Theorem 16.3. *Let f and g be parametric representations of the same arc C. Then*

$$l(f) = l(g).$$

PROOF. Let S be the function obtained in Theorem 16.1. Set

$$L_f = \sum_{i=1}^m |f(t_i) - f(t_{i-1})|, \qquad L_g = \sum_{i=1}^m |g(\tau_i) - g(\tau_{i-1})| \qquad (16.5)$$

where $\Delta: a = t_0 < t_1 < \cdots < t_m = b$ is a subdivision for f and $\Delta': c = \tau_0 < \tau_1 < \cdots < \tau_m = d$ is the subdivision of g which we obtain by setting $\tau_i = S(t_i)$ if S is an increasing function or by setting $\tau_i = S(t_{m-i})$ if S is a decreasing function. In the first case, we see that $g(\tau_i) = f(t_i)$ for all i and in the second case $g(\tau_i) = f(t_{m-i})$. In both cases the sums L_f and L_g in (16.5) are equal. Hence $l(f) = l(g)$. $\qquad\square$

Theorem 16.4. *Let $f: I \to V_N$ be a continuous function. Set*

$$f(t) = f_1(t)e_1 + f_2(t)e_2 + \cdots + f_N(t)e_N.$$

Then the path f is rectifiable if and only if all the f_i, $i = 1, 2, \ldots, N$, are of bounded variation on I.

PROOF. Let $\Delta: a = t_0 < t_1 < \cdots < t_m = b$ be a subdivision of I. We have

$$|f_i(t_k) - f_i(t_{k-1})| \leqslant |f(t_k) - f(t_{k-1})| \leqslant \sum_{i=1}^N |f_i(t_k) - f_i(t_{k-1})|.$$

Therefore

$$V_a^b f_i \leqslant l(f) \leqslant \sum_{i=1}^N V_a^b f_i. \qquad\square$$

When the parametric representation of an arc is sufficiently differentiable, methods of calculus can be used to compute the length, the tangent vector at any point, and other geometric quantities associated with a curve in $\mathbb{R}^N$.

Definitions. An arc Γ in $\mathbb{R}^N$ is said to be **smooth** if and only if Γ possesses a parametric representation f on some closed interval $I = \{x: a \leqslant x \leqslant b\}$ such that $f: I \to V_N$ is continuous and f' is uniformly continuous on the interior of I with the limits of f' at a and b different from zero. An arc Γ in $\mathbb{R}^N$ is said to be **piecewise smooth** if and only if Γ has a parametric representation f on I such that (i) f is continuous on I and (ii) there is a subdivision $a = T_0 < T_1 < \cdots < T_n = b$ of I such that for $k = 1, 2, \ldots, n$ the restriction f_k of f to the subinterval $I_k = \{x: T_{k-1} \leqslant x \leqslant T_k\}$ is smooth. We do not assume that the right and left limits of f' are equal at T_k. If they are unequal at a particular T_i, we suppose that the components are not proportional. A **smooth path** and a **piecewise smooth path** are defined similarly.

Theorem 16.5 (Length of a piecewise smooth path). *Let $f: I \to V_N$ be a piecewise smooth path such that f is smooth on the subintervals $a = T_0 < T_1 < \cdots < T_n = b$. Then the length $l(f)$ of the path, is given by*

$$l(f) = \int_a^b |f'(t)| \, dt = \sum_{k=1}^n \int_{T_{k-1}}^{T_k} |f'(t)| \, dt. \tag{16.6}$$

PROOF. From Theorem 16.2 it is sufficient to prove the result for an interval I on which f is smooth. For this purpose, we introduce a subdivision of $I: a = \tau_0 < \tau_1 < \cdots < \tau_m = b$. Then the length of f is the supremum over all subdivisions of

$$\sum_{i=1}^m |f(\tau_i) - f(\tau_{i-1})| = \sum_{i=1}^m \left| \frac{f(\tau_i) - f(\tau_{i-1})}{\tau_i - \tau_{i-1}} \right| \cdot |\tau_i - \tau_{i-1}|.$$

Let $\varepsilon > 0$ be given. From the definition of length of a path, there is a subdivision $\Delta: a = u_1 < u_2 < \cdots < u_p = b$ such that

$$l(f) - \frac{1}{4}\varepsilon < \sum_{i=1}^p |f(u_i) - f(u_{i-1})| \leqslant l(f).$$

From the definition of integral, there is a $\delta > 0$ such that if $\Delta_1: a = t_0 < t_1 < \cdots < t_m = b$ is any subdivision of mesh less than δ and if $\xi_i \in [t_{i-1}, t_i]$ for each i. then

$$\left| \sum_{i=1}^m |f'(\xi_i)|(t_i - t_{i-1}) - \int_a^b |f'(x)| \, dx \right| < \frac{1}{4}\varepsilon.$$

We may assume that Δ_1 is a refinement of Δ. From the Mean-value theorem applied to the components $f_1, f_2, \ldots, f_N$ of f, there are numbers ξ_{ki} in the interval (t_{i-1}, t_i) such that

$$f_k(t_i) - f_k(t_{i-1}) = f_k'(\xi_{ki})(t_i - t_{i-1}), \qquad i = 1, 2, \ldots, m, \, k = 1, 2, \ldots, N.$$

Set $\Delta t_i = t_i - t_{i-1}$ and then

$$\sum_{i=1}^m |f(t_i) - f(t_{i-1})| = \sum_{i=1}^m \left(\sum_{k=1}^N [f_k'(\xi_{ki})]^2 \right)^{1/2} \Delta t_i. \tag{16.7}$$

If for each fixed i the numbers ξ_{ki} all have the same value, say ξ_i, then the sum on the right in (16.7) would be

$$\sum_{i=1}^{m} |f'(\xi_i)|\Delta t_i, \tag{16.8}$$

which is a Riemann sum. Then proceeding to the limit we get (16.6). Since f' is uniformly continuous on I, it is possible to show that the sum in (16.7) yields the same limit as the Riemann sum (16.8). This is known as **Duhamel's principle**. To establish this result, set $\xi'_{ki} = \xi_{1i}$, $k = 2, 3, \ldots, N$, $i = 1, 2, \ldots, m$, and consider the quantity

$$\sum_{i=1}^{m} (|f'_k(\xi_{ki})| - |f'_k(\xi'_{ki})|)\Delta t_i.$$

For each fixed k, the uniform continuity of f'_k shows that the above sum tends to zero as the maximum of the Δt_i tends to zero. We leave the remaining details to the reader. $\qquad\square$

Suppose that $f: \mathbb{R}^1 \to V_N$ is a smooth function; we define

$$s(t) = \int_a^t |f'(\tau)|\, d\tau. \tag{16.9}$$

Then s represents the "directed distance" along the curve Γ in $\mathbb{R}^N$, where Γ is the graph of the point P of the radius vector $\overrightarrow{OP}$ from the origin to a point P in $\mathbb{R}^N$. That is, Γ is the arc $f(\overrightarrow{OP})$ (see Figure 16.1). Distance is measured from the point corresponding to $f(a)$. By differentiating (16.9), we have $s'(t_0) = |f'(t_0)|$ and the unit vector $T(t_0)$ is defined by

$$T = \frac{f'(t_0)}{|f'|} = \frac{f'(t_0)}{s'(t_0)}.$$

The vector T is the **unit tangent vector** to Γ at each point of the curve. Since s is a monotone function of t we note that f and T are functions of s. Then it follows that

$$\frac{df}{ds} = T.$$

For any two vectors c, d the **scalar product** $c \cdot d$ is defined by the formula

$$c \cdot d = |c||d| \cos \theta$$

where θ is the angle between two directed line segments representing the vectors c and d having the same point as base. If c and d have the representations $c = c_1 e_1 + \cdots + c_N e_N$, $d = d_1 e_1 + \cdots + d_N e_N$, then

$$c \cdot d = c_1 d_1 + c_2 d_2 + \cdots + c_N d_N.$$

If $c \cdot d = 0$, the vectors are said to be **orthogonal**.

We now suppose that $f = f_1(s)e_1 + f_2(s)e_2 + \cdots + f_N(s)e_N$ is a C^2 function.

That is, each component $f_i(s)$ of f has a continuous second derivative. Then since $T \cdot T = 1$ for all s, we may differentiate and obtain the relation

$$T \cdot \frac{dT}{ds} + \frac{dT}{ds} \cdot T = 0 \quad \text{or} \quad T \cdot \frac{dT}{ds} = 0.$$

Hence T and dT/ds are orthogonal vectors. The **scalar curvature** κ of Γ is defined by

$$\kappa = \left| \frac{dT}{ds} \right|.$$

The quantity κ measures the rate of change of the direction of Γ with respect to the arc length. We define the **principal normal** N to Γ as the unit vector in the direction of dT/ds. We have

$$\frac{dT}{ds} = \kappa N.$$

The quantity $R = 1/\kappa$ is called the **radius of curvature** of Γ.

If f is sufficiently differentiable, we may differentiate the relation $N \cdot N = 1$ and find that dN/ds is orthogonal to N. We express dN/ds in the form

$$\frac{dN}{ds} = \kappa_2 N_1 + \beta T$$

where N_1 is a unit vector orthogonal to the subspace determined by T and N, and κ_2 is called the **second curvature** of Γ. By differentiating the relation $T \cdot N = 0$, we see that $\beta = -\kappa$. We may continue this process by computing dN_1/ds and obtain $N - 1$ **scalar curvatures** $\kappa, \kappa_2, \ldots, \kappa_{N-1}$ corresponding to the $N - 1$ normals $N_1, \ldots, N_{N-1}$ to the curve Γ.

The above discussion is of greatest interest in ordinary Euclidean three-space. In this case the vector N_1 is called the **binormal** and is usually denoted by B. The vectors T, N, and B form an orthonormal set of vectors at each point of Γ. This set is called the **moving trihedral**. The quantity κ_2 is usually designated by $-\tau$, and τ is called the **torsion** of Γ. It is easy to verify that $\tau = 0$ for a plane curve. We choose an orientation in three-space so that

$$B = T \times N$$

where $\times$ denotes the usual vector product of two vectors in V_3. Differentiation of the above formula yields

$$\frac{dB}{ds} = \frac{dT}{ds} \times N + T \times \frac{dN}{ds} = T \times (-\tau B - \kappa T) = \tau N.$$

We obtain in this way the **Frenet formulas**

$$\frac{dT}{ds} = \kappa N, \qquad \frac{dN}{ds} = -\kappa T + \tau B, \qquad \frac{dB}{ds} = +\tau N. \tag{16.10}$$

EXAMPLE. Given the function $f: \mathbb{R}^1 \to V_3$ defined by

$$f(t) = te_1 + t^2 e_2 + \tfrac{2}{3}t^3 e_3,$$

find T, N, B, κ, and τ.

Solution. We have $f'(t) = e_1 + 2te_2 + 2t^2 e_3$. Therefore $|f'(t)| = (1 + 2t^2)$ and $s(t) = \int_0^t |f'(t)|\, dt = t + \tfrac{2}{3}t^3$, where distance is measured from $t = 0$. We compute

$$T = \frac{f'}{s'} = (1 + 2t^2)^{-1}[e_1 + 2te_2 + 2t^2 e_3].$$

Continuing the process, we obtain

$$\frac{dT}{ds} = \frac{dT}{dt}(1 + 2t^2)^{-1} = (1 + 2t^2)^{-3}[-4te_1 + (2 - 4t^2)e_2 + 4te_3].$$

Consequently,

$$\kappa = \left|\frac{dT}{ds}\right| = \frac{2}{(1 + 2t^2)^2}$$

and $N = (1 + 2t^2)^{-1}[-2te_1 + (1 - 2t^2)e_2 + 2te_3]$. To obtain B we compute $dN/ds = (dN/dt)(1 + 2t^2)^{-1}$ and then formula $dN/ds = -\kappa T - \tau B$ yields the torsion τ. We leave the remaining details to the reader. $\qquad\square$

PROBLEMS

1. Given $f: \mathbb{R}^1 \to V_N$ with $f = (f_1, f_2, \ldots, f_N)$ and $x_i = f_i(t) = t^i$, $i = 1, 2, \ldots, N$, eliminate the parameter by expressing $x_1, x_3, \ldots, x_N$ in terms of x_2. In what domain of $\mathbb{R}^1$ is this elimination valid?

2. Let $f(t): I \to V_2$ be given by $f_1(t) = t^2 - t$, $f_2(t) = t^3 - 3t$, with $I = \{t: -3 \leqslant t \leqslant 3\}$. Decide whether or not the path Γ in $\mathbb{R}^2$ represented by f is an arc.

3. Let S be a metric space and let I, J be intervals of $\mathbb{R}^1$. Suppose that $f: I \to S$ and $g: J \to S$ are continuous functions which have the same arc C as image. Prove that there is a continuous monotone function $h: I \to J$ such that $f(t) = g[h(t)]$ for $t \in I$.

4. Let S be a metric space and $f: I \to S$ a continuous function with I an interval of $\mathbb{R}^1$. Define the length of the path of f. State and prove the analog of Theorem 16.2 for such a function f.

5. State and prove the analog of Theorem 16.3 for functions from an interval I in $\mathbb{R}^1$ to a metric space S.

6. Given $f: I \to V_3$ where $I = \{t: 0 \leqslant t \leqslant 2\pi\}$ and $f = f_1 e_1 + f_2 e_2 + f_3 e_3$ with $f_1 = 3\cos t$, $f_2 = 3\sin t$, $f_3 = 2t$, find the length of the arc represented by f.

7. Let $f: I \to V_2$ where $I = \{t: 0 \leqslant t \leqslant 1\}$ and $f = f_1 e_1 + f_2 e_2$ with $f_1 = t$, and

$$f_2 = \begin{cases} \dfrac{1}{\sqrt{t}}\sin\dfrac{\pi}{t}, & t > 0, \\[2mm] 0, & t = 0. \end{cases}$$

Decide whether or not f is rectifiable.

8. Complete the proof of Theorem 16.5.

In each of Problems 9 through 12 find T, N, B, κ, and τ at the given value of t for the given function $f: \mathbb{R}^1 \to V_3$.

9. $f(t) = e^t[(\cos t)e_1 + (\sin t)e_2 + e_3]$, $t = 0$.

10. $f(t) = (1/3)t^3 e_1 + 2te_2 + (2/t)e_3$, $t = 2$.

11. $f(t) = te_1 + (3/2)t^2 e_2 + (3/2)t^3 e_3$, $t = 2$.

12. $f(t) = (t \cos t)e_1 + (t \sin t)e_2 + te_3$, $t = 0$.

13. Let $f: I \to V_N$ be the representation of a curve Γ in $\mathbb{R}^N$. Suppose that f is a C^N function. Let T be the tangent vector and let s denote arc length. Then $dT/ds = \kappa N$ where N is orthogonal to T. Differentiate N with respect to s and obtain a unit vector N_1 such that

$$\frac{dN}{ds} = -\kappa T + \kappa_1 N_1.$$

Differentiate N_1 with respect to s and obtain a unit vector N_2 such that

$$\frac{dN_1}{ds} = -\kappa_1 N + \kappa_2 N_2.$$

Continue this process and obtain a sequence of N mutually orthogonal unit vectors $T, N, N_1, \ldots, N_{N-2}$ and the formulas

$$\frac{dN_k}{ds} = -\kappa_k N_{k-1} + \kappa_{k+1} N_{k+1}, \qquad k = 2, 3, \ldots, N - 3.$$

Finally, show that $dN_{N-2}/ds = -\kappa_{N-2} N_{N-3}$. The quantities $\kappa, \kappa_1, \ldots, \kappa_{N-2}$ are called the **curvatures** of Γ.

14. Given $f: I \to V_3$ where $I = \{t: a \leqslant x \leqslant b\}$ and suppose that f is a C^3 function.
    (a) Prove that

$$\kappa^2 = \frac{(f' \times f'') \cdot (f' \times f'')}{|f'|^2}.$$

    (b) Prove that

$$\tau = \frac{f' \cdot (f'' \times f''')}{|f' \times f''|^2}.$$

In each of Problems 15 through 18 find the length of the given arc.

15. $f(t) = te_1 + \frac{3}{2}t^2 e_2 + \frac{3}{2}t^3 e_3$, $I = \{t: 0 \leqslant t \leqslant 2\}$.

16. $f(t) = te_1 + \frac{1}{2}\sqrt{2}t^2 e_2 + \frac{1}{3}t^3 e_3$, $I = \{t: 0 \leqslant t \leqslant 2\}$.

17. $f(t) = t \cos te_1 + t \sin te_2 + te_3$, $I = \{t: 0 \leqslant t \leqslant \pi/2\}$.

18. $f(t) = te_1 + \log(\sec t + \tan t)e_2 + \log \sec te_3$, $I = \{t: 0 \leqslant t \leqslant \pi/4\}$.

19. State and prove the analog of Theorem 16.1 when the range of f and g are in a metric space.

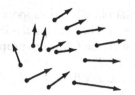

Figure 16.3. A vector field.

16.2. Vector Functions and Fields on $\mathbb{R}^N$

In this section we develop the basic properties of differential and integral calculus for functions whose domain is a subset of $\mathbb{R}^M$ and whose range is in a vector space V_N. We shall emphasize the coordinate-free character of the definitions and results, but we shall introduce coordinate systems and use them in proofs and computations whenever convenient.

Definition. A **vector function** from a domain D in $\mathbb{R}^M$ into V_N is a mapping $f: D \to V_N$. If the image is V_1 (which we identify with $\mathbb{R}^1$ in the natural way) then f is called a **scalar function**. Vector or scalar functions are often called **vector** or **scalar fields**. The term field is used because we frequently visualize a vector function as superimposed on $\mathbb{R}^M$. At each point of the domain D of $\mathbb{R}^M$ a representative directed line segment of f is drawn with base at that point. We obtain a **field** of vectors (see Figure 16.3 where $M = N = 2$).

Let $f: D \to V_N$ be a vector function where D is a domain in $\mathbb{R}^M$. An **orthonormal basis** for V_N is a set of N unit vectors $e_1, e_2, \ldots, e_N$ such that $e_i \cdot e_j = 0$ for $i \neq j$, $i, j = 1, 2, \ldots, N$. Then for each point $P \in D$, we may represent f by the formula

$$f(P) = \sum_{i=1}^{N} f_i(P)e_i \qquad (16.11)$$

where $f_i: D \to \mathbb{R}^1$, $i = 1, 2, \ldots, N$, are scalar functions from D into $\mathbb{R}^1$. We can consider vector fields with domain in any space such as V_M or E_M (Euclidean space). Equation (16.11) illustrates what is meant in that case. We shall ordinarily assume that D is in $\mathbb{R}^M$. The function f is **continuous at** P if and only if each f_i, $i = 1, 2, \ldots, N$, is continuous at P. We recall that the basic properties of functions from $\mathbb{R}^M$ into $\mathbb{R}^1$ were developed in Chapters 6, 7, and 8. Properties developed there which involve linear processes are easily transferred to functions f from $\mathbb{R}^M$ into V_N.

For simplicity we shall usually consider functions f from $\mathbb{R}^N$ into V_N, although many of the results developed below are valid with only modest changes if the dimension of the domain of f is different from that of its range.

The definite integral of a vector function, defined below, is similar to the integral of a scalar function on $\mathbb{R}^N$ as given in Chapter 8.

Definition. Let D be a domain in $\mathbb{R}^N$ and suppose that $u: D \to V_N$ is a vector function. Let F be a figure in D. The function u is **integrable over** F if and only if there is a vector L in V_N with the following property: for every $\varepsilon > 0$, there is a $\delta > 0$ such that if $\Delta = \{F_1, F_2, \ldots, F_n\}$ is any subdivision of F with mesh less than δ and $p_1, p_2, \ldots, p_n$ is any finite set of points with p_i in F_i, then

$$\left| \sum_{i=1}^{n} u(p_i)V(F_i) - L \right| < \varepsilon,$$

where $V(F_i)$ is the volume of F_i. We call L the **integral of u over F** and we write

$$L = \int_F u \, dV.$$

If, for example, u is given in the form (16.11) so that

$$u(P) = \sum_{i=1}^{N} u_i(P)e_i, \tag{16.12}$$

then it is clear that u is integrable over F if and only if each scalar function u_i is integrable over F. Therefore if L exists it must be unique. In fact, if u is integrable and given by (16.12), we have

$$\int_F u \, dV = \sum_{i=1}^{N} \left(\int_F u_i \, dV \right) e_i. \tag{16.13}$$

Each coefficient on the right in (16.13) is an integral of a function from $\mathbb{R}^N$ into $\mathbb{R}^1$ as defined in Chapter 8.

Let v be a vector in V_N and denote by $\overrightarrow{p_0 p}$ a directed line segment in $\mathbb{R}^N$ which represents v. As usual, we use the notation $v(\overrightarrow{p_0 p})$ for this vector.

Definitions. Let p_0 be a point in $\mathbb{R}^N$ and let a be a unit vector in V_N. Suppose that $w: D \to V_N$ is a vector function with domain D in $\mathbb{R}^N$. We define the **directional derivative of w in the direction a** at p_0, denoted by $D_a w(p_0)$, by the formula

$$D_a w(p_0) = \lim_{h \to 0} \frac{w(p) - w(p_0)}{h}$$

where the point $p \in D$ is chosen so that $v(\overrightarrow{p_0 p}) = ha$ (see Figure 16.4). The vector function w is **continuously differentiable** in D if and only if w and $D_a w$ are continuous on D for every unit vector a in V_N. We write $w \in C^1(D)$.

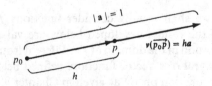

Figure 16.4

In order to derive formulas for computing directional derivatives we require a coordinate system in $\mathbb{R}^N$. When we write $w(p)$, $p \in \mathbb{R}^N$ we indicate the coordinate-free character of the function. If a Cartesian coordinate system $(x_1, x_2, \ldots, x_N)$ is introduced, then we write $w(x_1, x_2, \ldots, x_N)$ or $w(x)$. If e_1, $e_2, \ldots, e_N$ is an orthonormal basis in V_N, we use coordinates to write

$$w(x) = \sum_{i=1}^{N} w_i(x)e_i,$$

where each of the scalar functions $w_i(x)$ is now expressed in terms of a coordinate system. Strictly speaking we should use different symbols for a function w when expressed in terms of a particular coordinate system as compared with w defined in a purely geometric manner. However, it will be clear from the context whenever coordinates are used. Furthermore, we omit the statement and proofs of the theorems which show, for example, that a vector function w has a directional derivative if and only if w has one is an arbitrary Cartesian coordinate system, that w is continuous if and only if $w(x)$ is, and so forth.

The following theorem establishes the formula for obtaining the directional derivative of scalar and vector fields.

Theorem 16.6

(i) *Let a be a unit vector in V_N. Suppose that $w: D \to \mathbb{R}^1$ is a continuously differentiable scalar field with domain $D \subset \mathbb{R}^N$. If a has the representation $a = a_1 e_1 + a_2 e_2 + \cdots + a_N e_N$, then*

$$D_a w = \sum_{i=1}^{N} \frac{\partial w}{\partial x_i} a_i. \tag{16.14}$$

(ii) *If $w: D \to V_N$ is a vector field and $w = \sum_{i=1}^{N} w_i e_i$ then*

$$D_a w = \sum_{i=1}^{N} D_a w_i e_i = \sum_{i,j=1}^{N} w_{i,j} a_j e_i. \tag{16.15}$$

PROOF. (i) If a is a unit vector in one of the coordinate directions, say x_j, then (16.14) holds since $D_a w$ is the partial derivative with respect to x_j. In the general case, we fix a point $x^0 = (x_1^0, \ldots, x_N^0)$ and define

$$\varphi(t) = w(x_1^0 + a_1 t, \ldots, x_N^0 + a_N t).$$

Then $D_a w(x^0)$ is obtained by computing $\varphi'(0)$ according to the Chain rule, Theorem 7.3. The proof of (ii) is an immediate consequence of formula (16.14) and the representation of w in the form $\sum w_i e_i$. $\square$

The next theorem shows that the directional derivative of a scalar function may be expressed in terms of the scalar product of the given direction a with a uniquely determined vector field. Observe that the result is independent of the coordinate system, although coordinates are used in the proof.

Theorem 16.7. *Let* $f: D \to \mathbb{R}^1$ *be a continuously differentiable scalar field with domain D in $\mathbb{R}^N$. Then there is a unique vector field* $w: D \to V_N$ *such that for each unit vector $a \in V_N$ and for each $p \in D$, the directional derivative of f is given by the scalar product of a and w:*

$$D_a f(p) = a \cdot w(p). \tag{16.16}$$

If $e_1, e_2, \ldots, e_N$ is an orthonormal set of unit vectors in V_N, then $w(x)$ may be computed by the formula

$$w(x) = \sum_{j=1}^{N} f_{,j}(x)e_j \quad \text{for each } x \in D. \tag{16.17}$$

PROOF. From Formula (16.14), we find

$$D_a f(x) = \sum_{j=1}^{N} f_{,j}(x)a_j. \tag{16.18}$$

We now define $w(x)$ by (16.17) and therefore the scalar product of w and a yields (16.18). To show uniqueness, assume that w' is another vector field satisfying (16.16). Then $(w - w') \cdot a = 0$ for all unit vectors a. If $w - w' \neq 0$, choose a to be the unit vector in the direction of $w - w'$, in which case $(w - w') \cdot a = |w - w'| \neq 0$, a contradiction. $\qquad\square$

Definition. The vector w defined by (16.16) in Theorem 16.7 is called the **gradient** *of f* and is denoted by grad f or ∇f.

Remark. If f is a scalar field from $\mathbb{R}^N$ into $\mathbb{R}^1$, then in general $f(p) = $ const. represents a hypersurface in $\mathbb{R}^N$. At any point p, a vector a tangent to this hypersurface at p has the property that $\sum_{i=1}^{N} f_{,i}(p) \cdot a_i = 0$ where $a = \sum_{i=1}^{N} a_i e_i$. We conclude that ∇f *is orthogonal to the hypersurface $f(p) = $ constant at each point p on the hypersurface.*

EXAMPLE 1. Let e_1, e_2, e_3 be an orthonormal set of vectors in $V_3(\mathbb{R}^3)$. Given the scalar and vector functions

$$f(x_1, x_2, x_3) = x_1^3 + x_2^3 + x_3^3 - 3x_1 x_2 x_3$$

$$u(x_1, x_2, x_3) = (x_1^2 - x_2 + x_3)e_1 + (2x_2 - 3x_3)e_2 + (x_1 + x_3)e_3$$

and the vector $a = \lambda e_1 + \mu e_2 + \nu e_3$, $\lambda^2 + \mu^2 + \nu^2 = 1$, find ∇f and $D_a u$ in terms of $x_1, x_2, x_3, e_1, e_2, e_3$.

Solution. According to (16.17), we find

$$\nabla f = 3(x_1^2 - x_2 x_3)e_1 + 3(x_2^2 - x_1 x_3)e_2 + 3(x_3^2 - x_1 x_2)e_3.$$

Employing (16.15), we obtain

$$D_a w = (2x_1 \lambda - \mu + \nu)e_1 + (2\mu - 3\nu)e_2 + (\lambda + \nu)e_3. \qquad\square$$

Theorem 16.8. *Suppose that $f, g,$ and u are C^1 scalar fields with domain $D \subset \mathbb{R}^N$.*

Let $h: \mathbb{R}^1 \to \mathbb{R}^1$ be a C^1 function with the range of u in the domain of h. Then

$$\nabla(f + g) = \nabla f + \nabla g, \qquad \nabla(fg) = f\nabla g + g\nabla f,$$

$$\nabla\left(\frac{f}{g}\right) = \frac{1}{g^2}(g\nabla f - f\nabla g) \quad \text{if } g \neq 0, \qquad \nabla h(u) = h'(u)\nabla u.$$

PROOF. We prove the second formula, the remaining proofs being left to the reader. Let $e_1, e_2, \ldots, e_N$ be an orthonormal set in V_N. Then from the formula for $\nabla(fg)$ given by (16.17) it follows that

$$\nabla(fg) = \sum_{j=1}^{N} (fg)_{,j} e_j.$$

Performing the differentiations, we find

$$\nabla(fg) = \sum_{j=1}^{N} (fg_{,j} + f_{,j}g)e_j = f\nabla g + g\nabla f. \qquad \square$$

The operator ∇ carries a continuously differentiable scalar field from $\mathbb{R}^N$ to $\mathbb{R}^1$ into a continuous vector field from $\mathbb{R}^N$ to V_N. In a Cartesian coordinate system, we may write ∇ symbolically according to the formula

$$\nabla = \sum_{j=1}^{N} \frac{\partial}{\partial x_j} e_j.$$

However, the operator ∇ has a significance independent of the coordinate system. Suppose that w is a vector field from $\mathbb{R}^N$ into V_N which in coordinates may be written $w = \sum_{j=1}^{N} w_j e_j$. We define the operator $\nabla \cdot w$ by the formula

$$\nabla \cdot w = \sum_{j=1}^{N} \frac{\partial w_j}{\partial x_j},$$

and we shall show that this operator, called the **divergence operator**, is independent of the coordinate system (Theorem 16.9 below).

Definitions. The **support** of a scalar function f with domain D is the closure of the set of points p in D where $f(p) \neq 0$. If the support of a function is a compact subset of D, we say that f has **compact support in** D. If f is a C^n function for some nonnegative integer n, with compact support, we use the symbol $f \in C_0^n(D)$ to indicate this fact.

Lemma 16.2. *Let D be an open figure in $\mathbb{R}^N$ and suppose that $f \in C_0^1(D)$. Then*

$$\int_D f_{,j}(x) \, dV = 0, \qquad j = 1, 2, \ldots, N.$$

PROOF. Let S be the set of compact support of f. Then $f = 0$ and $\nabla f = 0$ on $D - S$. Let R be any hypercube which contains $\bar{D}$ in its interior. We extend the definition of f to be zero on $R - D$. Then integrating with respect to x_j

first and the remaining $N - 1$ variables next, we find (in obvious notation)

$$\int_D f_{,j}(x)\, dV_N = \int_R f_{,j}(x)\, dV_N = \int_{R'} \left[\int_{a_j}^{b_j} f_{,j}(x)\, dx_j \right] dV_{N-1}$$

$$= \int_{R'} [f(b_j, x') - f(a_j, x')]\, dV_{N-1} = 0. \qquad \square$$

The next lemma, useful in many branches of analysis, shows that a continuous function f must vanish identically if the integral of the product of f and all arbitrary smooth functions with compact support is always zero.

Lemma 16.3. *Let D be an open figure in $\mathbb{R}^N$ and suppose that the scalar function f is continuous on D. If*

$$\int_D fg\, dV_N = 0 \qquad \text{for all} \quad g \in C_0^1(D), \tag{16.19}$$

then $f \equiv 0$ on D.

PROOF. We prove the result by contradiction. Suppose there is an $x^0 \in D$ such that $f(x^0) \neq 0$. We may assume $f(x^0) > 0$, otherwise we consider $-f$. Since f is continuous and D is open, there is a ball in D of radius $3r_0$ and center x^0 on which $f > 0$. Denoting distance from x^0 by r, we define the function

$$g(x) = \begin{cases} 1 & \text{for } 0 \leqslant r \leqslant r_0, \\ \dfrac{1}{r_0^3}(2r - r_0)(r - 2r_0)^2 & \text{for } r_0 \leqslant r \leqslant 2r_0, \\ 0 & \text{for } r \geqslant 2r_0. \end{cases}$$

It is easily verified that $g(x) \in C_0^1(D)$ with $g(x) \geqslant 0$. We have

$$\int_D fg\, dV_N = \int_{B(x^0, 3r_0)} fg\, dV \geqslant \int_{B(x^0, r_0)} fg\, dV > 0,$$

which contradicts (16.19). Hence $f \equiv 0$ on D. $\qquad \square$

Remark. Lemma 16.3 remains valid if the functions g in formula (16.19) are restricted to the class $C_0^\infty(D)$. To see this merely replace g by its mollifier as defined in Chapter 15 and proceed with the same method of proof.

We now show that the divergence operator is independent of the coordinate system.

Theorem 16.9. *Let D be an open figure in $\mathbb{R}^N$ and suppose that $w \in C^1(D)$ is a vector field. Then there is a unique scalar field v, continuous on D, such that for all $u \in C_0^1(D)$, we have*

$$\int_D (uv + \nabla u \cdot w)\, dV = 0. \tag{16.20}$$

Furthermore, if $e_1, e_2, \ldots, e_N$ is an orthonormal basis and if, for each point $x \in D$, w is given by

$$w(x) = \sum_{j=1}^{N} w_j(x)e_j,$$

then

$$v(x) = \sum_{j=1}^{N} \frac{\partial}{\partial x_j} w_j(x). \tag{16.21}$$

PROOF. Suppose that w is given and that $e_1, \ldots, e_N$ is an orthonormal basis in some coordinate system. We *define* v by (16.21). Let u be any function in class $C_0^1(D)$. Then, according to Lemma 16.2,

$$\int_D (uv + \nabla u \cdot w) \, dV = \int_D \sum_{j=1}^{N} \left(u \frac{\partial}{\partial x_j} w_j + \frac{\partial u}{\partial x_j} w_j \right) dV$$

$$= \sum_{j=1}^{N} \int_D \frac{\partial}{\partial x_j} (uw_j) \, dV = 0.$$

To show that v is unique, suppose that v' is another scalar field which satisfies (16.20). By subtraction we get

$$\int_D (v - v')u \, dV = 0 \qquad \text{for all} \quad u \in C_0^1(D).$$

Thus $v' \equiv v$ according to Lemma 16.3. $\qquad\square$

Definition. The scalar field v determined by (16.20) in Theorem 16.9 and defined in any coordinate system by (16.21) is called the **divergence of** w. We use the notation $v = \text{div } w$ or $v = \nabla \cdot w$. We note that v is determined in (16.20) without reference to a coordinate system, although (16.21) is used for actual computations.

Theorem 16.10. *Let D be a domain in $\mathbb{R}^N$ and suppose that $w, w_1, w_2, \ldots, w_n$ are C^1 vector fields on D. Let f be a C^1 scalar field on D and suppose that c_1, $c_2, \ldots, c_n$ are real numbers. Then*

(a) $\text{div}(\sum_{j=1}^{n} c_j w_j) = \sum_{j=1}^{n} c_j \text{ div } w_j$.
(b) $\text{div}(fw) = f \text{ div } w + \nabla f \cdot w$.

We leave the proof to the reader.

EXAMPLE 2. Suppose that the origin of a coordinate system is at the center of the earth and R is its radius. Denote by g the acceleration due to gravity at the surface of the earth. For points p in $\mathbb{R}^3$ let r be the vector having $\overrightarrow{Op}$ as a representative, and set $r = |r|$. From classical physics it is known that the vector field of force due to gravity, denoted $v(p)$ and called the **gravitational**

field of the earth, is given approximately by

$$v(p) = -\frac{gR^2}{r^3}r \qquad \text{for} \quad r > R.$$

Show that div $v(p) = 0$ for $r > R$.

Solution. We use Part (b) of Theorem 16.10 to obtain

$$\text{div } v = \text{div}\left(\frac{-gR^2}{r^3}r\right) = -\frac{gR^2}{r^3}\text{div } r - \left(\nabla\frac{gR^2}{r^3}\right)\cdot r.$$

Next, we introduce the unit vectors e_1, e_2, e_3 and obtain

$$r = x_1 e_1 + x_2 e_2 + x_3 e_3, \qquad \text{div } r = 1 + 1 + 1 = 3,$$

$$r = (x_1^2 + x_2^2 + x_3^2)^{1/2},$$

$$\nabla r = (x_1^2 + x_2^2 + x_3^2)^{-1/2}(x_1 e_1 + x_2 e_2 + x_3 e_3) = \frac{r}{r}.$$

Using Theorem 16.8 and the above formulas, we find

$$\text{div } v = -\frac{3gR^2}{r^3} - \left(\frac{-3gR^2}{r^4}\right)\frac{r}{r}\cdot r = 0. \qquad \Box$$

The vector or cross product of two vectors in a three-dimensional vector space allows us to introduce a new differential operator acting on smooth vector fields. We shall define the operator without reference to a coordinate system although coordinates are used in all the customary computations. If e_1, e_2, e_3 is any orthonormal basis for V_3 we may construct the *formal operator*

$$\nabla \times u = \begin{vmatrix} e_1 & e_2 & e_3 \\ \dfrac{\partial}{\partial x_1} & \dfrac{\partial}{\partial x_2} & \dfrac{\partial}{\partial x_3} \\ u_1 & u_2 & u_3 \end{vmatrix}$$

$$= \left(\frac{\partial u_3}{\partial x_2} - \frac{\partial u_2}{\partial x_3}\right)e_1 + \left(\frac{\partial u_1}{\partial x_3} - \frac{\partial u_3}{\partial x_1}\right)e_2 + \left(\frac{\partial u_2}{\partial x_1} - \frac{\partial u_1}{\partial x_1}\right)e_3. \qquad (16.22)$$

We note that if the symbol ∇ is replaced by a vector v with components v_1, v_2, v_3 and if the partial derivatives in (16.22) are replaced by these components, then we obtain the usual formula for the vector product of two vectors.

Theorem 16.11. *Let D be any set in $\mathbb{R}^3$ and let $u \in C^1(D)$ be a vector field into V_3. Suppose that $\mathbb{R}^3$ is given one of its two possible orientations. Then there is a unique continuous vector field w from D into V_3 such that*

$$\text{div}(u \times a) = w \cdot a \qquad (16.23)$$

for every constant vector a. If e_1, e_2, e_3 is any orthonormal basis of a coordinate system consistent with the orientation of $\mathbb{R}^3$, then the vector w is given by (16.22).

PROOF. With the orthonormal basis e_1, e_2, e_3 given, we write $a = a_1 e_1 + a_2 e_2 + a_3 e_3$, $u = u_1 e_1 + u_2 e_2 + u_3 e_3$. The formula for the vector product yields

$$u \times a = (u_2 a_3 - u_3 a_2)e_1 + (u_3 a_1 - u_1 a_3)e_2 + (u_1 a_2 - u_2 a_1)e_3.$$

Using (16.21) for the divergence formula, we find

$$\operatorname{div}(u \times a) = a_3 \frac{\partial u_2}{\partial x_1} - a_2 \frac{\partial u_3}{\partial x_1} + a_1 \frac{\partial u_3}{\partial x_2} - a_3 \frac{\partial u_1}{\partial x_2} + a_2 \frac{\partial u_1}{\partial x_3} - a_1 \frac{\partial u_2}{\partial x_3}.$$

If we denote by w_1, w_2, w_3 the coefficients in the right side of (16.22), then

$$\operatorname{div}(u \times a) = w_1 a_1 + w_2 a_2 + w_3 a_3.$$

That is, $\operatorname{div}(u \times a) = w \cdot a$ where $w = w_1 e_1 + w_2 e_2 + w_3 e_3$. The vector w is unique since if w' were another such vector we would have $w \cdot a = w' \cdot a$ for all unit vectors a. This fact implies that $w = w'$.   □

Definition. The vector w in Theorem 16.11 is called the **curl of u** and is denoted by curl u and $\nabla \times u$.

The elementary properties of the curl operator are given in the next theorem.

Theorem 16.12. *Let D be a domain in $\mathbb{R}^3$ and suppose that $u, v, u_1, \ldots, u_n$ are C^1 vector fields from D into V_3. Let f be a C^1 scalar field on D and $c_1, \ldots, c_n$ real numbers. Then*

(a) $\operatorname{curl}(\sum_{j=1}^{n} c_j u_j) = \sum_{j=1}^{n} c_j(\operatorname{curl} u_j)$.
(b) $\operatorname{curl}(fu) = f \operatorname{curl} u + \nabla f \times u$.
(c) $\operatorname{div}(u \times v) = v \cdot \operatorname{curl} u - u \cdot \operatorname{curl} v$.
(d) $\operatorname{curl} \nabla f = 0$ *provided that* $\nabla f \in C^1(D)$.
(e) $\operatorname{div} \operatorname{curl} v = 0$ *provided that* $v \in C^2(D)$.

PROOF. We shall establish Part (b) and leave the remaining proofs to the reader. Let e_1, e_2, e_3 be an orthonormal basis and set $u = u_1 e_1 + u_2 e_2 + u_3 e_3$. Then

$$\operatorname{curl}(fu) = \left(\frac{\partial(fu_3)}{\partial x_2} - \frac{\partial(fu_2)}{\partial x_3} \right) e_1$$
$$+ \left(\frac{\partial(fu_1)}{\partial x_3} - \frac{\partial(fu_3)}{\partial x_1} \right) e_2 + \left(\frac{\partial(fu_2)}{\partial x_1} - \frac{\partial(fu_1)}{\partial x_2} \right) e_3.$$

Therefore

$$\operatorname{curl}(fu) = f \operatorname{curl} u + \left(u_3 \frac{\partial f}{\partial x_2} - u_2 \frac{\partial f}{\partial x_3} \right) e_1$$
$$+ \left(u_1 \frac{\partial f}{\partial x_3} - u_3 \frac{\partial f}{\partial x_1} \right) e_2 + \left(u_2 \frac{\partial f}{\partial x_1} - u_1 \frac{\partial f}{\partial x_2} \right) e_3. \quad (16.24)$$

Recalling that $\nabla f = (\partial f/\partial x_1)e_1 + (\partial f/\partial x_2)e_2 + (\partial f/\partial x_3)e_3$ and using the formula for the vector product, we see that (16.24) is precisely

$$\text{curl}(f\boldsymbol{u}) = f \,\text{curl}\, \boldsymbol{u} + \nabla f \times \boldsymbol{u}. \qquad \square$$

Remark. Part (d) in Theorem 16.12 states that if a vector $\boldsymbol{u}$ is the gradient of a scalar function f, that is if $\boldsymbol{u} = \nabla f$, then curl $\boldsymbol{u} = 0$. It is natural to ask whether or not the condition curl $\boldsymbol{u} = 0$ implies that $\boldsymbol{u}$ is the gradient of a scalar function f. Under certain conditions this statement is valid. In any Cartesian coordinate system, the condition $\boldsymbol{u} = \nabla f$ becomes

$$u_1 = \frac{\partial f}{\partial x_1}, \qquad u_2 = \frac{\partial f}{\partial x_2}, \qquad u_3 = \frac{\partial f}{\partial x_3}$$

or

$$df = u_1 dx_1 + u_2 dx_2 + u_3 dx_3. \qquad (16.25)$$

Then the statement curl $\boldsymbol{u} = 0$ asserts that the right side of (16.25) is an exact differential.

EXAMPLE 3. Let the vector field $\boldsymbol{u}$ be given (in coordinates) by

$$\boldsymbol{u} = 2x_1 x_2 x_3 \boldsymbol{e}_1 + (x_1^2 x_3 + x_2)\boldsymbol{e}_2 + (x_1^2 x_2 + 3x_3^2)\boldsymbol{e}_3.$$

Verify that curl $\boldsymbol{u} = 0$ and find the function f such that $\nabla f = \boldsymbol{u}$.

Solution. Computing $\nabla \times \boldsymbol{u}$ by Formula (16.22), we get

$$\text{curl}\, \boldsymbol{u} = \begin{vmatrix} \boldsymbol{e}_1 & \boldsymbol{e}_2 & \boldsymbol{e}_3 \\ \dfrac{\partial}{\partial x_1} & \dfrac{\partial}{\partial x_2} & \dfrac{\partial}{\partial x_3} \\ 2x_1 x_2 x_3 & x_1^2 x_3 + x_2 & x_1^2 x_2 + 3x_3^2 \end{vmatrix}$$

$$= (x_1^2 - x_1^2)\boldsymbol{e}_1 + (2x_1 x_2 - 2x_1 x_2)\boldsymbol{e}_2 + (2x_1 x_3 - 2x_1 x_3)\boldsymbol{e}_3 = 0.$$

We seek the function f such that

$$\frac{\partial f}{\partial x_1} = 2x_1 x_2 x_3, \qquad \frac{\partial f}{\partial x_2} = x_1^2 x_3 + x_2, \qquad \frac{\partial f}{\partial x_3} = x_1^2 x_2 + 3x_3^2.$$

Integrating the first equation, we find

$$f(x_1, x_2, x_3) = x_1^2 x_2 x_3 + C(x_2, x_3).$$

Differentiating this expression with respect to x_2 and x_3, we obtain

$$\frac{\partial f}{\partial x_2} = x_1^2 x_3 + \frac{\partial C}{\partial x_2} = x_1^2 x_3 + x_2 \quad \text{and} \quad \frac{\partial f}{\partial x_3} = x_1^2 x_2 + \frac{\partial C}{\partial x_3} = x_1^2 x_2 + 3x_3^2.$$

Thus it follows that

$$\frac{\partial C}{\partial x_2} = x_2, \qquad \frac{\partial C}{\partial x_3} = 3x_3^2 \quad \text{or} \quad C = \frac{1}{2}x_2^2 + K(x_3)$$

where $K'(x_3) = 3x_3^2$. Therefore $C(x_2, x_3) = \frac{1}{2}x_2^2 + x_3^3 + K_1$ with K_1 a constant. Hence

$$f(x_1, x_2, x_3) = x_1^2 x_2 x_3 + \frac{1}{2}x_2^2 + x_3^3 + K_1. \qquad \square$$

PROBLEMS

1. Let D be a domain in $\mathbb{R}^N$ and suppose that $u: D \to V_N$ is a vector function. If F is a figure in D and if $\int_F u\,dV$ exists, prove that the value is unique.

2. Suppose that $u: D \to V_N$ is integrable over some figure F in D. If $u(x) = \sum_{j=1}^N u_j(x)e_j$ for some orthonormal basis $e_1, \ldots, e_N$, write a detailed proof of the formula

$$\int_F u\,dV = \sum_{j=1}^N \left(\int_F u_j\,dV \right) e_j.$$

3. Suppose that $u: \mathbb{R}^2 \to V_2$ is given by $u(x_1, x_2) = (x_1^2 - x_2^2)e_1 + 2x_1 x_2 e_2$. Find $\int_F u\,dV$ where $F = \{(x_1, x_2): x_1^2 + x_2^2 \leq 1\}$.

In each of the following Problems 4 through 7 express $\nabla f(x)$ in terms of (x_1, x_2, x_3) and (e_1, e_1, e_3) and compute $D_a f(\bar{x})$ where a is the given unit vector and $\bar{x}$ is the given point.

4. $f(x) = 2x_1^2 + x_2^2 - x_1 x_3 - x_3^2$, $a = (1/3)(2e_1 - 2e_2 - e_3)$, $\bar{x} = (1, -1, 2)$.

5. $f(x) = x_1^2 + x_1 x_2 - x_2^2 + x_3^2$, $a = (1/7)(3e_1 + 2e_2 - 6e_3)$, $\bar{x} = (2, 1, -1)$.

6. $f(x) = e^{x_1} \cos x_2 + e^{x_2} \cos x_3$, $a = (1/\sqrt{3})(e_1 - e_2 + e_3)$, $\bar{x} = (1, \pi, -1/2)$.

7. $f(x) = x_1^2 \log(1 + x_2^2) - x_3^3$, $a = (1/\sqrt{10})(3e_1 + e_3)$, $\bar{x} = (1, 0, -2)$.

In each of Problems 8 and 9 express $D_a w(x)$ in terms of (x_1, x_2, x_3) and (e_1, e_2, e_3). Find the value at $\bar{x}$ as given.

8. $w(x) = x_2 x_3 e_1 + x_1 x_3 e_2 + x_1 x_2 e_3$, $a = (1/3)(e_1 + 2e_2 - 2e_3)$, $\bar{x} = (1, 2, -1)$.

9. $w(x) = (x_1 - 2x_2)e_1 + x_2 x_3 e_2 - (x_2^2 - x_3^2)e_3$, $a = (1/\sqrt{14})(3e_1 + 2e_2 - e_3)$, $\bar{x} = (2, -1, 3)$.

10. Prove the first, third, and fourth formulas in Theorem 16.8.

11. Show that Lemma 16.3 remains valid if the functions g in Formula (16.19) are restricted to the class $C_0^\infty(D)$.

12. Prove Theorem 16.10.

In each of Problems 13 through 17 find the value of div $v(x)$ at the given point $\bar{x}$.

13. $v(x) = x_1 x_2 e_1 + x_3^2 e_2 - x_1^2 e_3$, $\bar{x} = (1, 0, 1)$.

14. $v(x) = (x_1^2 - x_2 x_3)e_1 + (x_2^2 - x_1 x_3)e_2 + (x_3^2 - x_1 x_2)e_3$, $\bar{x} = (2, -1, 1)$.

15. $v(x) = \nabla u$, $u(x) = 3x_1 x_2^2 - x_2^3 + x_3$, $\bar{x} = (-1, 1, 2)$.

16. $v(x) = r^{-n}r$, $r = x_1 e_1 + x_2 e_2 + x_3 e_3$, $r = |r|$, $\bar{x} = (2, 1, -2)$.

17. $v = a \times r$, $\quad r = x_1 e_1 + x_2 e_2 + x_3 e_3$, $\quad a = $ constant vector, $\quad \bar{x} = (x_1^0, x_2^0, x_3^0)$.

In each of Problems 18 through 20, find curl v in terms of (x_1, x_2, x_3) and (e_1, e_2, e_3). If curl $v = 0$ find the function f such that $\nabla f = v$.

18. $v(x) = (x_1^2 + x_2^2 + x_3^2)^{-1}(x_1 e_1 + x_2 e_2 + x_3 e_3)$.

19. $v(x) = (x_1^2 + x_2 x_3)e_1 + (x_2^2 + x_1 x_3)e_2 + (x_3^2 + x_1 x_2)e_3$.

20. $v(x) = e^{x_1}(\sin x_2 \cos x_3 e_1 + \sin x_2 \sin x_3 e_2 + \cos x_2 e_3)$.

21. Find curl v where v is the gravitation field given in Example 2.

22. Prove Parts (a), (c), (d), and (e) of Theorem 16.12.

23. Find a formula for curl(curl u) in terms of (x_1, x_2, x_3) and (e_1, e_2, e_3) if $u = u_1 e_1 + u_2 e_2 + u_3 e_3$ and each of the u_i is a C^2 function on $\mathbb{R}^3$.

16.3. Line Integrals in $\mathbb{R}^N$

Let $I = \{t: a \leqslant t \leqslant b\}$ be an interval and f a vector function with domain I and range D, a subset of V_N. We consider in $\mathbb{R}^N$ the directed line segments having base at the origin 0 which represent the vectors f in D. The heads of these directed line segments trace out a curve in $\mathbb{R}^N$ which we denote by C. In Figure 16.5, $\overrightarrow{0P}$ represents a vector in D, the range of f. We recall that if f is continuous and the curve C has finite length, then we say that C is a rectifiable path (see Section 16.1).

Let g be a continuous vector function from C into V_N. If f, as defined above, is rectifiable then we can determine the *Riemann–Stieltjes* integral of g with respect to f. To do so we introduce a coordinate system with an orthonormal basis $e_1, e_2, \ldots, e_N$ in V_N, although the formula we shall obtain will be independent of the coordinate system. We write

$$g(x) = g_1(x)e_1 + \cdots + g_N(x)e_N,$$

$$f(t) = f_1(t)e_1 + \cdots + f_N(t)e_N, \qquad x \in C, t \in I. \qquad (16.26)$$

If f is rectifiable, then the functions $f_i: I \to \mathbb{R}^1$, $i = 1, 2, \ldots, N$, are continuous and of bounded variation. Also, if g is continuous, then the functions $g_i: C \to \mathbb{R}^N$, $i = 1, 2, \ldots, N$, are continuous. Using the notation $g_i[f(t)]$ for $g_i[f_1(t)$,

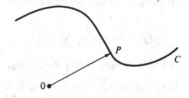

Figure 16.5. The curve $f(t)$.

..., $f_N(t)$], we observe that the following Riemann–Stieltjes integrals exist:

$$\int_a^b g_i[f(t)] \, df_i(t), \qquad i = 1, 2, \ldots, N. \tag{16.27}$$

We now establish the basic theorem for the existence of a Riemann–Stieltjes integral of a vector function g with respect to a vector function f.

Theorem 16.13. *Suppose that $f: I \to V_N$ is a vector function and that the range C in $\mathbb{R}^N$ as defined above by the radius vector $r(t) = v(\overrightarrow{OP})$ is a rectifiable path. Let g be a continuous vector field from C into V_N. Then there is a number L with the following property: for every $\varepsilon > 0$, there is a $\delta > 0$ such that for any subdivision $\Delta: a = t_0 < t_1 < \cdots < t_n = b$ with mesh less than δ and any choices of ξ_i with $t_{i-1} \leqslant \xi_i \leqslant t_i$, it follows that*

$$\left| \sum_{i=1}^n g[f(\xi_i)] \cdot [r(t_i) - r(t_{i-1})] - L \right| < \varepsilon. \tag{16.28}$$

The number L is unique.

PROOF. We introduce an orthonormal set $e_1, e_2, \ldots, e_N$ in V_N and write g and f in the form (16.26). Then we set

$$L = \sum_{i=1}^N \int_a^b g_i[f(t)] \, df_i(t),$$

and it is clear that each of the integrals exists. Replacing each integral by its Riemann sum, we obtain Inequality (16.28). $\qquad\qquad\square$

Definition. We write $L = \int_a^b g[f(t)] \cdot df(t)$ and we call this number the **Riemann–Stieltjes integral of g with respect to f.**

EXAMPLE 1. Given f and g defined by

$$f(t) = t^2 e_1 + 2te_2 - te_3, \qquad g(x) = (x_1^2 + x_3)e_1 + x_1 x_3 e_2 + x_1 x_2 e_3.$$

Find $\int_0^1 g \cdot df$.

Solution. We have

$$\int_0^1 g[f(t)] \cdot df(t)$$

$$= \int_0^1 [(t^4 - t)e_1 + t^2(-t)e_2 + t^2(2t)e_3] \cdot [d(t^2 e_1 + 2te_2 - te_3)]$$

$$= \int_0^1 [(t^4 - t) \cdot 2t + t^2(-t) \cdot 2 + t^2(2t)(-1)] \, dt = -\frac{4}{3}. \qquad\square$$

Let $I_1 = \{t: a_1 \leqslant t \leqslant b_1\}$ and $I_2 = \{t: a_2 \leqslant t \leqslant b_2\}$ be any intervals and suppose that $f_1: I_1 \to C$ and $f_2: I_2 \to C$ are one-to-one mappings onto a path C in $\mathbb{R}^N$, as described at the beginning of this section. We saw in Theorem 16.1 that there is a continuous function $U: I_1 \to I_2$ such that

$$f_1(t) = f_2[U(t)] \quad \text{for } t \in I_1, \tag{16.29}$$

and that U is either increasing on I_1 or decreasing on I_1. If C is rectifiable and $g: C \to V_N$ is continuous, then

$$\int_{a_1}^{b_1} g[f_1(t)] \cdot df_1(t) = \pm \int_{a_2}^{b_2} g[f_2(u)] \cdot df_2(u),$$

the plus sign corresponding to U increasing and the minus sign to U decreasing on I_1.

Definitions. Let $f: I \to V_N$ define an arc C in $\mathbb{R}^N$ by means of the radius vector $r(t) = v(\overrightarrow{OP})$, where $\overrightarrow{OP}$ represents f. We consider the collection $\mathcal{A}$ of all functions $f_\alpha: I_\alpha \to V_N$ such that the range of f_α is C and f_α is related to f by an equation such as (16.29) with U_α *increasing*. We define a **directed arc**, denoted $\vec{C}$, as the ordered pair $(C, \mathcal{A})$. Any function f_α in $\mathcal{A}$ is a **parametric representation** of $\vec{C}$. From Theorem 16.1 it follows that there are exactly two directed arcs $\vec{C}_1$ and $\vec{C}_2$ corresponding to an **undirected arc** C. We write $|\vec{C}_1| = |\vec{C}_2| = C$. If f_1 is a parametric representation of C_1 and f_2 is one of $\vec{C}_2$, then f_1 and f_2 are related according to (16.29) with U *decreasing*. It is therefore appropriate to write $\vec{C}_1 = -\vec{C}_2$ (see Figure 16.6).

The Riemann–Stieltjes integral along a directed arc $\vec{C}$ with radius vector r is defined by the formula

$$\int_{\vec{C}} g(r) \cdot dr = \int_a^b g[f(t)] \cdot df(t) \tag{16.30}$$

where f is a parametric representation of $\vec{C}$. It is not difficult to see that the integral along a directed arc as in (16.30) depends only on g and $\vec{C}$ and not on the particular parametric representation of $\vec{C}$. Also, it follows at once that

$$\int_{-\vec{C}} g(r) \cdot dr = -\int_{\vec{C}} g(r) \cdot dr.$$

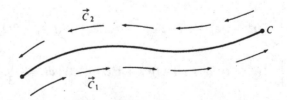

Figure 16.6. Directed and undirected arcs.

A directed arc may be decomposed into the union of directed subarcs. Let $I = \{t: a \leqslant t \leqslant b\}$ be decomposed into the subdivision $I_k = \{t: t_{k-1} \leqslant t \leqslant t_k\}$, $k = 1, 2, \ldots, n$, with $t_0 = a$ and $t_n = b$. If $\vec{C}$ has the representation $f: I \to V_n$, define $f_k: I_k \to V_N$ as the restriction of f to the interval I_k. Then each function f_k determines a directed arc $\vec{C}_k$ and we may write

$$\vec{C} = \vec{C}_1 + \vec{C}_2 + \cdots + \vec{C}_k.$$

The following result is an immediate consequence of the basic properties of Riemann–Stieltjes integrals.

Theorem 16.14

(a) *Suppose that $\vec{C}$ is a rectifiable arc and that $\vec{C} = \vec{C}_1 + \vec{C}_2 + \cdots + \vec{C}_n$. If g is continuous on $\vec{C}$, then*

$$\int_{\vec{C}} g(r) \cdot dr = \sum_{k=1}^{n} \int_{\vec{C}_k} g(r) \cdot dr. \tag{16.31}$$

(b) *Suppose that $\vec{C}$ has a piecewise smooth representation f and that g is continuous on $|\vec{C}|$. Then*

$$\int_{\vec{C}} g(r) \cdot dr = \int_a^b g[f(t)] \cdot f'(t)\, dt. \tag{16.32}$$

Remark. If f is piecewise smooth then the integral on the right in (16.32) may be evaluated by first decomposing $\vec{C}$ into subarcs, each of which has a smooth representation, and then using (16.3) to add up the integrals evaluated by a smooth f on the individual subarcs. We give an illustration.

EXAMPLE 2. In V_3, let $g = 2x_1 e_1 - 3x_2 e_2 + x_3 e_3$ and define the arc $\vec{C}$ as the union $\vec{C}_1 + \vec{C}_2$ where $\vec{C}_1$ is the directed line segment from $(1, 0, 1)$ to $(2, 0, 1)$, and $\vec{C}_2$ is the directed line segment from $(2, 0, 1)$ to $(2, 0, 4)$ (see Figure 16.7). Find the value of $\int_{\vec{C}} g(r) \cdot dr$.

Solution. $\vec{C}_1 = \{r: r = x_1 e_1 + e_3, 1 \leqslant x_1 \leqslant 2\}$ and $\vec{C}_2 = \{r: r = 2e_1 + x_3 e_3, 1 \leqslant x_3 \leqslant 4\}$. On $\vec{C}_1$, we have

$$dr = (dx_1)e_1$$

and on $\vec{C}_2$,

$$dr = (dx_3)e_3.$$

Therefore

$$\int_{\vec{C}_1} g \cdot dr = \int_1^2 2x_1\, dx_1 = 3 \quad \text{and} \quad \int_{\vec{C}_2} g \cdot dr = \int_1^4 x_3\, dx_3 = \frac{15}{2}.$$

We conclude that

$$\int_{\vec{C}} g \cdot dr = \frac{21}{2}. \qquad \square$$

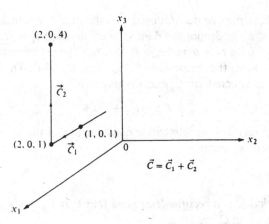

Figure 16.7

If a function g is continuous on a domain D in $\mathbb{R}^N$, then $\int_{\vec{C}} g \cdot dr$ will depend on the path $\vec{C}$ chosen in D. However, there are certain situations in which the value of the integral will be the same for all paths $\vec{C}$ in D which have the same endpoints. Under such circumstances we say *the* **integral is independent of the path**. The next result illustrates this fact.

Theorem 16.15. *Suppose that u is a continuously differentiable scalar field on a domain D in $\mathbb{R}^N$ and that p, q are points of D. Let $\vec{C}$ be any piecewise smooth path with representation f such that $f(t) \in D$ for all t on $I = \{t: a \leqslant t \leqslant b\}$, the domain of f. Suppose that $f(a) = p$, $f(b) = q$. Then*

$$\int_{\vec{C}} \nabla u \cdot dr = u(q) - u(p).$$

PROOF. Because of Part (a) in Theorem 16.14 it suffices to prove the result for f smooth rather than piecewise smooth. Let $e_1, e_2, \ldots, e_N$ be an orthonormal basis. Define

$$G(t) = u[f(t)] \qquad \text{for } t \in I,$$

with $f(t) = f_1(t)e_1 + \cdots + f_N(t)e_N$. Using the Chain rule we find

$$G'(t) = \sum_{i=1}^{N} u_{,i}[f(t)] f_i'(t) = \nabla u[f(t)] \cdot f'(t).$$

Hence

$$\int_{\vec{C}} \nabla u \cdot dr = \int_a^b \nabla u[f(t)] \cdot f'(t) \, dt = \int_a^b G'(t) \, dt = G(b) - G(a)$$

$$= u(q) - u(p). \qquad \qquad \square$$

Next we establish a converse of Theorem 16.15.

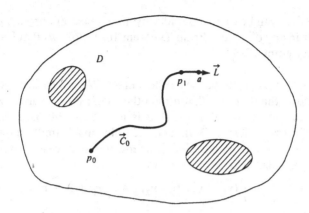

Figure 16.8

Theorem 16.16. *Let v be a continuous vector field with domain D in $\mathbb{R}^N$ and range in V_N. Suppose that for every smooth arc $\vec{C}$ lying entirely in D, the value of*

$$\int_{\vec{C}} v \cdot dr$$

is independent of the path. Then there is a continuously differentiable scalar field u on D such that $\nabla u(p) = v(p)$ for all $p \in D$.

PROOF. Let p_0 be a fixed point of D and suppose that $\vec{C}$ is any smooth path in D from p_0 to a point p. Define

$$u(p) = \int_{\vec{C}} v \cdot dr,$$

which, because of the hypothesis on v, does not depend on $\vec{C}$. Let p_1 be any point of D and $\vec{C}_0$ and arc from p_0 to p_1. Extend $\vec{C}_0$ at p_1 by adding a straight line segment $\vec{L}$ which begins at p_1 in such a way that the extended arc $\vec{C}_0 + \vec{L}$ is smooth. Denote by a the unit vector in the direction $\vec{L}$ (see Figure 16.8). We introduce a coordinate system in $\mathbb{R}^N$ and designate the coordinates of p_1 by x^0. Then any point q on $\vec{L}$ will have coordinates $x^0 + ta$ for $t \in \mathbb{R}^1$. Thus, if $h > 0$ we find

$$\frac{1}{h}[u(x^0 + ha) - u(x^0)] = \frac{1}{h}\int_{x^0}^{x^0 + ha} v(t) \cdot a \, dt.$$

Therefore

$$\lim_{h \to 0^+} \frac{1}{h}[u(x^0 + ha) - u(x^0)] = v(x^0) \cdot a$$

$$\lim_{h \to 0^-} \frac{1}{h}[u(x^0 + ha) - u(x^0)] = -\lim_{k \to 0} \frac{1}{k}[u(x^0 - ka) - u(x^0)]$$

$$= -v(x^0) \cdot (-a) = v(x^0) \cdot a.$$

This procedure can be carried out for any arc $\vec{C}_0$ passing through p_1 with a a unit vector in any direction. From Theorem 16.7 it follows that $\nabla u = v$ at x^0, an arbitrary point of D.   □

Suppose that u is a smooth scalar field in $\mathbb{R}^3$. We saw earlier that if $v = \nabla u$ then it follows that curl $v = 0$. On the other hand if v is a given vector field such that curl $v = 0$ in a domain D, it is not necessarily true that v is the gradient of a scalar field in D. In fact, the following example shows that the integral of v may not be path-independent, in which case v cannot be the gradient of a scalar field. To see this, we set

$$v = (x_1^2 + x_2^2)^{-1}(-x_2 e_1 + x_1 e_2 + 0 \cdot e_3)$$

with $D = \{(x_1, x_2, x_3): \frac{1}{2} \leqslant x_1^2 + x_2^2 \leqslant \frac{3}{2}, -1 \leqslant x_3 \leqslant 1\}$. We choose for $\vec{C}$ the path $r(t) = (\cos t)e_1 + (\sin t) \cdot e_2 + 0 \cdot e_3, 0 \leqslant t \leqslant 2\pi$. Then

$$\int_{\vec{C}} v \cdot dr = \int_0^{2\pi} dt = 2\pi.$$

However, if the integral were independent of the path its value should be zero since the initial and terminal points of $\vec{C}$ are the same. The difficulty arises because the path $\vec{C}$ encloses a singularity of v on the line $x_1 = x_2 = 0$ whereas the cylindrical domain D does not contain this line.

If a smooth vector field v is defined in a domain which is not merely connected but also simply connected (as defined below), then we shall prove that if curl $v = 0$ then v is the gradient of a scalar field u.

Definition. Let D be a domain in $\mathbb{R}^N$. Then D is **simply connected** if and only if whenever f_0 and f_1 are paths from $I = \{t: a \leqslant t \leqslant b\}$ into D with $f_0(a) = f_1(a)$ and $f_0(b) = f_1(b)$, there exists a function $f(t, s)$ which is continuous for $t \in I$ and $s \in J = \{s: 0 \leqslant s \leqslant 1\}$, has range in D, and has the properties:

$$f(t, 0) = f_0(t),$$

$$f(t, 1) = f_1(t),$$

$$f(a, s) = f_0(a) = f_1(a) \qquad \text{for all} \quad s \in J,$$

$$f(b, s) = f_0(b) = f_1(b) \qquad \text{for all} \quad s \in J. \qquad (16.33)$$

In other words, a domain D is simply connected if any two paths situated in D which have the same starting and ending points can be deformed continuously one into the other without leaving D (see Figure 16.9).

The following technical lemma shows that in general the function $f(t, s)$ used for deforming one path into another in a simply connected domain can be chosen so that it has certain smoothness properties. For the proof, see Problem 17 at the end of this section and the hints given there.

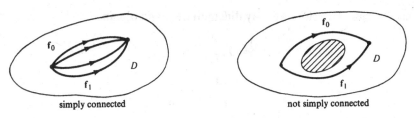

simply connected   not simply connected

Figure 16.9

Lemma 16.4. *Let $I = \{t: a \leqslant t \leqslant b\}$ and $J = \{s: 0 \leqslant s \leqslant 1\}$ be intervals. Let D be a simply connected domain in $\mathbb{R}^N$ and suppose that f_0 and f_1 with domain I are smooth paths in D for which $f_0(a) = f_1(a)$ and $f_0(b) = f_1(b)$. Then there is a continuous function $f(t, s)$ with domain $R = I \times J$ and with range in D satisfying conditions (16.33) and also the conditions: (i) $(\partial/\partial t) f$ is continuous in R; and (ii), there is a sufficiently large integer n so that, for each $t \in I$, the function f is linear in s on $J_i = \{s: (1/n)(i - 1) \leqslant s \leqslant (1/n)i\}, i = 1, 2, \ldots, n$, with $\partial f/\partial s$ and $\partial^2 f/\partial t \partial s$ uniformly continuous on each rectangle $I \times J_i$.*

Theorem 16.17. *Suppose that v is a continuously differentiable vector field on a simply connected domain D in $\mathbb{R}^3$ and that curl $v = 0$ on D. Then there is a continuously differentiable scalar field u on D such that $v = \nabla u$.*

PROOF. We need only show that $\int_{\vec{C}} v \cdot dr$ is independent of the path. To do this let p and q be two points of D and let f_0 and f_1 represent two smooth paths $\vec{C}_0$ and $\vec{C}_1$ in D such that $f_0(a) = f_1(a) = p$ and $f_0(b) = f_1(b) = q$. We choose a coordinate system with orthonormal basis e_1, e_2, e_3 and write $v = v_1 e_1 + v_2 e_2 + v_3 e_3$. From Lemma 16.4 there is a function $f(t, s) = f_1 e_1 + f_2 e_2 + f_3 e_3$ which is differentiable in t, piecewise linear in s and satisfies conditions (16.33). We define the function

$$\varphi(s) = \int_a^b [v_1(f)(\partial f_1/\partial t) + v_2(f)(\partial f_2/\partial t) + v_3(f)(\partial f_3/\partial t)] \, dt.$$

Then from (16.33) it follows that

$$\varphi(0) = \int_a^b v(f_0) \cdot f_0' \, dt = \int_{\vec{C}_0} v \cdot dr,$$

$$\varphi(1) = \int_a^b v(f_1) \cdot f_1' \, dt = \int_{\vec{C}_1} v \cdot dr.$$

We shall show that $\varphi'(s) \equiv 0$, which implies that the desired integral is independent of the path. Since v, f, and $\partial f/\partial t$ are continuous, it follows that φ is continuous for $s \in J = \{s: 0 \leqslant s \leqslant 1\}$. On each subinterval $J_i = \{s: (1/n)(i - 1) \leqslant s \leqslant (1/n)i\}$, we know that $\partial f/\partial s$ and $\partial^2 f/\partial t \partial s$ are uniformly continuous. Therefore we may differentiate φ and obtain

$$\varphi'(s) = \int_a^b \sum_{i=1}^3 \left\{ v_i(f) \frac{\partial^2 f_i}{\partial s \partial t} + \sum_{j=1}^3 v_{i,j}(f) \frac{\partial f_i}{\partial t} \frac{\partial f_j}{\partial s} \right\} dt,$$

$$\frac{i-1}{n} < s < \frac{i}{n}.$$

Integrating by parts in the terms containing $\partial f_i/\partial t \partial s$, we find

$$\varphi'(s) = \sum_{i=1}^3 \left\{ v_i[f(b, s)] \frac{\partial f_i(b, s)}{\partial s} - v_i[f(a, s)] \frac{\partial f_i(a, s)}{\partial s} \right.$$
$$\left. + \int_a^b \sum_{i,j=1}^3 \left(v_{i,j} \frac{\partial f_i}{\partial t} \frac{\partial f_j}{\partial s} - v_{i,j} \frac{\partial f_i}{\partial s} \frac{\partial f_j}{\partial t} \right) dt \right\}. \qquad (16.34)$$

Since $f(a, s)$ and $f(b, s)$ are independent of s (according to (16.33)), it is clear that $\partial f_i(a, s)/\partial s = \partial f_i(b, s)/\partial s = 0$, $i = 1, 2, 3$. Therefore the first sum in (16.34) is zero. In the second sum we interchange the indices i and j and obtain

$$\varphi'(s) = \int_a^b \sum_{i,j=1}^3 (v_{i,j} - v_{j,i}) \frac{\partial f_i}{\partial t} \frac{\partial f_j}{\partial s} dt.$$

The hypothesis curl $v = 0$ implies that the above integrand vanishes, and so $\varphi'(s) = 0$ for $s \in J$. $\qquad \square$

Remark. We can easily extend Theorem 16.17 to functions v from a domain D in $\mathbb{R}^N$ to V_N. If curl $v = 0$ is replaced by the condition

$$v_{i,j} - v_{j,i} = 0, \qquad i, j = 1, 2, \ldots, N,$$

where $v = v_1 e_1 + \cdots + v_N e_N$, then v is the gradient of a scalar field u provided that the domain D is simply connected.

It is useful to have criteria which establish the simple-connectivity of domains so that Theorem 16.17 can be employed. The reader should not find it difficult to prove the following result.

Theorem 16.18

(a) If D is a convex[2] domain in $\mathbb{R}^N$, then D is simply connected.
(b) Let h be a one-to-one continuous mapping of a domain D in $\mathbb{R}^N$ onto a domain D_1 in $\mathbb{R}^N$. If D is simply connected then so is D_1.

PROBLEMS

In each of Problems 1 through 8 assume that g is a continuous vector field from a domain D in $\mathbb{R}^N$ into V_N for the appropriate value of N. The vectors

---

[2] Recall that a set S is convex if, whenever p and q are in S, the line segment joining p and q is also in S.

$e_1, \ldots, e_N$ form an orthonormal basis for a coordinate system. Compute $\int_{\vec{C}} g \cdot dr$.

1. $g = x_1 x_3 e_1 - x_2 e_2 + x_1 e_3$; $\vec{C}$ is the directed line segment from $(0, 0, 0)$ to $(1, 1, 1)$.

2. $g = x_1 x_3 e_1 - x_2 e_2 + x_1 e_3$; $\vec{C}$ is the directed arc given by $f(t) = t e_1 + t^2 e_2 + t^3 e_3$, $0 \leqslant t \leqslant 1$, from 0 to 1.

3. $g = -x_1 e_1 + x_2 e_2 - x_3 e_3$; $\vec{C}$ is the helix given by $f(t) = (\cos t)e_1 + (\sin t)e_2 + (t/\pi)e_3$, $0 \leqslant t \leqslant 2\pi$, from 0 to 2π.

4. $g = -x_1 e_1 + x_2 e_2 - x_3 e_3$; $\vec{C}$ is the directed line segment from $(1, 0, 0)$ to $(1, 0, 2)$.

5. $g = x_1^2 e_1 + x_1 x_2 e_2 + 0 \cdot e_3$; $\vec{C} = \{(x_1, x_2, x_3): x_2 = x_1^2, x_3 = 0, 0 \leqslant x_1 \leqslant 1$, from $(0, 0, 0)$ to $(1, 1, 0)\}$.

6. $g = 2x_2 x_3 e_2 + (x_3^2 - x_2^2)e_3$; $\vec{C}$ is the shorter circular arc given by $x_1 = 0$, $x_2^2 + x_3^2 = 4$, from $(0, 2, 0)$ to $(0, 0, 2)$.

7. $g = x_1 x_4 e_1 + x_2 e_2 - x_2 x_4 e_3 + x_3 e_4$; $\vec{C}$ is the directed straight line segment from $(0, 0, 0, 0)$ to (a_1, a_2, a_3, a_4).

8. $g = \sum_{i=1}^{N} x_i^2 e_i$; $\vec{C}$ is the directed line segment given by $f(t) = \sum_{i=1}^{N} \alpha_i t e_i$, $0 \leqslant t \leqslant 1$ from 0 to 1; $\{\alpha_i\}$ are constants.

9. Given the scalar function $u(x) = x_1^2 + 2x_2^2 - x_3^2 + x_4$ in $\mathbb{R}^4$. Verify that $\int_{\vec{C}} \nabla u \cdot dr$ is independent of the path by computing the value of the integral along $\vec{C}_1$, the straight line segment from $(0, 0, 0, 0)$ to $(1, 1, 1, 1)$, and the along $\vec{C}_2$, the straight segment from $(0, 0, 0, 0)$ to $(1, 0, 0, 0)$ followed by the straight segment from $(1, 0, 0, 0)$ to $(1, 1, 1, 1)$. Show that the two values are the same.

10. Given the scalar function $u(x) = \sum_{i=1}^{N} x_i^2$ in $\mathbb{R}^N$. Let $\vec{C}_1$ be the line segment from $(0, 0, \ldots, 0)$ to $(1, 1, \ldots, 1)$ and $\vec{C}_2$ the path $f(t) = t^2 e_1 + \sum_{i=2}^{N} t e_i$, $0 \leqslant t \leqslant 1$. Verify that $\int_{\vec{C}} \nabla u \cdot dr$ is independent of the path by computing this integral along $\vec{C}_1$ and $\vec{C}_2$ and showing that the values are the same.

11. In $\mathbb{R}^2$, let v be the vector field given by $v = (x_1^2 + x_2^2)^{-1}(-x_2 e_1 + x_1 e_2)$. Let $\vec{C}$ be the directed arc given by $f(t) = (\cos t)e_1 + (\sin t)e_2$, $0 \leqslant t \leqslant 2\pi$. Use Theorem 16.17 to show that $\mathbb{R}^2$ with the origin removed is not simply connected.

12. Prove Theorem 16.18.

13. (a) Show that any half-plane in $\mathbb{R}^2$ is convex.
    (b) Show that in $\mathbb{R}^N$ the half-space given by $x_N \geqslant 0$ is convex.

14. Show that the set S in $\mathbb{R}^2$ given by $S = \{(x_1, x_2): x_1 \geqslant 0, -\pi \leqslant x_2 \leqslant \pi\}$ is convex.

15. Let $I = \{x_1: a \leqslant x_1 \leqslant b\}$ and suppose that $f: I \to \mathbb{R}^1$ is a continuous, positive function. Show that the set $S = \{(x_1, x_2): a < x_1 < b, 0 < x_2 < f(x_1)\}$ is a simply connected set in $\mathbb{R}^2$.

16. A **torus** is obtained by revolving a circle about an axis in the plane of the circle provided the axis does not intersect the circle (see Figure 16.10). Let a be the radius of the circle and let b be the distance from the axis of revolution to the center of

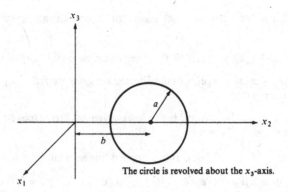

The circle is revolved about the x_3-axis.

Figure 16.10

the circle. The interior of the torus is the set S given by

$$S = \{(x_1, x_2, x_3): (\sqrt{x_1^2 + x_2^2} - b)^2 + x_3^2 < a^2\}.$$

Use the method of Problem 11 to show that the set S is not simply connected.

17. Employ the following steps to prove Lemma 16.4.
   (a) Let $f^*(t, s)$ be a continuous function satisfying the conditions (16.33) for the definition of simple connectivity. Let R be the rectangle $\{(t, s): a \leqslant t \leqslant b, 0 \leqslant s \leqslant 1\}$. Show that there is a number $\rho > 0$ such that the ball with center at $f^*(t, s)$ and radius ρ is in D for all $(t, s) \in R$.
   (b) Define a sequence $\{g_{ni}\}$, $i = 0, 1, 2, \ldots, n$, $n = 1, 2, \ldots$, such that $g_{n0}(t) = f^*(t, s) = f_0(t)$, $g_{nn}(t) = f^*(t, 1) = f_1(t)$, and such that, for $1 \leqslant i \leqslant n - 1$, we set g_{ni}^* equal to a polynomial with the property that $|g_{ni}^*(t) - f^*(t, i/n)| < \rho/8$ for $a \leqslant t \leqslant b$. Then let

$$g_{ni}(t) = g_{ni}^*(t) - l_{ni}(t)$$

   where $l_{ni}(t)$ is the linear function coinciding with $g_{ni}^*(t) - f^*(t, i/n)$ for $t = a$ and $t = b$. Show that $|l_{ni}(t)| < \rho/8$ on $a \leqslant t \leqslant b$ and that

$$|g_{ni}(t) - f^*(t, i/n)| < \frac{\rho}{4} \qquad \text{for} \quad a \leqslant t \leqslant b, 0 \leqslant i \leqslant n.$$

   (c) Show that there is an n so large that

$$|f^*(t, s_1) - f^*(t, s_2)| < \frac{\rho}{4} \qquad \text{if} \quad |s_1 - s_2| \leqslant \frac{1}{n}.$$

   (d) With the n chosen as in (c), define

$$f(t, s) = (i - ns)g_{ni-1}(t) + (ns - i + 1)g_{ni}(t),$$

$$a \leqslant t \leqslant b, \quad \frac{i - 1}{n} \leqslant s \leqslant \frac{i}{n}, \quad i = 1, 2, \ldots, n.$$

   Show that f and $\partial f/\partial t$ are continuous on R and observe that f is linear in s for $(i - 1)/n \leqslant s \leqslant i/n$, $i = 1, 2, \ldots, n$, with $f(t, (i - 1)/n) = g_{ni-1}(t)$ and $f(t, i/n) = g_{ni}(t)$.

(e) By considering the inequalities

$$\left| f(t, s) - f^*\left(t, \frac{i}{n}\right) \right| \leq |f(t, s) - g_{ni}(t)| + \left| g_{ni}(t) - f^*\left(t, \frac{i}{n}\right) \right|$$

$$\leq |g_{ni-1}(t) - g_{ni}(t)| + \frac{\rho}{4} \leq \left| g_{ni-1}(t) - f^*\left[t, \frac{(i-1)}{n}\right] \right|$$

$$+ \left| f^*\left[t, \frac{(i-1)}{n}\right] - f^*\left(t, \frac{i}{n}\right) \right|$$

$$+ \left| f^*\left(t, \frac{i}{n}\right) - g_{ni}(t) \right| + \frac{\rho}{4},$$

show that

$$\left| f(t, s) - f^*\left(t, \frac{i}{n}\right) \right| < \rho \quad \text{for } \frac{i-1}{n} \leq s \leq \frac{i}{n}.$$

Conclude that $f(t, s) \in D$ for all (t, s) in R.

16.4. Green's Theorem in the Plane

Green's theorem is an extension to the plane of the Fundamental theorem of calculus. We recall that this Fundamental theorem states that if I is an interval in $\mathbb{R}^1$ and $f: I \to \mathbb{R}^1$ is a continuously differentiable function then for any points a and b in I we have the formula

$$\int_a^b f'(x) \, dx = f(b) - f(a). \tag{16.35}$$

In the extension of this theorem to the plane we suppose that F is a figure in $\mathbb{R}^2$ and C is its boundary. Then Green's theorem is a formula which connects the line integral of a vector function over C with the double integral of the derivative of the function taken over the figure F.

In order to achieve the appropriate precision in the statement and proof of the theorem we first develop lemmas and theorems which deal with geometry in the plane and the representation of vector functions defined on figures. The reader who is interested only in the basic formula may ship all the introductory material and start with the definitions on page 447.

Definitions. Let $I = \{t: a \leq t \leq b\}$ be the domain of a path $f: I \to V_N$. The function f represents a curve C in $\mathbb{R}^N$. Then f is said to be a **closed path** if and only if $f(a) = f(b)$. Also, f is a **simple closed path** if and only if f is closed and, whenever $a < t_1, t_2 < b$, then $f(t_1) = f(t_2)$ implies that $t_1 = t_2$. The curve C which is the range of the path f is called a **simple closed curve** or a **closed Jordan curve**.

The next theorem is an extension to simple paths of Theorem 16.1 which connects any two representations of a curve in $\mathbb{R}^N$.

Theorem 16.19*. *Let $I = \{t: a \leqslant t \leqslant b\}$ and $J = \{t: c \leqslant t \leqslant d\}$ be any intervals and suppose that $f: I \to V_N, g: J \to V_N$ have as range the same simple closed curve C.*

(a) *If $f(a) = f(b) = g(c) = g(d)$, then there is a continuous function U from I onto J such that $f(t) = g[U(t)]$ for $t \in I$, and either U is increasing ith $U(a) = c$, $U(b) = d$ or U is decreasing with $U(a) = d$, $U(b) = c$.*
(b) *If f is rectifiable, then g is rectifiable and the length of the path f equals the length of the path g.*

PROOF

(a) Let $\varepsilon > 0$ be given and define $I_\varepsilon = \{t: a + \varepsilon \leqslant t \leqslant b - \varepsilon\}$. Then f restricted to I_ε has as range an arc C_ε. The arc C_ε will be the range of g restricted to J_δ, a subinterval of J. According to Theorem 16.1 there is a function $U_\varepsilon: I_\varepsilon \to J_\delta$ which is continuous on I_ε and is either increasing on I_ε or decreasing on I_ε. Furthermore, $f(t) = g[U_\varepsilon(t)]$ for $t \in I_\varepsilon$. We let ε tend to zero and obtain in the limit a function U defined from $I' = \{t: a < t < b\}$ onto $J' = \{t: c < t < d\}$. If we define $U(a) = c$ and $U(b) = d$ when U is increasing, then U will be continuous from I onto J. Similarly, if U is decreasing, we set $U(a) = d$, $U(b) = c$ and U is a continuous decreasing function from I onto J. $\qquad\square$

We leave to the reader the proof of Part (b) (see Theorem 16.3).

It may happen that $f: I \to V_N$ and $g: J \to V_N$ have the same simple closed curve C as range but that $g(c)$ and $g(d)$ are equal to $f(\beta)$ where β is an interior point of I. The following corollary shows that if the definition of f is extended by periodicity, we get the same result as in Theorem 16.19(a).

Corollary*. *Suppose that the domain of f in Theorem 16.19 is extended from I to $I_1 = \{t: a \leqslant t \leqslant b + (b - a)\}$ by the formula $f(t) = f[t + (b - a)]$ for $a \leqslant t \leqslant b$. Suppose that $g(c) = g(d) = f(a + \alpha) = f(b + \alpha)$ for some α such that $0 \leqslant \alpha \leqslant (b - a)$. Define $f_\alpha(t) = f(t + \alpha)$ for $t \in I$. Then there is a function $U_\alpha: I \to J$ with the properties of U in Theorem 16.19(a) such that $f_\alpha(t) = g[U_\alpha(t)]$ for $t \in I$.*

Suppose that f and g are as in the Corollary to Theorem 16.19 with range C, a simple closed rectifiable curve. Let $F: C \to V_N$ be a continuous function on C. Then

$$\int_a^b F[f_\alpha(t)] \cdot df_\alpha(t) = \pm \int_c^d F[g(s)] \cdot dg(s)$$

with the plus sign taken if U_α is increasing and the minus sign if U_α is decreasing.

---

* Material marked with an asterisk may be omitted on a first reading.

The proof of the above formula is virtually unchanged from the proof when C is a rectifiable arc.

The following definitions which give an orientation to simple closed curves in $\mathbb{R}^N$ are analogous to those for directed curves (see page 436).

Definition*. Let $f: I \to V_N$ have as range a simple closed curve C in $\mathbb{R}^N$. An **oriented simple closed curve** is a pair $(C, \mathscr{A})$ where $\mathscr{A}$ is the class of vector functions consisting of all paths f_β having range C which are related to f by a function U_α as in the Corollary to Theorem 16.19 with U_α *increasing*. We call any such f_β a **parametric representation** of $(C, \mathscr{A})$. We denote $(C, \mathscr{A})$ by the symbol $\vec{C}$. The **unoriented simple closed curve** will be designated by C or $|\vec{C}|$.

An unoriented simple closed curve C has two possible orientations $\vec{C}_1$ and $\vec{C}_2$ according as U_α is increasing or decreasing. We often write $\vec{C}_1 = -\vec{C}_2$. Suppose that $f: I \to V_N$ is a parametric representation of $\vec{C}$; we decompose $I = \{t: a \leqslant t \leqslant b\}$ into subintervals $I_k = \{t: t_{k-1} \leqslant t \leqslant t_k\}, k = 1, 2, \ldots, n$, with $t_0 = a, t_n = b$ and denote by $f_k: I_k \to V_N$ the restriction of f to I_k. Then the directed arc $\vec{C}_k$ is the range of f_k. We write $\vec{C} = \vec{C}_1 + \vec{C}_2 + \cdots + \vec{C}_n$. As in Theorem 16.14 an integral over $\vec{C}$ may be decomposed into a sum of integrals over the $\vec{C}_k, k = 1, 2, \ldots, n$.

Let D be a region in $\mathbb{R}^2$ with a boundary which is a simple closed curve. We denote the unoriented boundary by ∂D and the boundary oriented in a counterclockwise direction by $\overrightarrow{\partial D}$. The symbol $-\overrightarrow{\partial D}$ is used for the clockwise orientation.

We shall first establish Green's theorem for regions in $\mathbb{R}^2$ which have a special simple shape. Then we shall show how the result for these special regions can be used to yield the same formula for general regions in the plane.

Definitions. Let D be a region in $\mathbb{R}^2$ and $v: D \to V_2$ a vector function. We introduce a coordinate system in V_2 with orthonormal basis e_1 and e_2. There are two possible orientations for such a basis: e_2 is obtained from e_1 by a 90° rotation in a counterclockwise direction or e_2 is obtained from e_1 by a 90° rotation in a clockwise direction. We call the plane $\mathbb{R}^2$ **oriented** when one of these systems is introduced in $V_2(\mathbb{R}^2)$ and we use the notation $\vec{\mathbb{R}}^2$ and $-\vec{\mathbb{R}}^2$ for the orientations (see Figure 16.11). Generally, we shall consider the oriented plane $\vec{\mathbb{R}}^2$. In coordinates we write $v = P(x_1, x_2)e_1 + Q(x_1, x_2)e_2$ where P and Q are functions from D into $\mathbb{R}^1$. The **scalar curl** of v is the function $Q_{,1} - P_{,2}$ and we write

$$\text{curl } v = Q_{,1} - P_{,2}.$$

We recall that in V_3 the curl of a vector function is a vector function whereas the above curl is a scalar function. To justify the above terminology observe

---

* See the footnote on page 446.

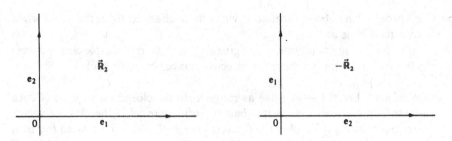

Figure 16.11. Two orientations of the plane.

that if v is a vector function from $\mathbb{R}^3$ to V_3 in the special form

$$v = Pe_1 + Qe_2 + 0 \cdot e_3$$

and $P = P(x_1, x_2)$, $Q = Q(x_1, x_2)$, then

$$\operatorname{curl} v = (Q_{,1} - P_{,2})e_3.$$

Lemma 16.5. *Let f be a continuously differentiable function with domain $I = \{x_1 : a \leqslant x_1 \leqslant b\}$ and range in $\mathbb{R}^1$. Let c be a constant with $f(x_1) \geqslant c$ for $x_1 \in I$. Define $D = \{(x_1, x_2) : a \leqslant x_1 \leqslant b, c \leqslant x_2 \leqslant f(x_1)\}$. (See Figure 16.12) Suppose that G is any region in $\vec{\mathbb{R}}^2$ with $\bar{D} \subset G$ and let $v : G \to V_2(\mathbb{R}^2)$ be a continuously differentiable vector function. Then*

$$\int_D \operatorname{curl} v \, dA = \int_{\overrightarrow{\partial D}} v \cdot dr, \tag{16.36}$$

where $\overrightarrow{\partial D}$ is oriented in a counterclockwise sense.

PROOF. Writing $v(x_1, x_2) = P(x_1, x_2)e_1 + Q(x_1, x_2)e_2$, we see that it is sufficient to prove (16.36) for the functions $P(x_1, x_2)e_1$ and $Q(x_1, x_2)e_2$ separately. Letting $v_1 = Pe_1$, we have

$$\int_D \operatorname{curl} v_1 \, dA = \int_D P_{,2}(x_1, x_2) \, dA = -\int_a^b \{P(x_1, f(x_1)) - P(x_1, c)\} \, dx_1.$$

We write $\overrightarrow{\partial D} = \vec{C}_1 + \vec{C}_2 + \vec{C}_3 + \vec{C}_4$ as shown in Figure 16.12. Then

$$\int_D \operatorname{curl} v_1 \, dA = \int_{\vec{C}_3} P \, dx_1 + \int_{\vec{C}_1} P \, dx_1.$$

Since x_1 is constant along $\vec{C}_2$ and $\vec{C}_4$, it follows that $\int_{\vec{C}_2} P \, dx_1 = \int_{\vec{C}_4} P \, dx_1 = 0$. Therefore

$$\int_D \operatorname{curl} v_1 \, dA = \int_{\sum_{i=1}^4 \vec{C}_i} P \, dx_1 = \int_{\overrightarrow{\partial D}} P \, dx_1. \tag{16.37}$$

Next, let $v_2 = Qe_2$ and define

$$U(x_1, x_2) = \int_c^{x_2} Q(x_1, \eta) \, d\eta.$$

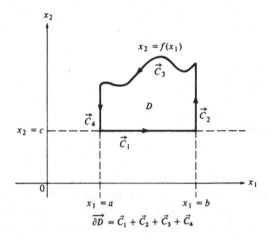

$$\overrightarrow{\partial D} = \vec{C}_1 + \vec{C}_2 + \vec{C}_3 + \vec{C}_4$$

Figure 16.12

Then

$$U_{,1}(x_1, x_2) = \int_c^{x_2} Q_{,1}(x_1, \eta)\, d\eta, \qquad U_{,2}(x_1, x_2) = Q(x_1, x_2),$$

and

$$U_{,1,2} = U_{,2,1} = Q_{,1}.$$

Since U is smooth in G, it is clear that

$$\int_{\overrightarrow{\partial D}} (U_{,1}\, dx_1 + U_{,2}\, dx_2) = 0 = \int_{\overrightarrow{\partial D}} U_{,1}\, dx_1 + \int_{\overrightarrow{\partial D}} Q\, dx_2, \qquad (16.38)$$

and, using Formula (16.37) with $U_{,1}$ in place of P, we obtain

$$\int_D \operatorname{curl} v_2\, dA = \int_D U_{,1,2}\, dA = -\int_{\overrightarrow{\partial D}} U_{,1}\, dx_1.$$

Taking (16.38) into account we conclude that

$$\int_D \operatorname{curl} v_2\, dA = \int_{\overrightarrow{\partial D}} Q\, dx_2. \qquad (16.39)$$

Adding (16.37) and (16.39) we get the result. $\qquad\square$

Definitions. Let C_1 and C_2 be two smooth arcs which have as their only point in common an end point of each of them. This point, denoted P, is called a **corner of C_1 and C_2** if each of the arcs has a (one-sided) tangent line at P and if the two tangent lines make a *positive angle* (see Figure 16.13). A **piecewise smooth simple closed curve** is a simple closed curve which is made up of a finite number of smooth arcs which are joined at corners.

Remark. In Lemma 16.5 we may suppose that f is a piecewise continuously differentiable (i.e., piecewise smooth) function on I consisting of a finite number of smooth arcs joined at corners. The proof is unchanged.

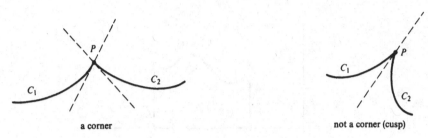

a corner   not a corner (cusp)

Figure 16.13

Definition. A region D in $\mathbb{R}^2$ is said to be **regular** if and only if (i) D is bounded, (ii) ∂D consists of a finite number of piecewise smooth simple closed curves, and (iii) at each point P of ∂D a Cartesian coordinate system can be constructed with P as origin with the following properties: for sufficiently small a and b a rectangle

$$R = \{(x_1, x_2): -a \leqslant x_1 \leqslant a, \ -b \leqslant x_2 \leqslant b\}$$

can be found so that the part of ∂D in R has the form $x_2 = f(x_1)$ for x_1 on $I = \{x_1: -a \leqslant x_1 \leqslant a\}$ and with range on $J = \{x_2: -b < x_2 < b\}$, where f is a piecewise smooth function (see Figure 16.14).

Of course, the values of a, b, and the function f will change with the point P. If a rectangle R is determined for a point P, then it is clear that any rectangle with the same value for b and a smaller value for a is also adequate. See Figure 16.15 for examples of regions which are not regular.

If D is a regular region in $\vec{\mathbb{R}}^2$, a tangent vector can be drawn at any point of ∂D which is not a corner. Let $\vec{T}$ be the tangent vector at a point $P \in \partial D$ and construct a coordinate system in $\mathbb{R}^2$ so that e_1 is parallel to $\vec{T}$. Then ∂D may be oriented at P in two ways: the vector e_2 may point into D or the vector e_2 may point outward from D. In the first case we say ∂D is oriented at P so that D is

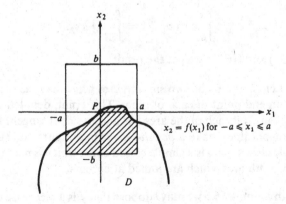

Figure 16.14

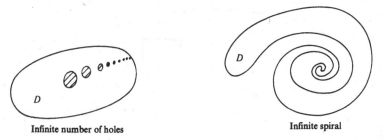

Infinite number of holes   Infinite spiral

Regions which are not regular

Figure 16.15

to the left; in the second case, we say *D is to the right*. It can be shown that the oriented piecewise smooth simple closed curve $\overrightarrow{\partial D}$ has the property that D is always on the left or always on the right for all points at which a tangent vector can be drawn.

Theorem 16.20 (Green's theorem in the plane). *Let D be a regular region in $\vec{\mathbb{R}}^2$. Suppose that G is a region in $\mathbb{R}^2$ with $\bar{D} \subset G$, and let $v: \underrightarrow{G} \to V_2(\vec{\mathbb{R}}^2)$ be a continuously differentiable vector function on G. Assume that $\overrightarrow{\partial D}$ is oriented so that D is on the left and that $\overrightarrow{\partial D} = \vec{C}_1 + \vec{C}_2 + \cdots + \vec{C}_k$ where each $\vec{C}_i$ is a smooth arc. Then*

$$\int_D \operatorname{curl} v \, dA = \sum_{i=1}^{k} \int_{\vec{C}_i} v \cdot dr = \int_{\overrightarrow{\partial D}} v \cdot dr. \tag{16.40}$$

PROOF. For each point of ∂D, choose a rectangle R_P and a function f_P as prescribed in the definition of a regular region. For each interior point P of D choose a coordinate system with origin at P and an open rectangle R_P with sides parallel to the coordinate axes so that $\bar{R}_P \subset D$. Since $\bar{D}$ is compact, a finite number of the rectangles $\{R_P\}$ as described for $P \in D \cup \partial D$ cover $\bar{D}$ (see Figure 16.16). We denote these rectangles by $S_1, S_2, \ldots, S_n$ and their centers by P_i, $i = 1, 2, \ldots, n$. According to the process described in Section 15.4, we can find mollifiers $\psi_1, \psi_2, \ldots, \psi_n$ of class C^∞ on an open set $G_0 \supset \bar{D}$ such that each ψ_i vanishes on $G_0 - F_i$ where F_i is a compact subset of the open rectangle S_i containing P_i. For each point P define

$$\varphi_i(P_\Lambda) = \frac{\psi_i(P_\Lambda)}{\sum\limits_{i=1}^{n} \psi_i(P_\Lambda)}.$$

Then each φ_i is of class C^∞ and, for every $P_\Lambda \in G_0$, it follows that $\sum_{i=1}^{n} \varphi_i(P) = 1$. The functions $\varphi_1, \varphi_2, \ldots, \varphi_n$ are called a **partition of unity** (see Problem 2 of Section 15.4). We define

$$v_i = \varphi_i v, \qquad i = 1, 2, \ldots, n.$$

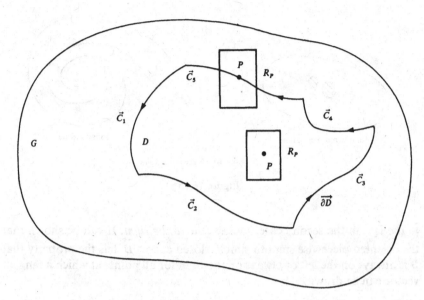

Figure 16.16. A finite number of rectangles covers $\bar{D}$.

Then each v_i is continuously differentiable on G_0 and

$$v = v_1 + v_2 + \cdots + v_n.$$

Hence it suffices to establish (16.40) for each v_i. If $\bar{S}_i$ is contained in D, we have

$$\int_{\overrightarrow{\partial D}} v_i \cdot dr = 0$$

since v_i is zero at every point of $\overrightarrow{\partial D}$. Also, by Lemma 16.5

$$\int_D \text{curl } v_i \cdot dA = \int_{S_i} \text{curl } v_i \, dA = \int_{\overrightarrow{\partial S_i}} v_i \cdot dr = 0,$$

since $v_i = 0$ except in F_i, and $\overrightarrow{\partial S_i}$ is disjoint from F_i. Finally, suppose S_i is a boundary rectangle. Let $D_i = D \cap S_i$ and $\vec{C}_i = \partial D \cap S_i$. Then, since $v_i = 0$ on $\partial D - \vec{C}_i$, and on $D - D_i$, it follows from Lemma 16.5 that

$$\int_{\overrightarrow{\partial D}} v_i \cdot dr = \int_{\vec{C}_i} v_i \cdot dr = \int_{\overrightarrow{\partial D_i}} v_i \cdot dr = \int_{D_i} \text{curl } v_i \, dA = \int_D \text{curl }_i \, dA.$$

Performing this process for each i and adding the results, we obtain (16.40).
                                                                                                    □

If we set $v = P(x_1, x_2)e_1 + Q(x_1, x_2)e_2$ and express (16.40) in terms of coordinates we have the following form of Green's theorem.

Corollary. *If P and Q are smooth in a domain G in $\mathbb{R}^2$ which contains a regular region D, then*

$$\int_D \left(\frac{\partial Q}{\partial x_1} - \frac{\partial P}{\partial x_2}\right) dA = \oint_{\partial D} (P \, dx_1 + Q \, dx_2) \tag{16.41}$$

where the line integral is taken in a counterclockwise direction.

EXAMPLE 1. Given the disk $K = \{(x_1, x_2): x_1^2 + x_2^2 < 1\}$, use Green's theorem to evaluate

$$\int_{\overrightarrow{\partial K}} [(2x_1 - x_2^3) \, dx_1 + (x_1^3 + 3x_2^2) \, dx_2].$$

Solution. Setting $P = 2x_1 - x_2^3$, $Q = x_1^3 + 3x_2^2$, we see from (16.41) that

$$\oint_{\partial K} (P \, dx_1 + Q \, dx_2) = \int_K 3(x_1^2 + x_2^2) \, dA = 3 \int_0^{2\pi} \int_0^1 r^2 \cdot r \, dr \, d\theta = \frac{3}{2}\pi.$$

$\square$

EXAMPLE 2. Let $K = \{(x_1, x_2): x_1^2 + x_2^2 \leqslant 1\}$ be the unit disk and let D be the region outside K which is bounded on the left by the parabola $x_2^2 = 2(x_1 + 2)$ and on the right by the line $x_1 = 2$ (see Figure 16.17). Use Green's theorem to evaluate

$$\int_{\vec{C}_1} \left(-\frac{x_2}{x_1^2 + x_2^2} \, dx_1 + \frac{x_1}{x_1^2 + x_2^2} \, dx_2\right)$$

where $\vec{C}_1$ is the outer boundary of D.

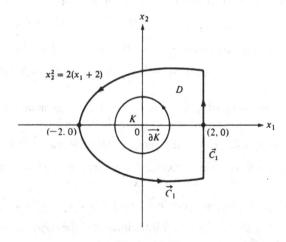

Figure 16.17

Solution. We write $P = -x_2/(x_1^2 + x_2^2)$, $Q = x_1/(x_1^2 + x_2^2)$ and note that $Q_{,1} - P_{,2} = 0$ in D. Hence with $\overrightarrow{\partial K}$ oriented in the counterclockwise sense, Green's theorem implies that

$$0 = \int_{\overrightarrow{\partial G}} (P\, dx_1 + Q\, dx_2) = \int_{\vec{C}_1} (P\, dx_1 + Q\, dx_2) - \int_{\overrightarrow{\partial K}} (P\, dx_1 + Q\, dx_2).$$

Using the representation $x_1 = \cos\theta$, $x_2 = \sin\theta$, $-\pi \leqslant \theta \leqslant \pi$, for the last integral on the right, we obtain

$$\int_{\vec{C}_1} (P\, dx_1 + Q\, dx_2) = \int_{-\pi}^{\pi} (\sin^2\theta + \cos^2\theta)\, d\theta = 2\pi. \qquad \square$$

EXAMPLE 3. Let D be a regular region with area A. Let $v = -\frac{1}{2}(x_2 e_1 - x_1 e_2)$. Show that

$$A = \int_{\overrightarrow{\partial D}} v \cdot dr.$$

Solution. We apply Green's theorem and find that

$$\int_{\overrightarrow{\partial D}} v \cdot dr = \int_{D} \operatorname{curl} v\, dA = \int_{D} \left(\frac{1}{2} + \frac{1}{2}\right) dA = A. \qquad \square$$

Example 3 shows that the area of any regular region may be expressed as an integral over the boundary of that region.

PROBLEMS

In each of Problems 1 through 8 verify Green's theorem.

1. $P(x_1, x_2) = -x_2$, $Q(x_1, x_2) = x_1$; $D = \{(x_1, x_2): 0 \leqslant x_1 \leqslant 1, 0 \leqslant x_2 \leqslant 1\}$.

2. $P(x_1, x_2) = x_1 x_2$, $Q(x_1, x_2) = -2x_1 x_2$; $D = \{(x_1, x_2, : 1 \leqslant x_1 \leqslant 2, 0 \leqslant x_2 \leqslant 3\}$.

3. $P(x_1, x_2) = 2x_1 - 3x_2$, $Q(x_1, x_2) = 3x_1 + 2x_2$, $D = \{(x_1, x_2): 0 \leqslant x_1 \leqslant 2, 0 \leqslant x_2 \leqslant 1\}$.

4. $P(x_1, x_2) = 2x_1 - x_2$, $Q(x_1, x_2) = x_1 + 2x_2$; D is the region outside the unit disk, above the curve $x_2 = x_1^2 - 2$, and below the line $x_2 = 2$.

5. $P = x_1^2 - x_2^2$, $Q = 2x_1 x_2$; D is the triangle with vertices at $(0, 0)$, $(2, 0)$, and $(1, 1)$.

6. $P = -x_2$, $Q = 0$; D is the region inside the circle $x_1^2 + x_2^2 = 4$ and outside the circles $x_1^2 + (x_2 - 1)^2 = 1/4$ and $x_1^2 + (x_2 + 1)^2 = 1/4$.

7. $v = (x_1^2 + x_2^2)^{-1}(-x_2 e_1 + x_1 e_2)$; $D = \{(x_1, x_2): 1 < x_1^2 + x_2^2 < 4\}$.

8. $P = 4x_1 - 2x_2$, $Q = 2x_1 + 6x_2$; D is the interior of the ellipse: $x_1 = 2\cos\theta$, $x_2 = \sin\theta$, $-\pi \leqslant \theta \leqslant \pi$.

9. Prove Part (b) of Theorem 16.19.

10. Given $D = \{(x_1, x_2): 0 \leqslant x_1^{2/3} + x_2^{2/3} < 1\}$. Decide whether or not D is a regular region.

In Problems 11 through 14 compute $\int_{\overrightarrow{\partial D}} v \cdot dr$ by means of Green's theorem.

11. $v(x_1, x_2) = (4x_1 e^{x_2} + 3x_1^2 x_2 + x_2^3)e_1 + (2x_1^2 e^{x_2} - \cos x_2)e_2$; $D = \{(x_1, x_2): x_1^2 + x_2^2 \leqslant 4\}$.

12. $v(x_1, x_2) = \arctan(x_2/x_1)e_1 + \frac{1}{2} \log(x_1^2 + x_2^2)e_2$; $D = \{(x_1, x_2): 1 \leqslant x_1 \leqslant 3, -2 \leqslant x_2 \leqslant 2\}$.

13. $v(x_1, x_2) = -x_2 e_1 + x_1 e_2$; $D = \{(x_1, x_2): (x_1 - 1)^2 + x_2^2 < 1\}$.

14. $v(x_1, x_2) = -3x_1^2 x_2 e_1 + 3x_1 x_2^2 e_2$;

$$D = \{(x_1, x_2): -a \leqslant x_1 \leqslant a, 0 \leqslant x_2 \leqslant \sqrt{a^2 - x_1^2}\}.$$

15. Given $P = -x_1(x_1^2 + x_2^2)^{-1}$, $Q = x_2(x_1^2 + x_2^2)^{-1}$. Use Green's theorem to find the value of $\int_C (P \, dx_1 + Q \, dx_2)$ where C is the arc of the parabola $x_2 = x_1^2 - 1$, $-1 \leqslant x_1 \leqslant 2$, followed by the line segment from $(2, 3)$ to $(-1, 0)$. Let D be the region outside a small disk of radius ρ center at $(0, 0)$ and inside of $|\vec{C}|$.

16. Suppose $u(x_1, x_2)$ satisfies $u_{x_1 x_1} + u_{x_2 x_2} = 0$ in a region G. Show that $\oint_{G^*} (u_{x_2} \, dx_1 - u_{x_1} \, dx_2) = 0$ where G^* is any region interior to G.

17. If $u_{x_1 x_1} + u_{x_2 x_2} = 0$ in a region G and $v(x_1, x_2)$ is a smooth function, use the identity $(vu_{x_1})_{x_1} = vu_{x_1 x_1} + v_{x_1} u_{x_1}$ and a similar one for $(vu_{x_2})_{x_2}$ to prove that

$$\int_{\partial G^*} v(u_{x_2} \, dx_1 - u_{x_1} \, dx_2) = -\iint_{G^*} (v_{x_1} u_{x_1} + v_{x_2} u_{x_2}) \, dV$$

where G^* is any region interior to G.

16.5. Surfaces* in $\mathbb{R}^3$; Parametric Representation

Until now we have considered a surface in $\mathbb{R}^3$ as the graph (in a Cartesian coordinate system) of an equation of the form $x_3 = f(x_1, x_2)$ or $F(x_1, x_2, x_3) = 0$. Now we are interested in studying surfaces which are more complicated than those which can be described by a single equation in x_1, x_2, x_3.

We simplify the study of a complicated surface by decomposing it into a number of small pieces and by examining each piece separately. It may happen that if the decomposition is fine enough, each individual piece will have a simple structure even when the entire surface is unusual or bizarre. For most surfaces we shall study, each small piece will have a structure like that of a small section of a sphere, a cylinder, a hyperboloid, or similar smooth surface.

A sphere, an ellipsoid, and the surface of a parallelepiped are examples of *surfaces without boundary*. On the other hand, a hemisphere has as a boundary

---

* Those readers interested only in the basic form of Stokes' theorem may omit this section without loss of continuity.

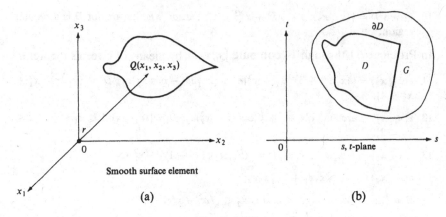

Figure 16.18

the circle consisting of its (equatorial) rim. When dividing a surface into small pieces we must be careful to distinguish those pieces which contain a portion of the boundary from those which are entirely interior to the surface.

Definitions. A **smooth surface element** is the graph of a system of equations of the form

$$x_1 = x_1(s, t), \qquad x_2 = x_2(s, t), \qquad x_3 = x_3(s, t), \qquad (s, t) \in D \cup \partial D \qquad (16.42)$$

in which x_1, x_2, x_3 are C^1 functions with domain a region G in $\mathbb{R}^2$ containing $D \cup \partial D$. We suppose that D is a region whose boundary consists of a finite number of piecewise smooth simple closed curves. In vector notation Equations (16.42) may be written

$$v(\overrightarrow{OQ}) = r(s, t), \qquad (s, t) \in D \cup \partial D, \qquad O \text{ given.} \qquad (16.43)$$

In Figure 16.18 we exhibit a smooth surface element and a domain D whose boundary ∂D consists of a single piecewise smooth curve. We shall assume that

$$r_s \times r_t \neq 0 \quad \text{for } (s, t) \in G \qquad (16.44)$$

and that $r(s_1, t_1) \neq r(s_2, t_2)$ whenever $(s_1, t_1) \neq (s_2, t_2)$. In other words, Equations (16.42) define a one-to-one C^1 transformation from $D \cup \partial D$ onto the points of the surface element. Observe that if

$$r(s, t) = x_1(s, t)e_1 + x_2(s, t)e_2 + x_3(s, t)e_3,$$

then

$$r_s \times r_t = \left(\frac{\partial x_2}{\partial s}\frac{\partial x_3}{\partial t} - \frac{\partial x_2}{\partial t}\frac{\partial x_3}{\partial s}\right)e_1 + \left(\frac{\partial x_3}{\partial s}\frac{\partial x_1}{\partial t} - \frac{\partial x_3}{\partial t}\frac{\partial x_1}{\partial s}\right)e_2$$
$$+ \left(\frac{\partial x_1}{\partial s}\frac{\partial x_3}{\partial t} - \frac{\partial x_1}{\partial t}\frac{\partial x_3}{\partial s}\right)e_3.$$

Employing the Jacobian notation which we introduced on page 359

$$J\left(\frac{\varphi, \psi}{u, v}\right) = \begin{vmatrix} \dfrac{\partial \varphi}{\partial u} & \dfrac{\partial \varphi}{\partial v} \\[2mm] \dfrac{\partial \psi}{\partial u} & \dfrac{\partial \psi}{\partial v} \end{vmatrix} = \left(\frac{\partial \varphi}{\partial u}\frac{\partial \psi}{\partial v} - \frac{\partial \varphi}{\partial v}\frac{\partial \psi}{\partial u}\right),$$

we may write

$$r_s \times r_t = J\left(\frac{x_2, x_3}{s, t}\right)e_1 + J\left(\frac{x_3, x_1}{s, t}\right)e_2 + J\left(\frac{x_1, x_2}{s, t}\right)e_3. \tag{16.45}$$

We say that the equations (16.42) or (16.43) form a **parametric representation** of the smooth surface element. The quantities s and t are called **parameters**.

The next result shows the relation between two parametric representations of the same surface element.

Theorem 16.21. *Suppose that the transformation given by (16.43) satisfies Condition (16.44) and that (16.43) is C^1 on a region G which contains $D \cup \partial D$. Let S and S_1 denote the images of $D \cup \partial D$ and G, respectively.*

(a) *If $(\bar{s}, \bar{t})$ is any point of $D \cup \partial D$, then there is a positive number ρ such that the part of S corresponding to the disk*

$$(s - \bar{s})^2 + (t - \bar{t})^2 < \rho^2$$

has one of the forms

$$x_3 = f(x_1, x_2), \qquad x_1 = g(x_2, x_3) \qquad or \qquad x_2 = h(x_1, x_3),$$

where $f, g,$ or h is smooth near the point $(\bar{x}_1, \bar{x}_2, \bar{x}_3)$ corresponding to $(\bar{s}, \bar{t})$.

(b) *Suppose that another parametric representation of S and S_1 is given by*

$$v(\overrightarrow{OP}) = r_1(s', t'), \qquad (s', t') \in D_1 \cup \partial D_1,$$

in which S and S_1 are the images of $D_1 \cup \partial D_1$ and G_1 respectively, in the (s', t') plane. Assume that D_1, G_1 and r_1 have all the properties which D, G, and r have. Then there is a one-to-one transformation

$$T: s = U(s', t'), \qquad t = V(s', t'), \qquad (s', t') \in G_1, \tag{16.46}$$

form G_1 to G such that $T(D_1) = D$, $T(\partial D_1) = \partial D$ and

$$r[U(s', t'), V(s', t')] = r_1(s', t') \quad for \ (s', t') \in G_1.$$

PROOF

(a) Since $r_s \times r_t \neq 0$ at a point $(\bar{s}, \bar{t})$, it follows that at least one of the three Jacobians in (16.45) is not zero at $(\bar{s}, \bar{t})$. Suppose, for instance, that

$$J\left(\frac{x_1, x_2}{s, t}\right) \neq 0$$

at $(\bar{s}, \bar{t})$. We set

$$r(s, t) = X_1(s, t)e_1 + X_2(s, t)e_2 + X_3(s, t)e_3$$

and denote $\bar{x}_1 = X_1(\bar{s}, \bar{t}), \bar{x}_2 = X_2(\bar{s}, \bar{t}), \bar{x}_3 = X_3(\bar{s}, \bar{t})$. Then from the Implicit function theorem it follows that there are positive numbers α and β such that all numbers x_1, x_2, s, t which satisfy

$$(x_1 - \bar{x}_1)^2 + (x_2 - \bar{x}_2)^2 < \alpha^2, \qquad (s - \bar{s})^2 + (t - \bar{t})^2 < \beta^2,$$

and

$$x_1 = X_1(s, t), \qquad x_2 = X_2(s, t),$$

lie on the graph of

$$s = \varphi(x_1, x_2), \qquad t = \psi(x_1, x_2),$$

where φ and ψ are smooth functions in the disk $(x - \bar{x}_1)^2 + (x_2 - \bar{x}_2)^2 < \alpha^2$. In this case the part of S_1 near $(\bar{x}_1, \bar{x}_2, \bar{x}_3)$ is the graph of

$$x_3 = X_3[\varphi(x_1, x_2), \psi(x_1, x_2)] \quad \text{for } (x_1 - \bar{x}_1)^2 + (x_2 - \bar{x}_2)^2 < \alpha^2.$$

The conclusion (a) follows when we select $\rho > 0$ so small that the image of the disk $(s - \bar{s})^2 + (t - \bar{t})^2 < \rho^2$ lies inside the disk $(x_1 - \bar{x}_1)^2 + (x_2 - \bar{x}_2)^2 < \alpha^2$.

(b) Since $r(s, t)$ and $r_1(s', t')$ are one-to-one, it follows that to each (s', t') in G_1 there corresponds a unique P on S_1 which comes from a unique pair (s, t) in G. The relationship is shown in Figure 16.19. The transformation T in (16.46) is defined by this correspondence: $s = U(s', t'), t = V(s', t')$. Then T is clearly one-to-one. To see that T is smooth, let (s_0', t_0') be any point in G_1 and let $s_0 = U(s_0', t_0'), t_0 = V(s_0', t_0')$. Denote by P_0 the point of S corresponding to both (s_0, t_0) and (s_0', t_0'). At least one of the three Jacobians in (16.45) does not vanish at (s_0, t_0). If, for example, the last one does not vanish, we can solve for s and t in terms of x_1 and x_2 as in Part (a). We now set

$$r_1(s', t') = X_1'(s', t')e_1 + X_2'(s', t')e_2 + X_3'(s', t')e_3.$$

At points P of S near P_0 the surface can be represented as a function of (x_1, x_2) so that there is a one-to-one correspondence between the range and domain

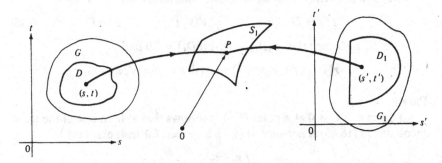

Figure 16.19. Defining the transformation T.

of this function. Therefore we find

$$U(s', t') = \varphi[X_1'(s', t'), X_2'(s', t')], \qquad V(s', t') = \psi[X_1'(s', t'), X_2'(s', t')].$$

Hence U and V are smooth near (s_0', t_0'), an arbitrary point of $D_1 \cup \partial D_1$. $\square$

Remark. The inverse transformation, T^{-1}, has the same smoothness properties that T does.

Definition. Suppose that S is a smooth surface element given by

$$r(s, t) = x_1(s, t)e_1 + x_2(s, t)e_2 + x_3(s, t)e_3, \qquad (s, t) \in D \cup \partial D.$$

Then the **boundary** *of* S is the image of ∂D in the above parametric representation. From Part (b) of Theorem 16.21 the boundary of a smooth surface element is independent of the particular parametric representation. Also, it is not difficult to see that such a boundary consists of a finite number of piecewise smooth, simple closed curves, no two of which intersect.

From Part (a) of Theorem 16.21 it follows that a neighborhood of any point of a smooth surface element is the graph of an equation of one of the three forms

$$x_3 = f(x_1, x_2), \qquad x_1 = g(x_2, x_3), \qquad x_2 = h(x_1, x_3),$$

where f, g, or h is a smooth function on its domain.

The set of points

$$\{(x_1, x_2, x_3): x_3 = f(x_1, x_2), (x_1, x_2) \in D \cup \partial D\}$$

where f is smooth on a region G containing $D \cup \partial D$ is a smooth surface element S. To see this, observe that S is the graph of the parametric equations

$$x_1 = s, \qquad x_2 = t, \qquad x_3 = f(s, t), \qquad (s, t) \in D \cup \partial D.$$

Definition. A **piecewise smooth surface** S is the union $S_1 \cup S_2 \cup \cdots \cup S_n$ of a finite number of smooth surface elements $S_1, S_2, \ldots, S_n$ satisfying the following five conditions:

(i) No two S_i have common interior points.
(ii) The intersection of the boundaries of two elements $\partial S_i \cap \partial S_j$, $i \neq j$, is either empty, or a single point, or a piecewise smooth arc. See Figure 16.20 for an example of a surface S.
(iii) The boundaries of any three distinct elements have at most one point in common.
(iv) Any two points of S can be joined by a path in S.
(v) The union of all arcs each of which is on the boundary of only one of the S_i form a finite number of disjoint piecewise smooth simple closed curves.

The set of points in (v) constitute the **boundary** of S, denoted ∂S. If this set is empty then S is a piecewise smooth **surface without boundary**.

A piecewise smooth surface S has many decompositions into the finite

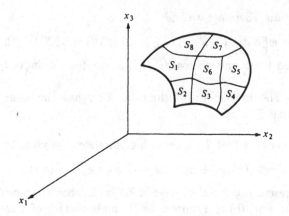

Figure 16.20. A piecewise smooth surface.

union of smooth surface elements. An arc which is part of the boundary of two of the S_i is called an **edge** of the decomposition. A corner of one of the boundaries of an S_i is called a **vertex**.

A piecewise smooth surface S is said to be a **smooth surface** if and only if for every point P not on ∂S a decomposition of S into piecewise smooth surface elements can be found such that P is an interior point of one of the surface elements.

It can be shown that if F is a smooth scalar field on some domain in $\mathbb{R}^3$ and $F(x_1, x_2, x_3)$ and $\nabla F(x_1, x_2, x_3)$ are never simultaneously zero, then the graph of the equation $F = 0$ is a smooth surface provided that the graph is a closed, bounded set. We leave the details of the proof to the reader. For example, the functions

$$F(x_1, x_2, x_3) = x_1^2 + x_2^2 + x_3^2 - a^2,$$

and

$$F_1(x_1, x_2, x_3) = \frac{x_1^2}{a^2} + \frac{x_2^2}{b^2} + \frac{x_3^2}{c^2} - 1,$$

exhibit the fact that spheres and ellipsoids are smooth surfaces. Any polyhedron can be shown to be a piecewise smooth surface.

EXAMPLE. Suppose that a smooth scalar field is defined in all of $\mathbb{R}^3$ by the formula

$$F(x_1, x_2, x_3) = x_1^2 + 4x_2^2 + 9x_3^2 - 44.$$

Show that the set $S = \{(x_1, x_2, x_3): F = 0\}$ is a smooth surface.

Solution. We compute the gradient:

$$\nabla F = 2x_1 e_1 + 8x_2 e_2 + 18x_3 e_3.$$

Then $\nabla F = 0$ only at $(0, 0, 0)$. Since S is not void (e.g., the point $(\sqrt{44}, 0, 0)$ is on it), since $(0, 0, 0)$ is not a point of S, and since $F = 0$ is a closed, bounded set, the surface is smooth.   □

PROBLEMS

1. Find two parametric representations of the hemisphere $S = \{(x_1, x_2, x_3): x_1^2 + x_2^2 + x_3^2 = 1, x_3 > 0\}$ and find the transformation T as given by (16.46) relating the two representations.

2. Same as Problem 1 for the surface $S = \{(x_1, x_2, x_3): x_1^2 + x_2^2 + 2x_3^2 = 1, x_3 > 0\}$.

3. Given the cube with vertices $(0, 0, 0)$, $(1, 0, 0)$, $(0, 1, 0)$, $(1, 1, 0)$, $(0, 0, 1)$, $(0, 1, 1)$ $(1, 0, 1)$, $(1, 1, 1)$. Find a decomposition of the surface of the cube into smooth surface elements, and verify that the boundary is empty.

4. Given the sphere $S = \{(x_1, x_2, x_3): x_1^2 + x_2^2 + x_3^2 = 1\}$. Decompose S into smooth surface elements and verify that S is a smooth surface without boundary.

5. Given the conical surface $S = \{(x_1, x_2, x_3): x_1^2 + x_2^2 - x_3^2 = 0, -1 < x_3 < 1\}$. Decide whether or not S is a smooth surface.

6. A torus S is given by $S = \{(x_1, x_2, x_3): (\sqrt{x_1^2 + x_2^2} - b)^2 + x_3^2 = a^2\}$ with $0 < a < b$. Find a decomposition of S into smooth surface elements and show that S is a surface without boundary.

7. Let $F: D \to \mathbb{R}^1$ be a smooth scalar function where D is a region in $\mathbb{R}^3$. Suppose that F and ∇F are never simultaneously zero. Suppose that $S = \{(x_1, x_2, x_3): F(x_1, x_2, x_3) = 0\}$ is a closed, bounded set. Show that S is a smooth surface.

8. Let $S_0 = \{(x_1, x_2, x_3): 1 \leqslant x_1^2 + x_2^2 + x_3^2 \leqslant 4\}$ and $S_1 = \{(x_1, x_2, x_3): 2(x_1^2 + x_2^2) = 1, x_3 > 0\}$ be given. Show that $S_1 \cap S_0 \equiv S$ is a piecewise smooth surface. Describe ∂S.

16.6. Area of a Surface in $\mathbb{R}^3$; Surface Integrals

Let a surface S in $\mathbb{R}^3$ be given by

$$S = \{(x_1, x_2, x_3): x_3 = f(x_1, x_2), (x_1, x_2) \in D \cup \partial D\}$$

where D is a bounded region in $\mathbb{R}^2$. We know that if f has continuous first derivatives the area of the surface, $A(S)$, may be computed by the formula developed in calculus. In fact, we recall that

$$A(S) = \iint_D \sqrt{1 + \left(\frac{\partial f}{\partial x_1}\right)^2 + \left(\frac{\partial f}{\partial x_2}\right)^2}\, dA.$$

In order to develop methods for finding the area of more complicated surfaces we first define the area of a smooth surface element (see p. 456 for the definition of a smooth surface element).

Let σ be a smooth surface element given by

$$\sigma: v(\overrightarrow{0Q}) = r(s, t), \qquad (s, t) \in D \cup \partial D, \qquad (16.47)$$

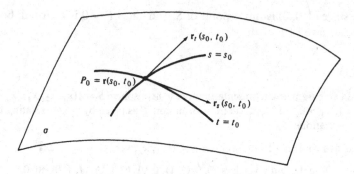

Figure 16.21. r_s and r_t are tangent to σ.

with r and D satisfying the conditions stated in the definition of a smooth surface element. If t is held constant, say equal to t_0, then the graph of (16.47) is a smooth curve on σ. Therefore $r_s(s, t_0)$, the partial derivative of r with respect to s, is a vector tangent to this curve on the surface. Similarly, the vector $r_t(s_0, t)$ is tangent to the curve on σ obtained when we set $s = s_0$ (see Figure 16.21). The vectors $r_s(s_0, t_0)$ and $r_t(s_0, t_0)$ lie in the plane tangent to the surface element σ at the point $P_0 = r(s_0, t_0)$.

Definition. The **tangent linear transformation** of the surface given by (16.47) at the point P_0 on σ is given by

$$v(\overrightarrow{OP}) = r(s_0, t_0) + r_s(s_0, t_0)(s - s_0) + r_t(s_0, t_0)(t - t_0) \qquad (16.48)$$

for $(s, t) \in \mathbb{R}^2$. Observe that the graph of the point P in (16.48) is the plane tangent to σ at P_0. The image of a rectangle in $\mathbb{R}^2$ such as

$$R = \{(s, t): s_1 \leqslant s \leqslant s_2, t_1 \leqslant t \leqslant t_2\}$$

under the transformation (16.48) is a parallelogram $ABCD$ in the plane tangent to σ at P_0 (see Figure 16.22). We set $a = r_s(s_0, t_0)$ and $b = r_t(s_0, t_0)$. Then the

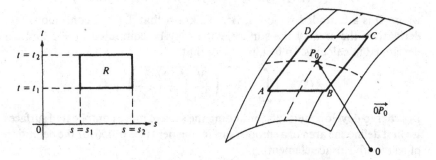

Figure 16.22. Image of R is a parallelogram.

points A, B, C, D are determined by

$$v(\overrightarrow{OA}) = r(s_0, t_0) + a(s_1 - s_0) + b(t_1 - t_0),$$
$$v(\overrightarrow{OB}) = r(s_0, t_0) + a(s_2 - s_0) + b(t_1 - t_0),$$
$$v(\overrightarrow{OC}) = r(s_0, t_0) + a(s_2 - s_0) + b(t_2 - t_0),$$
$$v(\overrightarrow{OD}) = r(s_0, t_0) + a(s_1 - s_0) + b(t_2 - t_0).$$

By subtraction of $v(\overrightarrow{OA})$ from $v(\overrightarrow{OB})$ and $v(\overrightarrow{OA})$ from $v(\overrightarrow{OD})$ we find that

$$v(\overrightarrow{AB}) = (s_2 - s_1)a, \qquad v(\overrightarrow{AD}) = (t_2 - t_1)b.$$

Denoting the area of R by $A(R)$, we use the vector product to obtain the area of the parallelogram $ABCD$. Hence

$$\text{Area } ABCD = |v(\overrightarrow{AB}) \times v(\overrightarrow{AD})| = |(s_2 - s_1)(t_2 - t_1)||a \times b|$$
$$= A(R)|r_s(s_0, t_0) \times r_t(s_0, t_0)|.$$

In defining the area of a surface element we use the fact that a very small piece of surface is approximated by a small parallelogram in the plane tangent to this small piece.

Definition. Let σ be a smooth surface element given by (16.47). We subdivide D into a number of subregions $D_1, D_2, \ldots, D_n$ and define the mesh size $\|\Delta\|$ as the maximum diameter of all the D_i. The **area of the surface element** σ, denoted $A(\sigma)$, is defined by the formula

$$A(\sigma) = \lim_{\|\Delta\| \to 0} \sum_{i=1}^{n} A(D_i)|r_s(s_i, t_i) \times r_t(s_i, t_i)|$$

where $A(D_i)$ is the area of D_i, and (s_i, t_i) is any point of D_i, and where the limit exists in the same manner as that determined in the definition of a definite integral. From this definition of area we obtain at once the formula

$$A(\sigma) = \iint_D |r_s(s, t) \times r_t(s, t)| \, dA. \tag{16.49}$$

Suppose a surface element σ has another parametric representation in addition to (16.47). That is, suppose σ is given by

$$\sigma : v(\overrightarrow{OP}) = r'(s', t') \quad \text{for } (s', t') \in D' \cup \partial D'.$$

Then the area $A'(\sigma)$ in this representation is given by

$$A'(\sigma) = \iint_{D'} |r'_{s'}(s', t') \times r'_{t'}(s', t')| \, dA. \tag{16.50}$$

From the rule for multiplying Jacobians and the rule for change of variables in a multiple integral, it follows that $A(\sigma) = A'(\sigma)$. In fact, we have

$$|r'_{s'}(s', t') \times r'_{t'}(s', t')| = |r_s(s, t) \times r_t(s, t)| \cdot \left| J\left(\frac{s, t}{s', t'}\right) \right|,$$

and so (16.49) and (16.50) yield the same value.

If the representation of σ is of the form $x_3 = f(x_1, x_2)$ discussed at the beginning of the section, we may set $x_1 = s$, $x_2 = t$, $x_3 = f(s, t)$ and find that

$$J\left(\frac{x_2, x_3}{s, t}\right) = -f_s, \qquad J\left(\frac{x_3, x_1}{s, t}\right) = -f_t, \qquad J\left(\frac{x_1, x_2}{s, t}\right) = 1.$$

Then (16.49) becomes the familiar formula

$$A(\sigma) = \iint_D \sqrt{1 + f_s^2 + f_t^2} \, dA. \tag{16.51}$$

When a surface σ is described by a single equation such as $x_3 = f(x_1, x_2)$ we say that σ is given in **nonparametric form**.

Unfortunately most surfaces cannot be described in nonparametric form, and so (16.51) cannot be used in general for the computation of surface area. In fact, it can be shown that a simple surface such as a sphere cannot be part of a single smooth surface element.

If S is a piecewise continuous surface, we can define area by decomposition. First, if σ is a smooth surface element, a set F is a **figure in** σ if and only if F is the image under (16.47) of a figure E in the plane region $D \cup \partial D$. The area of F is defined by Formula (16.49). If $r_1 = r_1(s', t')$ is another representation of σ and E_1 is the set in $D_1 \cup \partial D_1$ corresponding to F, it follows from Theorem 16.21(b) and the rule for multiplying Jacobians that E_1 is a figure and hence the area of F is also given by (16.49).

Definitions. Let S be a piecewise continuous surface. A set F contained in S is a **figure** if and only if $F = F_1 \cup F_2 \cup \cdots \cup F_k$ where each F_i is a figure contained in a single smooth surface element σ_i of S and no two F_i have common interior points. The **area** of F is defined by the formula

$$A(F) = A(F_1) + \cdots + A(F_k).$$

It is important to know that any two such decompositions of a set F yield the same value for $A(F)$. We omit the proof of the theorem which establishes this fact.

We now discuss integration of a function f defined on a surface F. Suppose that F is a closed figure on a piecewise smooth surface. We write $F = F_1 \cup F_2 \cup \cdots \cup F_k$ where each F_i is a figure contained in one smooth surface element. Then each F_i is the image of a figure E_i in $\mathbb{R}^2$, $i = 1, 2, \ldots, k$, under the map

$$v(\overrightarrow{OQ}) = r_i(s, t) \quad \text{for } (s, t) \in D_i \cup \partial D_i \tag{16.52}$$

with $E_i \subset (D_i \cup \partial D_i)$. Let $f: F \to \mathbb{R}^1$ be a continuous scalar field. We define

$$\iint_F f \, dS = \sum_{i=1}^k \iint_{F_i} f \, dS \tag{16.53}$$

where each term in the right side of (16.53) is defined by the formula

$$\iint_{F_i} f \, dS = \iint_{E_i} f[r_i(s, t)] \cdot |r_{is} \times r_{it}| \, dA. \tag{16.54}$$

The result is independent of the particular subdivision $\{F_i\}$ and the particular parametric representation (16.52).

· If a surface element σ has a nonparametric representation $x_3 = \varphi(x_1, x_2)$, then (16.54) becomes

$$\iint_{F_i} f \, dS = \iint_{E_i} f[x_1, x_2, \varphi(x_1, x_2)] \sqrt{1 + \left(\frac{\partial \varphi}{\partial x_1}\right)^2 + \left(\frac{\partial \varphi}{\partial x_2}\right)^2} \, dA. \tag{16.55}$$

The evaluation of the integral in (16.55) follows the usual rule for evaluation of ordinary double and iterated integrals. We show the technique and give several applications in the following examples.

EXAMPLE 1. Find the value of $\iint_F x_3^2 \, dS$ where F is the part of the lateral surface of the cylinder $x_1^2 + x_2^2 = 4$ between the planes $x_3 = 0$ and $x_3 = x_2 + 3$ (see Figure 16.23).

Solution. When we transform to the cylindrical coordinate system $x_1 = r \sin \theta$, $x_2 = r \cos \theta$, $x_3 = z$, then F lies on the surface $r = 2$. We choose θ and z as

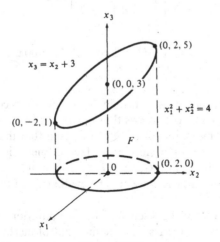

Figure 16.23

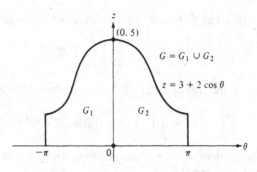

Figure 16.24

parametric coordinates on F and set

$$G = \{(\theta, z): -\pi \le \theta \le \pi, 0 \le z \le 3 + 2\cos\theta\},$$

$$F = \{(x_1, x_2, x_3): x_1 = 2\sin\theta, x_2 = 2\cos\theta, x_3 = z, (\theta, z) \in G\}.$$

The element of surface area dS is given by $dS = |r_\theta \times r_z| \, dA_{\theta z}$, (see Figure 16.24) and since

$$r_\theta \times r_z = J\left(\frac{x_2, x_3}{\theta, z}\right)e_1 + J\left(\frac{x_3, x_1}{\theta, z}\right)e_2 + J\left(\frac{x_1, x_2}{\theta, z}\right)e_3$$

$$= -(2\sin\theta)e_1 - (2\cos\theta)e_2 + 0\cdot e_3,$$

we find

$$dS = 2dA_{\theta z}.$$

Therefore

$$\iint_F x_3^2 \, dS = 2\iint_G z^2 \, dA_{\theta z} = 2\int_{-\pi}^{\pi}\int_0^{3+2\cos\theta} z^2 \, dz \, d\theta$$

$$= \frac{2}{3}\int_{-\pi}^{\pi} (3 + 2\cos\theta)^3 \, d\theta = 60\pi. \tag{16.56}$$

□

Remark. The surface F is not a single smooth surface element since the transformation from G to F shows that the points $(-\pi, z)$ and (π, z) of G are carried into the same points of F. The condition that the transformation be one-to-one, necessary for a smooth surface element, is therefore violated. However, if we divide G into G_1 and G_2 as shown in Figure 16.24, the image of each is a smooth surface element. The evaluation of the integral (16.56) is unchanged.

Surface integrals can be used for the computation of various physical quantities. The center of mass and the moment of inertia of thin curvilinear plates are sometimes computable in terms of surface integrals. Also, the potential of a distribution of an electric charge on a surface may be expressed

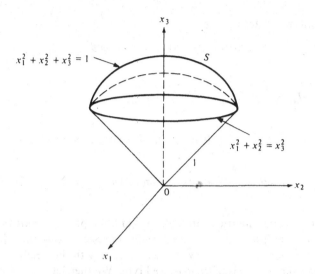

Figure 16.25. A cone surmounted by part of a sphere.

as a surface integral. We may associate a mass with any surface F by assuming it to be made of thin material having density δ. The density is assumed to be continuous but not necessarily constant. The total mass of such a surface F, denoted $M(F)$, is given by

$$M(F) = \iint_F \delta(P)\,dS.$$

The moment of inertia I_{x_3} (of a surface F) about the x_3 axis is given by

$$I_{x_3}(F) = \iint_F \delta(P)(x_1^2(P) + x_2^2(P))\,dS,$$

with analogous formulas for $I_{x_1}(F)$ and $I_{x_2}(F)$.

EXAMPLE 2. Find the moment of inertia about the x_1 axis of the surface $S = \{(x_1, x_2, x_3): x_1^2 + x_2^2 + x_3^2 = 1, x_3 \geqslant \sqrt{x_1^2 + x_2^2}\}$. Assume the density δ is a constant (see Figure 16.25).

Solution. We introduce spherical coordinates $x_1 = \rho \cos \theta \sin \varphi$, $x_2 = \rho \sin \theta \sin \varphi$, $x_3 = \rho \cos \varphi$ and set

$$D = \{(\theta, \varphi): 0 \leqslant \theta \leqslant 2\pi, 0 \leqslant \varphi \leqslant \pi/4\}.$$

Then the surface S is given by $S = \{(\rho, \theta, \varphi): \rho = 1, (\theta, \varphi) \in D\}$. We compute $r_\varphi \times r_\theta$ and find

$$r_\varphi \times r_\theta = J\left(\frac{x_2, x_3}{\varphi, \theta}\right)e_1 + J\left(\frac{x_3, x_1}{\varphi, \theta}\right)e_2 + J\left(\frac{x_1, x_2}{\varphi, \theta}\right)e_3$$

$$= (\sin^2 \varphi \cos \theta)e_1 + (\sin^2 \varphi \sin \theta)e_2 + (\sin \varphi \cos \varphi)e_3.$$

The element of surface area is

$$dS = |r_\varphi \times r_\theta| \, dA_{\varphi\theta} = \sin \varphi \, dA_{\varphi\theta}.$$

Then

$$I_{x_1} = \iint_D \delta(x_2^2 + x_3^2) \sin \varphi \, dA_{\varphi\theta}$$

$$= \delta \int_0^{\pi/4} \int_0^{2\pi} (\sin^2 \varphi \sin^2 \theta + \cos^2 \varphi) \sin \varphi \, d\theta \, d\varphi$$

$$= \pi\delta \int_0^{\pi/4} (1 + \cos^2 \varphi) \sin \varphi \, d\varphi = \frac{\pi\delta}{12}(16 - 7\sqrt{2}). \qquad \square$$

Remark. The parametric representation of S in spherical coordinates does not fulfill the required conditions for such representations since $|r_\varphi \times r_\theta| = \sin \varphi$ vanishes for $\varphi = 0$. However, the surface S with the deletion of a small hole around the x_3 axis is of the required type. We then let the size of the hole tend to zero and obtain the above result for I_{x_1}.

EXAMPLE 3. Given $R = \{(x_1, x_2, x_3): x_1^2 + x_2^2 \leqslant 1, 0 \leqslant x_3 \leqslant x_1 + 2\}$ and $S = \partial R$. If S has uniform density δ, find its mass. Find the value of

$$\iint_S x_1 \, dS$$

and obtain $\bar{x}_1$, the x_1-coordinate of the center of mass of S (see Figure 16.26).

Solution. The surface is composed of three parts: S_1, the disk in the $x_1 x_2$-plane; S_2, the lateral surface of the cylinder; and S_3, the part of the plane $x_3 = x_1 + 2$

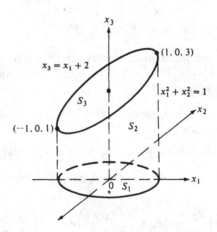

Figure 16.26

inside the cylinder. We have (since x_1 is an odd function)

$$\iint_{S_1} x_1 \, dS = 0.$$

On S_3 we see that $x_3 = x_1 + 2$ so that $dS = \sqrt{2} \, dA$ is the element of area on S_1. Hence

$$\iint_{S_3} x_1 \, dS = \sqrt{2} \iint_{S_1} x_1 \, dA = 0.$$

On S_2 we choose coordinates (θ, z) with $x_1 = \cos \theta$, $x_2 = \sin \theta$, $x_3 = z$, $dS = dA_{\theta z}$. We write

$$D_2 = \{(\theta, z): -\pi \leqslant \theta \leqslant \pi, \quad 0 \leqslant z \leqslant 2 + \cos \theta\}$$

and find

$$\iint_{S_2} x_1 \, dS = \iint_{D_2} \cos \theta \, dA_{\theta z} = \int_{-\pi}^{\pi} \int_0^{2+\cos\theta} \cos \theta \, dz \, d\theta = \pi.$$

Therefore

$$\iint_S x_1 \, ds = \iint_{S_1 \cup S_2 \cup S_3} x_1 \, dS = \pi.$$

The surface S is a piecewise smooth surface without boundary. The intersections of $x_1^2 + x_2^2 = 1$ with the planes $x_3 = 0$, $x_3 = x_1 + 2$ form two smooth edges. There are no vertices on S. To find the mass $M(S)$ of S, we observe that

$$M(S) = \iint_S \delta \, dS = \delta A(S) = \delta[A(S_1) + A(S_2) + A(S_3)].$$

Clearly, $A(S_1) = \pi$, $A(S_3) = \pi\sqrt{2}$. Also,

$$A(S_2) = \iint_{D_2} dA_{\theta z} = \int_{-\pi}^{\pi} \int_0^{2+\cos\theta} dz \, d\theta = 4\pi.$$

Therefore $M(S) = \pi\delta(5 + \sqrt{2})$. To obtain the x_1 coordinate of the center of mass, note that

$$\bar{x}_1 = \frac{\delta \iint_S x_1 \, dA}{M(S)} = \frac{\delta\pi}{\delta\pi(5 + \sqrt{2})} = \frac{1}{5 + \sqrt{2}}. \qquad \square$$

PROBLEMS

In each of Problems 1 through 9, find the value of

$$\iint_S f(x_1, x_2, x_3) \, dS.$$

1. $f(x_1, x_2, x_3) = x_1, S = \{(x_1, x_2, x_3): x_1 + x_2 + x_3 = 1, x_1 \geqslant 0, x_2 \geqslant 0, x_3 \geqslant 0\}.$

2. $f(x_1, x_2, x_3) = x_1^2$, $S = \{(x_1, x_2, x_3): x_3 = x_1, x_1^2 + x_2^2 \leqslant 1\}$.

3. $f(x_1, x_2, x_3) = x_1^2$, $S = \{(x_1, x_2, x_3): x_3^2 = x_1^2 + x_2^2, 1 \leqslant x_3 \leqslant 2\}$.

4. $f(x_1, x_2, x_3) = x_1^2$, S is the part of the cylinder $x_3 = x_1^2/2$ cut out by the planes $x_2 = 0$, $x_1 = 2$, and $x_1 = x_2$.

5. $f(x_1, x_2, x_3) = x_1 x_3$, $S = \{(x_1, x_2, x_3): x_1^2 + x_2^2 = 1, 0 \leqslant x_3 \leqslant x_1 + 2\}$.

6. $f(x_1, x_2, x_3) = x_1$, S is the part of the cylinder $x_1^2 - 2x_1 + x_2^2 = 0$ between the two nappes of the cone $x_1^2 + x_2^2 = x_3^2$.

7. $f(x_1, x_2, x_3) = 1$; using polar coordinates (r, θ) in the x_1, x_2-plane, S is the part of the vertical cylinder erected on the spiral $r = \theta$, $0 \leqslant \theta \leqslant \pi/2$, bounded below by the $x_1 x_2$-plane and above by the cone $x_1^2 + x_2^2 = x_3^2$.

8. $f(x_1, x_2, x_3) = x_1^2 + x_2^2 - 2x_3^2$, $S = \{(x_1, x_2, x_3): x_1^2 + x_2^2 + x_3^2 = a^2\}$.

9. $f(x_1, x_2, x_3) = x_1^2$, $S = \partial R$ where $R = \{(x_1, x_2, x_3): x_3^2 \geqslant x_1^2 + x_2^2, 1 \leqslant x_3 \leqslant 2\}$ (see Problem 3).

In each of Problems 10 through 14, find the moment of inertia of S about the indicated axis, assuming that the density δ is constant.

10. The surface S of Problem 3; x_1 axis.

11. The surface S of Problem 6; x_1 axis.

12. The surface S of Problem 7; x_3 axis.

13. The surface S which is the boundary of R, where $R = \{(x_1, x_2, x_3): x_1 + x_2 + x_3 < 1, x_1 > 0, x_2 > 0, x_3 > 0\}$; about the x_2 axis.

14. The torus $S = \{(x_1, x_2, x_3): (\sqrt{x_1^2 + x_2^2} - b)^2 + x_3^2 = a^2, 0 < a < b\}$; x_3 axis. [*Hint:* If the parameters θ, φ are introduced by the relations

$$x_1 = (b + a \cos \varphi) \cos \theta$$

$$x_2 = (b + a \cos \varphi) \sin \theta$$

$$x_3 = a \sin \varphi,$$

the torus S is described by $\{(\varphi, \theta): 0 \leqslant \varphi \leqslant 2\pi, 0 \leqslant \theta \leqslant 2\pi\}$.]

In each of Problems 15 through 17 find the center of mass assuming the density δ is constant.

15. S is the surface of Problem 2.

16. S is the surface of Problem 13.

17. $S = \{(x_1, x_2, x_3): x_1^2 + x_2^2 + x_3^2 = 4a^2; x_1^2 + x_2^2 \leqslant a^2\}$.

The electrostatic potential $E(Q)$ at a point Q due to a distribution of electric charge (with charge density ρ) on a surface S is given by

$$E(Q) = \iint_S \frac{\rho(P)\, dS}{d_{PQ}}$$

where d_{PQ} is the distance from a point Q in $\mathbb{R}^3 - S$ to a point $P \in S$.

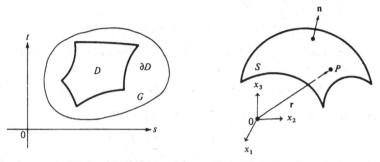

Figure 16.27. The unit normal to the surface S.

In Problems 18 through 21 find $E(Q)$ at the point given, assuming that ρ is constant.

18. $S = \{(x_1, x_2, x_3): x_1^2 + x_2^2 = 1, 0 \leqslant x_3 \leqslant 1\}; Q = (0, 0, 0)$.

19. $S = \{(x_1, x_2, x_3): x_1^2 + x_2^2 + x_3^2 = a^2\}; Q = (0, 0, c)$. Case 1: $c > a > 0$; Case 2: $a > c > 0$.

20. $S = \{(x_1, x_2, x_3): x_1^2 + x_2^2 + x_3^2 = a^2, x_3 \geqslant 0\}. Q = (0, 0, c); 0 < c < a$.

21. S is the surface of Problem 3; $Q = (0, 0, 0)$.

16.7. Orientable Surfaces

Suppose that S is a smooth surface element represented parametrically by

$$v(\overrightarrow{OP}) = r(s, t) \quad \text{with } (s, t) \in D \cup \partial D \tag{16.57}$$

where D is a region in the (s, t) plane with a piecewise smooth boundary ∂D. From the definition of a smooth surface element, we know that $r_s \times r_t \neq 0$ for (s, t) in a region G containing $D \cup \partial D$ (see Figure 16.27).

Definition. For a smooth surface element S, the **unit normal function to S** is defined by the formula

$$n = \frac{r_s \times r_t}{|r_s \times r_t|}, \quad (s, t) \in D.$$

Whenever S is a smooth surface element, the vector n is a continuous function of s and t. Using Jacobian notation, n may be written in coordinates

$$n = |r_s \times r_t|^{-1} \left[J\left(\frac{x_2, x_3}{s, t}\right) e_1 + J\left(\frac{x_3, x_1}{s, t}\right) e_2 + J\left(\frac{x_1, x_2}{s, t}\right) e_3 \right]. \tag{16.58}$$

Suppose now that the same surface element S has another parametric representation

$$v(\overrightarrow{OP}) = r'(s', t') \quad \text{with } (s', t') \in D' \cup \partial D'.$$

Then from Part (b) of Theorem 16.21 there is a one-to-one continuously differentiable transformation T given by

$$T: s = U(s', t'), \qquad t = V(s', t'), \qquad (s', t') \in G',$$

from G' to G such that $T(D') = D$ and $T(\partial D') = \partial D$. We also have

$$r[U(s', t'), V(s', t')] = r'(s', t') \quad \text{for } (s', t') \in D'.$$

We now define the unit normal function n' in terms of the second parametric representation:

$$n' = |r'_{s'} \times r'_{t'}|^{-1}\left[J\left(\frac{x_2, x_3}{s', t'}\right)e_1 + J\left(\frac{x_3, x_1}{s', t'}\right)e_2 + J\left(\frac{x_1, x_2}{s', t'}\right)e_3 \right]. \quad (16.59)$$

Using the law for multiplying Jacobians, we conclude from (16.58) and (16.59) that for all $P \in S$

$$n'(P) = n(P) \quad \text{or} \quad n'(P) = -n(P).$$

The choice of sign depends upon whether $J(s, t/s', t')$ is positive or negative on $D' \cup \partial D'$.

Definitions. A smooth surface S is **orientable** if and only if there exists a continuous unit normal function defined over all of S. Such a unit normal function is called an **orientation of** S.

Since the unit normal function to S at a point P is either $n(P)$ or $-n(P)$, each orientable surface possesses exactly two orientations, each of which is the negative of the other. An **oriented surface** is the pair (S, n) where n is one of the two orientations of S. We denote such an oriented surface by $\vec{S}$. Suppose that F is a smooth surface element of the oriented surface $\vec{S}$. The function n when restricted to F provides an orientation for F so that $(F, n) = \vec{F}$ is an oriented surface element. We say that the orientation of F agrees with the orientation of S if the parametric representations of F and S yield unit normal functions on F which are identical.

It is intuitively clear that smooth surfaces such as spheres, ellipsoids, toruses, and so forth, are all orientable surfaces. However, there are smooth surfaces for which there is no way to choose a continuous unit normal over the *entire* surface. One such surface is the **Möbius strip** shown in Figure 16.28. A model of this surface can be made from a long, narrow rectangular strip of paper by giving one end a half-twist (180°) and then gluing the ends together. A Möbius strip M can be represented parametrically on a rectangle (Figure 16.29)

$$R = \{(s, t): 0 \leqslant s \leqslant 2\pi, -h \leqslant t \leqslant h\}$$

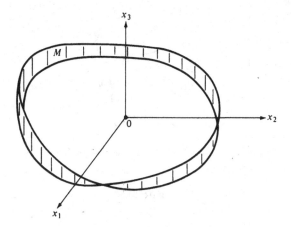

Figure 16.28. A Möbius strip.

by the equations $(0 < h < a)$:

$$x_1 = (a + t \sin\tfrac{1}{2}s) \cos s, \qquad x_2 = (a + t \sin\tfrac{1}{2}s) \sin s, \qquad x_3 = t \cos\tfrac{1}{2}s.$$
$$(16.60)$$

To find the unit normal to M, we first compute

$$J\left(\frac{x_2, x_3}{s, t}\right) = (a + t \sin\tfrac{1}{2}s) \cos\tfrac{1}{2}s \cos s + \tfrac{1}{2}t \sin s,$$

$$J\left(\frac{x_3, x_1}{s, t}\right) = (a + t \sin\tfrac{1}{2}s) \cos\tfrac{1}{2}s \sin s - \tfrac{1}{2}t \cos s,$$

$$J\left(\frac{x_1, x_2}{s, t}\right) = -(a + t \sin\tfrac{1}{2}s) \sin\tfrac{1}{2}s.$$

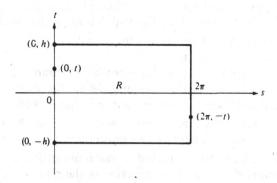

Figure 16.29

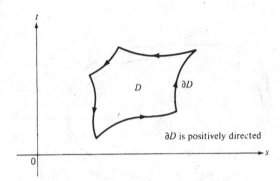

∂D is positively directed

Figure 16.30

Then

$$|r_s \times r_t|^2 = (a + t \sin\tfrac{1}{2}s)^2 + \tfrac{1}{4}t^2.$$

Therefore $r_s \times r_t$ is never $\mathbf{0}$, so that if we define $\mathbf{n}(s, t)$ by

$$\mathbf{n}(s, t) = |r_s \times r_t|^{-1}r_s \times r_t,$$

it follows that $\mathbf{n}(s, t)$ is continuous. However, $\mathbf{n}(0, 0) = \mathbf{e}_1$ and $\mathbf{n}(2\pi, 0) = -\mathbf{e}_1$, while the points $(0, 0)$, $(2\pi, 0)$ in R correspond to the same point on M. The transformation (16.60) is one-to-one except for the points of R given by $(0, t)$ and $(2\pi, -t)$, $-h \leqslant t \leqslant h$, which are carried into the same point of M (see Figure 16.29). The Möbius strip, called a "one-sided surface," is not a smooth surface element. It is not an orientable surface since, if a pencil line (with pencil perpendicular to the surface) is drawn down the center of the strip ($t = 0$) then, after one complete trip around, the pencil will be pointing to the "opposite side."

If S is a smooth surface element represented parametrically by

$$v(\overrightarrow{OP}) = r(s, t), \qquad (s, t) \in D \cup \partial D, \tag{16.61}$$

then the boundary ∂S is the image of ∂D. Since $\mathbb{R}^2$ is oriented, we say that ∂D is *positively directed* if it is oriented so that D is on the left as ∂D is traversed (see Figure 16.30). We write $\overrightarrow{\partial D}$ for this orientation and $-\overrightarrow{\partial D}$ when the boundary is oppositely directed.

Let $\vec{S}$ be an oriented surface and suppose that the parametric representation given by (16.61) agrees with the orientation of $\vec{S}$. Then every closed curve Γ on the boundary of $\vec{S}$ will have an orientation induced by the oriented closed curve $\vec{C}$ of $\overrightarrow{\partial D}$ which has Γ as its image. We write $\vec{\Gamma}$ for this oriented curve and say that it is **positively directed with respect to** $\vec{S}$ if $\vec{C}$ is positively directed. Geometrically, the curve $\vec{\Gamma}$ is directed so that if one proceeds along $\overrightarrow{\partial S}$ in an upright position with head in the direction of the positive normal $\mathbf{n}$ to the surface, then the surface is on the left (see Figure 16.31). In terms of positively and negatively oriented coordinate systems in $\mathbb{R}^3$, if t is tangent to $\overrightarrow{\partial S}$ pointing

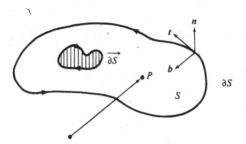

Figure 16.31. S is on the left as ∂S is traversed.

in the positive direction, if n is perpendicular to t and in the direction of the positive normal to $\vec{S}$, and if b is perpendicular to the vectors t and n and pointing *toward* the surface S, then the triple t, b, n is a positively-oriented triple. If ∂D and ∂S consist of several closed curves, the statement holds for each curve.

The notion of an orientable surface can be extended to *piecewise smooth surfaces*. Such surfaces have edges, and so a continuous unit normal vector field cannot be defined over the entirety of such a surface. However, a piecewise smooth surface can be divided into a finite number of smooth surface elements $F_1, F_2, \ldots, F_n$. Each surface element may be oriented and each boundary can be given a positive direction. Let γ_{ij} be a smooth arc which is the common boundary of the surface elements F_i and F_j.

Definitions. If, for all arcs γ_{ij}, the positive direction of γ_{ij} as part of $\overrightarrow{\partial F_i}$ is the negative of the positive direction of γ_{ij} as part of ∂F_j, the surface F is said to be an **orientable piecewise smooth surface**. In this case, the collection of oriented elements $\vec{F_i}$ forms an orientation of F and the unit normal function defined on the interior of each F_i is called the **positive unit normal function** of F.

Figure 16.32 exhibits a piecewise smooth, orientable surface and its decomposition into four smooth surface elements. Those boundary arcs of the F_i which are traversed only once comprise $\overrightarrow{\partial F}$ which is a positively directed,

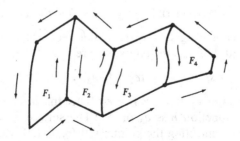

Figure 16.32. An orientable piecewise smooth surface.

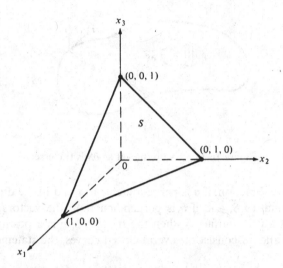

Figure 16.33. S is the surface of a tetrahedron.

closed, piecewise smooth curve. However $\overrightarrow{\partial F}$ may be empty (e.g., if S is the surface of a cube). From the above discussion, it follows that if a piecewise smooth surface is orientable according to one decomposition into smooth surface elements, then it is orientable according to any other such decomposition. A nonorientable surface, such as a Möbius strip, is not orientable even if it is treated as a piecewise smooth rather than as a smooth surface. That is, if a surface S is not orientable, then it can be shown that there is no decomposition into smooth surface elements such that all the γ_{ij} always have opposite orientations as parts of $\overrightarrow{\partial F_i}$ and $\overrightarrow{\partial F_j}$.

EXAMPLE. Let S be the surface of the tetrahedron with vertices at $(0, 0, 0)$, $(1, 0, 0)$, $(0, 1, 0)$, $(0, 0, 1)$. Divide S into smooth surface elements, find a parametric reprsentation of each, and find n, the unit normal function in terms of the parameters (see Figure 16.33).

Solution. The surface $S = S_1 \cup S_2 \cup S_3 \cup S_4$ where each S_i is one of the triangular faces. We define

$$D = \{(s, t): 0 < s < 1 - t, 0 \leqslant t \leqslant 1\}.$$

The surface S_1 consisting of the triangular region connecting the points $(0, 0, 0)$, $(1, 0, 0)$, $(0, 1, 0)$ has the parametric representation

$$r(s, t) = se_1 + te_2 + 0 \cdot e_3, \qquad (s, t) \in D.$$

Therefore $r_s = e_1, r_t = e_2$ and $r_s \times r_t = e_3$. Since this last vector has length 1, the unit normal function n is e_3 on S_1. The surface S_2 consisting of the triangular region connecting the points $(1, 0, 0)$, $(0, 1, 0)$, $(0, 0, 1)$ has the parametric representation

$$r(s, t) = se_1 + te_2 + (1 - s - t)e_3, \qquad (s, t) \in D.$$

Hence $r_s = e_1 - e_3$, $r_t = e_2 - e_3$ and $r_s \times r_t = e_1 + e_2 + e_3$. Therefore $n = |r_s \times r_t|^{-1} r_s \times r_t = (1/\sqrt{3})(e_1 + e_2 + e_3)$.

The remaining two surfaces of S are similar to S_1. $\qquad\square$

PROBLEMS

In each of Problems 1 through 6 a piecewise smooth (or smooth) surface S is described. Divide S into smooth surface elements (if necessary) and find a parametric representation for each element. Then express n, the unit normal function, in terms of these parameters. Describe $\partial\vec{S}$ in each case.

1. S is the surface of a cube of side length 1.

2. S is the pyramid with vertices at $(0, 0, 0)$, $(1, 0, 0)$, $(1, 1, 0)$, $(0, 1, 0)$, and $(0, 0, 1)$.

3. $S = \{(x_1, x_2, x_3): x_1^2 + x_2^2 + x_3^2 = 1, x_3 \geqslant 0\}$.

4. $S = \{(x_1, x_2, x_3): x_1^2 + x_2^2 - x_3^2 = 0, 1 \leqslant x_3 \leqslant 2\}$.

5. $S = S_1 \cup S_2$ where $S_1 = \{(x_1, x_2, x_3): x_1^2 + 4x_2^2 + x_3^2 = 12,\ 0 \leqslant x_2 \leqslant \frac{1}{2}\}$, $S_2 = \{(x_1, x_2, x_3): x_1^2 + 3x_2^2 + x_3^2 = 12, -\frac{1}{2} \leqslant x_2 \leqslant 0\}$.

6. $S = \{(x_1, x_2, x_3): |x_3| = x_1^2 + x_2^2, -1 \leqslant x_3 \leqslant 1\}$.

7. Let M be the Möbius strip given by $x_1 = (1 + t \sin\frac{1}{2}s) \cos s$, $x_2 = (1 + t \sin\frac{1}{2}s) \sin s$, $x_3 = t \cos\frac{1}{2}s$, with $(s, t) \in R = \{(s, t): 0 \leqslant s \leqslant 2\pi, -\frac{1}{2} \leqslant t \leqslant \frac{1}{2}\}$. Divide the rectangle R into n vertical strips of equal size and show that the portion of M corresponding to each strip is a smooth surface element. Also show that the curves γ_{ij}, the common boundary of the ith and jth surface element, are oriented in such a way that M cannot be oriented.

8. Let S be the surface of a pyramid with a square base. Show that it is an orientable piecewise smooth surface.

9. Let S be the surface of a pyramid with base which is a polygon with n sides, $n \geqslant 5$. Show that S is an orientable piecewise smooth surface.

16.8. The Stokes Theorem

Let $\vec{S}$ be a smooth oriented surface in $\mathbb{R}^3$ with boundary $\overrightarrow{\partial S}$ and suppose that v is a vector function defined on S and ∂S. Stokes's theorem is a generalization to surfaces of Green's theorem. We recall that Green's theorem establishes a relation between the integral of the derivative of a function in a domain D in $\mathbb{R}^2$ and the integral of the same function over ∂D (Theorem 16.20). The theorem of Stokes establishes an equality between the integral of curl $v \cdot n$ over a surface $\vec{S}$ and the integral of v over the boundary of $\vec{S}$. The principal result is given in Theorem 16.22.

Let S be represented parametrically by

$$r(s, t) = x_1(s, t)e_1 + x_2(s, t)e_2 + x_3(s, t)e_3 \quad \text{for } (s, t) \in D,$$

where D is a bounded region in the (s, t)-plane. Then the positive unit normal function is

$$n(s, t) = |r_s \times r_t|^{-1}\left[J\left(\frac{x_2, x_3}{s, t}\right)e_1 + J\left(\frac{x_3, x_1}{s, t}\right)e_2 + J\left(\frac{x_1, x_2}{s, t}\right)e_3\right].$$

Suppose that v is a continuous vector field defined on S and given in coordinates by

$$v(x_1, x_2, x_3) = v_1(x_1, x_2, x_3)e_1 + v_2(x_1, x_2, x_3)e_2 + v_3(x_1, x_2, x_3)e_3.$$

Since S is a smooth surface the scalar product $v \cdot n$ is continuous on S, and a direct computation yields

$$v \cdot n = |r_s \times r_t|^{-1}\left[v_1 J\left(\frac{x_2, x_3}{s, t}\right) + v_2 J\left(\frac{x_3, x_1}{s, t}\right) + v_3 J\left(\frac{x_1, x_2}{s, t}\right)\right].$$

Therefore it is possible to define the surface integral

$$\iint_S v \cdot n \, dS. \tag{16.62}$$

The surface element dS can be computed in terms of the parameters (s, t) by the formula

$$dS = |r_s \times r_t| \, dA_{st}$$

where dA_{st} is the element of area in the plane region D. Hence

$$\iint_S v \cdot n \, dS = \iint_D \left[v_1 J\left(\frac{x_2, x_3}{s, t}\right) + v_2 J\left(\frac{x_3, x_1}{s, t}\right) + v_3 J\left(\frac{x_1, x_2}{s, t}\right)\right] dA_{st}. \tag{16.63}$$

If $r_1(s', t')$ is another smooth representation of S with $(s', t') \in D_1$, it follows from Theorem 16.21 that there is a smooth one-to-one transformation

$$s = U(s', t'), \qquad t = V(s', t'),$$

such that

$$r[U(s', t'), V(s', t')] = r_1(s', t').$$

The rule for multiplying Jacobians implies that the representation $r_1(s', t')$ gives the same orientation as the representation $r(s, t)$ if and only if

$$J\left(\frac{s, t}{s', t'}\right) > 0.$$

Therefore if we replace (s, t) by (s', t') in (16.63) and integrate over D_1 we obtain the formula for

$$\iint_S v \cdot n \, dS$$

in terms of the parameters (s', t').

If S is piecewise smooth rather than smooth, it can be represented as the union of a finite number of smooth surface elements $S_1, S_2, \ldots, S_m$. Then it is natural to define

$$\iint_S v \cdot n \, dS = \sum_{i=1}^{m} \iint_{S_i} v \cdot n_i \, dS_i$$

where each integral on the right may be evaluated according to (16.63). In a similar way we can define the line integral over the boundary $\overrightarrow{\partial S}$ of a piecewise smooth oriented surface $\vec{S}$. If ∂S is made up of a finite number of smooth positively directed arcs $\vec{C}_1, \vec{C}_2, \ldots, \vec{C}_n$, and if v is a continuous vector field defined in a region of $\mathbb{R}^3$ which contains $\overrightarrow{\partial S}$, we define

$$\int_{\overrightarrow{\partial S}} v \cdot dr = \sum_{i=1}^{n} \int_{\vec{C}_i} v \cdot dr.$$

The following two lemmas are needed in the proof of the Stokes theorem.

Lemma 16.6. *Suppose that $\vec{S}$ is a smooth oriented surface element in $\mathbb{R}^3$ with a parametric representation*

$$r(s, t) = x_1(s, t)e_1 + x_2(s, t)e_2 + x_3(s, t)e_3 \quad for \quad (s, t) \in D \cup \partial D.$$

Let v be a continuous vector field defined on an open set G containing $\overrightarrow{\partial S}$ given by

$$v(x_1, x_2, x_3) = v_1(x_1, x_2, x_3)e_1 + v_2(x_1, x_2, x_3)e_2 + v_3(x_1, x_2, x_3)e_3$$

for $(x_1, x_2, x_3) \in G$. Then

$$\int_{\overrightarrow{\partial S}} v \cdot dr = \int_{\overrightarrow{\partial D}} \left[\left(v_1 \frac{\partial x_1}{\partial s} + v_2 \frac{\partial x_2}{\partial s} + v_3 \frac{\partial x_3}{\partial s} \right) ds \right.$$
$$\left. + \left(v_1 \frac{\partial x_1}{\partial t} + v_2 \frac{\partial x_2}{\partial t} + v_3 \frac{\partial x_3}{\partial t} \right) dt \right]. \tag{16.64}$$

PROOF. We establish the result when $\overrightarrow{\partial D}$ consists of a single piecewise smooth simple closed curve. The extension to several such curves is clear. Let ∂D be given parametrically by the equations

$$s = s(\tau), \qquad t = t(\tau), \qquad a \leqslant \tau \leqslant b.$$

Then the equations

$$x_1 = x_1[s(\tau), t(\tau)], \qquad x_2 = x_2[s(\tau), t(\tau)],$$
$$x_3 = x_3[s(\tau), t(\tau)], \qquad a \leqslant \tau \leqslant b,$$

give a parametric representation of $\overrightarrow{\partial S}$. Using the Chain rule we find

$$dr = (dx_1)e_1 + (dx_2)e_2 + (dx_3)e_3$$
$$= \left[\frac{\partial x_1}{\partial s} \frac{ds}{d\tau} + \frac{\partial x_1}{\partial t} \frac{dt}{d\tau} \right] d\tau e_1 + \left[\frac{\partial x_2}{\partial s} \frac{ds}{d\tau} + \frac{\partial x_2}{\partial t} \frac{dt}{d\tau} \right] d\tau e_2$$
$$+ \left[\frac{\partial x_3}{\partial s} \frac{ds}{d\tau} + \frac{\partial x_3}{\partial t} \frac{dt}{d\tau} \right] d\tau e_3.$$

Therefore

$$\int_{\overrightarrow{\partial s}} v \cdot dr = \int_a^b \left[\left(v_1 \frac{\partial x_1}{\partial s} + v_2 \frac{\partial x_2}{\partial s} + v_3 \frac{\partial x_3}{\partial s} \right) \frac{ds}{d\tau} \right.$$
$$\left. + \left(v_1 \frac{\partial x_1}{\partial t} + v_2 \frac{\partial x_2}{\partial t} + v_3 \frac{\partial x_3}{\partial t} \right) \frac{dt}{d\tau} \right] d\tau,$$

and (16.64) follows at once.   □

Let v be a smooth vector field defined in a region G of $\mathbb{R}^3$ given in coordinates

$$v(x_1, x_2, x_3) = v_1(x_1, x_2, x_3)e_1 + v_2(x_1, x_2, x_3)e_2 + v_3(x_1, x_2, x_3)e_3. \tag{16.65}$$

We recall that curl v is a vector field on G given by

$$\text{curl } v = \left(\frac{\partial v_3}{\partial x_2} - \frac{\partial v_2}{\partial x_3} \right) e_1 + \left(\frac{\partial v_1}{\partial x_3} - \frac{\partial v_3}{\partial x_1} \right) e_2 + \left(\frac{\partial v_2}{\partial x_1} - \frac{\partial v_1}{\partial x_2} \right) e_3. \tag{16.66}$$

Theorem 16.22 (The Stokes theorem). *Suppose that $\vec{S}$ is a bounded, closed, oriented piecewise smooth surface and that v is a smooth vector field on a region in $\mathbb{R}^3$ containing $\vec{S} \cup \partial S$. Then*

$$\iint_{\vec{S}} (\text{curl } v) \cdot n \, dS = \int_{\partial S} v \cdot dr. \tag{16.67}$$

Proof. We first suppose that $\vec{S}$ is a smooth oriented surface element and that the components of r have continuous second derivatives. From Equation (16.64) in Lemma 16.6, it follows that

$$\int_{\overrightarrow{\partial s}} v \cdot dr = \int_{\partial D} \left[\left(v_1 \frac{\partial x_1}{\partial s} + v_2 \frac{\partial x_2}{\partial s} + v_3 \frac{\partial x_3}{\partial s} \right) ds \right.$$
$$\left. + \left(v_1 \frac{\partial x_1}{\partial t} + v_2 \frac{\partial x_2}{\partial t} + v_3 \frac{\partial x_3}{\partial t} \right) dt \right].$$

We apply Green's theorem to the integral on the right and use the Chain rule to obtain

$$\iint_{\overrightarrow{\partial s}} v \cdot dr = \iint_D \left\{ \frac{\partial}{\partial s} \left[v_1 \frac{\partial x_1}{\partial t} + v_2 \frac{\partial x_2}{\partial t} + v_3 \frac{\partial x_3}{\partial t} \right] \right.$$
$$\left. - \frac{\partial}{\partial t} \left[v_1 \frac{\partial x_1}{\partial s} + v_2 \frac{\partial x_2}{\partial s} + v_3 \frac{\partial x_3}{\partial s} \right] \right\} dA_{st}$$
$$= \iint_D \left\{ \left(\frac{\partial v_1}{\partial x_1} \frac{\partial x_1}{\partial s} + \frac{\partial v_1}{\partial x_2} \frac{\partial x_2}{\partial s} + \frac{\partial v_1}{\partial x_3} \frac{\partial x_3}{\partial s} \right) \frac{\partial x_1}{\partial t} \right.$$

$$+ \left(\frac{\partial v_2}{\partial x_1} \frac{\partial x_1}{\partial s} + \frac{\partial v_2}{\partial x_2} \frac{\partial x_2}{\partial s} + \frac{\partial v_2}{\partial x_3} \frac{\partial x_3}{\partial s} \right) \frac{\partial x_2}{\partial t}$$

$$+ \left(\frac{\partial v_3}{\partial x_1} \frac{\partial x_1}{\partial s} + \frac{\partial v_3}{\partial x_2} \frac{\partial x_2}{\partial s} + \frac{\partial v_3}{\partial x_3} \frac{\partial x_3}{\partial s} \right) \frac{\partial x_3}{\partial t}$$

$$- \left(\frac{\partial v_1}{\partial x_1} \frac{\partial x_1}{\partial t} + \frac{\partial v_1}{\partial x_2} \frac{\partial x_2}{\partial t} + \frac{\partial v_1}{\partial x_3} \frac{\partial x_3}{\partial t} \right) \frac{\partial x_1}{\partial s}$$

$$- \left(\frac{\partial v_2}{\partial x_1} \frac{\partial x_1}{\partial t} + \frac{\partial v_2}{\partial x_2} \frac{\partial x_2}{\partial t} + \frac{\partial v_2}{\partial x_3} \frac{\partial x_3}{\partial t} \right) \frac{\partial x_2}{\partial s}$$

$$- \left(\frac{\partial v_3}{\partial x_1} \frac{\partial x_1}{\partial t} + \frac{\partial v_3}{\partial x_2} \frac{\partial x_2}{\partial t} + \frac{\partial v_3}{\partial x_3} \frac{\partial x_3}{\partial t} \right) \frac{\partial x_3}{\partial s} \Bigg\} dA_{st}.$$

All the terms containing $\partial^2 x_1 / \partial s \partial t$, $\partial^2 x_2 / \partial s \partial t$, $\partial^2 x_3 / \partial s \partial t$ in the expression on the right cancel. Collecting the terms on the right we find

$$\int_{\partial \vec{S}} v \cdot dr = \iint_D \left\{ \left(\frac{\partial v_3}{\partial x_2} - \frac{\partial v_2}{\partial x_3} \right) J\left(\frac{x_2, x_3}{s, t} \right) + \left(\frac{\partial v_1}{\partial x_3} - \frac{\partial v_3}{\partial x_1} \right) J\left(\frac{x_3, x_1}{s, t} \right) \right.$$

$$\left. + \left(\frac{\partial v_2}{\partial x_1} - \frac{\partial v_1}{\partial x_2} \right) J\left(\frac{x_1, x_2}{s, t} \right) \right\} dA_{st}.$$

The above expression is equivalent to (16.67). $\qquad \square$

If $x_1(s, t)$, $x_2(s, t)$, $x_3(s, t)$ are only continuously differentiable on an open set G containing $D \cup \partial D$, then it can be shown that there are sequences $\{x_{1n}\}$, $\{x_{2n}\}$, $\{x_{3n}\}$ such that x_{1n}, x_{2n}, x_{3n} and $\partial x_{1n}/\partial s, \partial x_{2n}/\partial s, \ldots, \partial x_{3n}/\partial t$ are smooth functions on an open set G containing $D \cup \partial D$ and such that all these sequences converge uniformly on $D \cup \partial D$ to $x_1, x_2, \ldots, \partial x_3 / \partial t$ as $n \to \infty$. Formula (16.67) holds for each n and, because of the uniform convergence, it also holds in the limit. Finally, if $\vec{S}$ is any piecewise smooth surface, it is the union $\vec{S}_1 \cup \vec{S}_2 \cup \cdots \cup \vec{S}_k$ of smooth surface elements $\vec{S}_i$ oriented in such a way that any piecewise smooth arc γ_{ij} which is the boundary of both $\vec{S}_i$ and $\vec{S}_j$ is directed oppositely on ∂S_j from the way it is directed on ∂S_i. Therefore

$$\iint_{\vec{S}} (\text{curl } v) \cdot n \, dS = \sum_{i=1}^{k} \int_{\partial \vec{S}_i} v \cdot dr = \int_{\partial \vec{S}} v \cdot dr. \qquad \square$$

Corollary. *Suppose that S is a bounded, closed, oriented piecewise smooth surface without boundary and that v is a smooth vector field defined on an open set containing S. Then*

$$\iint_{\vec{S}} (\text{curl } v) \cdot n \, dS = 0. \tag{16.68}$$

In the following example we show how an integral over a surface in $\mathbb{R}^3$ may be calculated by reducing it to an ordinary double integral in the plane.

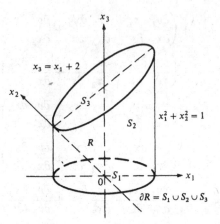

Figure 16.34

EXAMPLE 1. Let R be the region in $\mathbb{R}^3$ defined by $R = \{(x_1, x_2, x_3): x_1^2 + x_2^2 \leqslant 1, 0 \leqslant x_3 \leqslant x_1 + 2\}$. Let S be the boundary of R. Define $v(x_1, x_2, x_3) = 2x_1 e_1 - 3x_2 e_2 + x_3 e_3$. Find the value of

$$\iint_S v \cdot n \, dS \tag{16.69}$$

where n is the unit normal directed outward from S (see Figure 16.34).

Solution. S is a piecewise smooth surface and we divide it into three smooth surface elements as shown in Figure 16.34. The normal to S_1 is $-e_3$ and therefore $v \cdot n = -x_3$ on S_1. Since $x_3 = 0$ on S_1, the value of (16.69) over S_1 is 0. Since $x_3 = x_1 + 2$ on S_3 it follows that

$$n = \frac{1}{\sqrt{2}}(-e_1 + e_3)$$

on S_3. Hence on S_3:

$$v \cdot n = \frac{1}{\sqrt{2}}(-2x_1 + x_3) = \frac{1}{\sqrt{2}}(-x_1 + 2),$$

$$dS = \sqrt{2} \, dA_{st}$$

where dA_{st} is the element of area in the disk $D = \{(s, t): s^2 + t^2 < 1\}$. Using the parameters $x_1 = s$, $x_2 = t$, we find

$$\iint_{S_3} v \cdot n \, dS = \iint_D (-s + 2) \, dA_{st} = 2\pi.$$

To evaluate (16.69) over S_2 choose cylindrical coordinates:

$$x_1 = \cos s, \qquad x_2 = \sin s, \qquad x_3 = t, \qquad (s, t) \in D_1,$$

where $D_1 = \{(s, t): -\pi \leqslant s \leqslant \pi, 0 \leqslant t \leqslant 2 + \cos s\}$. The outward normal on S_2 is $n = (\cos s)e_1 + (\sin s)e_2$ and $v \cdot n = 2 \cos^2 s - 3 \sin^2 s$. Therefore

$$\iint_{S_2} v \cdot n \, dS = \iint_{D_1} (2 \cos^2 s - 3 \sin^2 s) \, dA_{st}$$

$$= \int_{-\pi}^{\pi} \int_0^{2+\cos s} (2 - 5 \sin^2 s) \, dt \, ds = -2\pi.$$

Finally,

$$\iint_S v \cdot n \, dS = \iint_{S_1 \cup S_2 \cup S_3} v \cdot n \, dS = 0 + 2\pi - 2\pi = 0. \qquad \square$$

EXAMPLE 2. Verify Stokes's theorem given that

$$v = x_2 e_1 + x_3 e_2 + x_1 e_3$$

and S_2 is the lateral surface in Example 1 with n pointing outward.

Solution. The boundary of $\vec{S}_2$ consists of the circle

$$C_1 = \{(x_1, x_2, x_3): x_1^2 + x_2^2 = 1, x_3 = 0\}$$

and the ellipse

$$C_2 = \{(x_1, x_2, x_3): x_1^2 + x_2^2 = 1, x_3 = x_1 + 2\}.$$

The curves $\vec{C}_1, \vec{C}_2$ are oriented as shown in Figure 16.35. We select cylindrical coordinates to describe $\vec{S}_2$:

$$\vec{S}_2 = \{(x_1, x_2, x_3): x_1 = \cos s, \quad x_2 = \sin s, \quad x_3 = t\}, \quad (s, t) \in D_1$$

where $D_1 = \{(s, t): -\pi \leqslant s \leqslant \pi, 0 \leqslant t \leqslant 2 + \cos s\}$. We decompose $\vec{S}_2$ into two smooth surface elements, one part corresponding to $x_2 \geqslant 0$ and the other to $x_2 \leqslant 0$. Then D_1 is divided into two parts E_1 and E_2 corresponding to $s \geqslant 0$ and $s \leqslant 0$ as shown in Figure 16.36. This subdivision is required because the representation of $\vec{S}$ by D_1 is not one-to-one ($s = \pi$ and $s = -\pi$ correspond to the same curve on $\vec{S}_2$). A computation yields

$$J\left(\frac{x_2, x_3}{s, t}\right) = \cos s, \quad J\left(\frac{x_3, x_1}{s, t}\right) = \sin s, \quad J\left(\frac{x_1, x_2}{s, t}\right) = 0,$$

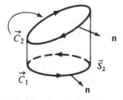

Figure 16.35

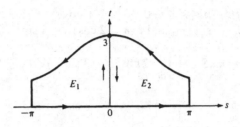

Figure 16.36

and $n = (\cos s)e_1 + (\sin s)e_2$. Also, $\text{curl } v = -e_1 - e_2 - e_3$, $dS = dA_{st}$. Therefore,

$$\iint_{\vec{S}} (\text{curl } v) \cdot n \, dS = \int_{-\pi}^{\pi} \int_{0}^{2 + \cos \theta} (-\cos \theta - \sin \theta) \, dt \, ds = -\pi. \quad (16.70)$$

The boundary integrals are

$$\int_{\overrightarrow{\partial s}} v \cdot dr = \int_{\vec{C}_1} v \cdot dr - \int_{-\vec{C}_2} v \cdot dr.$$

Using cylindrical coordinates on C_1 and C_2, we obtain

$$v = (\sin s)e_1 + (\cos s)e_3, \qquad dr = [(-\sin s)e_1 + (\cos s)e_2] \, ds \quad \text{on } \vec{C}_1;$$

and on $\vec{C}_2$

$$v = (\sin s)e_1 + (2 + \cos s)e_2 + (\cos s)e_3,$$

$$dr = [(-\sin s)e_1 + (\cos s)e_2 + (-\sin s)e_3] \, ds.$$

Taking scalar products, we find

$$\int_{\vec{C}_1} v \cdot dr = \int_{-\pi}^{\pi} (-\sin^2 s) \, ds = -\pi, \qquad (16.71)$$

$$\int_{\vec{C}_2} v \cdot dr = \int_{-\pi}^{\pi} (-\sin^2 \theta + 2 \cos \theta + \cos^2 \theta - \sin \theta \cos \theta) \, d\theta = 0. \quad (16.72)$$

Stokes's theorem is verified by comparing (16.70) with (16.71) and (16.72).

□

Problems

In each of Problems 1 through 6 compute

$$\iint_{\vec{S}} v \cdot n \, dS.$$

1. $v = (x_1 + 1)e_1 - (2x_2 + 1)e_2 + x_3 e_3$; $\vec{S}$ is the triangular region with vertices at $(1, 0, 0)$, $(0, 1, 0)$, $(0, 0, 1)$ and n is pointing away from the origin.

2. $v = x_1e_1 + x_2e_2 + x_3e_3$; $\vec{S} = \{(x_1, x_2, x_3): x_1^2 + x_2^2 = 2x_3, (x_1 - 1)^2 + x_2^2 \leqslant 1\}$, oriented so that $n \cdot e_3 > 0$.

3. $v = x_1^2e_1 + x_2^2e_2 + x_3^2e_3$; $\vec{S} = \{(x_1, x_2, x_3): x_1^2 + x_2^2 = x_3^2; 1 \leqslant x_3 \leqslant 2\}$, $n \cdot e_3 > 0$.

4. $v = x_1x_2e_1 + x_1x_3e_2 + x_2x_3e_3$; $\vec{S} = \{(x_1, x_2, x_3): x_2^2 = 2 - x_1; x_3^3 \leqslant x_2 \leqslant x_3^{1/2}\}$.

5. $v = x_2^2e_1 + x_3e_2 - x_1e_3$; $\vec{S} = \{(x_1, x_2, x_3): x_2^2 = 1 - x_1, 0 \leqslant x_3 \leqslant x_1; x_1 \geqslant 0\}$, $n \cdot e_1 > 0$.

6. $v = 2x_1e_1 - x_2e_2 + 3x_3e_3$; $\vec{S} = \{(x_1, x_2, x_3): x_3^2 = x_1; x_2^2 \leqslant 1 - x_1; x_2 \geqslant 0\}$, $n \cdot e_1 > 0$.

In each of Problems 7 through 12, verify the Stokes theorem.

7. $v = x_3e_1 + x_1e_2 + x_2e_3$; $\vec{S} = \{(x_1, x_2, x_3): x_3 = 1 - x_1^2 - x_2^2, x_3 \geqslant 0\}$, $n \cdot e_3 > 0$.

8. $v = x_2^2e_1 + x_1x_2e_2 - 2x_1x_3e_3$; $\vec{S} = \{(x_1, x_2, x_3): x_1^2 + x_2^2 + x_3^2 = 1, x_3 \geqslant 0\}$, $n \cdot e_3 > 0$.

9. $v = -x_2x_3e_3$; $\vec{S} = \{(x_1^2 + x_2^2 + x_3^2 = 4, x_1^2 + x_2^2 \geqslant 1\}$, n pointing outward from the sphere.

10. $v = -x_3e_2 + x_2e_3$; $\vec{S}$ is the surface of the cylinder given in cylindrical coordinates by $r = \theta, 0 \leqslant \theta \leqslant \pi/2$, which is bounded below by the plane $x_3 = 0$ and above by the surface of the cone $x_1^2 + x_2^2 = x_3^2$; $n \cdot e_1 > 0$ for $\theta > 0$.

11. $v = x_2e_1 + x_3e_2 + x_1e_3$; $\vec{S} = \{(x_1, x_2, x_3): x_3^2 = 4 - x_1, x_1 \geqslant x_2^2\}$, $n \cdot e_1 > 0$.

12. $v = x_3e_1 - x_1e_3$; $\vec{S}$ is the surface of the cylinder given in cylindrical coordinates by $r = 2 + \cos \theta$ above the plane $x_3 = 0$ and exterior to the cone $x_3^2 = x_1^2 + x_2^2$; n is pointing outward from the cylindrical surface.

In each of Problems 13 and through 15 use the Stokes theorem to compute $\int_{\partial S} v \cdot dr$.

13. $v = r^{-3}r$ where $r = x_1e_1 + x_2e_2 + x_3e_3$ and $r = |r|$; $\vec{S}$ is the surface $\vec{S}_2$ of Example 2.

14. $v = (e^{x_1} \sin x_2)e_1 + (e^{x_1} \cos x_2 - x_3)e_2 + x_2e_3$; $\vec{S}$ is the surface in Problem 3.

15. $v = (x_1^2 + x_3)e_1 + (x_1 + x_2^2)e_2 + (x_2 + x_3^2)e_3$; $\vec{S} = \{(x_1, x_2, x_3): x_1^2 + x_2^2 + x_3^2 = 1, x_3 \geqslant (x_1^2 + x_2^2)^{1/2}\}$; n points outward from the spherical surface.

16. Show that if $\vec{S}$ is given by $x_3 = f(x_1, x_2)$ for $(x_1, x_2) \in D = \{(x_1, x_2): x_1^2 + x_2^2 \leqslant 1\}$, if f is smooth, and if $v = (1 - x_1^2 - x_2^2)w(x_1, x_2, x_3)$ where w is any smooth vector field defined on an open set containing S, then

$$\iint_{\vec{S}} (\text{curl } v) \cdot n \, dS = 0.$$

17. Suppose that $v = r^{-3}(x_2e_1 + x_3e_2 + x_1e_3)$ where $r = x_1e_1 + x_2e_2 + x_3e_3$ and $r = |r|$, and $\vec{S}$ is the sphere $\{(x_1, x_2, x_3): x_1^2 + x_2^2 + x_3^2 = 1\}$ with n pointing outward. Show that

$$\iint_S (\text{curl } v) \cdot n \, dS = 0.$$

18. Suppose that a smooth surface $\vec{S}$ has two different smooth parametric representations, $r(s, t)$ for $(s, t) \in D$ and $r_1(s', t')$ for $(s', t') \in D_1$. Let v be a smooth vector field defined on $\vec{S}$. Find the relationship between the formulas for $(\text{curl } v) \cdot n$ in the two representations.

19. In the proof of the Stokes theorem, carry out the verification that all the second derivative terms cancel in the displayed formula before (16.68).

20. Let M be a Möbius strip. Where does the proof of the Stokes theorem break down for this surface?

16.9. The Divergence Theorem

Green's theorem establishes a relation between the line integral of a function over the boundary of a plane region and the double integral of the derivative of the same function over the region itself. Stokes's theorem extends this result to two-dimensional surfaces in three-space. In this section we establish another kind of generalization of the Fundamental theorem of calculus known as the Divergence theorem. This theorem determines the relationship between an integral of the derivative of a function over a three-dimensional region in $\mathbb{R}^3$ and the integral of the function itself over the boundary of that region. All three theorems (Geen, Stokes, Divergence) are special cases of a general formula which connects an integral over a set of points in $\mathbb{R}^N$ with another integral over the boundary of that set points. The integrand in the first integral is a certain derivative of the integrand in the boundary integral.

Let $v = v_1(x_1, x_2, x_3)e_1 + v_2(x_1, x_2, x_3)e_2 + v_3(x_1, x_2, x_3)e_3$ be a vector field defined for (x_1, x_2, x_3) in a region E in $\mathbb{R}^3$. We recall that div v is a scalar field given in coordinates by the formula

$$\text{div } v = \frac{\partial v_1}{\partial x_1} + \frac{\partial v_2}{\partial x_2} + \frac{\partial v_3}{\partial x_3}.$$

The Divergence theorem consists of proving the formula

$$\iiint_E \text{div } v \, dV = \iint_{\partial E} v \cdot n \, dS \tag{16.73}$$

where ∂E is oriented by choosing n as the exterior normal to ∂E. We first establish (16.73) in several special cases and then show that the formula holds generally, provided that the boundary of E is not too irregular and that v is smooth.

Lemma 16.7. *Let D be a domain in the (x_1, x_2)-plane with smooth boundary. Let $f: D \cup \partial D \to \mathbb{R}^1$ be a piecewise smooth function and define*

$$E = \{(x_1, x_2, x_3) : (x_1, x_2) \in D, c < x_3 < f(x_1, x_2)\}$$

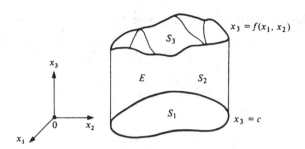

Figure 16.37. $\partial E = S_1 \cup S_2 \cup S_3$.

for some constant c. Suppose that $v = u(x_1, x_2, x_3)e_3$ is such that u and $\partial u/\partial x_3$ are continuous on an open set in $\mathbb{R}^3$ containing $E \cup \partial E$. Then

$$\iiint_E \operatorname{div} v \, dV = \iint_{\partial E} v \cdot n \, dS$$

where n is the outward unit normal of ∂E.

PROOF. Since $\operatorname{div} v = \partial u/\partial x_3$, it follows that

$$\iiint_E \operatorname{div} v \, dV = \iiint_E \frac{\partial u}{\partial x_3} \, dV = \iint_D \int_c^{f(x_1, x_2)} \frac{\partial u}{\partial x_3} \, dx_3 \, dA_{x_1 x_2}.$$

Performing the integration with respect to x_3, we find

$$\iiint_E \operatorname{div} v \, dV = \iint_D \{u[x_1, x_2, f(x_1, x_2)] - u(x_1, x_2, c)\} \, dA_{x_1 x_2}. \quad (16.74)$$

Let $\partial E = S_1 \cup S_2 \cup S_3$ where S_1 is the domain in the plane $x_3 = c$ which is congruent to D; S_2 is the lateral cylindrical surface of ∂E; and S_3 is the part of ∂E corresponding to $x_3 = f(x_1, x_2)$ (see Figure 16.37). We wish to show that (16.74) is equal to

$$\iint_{S_1 \cup S_2 \cup S_3} v \cdot n \, dS.$$

Along S_2, the unit normal n is parallel to the (x_1, x_2)-plane, and so $n \cdot e_3 = 0$. Therefore $v \cdot n = 0$ on S_2 and

$$\iint_{S_2} v \cdot n \, dS = 0. \quad (16.75)$$

The outer normal along S_1 is clearly $-e_3$ and therefore

$$\iint_{S_1} v \cdot n \, dS = -\iint_D u(x_1, x_2, c) \, dA_{x_1 x_2}. \quad (16.76)$$

As for S_3, the unit normal function is given by

$$n = \frac{-\dfrac{\partial f}{\partial x_1}e_1 - \dfrac{\partial f}{\partial x_2}e_2 + e_3}{\left[1 + \left(\dfrac{\partial f}{\partial x_1}\right)^2 + \left(\dfrac{\partial f}{\partial x_2}\right)^2\right]^{1/2}}.$$

Also, for $(x_1, x_2, x_3) \in S_3$,

$$dS = \left[1 + \left(\frac{\partial f}{\partial x_1}\right)^2 + \left(\frac{\partial f}{\partial x_2}\right)^2\right]^{1/2} dA_{x_1 x_2}.$$

Therefore, since $v = ue_3$, we find

$$\iint_{S_3} v \cdot n \, dS = \iint_D u[x_1, x_2, f(x_1, x_2)] \, dA_{x_1 x_2}. \tag{16.77}$$

The result follows when (16.74) is compared with (16.75), (16.76), and (16.77). $\qquad\square$

Lemma 16.8. *Suppose that the hypotheses of Lemma 16.7 hold, except that v has the form*

$$v = v_1(x_1, x_2, x_3)e_1 + u_2(x_1, x_2, x_3)e_2$$

with u_1, u_2 smooth function on an open set containing $E \cup \partial E$. Then

$$\iiint_E \operatorname{div} v \, dV = \iint_{\partial E} v \cdot n \, dS$$

where n is the outward unit normal of ∂E.

PROOF. Let $U_1(x_1, x_2, x_3)$, $U_2(x_1, x_2, x_3)$ be defined by

$$U_1(x_1, x_2, x_3) = -\int_c^{x_3} u_1(x_1, x_2, t) \, dt, \quad U_2(x_1, x_2, x_3) = \int_c^{x_3} u_2(x_1, x_2, t) \, dt.$$

In addition, we define

$$w = U_2 e_1 + U_1 e_2, \qquad U_3 = -\frac{\partial U_1}{\partial x_1} + \frac{\partial U_2}{\partial x_2}, \qquad u = -U_3 e_3.$$

Then w is a smooth vector field and U_3, $\partial U_3/\partial x_3$ are continuous, and hence

$$\operatorname{curl} w = -\frac{\partial U_1}{\partial x_3}e_1 + \frac{\partial U_2}{\partial x_3}e_2 + \left(\frac{\partial U_1}{\partial x_1} - \frac{\partial U_2}{\partial x_2}\right)e_3 = v + u,$$

$$\operatorname{div} v = \frac{\partial u_1}{\partial x_1} + \frac{\partial u_2}{\partial x_2} = \frac{\partial U_3}{\partial x_3} = -\operatorname{div} u. \tag{16.78}$$

Since ∂E is a piecewise smooth surface without boundary, the Corollary to

Stokes's theorem is applicable. Hence

$$\iint_{\partial E} (\mathrm{curl}\ w) \cdot n\ dS = 0 = \iint_{\partial E} (v + u) \cdot n\ dS.$$

Therefore using Lemma 16.7 for the function $u = -U_3 e_3$, we find

$$\iint_{\partial E} v \cdot n\ dS = -\iint_{\partial E} u \cdot n\ dS = -\iiint_{E} \mathrm{div}\ u\ dV. \qquad (16.79)$$

The result of the lemma follows by inserting (16.78) into (16.79). $\qquad \square$

The Divergence theorem will now be established for a wide class of regions which are called "regular". Intuitively, a region in $\mathbb{R}^3$ is regular if its boundary can be subdivided into small pieces in such a way that each piece has a piecewise smooth representation of the form $x_3 = f(x_1, x_2)$ if a suitable Cartesian coordinate system is introduced.

Definition. A region E in $\mathbb{R}^3$ is **regular** if and only if: (i) ∂E consists of a finite number of piecewise smooth surfaces, each without boundary; (ii) at each point P of ∂E a Cartesian coordinate system is introduced with P as origin. There is a cylindrical domain $\Gamma = \{(x_1, x_2, x_3) : (x_1, x_2) \in D, -\infty < x_3 < \infty\}$, with D a region in the plane $x_3 = 0$ containing the origin, which has the property that $\Gamma \cap \partial E$ is a surface which can be represented in the form $x_3 = f(x_1, x_2)$ for $(x_1, x_2) \in D \cup \partial D$. Furthermore, f is piecewise smooth (see Figure 16.38); (iii) the set

$$\Gamma_1 = \{(x_1, x_2, x_3) : (x_1, x_2) \in D, -c < x_3 < f(x_1, x_2)\}$$

for some positive constant c (depending on P) is contained entirely in E.

Remarks. If E is a regular region, then each point P of ∂E is interior to a smooth surface element except for those points on a finite number of arcs on ∂E which have zero surface area. If a line in the direction of the unit normal

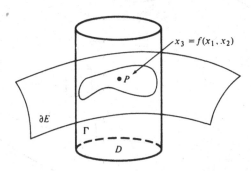

Figure 16.38. A regular region.

to ∂E is drawn through a point P interior to a smooth surface element on ∂E, then a segment of the line on one side of P will be in E, while a segment on the other side will be exterior to E. Finally, we observe that $\boldsymbol{n}$ varies continuously when P is in a smooth surface element of ∂E.

Theorem 16.23 (The Divergence theorem). *Suppose that E is a closed, bounded, regular region in $\mathbb{R}^3$ and that $\boldsymbol{v}$ is a continuously differentiable vector field on an open set G containing $E \cup \partial E$. Then*

$$\iiint_E \operatorname{div} \boldsymbol{v} \, dV = \iint_{\partial E} \boldsymbol{v} \cdot \boldsymbol{n} \, dS$$

where $\boldsymbol{n}$ is the unit normal function pointing outward from E.

PROOF. With each point $P \in \partial E$ associate a coordinate system and a cylindrical domain Γ as in the definition of a regular region. Let Γ_P be the bounded portion of Γ such that $-c < x_3 < c$ (where c is the constant in the definition of regular region; c depends on P). With each interior point P of E, introduce a Cartesian coordinate system and a cube Γ_P with P as origin, with the sides of Γ_P parallel to the axes, and with $\bar{\Gamma}_P$ entirely in E (see Figure 16.39). Since $E \cup \partial E$ is compact, a finite number $\Gamma_1, \Gamma_2, \ldots, \Gamma_n$ cover $E \cup \partial E$. As described in the proof of Green's theorem, there is a partition of unity $\varphi_1, \varphi_2, \ldots, \varphi_n$ of class C^∞ on an open set G containing $E \cup \partial E$ such that each φ_i vanishes on $G - F_i$ where F_i is a compact subset of Γ_i. We define

$$v_i = \varphi_i v, \qquad i = 1, 2, \ldots, n.$$

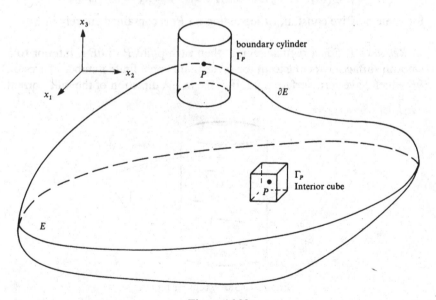

Figure 16.39

Then each v_i is continuously differentiable on G and vanishes on $G - F_i$. Also, $v = \sum_{i=1}^{n} v_i$. Therefore it is sufficient to establish the result for each v_i. Suppose first that P_i is an interior point of E. Then $\bar{\Gamma}_i \subset E$ and v_i vanishes on $\partial\Gamma_i$. Hence

$$\iiint_E \operatorname{div} v_i \, dV = \iiint_{\Gamma_i} \operatorname{div} v_i \, dV = 0$$

because of Lemma 16.8. Also, because $v_i = 0$ on ∂E, we have

$$\iint_{\partial E} v_i \cdot n \, dS = 0.$$

Thus the result is established for all such Γ_i. Suppose now that $P_i \in \partial E$. Define $E_i = \Gamma_i \cap E$. Then $v_i = 0$ on the three sets: (i) $E - E_i$; (ii) $\partial E - \gamma_i$ where $\gamma_i = \partial E \cap \Gamma_i$; (iii) $\partial E_i - \gamma_i$. Now, since E_i is a region of the type described in Lemmas 16.7 and 16.8, we find

$$\iiint_E \operatorname{div} v_i \, dV = \iiint_{E_i} \operatorname{div} v_i \, dV = \iint_{\partial E_i} v_i \cdot n \, dS$$

$$= \iint_{\gamma_i} v_i \cdot n \, dS = \iint_{\partial E} v_i \cdot n \, dS.$$

The result is established for all v_i and hence for v.   □

EXAMPLE 1. Let E be the region given by

$$E = \{(x_1, x_2, x_3) : 1 \leqslant x_1^2 + x_2^2 + x_3^2 \leqslant 9\}.$$

Let $r = x_1 e_1 + x_2 e_2 + x_3 e_3$ and $v = r^{-3} r$ where $r = |r|$. Verify the Divergence theorem.

Solution. Let $\vec{S}_1 = \{(x_1, x_2, x_3) : x_1^2 + x_2^2 + x_3^2 = 1\}$ and $\vec{S}_2 = \{(x_1, x_2, x_3) : x_1^2 + x_2^2 + x_3^2 = 9\}$ be the boundary spheres of E with n pointing outward from the origin 0. Then

$$\iint_{\overline{\partial E}} v \cdot n \, dS = \iint_{\vec{S}_2} v \cdot n \, dS - \iint_{\vec{S}_1} v \cdot n \, dS.$$

A computation shows that div $v = 0$ and so $\iiint_E \operatorname{div} v \, dV = 0$. On both $\vec{S}_1$ and $\vec{S}_2$ the normal n is in the radial direction and hence $n = r^{-1} r$. Therefore

$$\iint_{\vec{S}_2} v \cdot n \, dS - \iint_{\vec{S}_1} v \cdot n \, dS = \frac{1}{9} A(S_2) - A(S_1) = 0.$$ □

EXAMPLE 2. Let $E = \{(x_1, x_2, x_3) : x_1^2 + x_2^2 < 1, 0 < x_3 < x_1 + 2\}$ and define $v = \frac{1}{2}(x_1^2 + x_2^2) e_1 + \frac{1}{2}(x_2^2 + x_3 x_1^2) e_2 + \frac{1}{2}(x_3^2 + x_1^2 x_2) e_3$. Use the Divergence theorem to evaluate

$$\iint_{\partial E} v \cdot n \, dS.$$

Solution. A computation shows that div $v = x_1 + x_2 + x_3$. We denote the unit disk by F and obtain

$$\iint_{\partial E} v \cdot n \, dS = \iiint_E (x_1 + x_2 + x_3) \, dV$$

$$= \iint_F \left[(x_1 + x_2)x_3 + \frac{1}{2} x_3^2 \right]_0^{x_1+2} \, dA_{x_1 x_2}$$

$$= \frac{1}{2} \iint_F [2x_2(x_1 + 2) + 2x_1^2 + 4x_1 + (x_1^2 + 4x_1 + 4)] \, dA_{x_1 x_2}$$

$$= \frac{1}{2} \iint_F (3x_1^2 + 4) \, dA_{x_1 x_2}$$

$$= \frac{1}{2} \int_0^1 \int_0^{2\pi} (3r^2 \cos^2 \theta + 4]) r \, dr \, d\theta = \frac{19}{8} \pi. \qquad \square$$

PROBLEMS

In each of Problems 1 through 10 verify the Divergence theorem by computing $\iiint_E \text{div } v \, dV$ and $\iint_{\partial E} v \cdot n \, dS$ separately.

1. $v = x_1 x_2 e_1 + x_2 x_3 e_2 + x_3 x_1 e_3$; E is the tetrahedron with vertices at $(0, 0, 0)$, $(1, 0, 0)$, $(0, 1, 0)$, $(0, 0, 1)$.

2. $v = x_1^2 e_1 - x_2^2 e_2 + x_3^2 e_3$; $E = \{(x_1, x_2, x_3): x_1^2 + x_2^2 < 4, 0 < x_3 < 2\}$.

3. $v = 2x_1 e_1 + 3x_2 e_2 - 4x_3 e_3$; $E = \{(x_1, x_2, x_3): x_1^2 + x_2^2 + x_3^2 < 4\}$.

4. $v = x_1^2 e_1 + x_2^2 e_2 + x_3^2 e_3$; $E = \{(x_1, x_2, x_3): x_2^2 < 2 - x_1, 0 < x_3 < x_1\}$.

5. $v = x_1 e_1 + x_2 e_2 + x_3 e_3$; $E = E_1 \cap E_2$ where $E_1 = \{(x_1, x_2, x_3): x_1^2 + x_2^2 > 1\}$, $E_2 = \{(x_1, x_2, x_3): x_1^2 + x_2^2 + x_3^2 < 4\}$.

6. $v = x_1 e_1 - 2x_2 e_2 + 3x_3 e_3$; $E = E_1 \cap E_2$ where $E_1 = \{(x_1, x_2, x_3): x_2^2 < x_1\}$, $E_2 = \{(x_1, x_2, x_3): x_3^2 < 4 - x_1\}$.

7. $v = r^{-3}(x_3 e_1 + x_1 e_2 + x_2 e_3)$, $r = (x_1^2 + x_2^2 + x_3^2)^{1/2}$; $E = \{(x_1, x_2, x_3): 1 < x_1^2 + x_2^2 + x_3^2 < 4\}$.

8. $v = x_1 e_1 + x_2 e_2 + x_3 e_3$; $E = E_1 \cap E_2$ where $E_1 = \{(x_1, x_2, x_3): x_1^2 + x_2^2 < 4\}$; $E_2 = \{(x_1, x_2, x_3): x_1^2 + x_2^2 - x_3^2 > 1\}$.

9. $v = 2x_1 e_1 + x_2 e_2 + x_3 e_3$; $E = E_1 \cap E_2$ where $E_1 = \{(x_1, x_2, x_3): x_3 < x_1^2 + x_2^2\}$; $E_2 = \{(x_1, x_2, x_3): 2x_1 \leqslant x_3\}$.

10. $v = 3x_1 e_1 - 2x_2 e_2 + x_3 e_3$; $E = E_1 \cap E_2 \cap E_3$ where $E_1 = \{(x_1, x_2, x_3): x_2 \geqslant 0\}$; $E_2 = \{(x_1, x_2, x_3): x_1^2 + x_3^2 \leqslant 4\}$; $E_3 = \{(x_1, x_2, x_3): x_1 + x_2 + x_3 \leqslant 3\}$.

In each of Problems 11 through 13, use the Divergence theorem to evaluate $\iint_{\partial E} v \cdot n \, dS$.

11. $v = x_2 e^{x_3} e_1 + (x_2 - 2x_3 e^{x_1}) e_2 + (x_1 e^{x_2} - x_3) e_3$; $E = \{(x_1, x_2, x_3): [(x_1^2 + x_2^2)^{1/2} - 2]^2 + x_3^2 < 1\}$.

12. $v = x_1^3 e_1 + x_2^3 e_2 + x_3^3 e_3$; $E = \{(x_1, x_2, x_3): x_1^2 + x_2^2 + x_3^2 < 1\}$.

13. $v = x_1^3 e_1 + x_2^3 e_2 + x_3 e_3$; $E = \{(x_1, x_2, x_3): x_1^2 + x_2^2 < 1, 0 < x_3 < x_1 + 2\}$.

14. Let E be a regular region in $\mathbb{R}^3$. Suppose that u is a scalar field and v is a vector field, both smooth in an open region G containing $E \cup \partial E$. Show that

$$\iiint_E u \operatorname{div} v \, dV = \iint_{\partial E} uv \cdot n \, dS - \iiint_E (\operatorname{grad} u) \cdot v \, dV.$$

15. Let E and G be as in Problem 14 and u, grad u and v smooth functions in G. Let $\partial/\partial n$ denote the normal derivative on ∂E in the direction of n. If Δ denotes the Laplace operator: $(\partial^2/\partial x_1^2) + (\partial^2/\partial x_2^2) + (\partial^2/\partial x_3^2)$, show that

$$\iiint_E v \Delta u \, dV = \iint_{\partial E} u \frac{\partial u}{\partial n} \, dS - \iiint_E (\operatorname{grad} u) \cdot (\operatorname{grad} v) \, dV.$$

If u is any solution of $\Delta u = 0$ in E, prove that

$$\iint_{\partial E} \frac{\partial u}{\partial n} \, dS = 0.$$

Appendixes

Appendix 1. Absolute Value

If a is a real number, the **absolute value** of a, denoted by $|a|$, is defined by the conditions

$$|a| = a \quad \text{if } a > 0,$$

$$|a| = -a \quad \text{if } a < 0,$$

$$|0| = 0.$$

Algebraic manipulations with absolute values are described in the following theorem.

Theorem A.1

(i) $|a| \geqslant 0$, $|-a| = |a|$, and $|a|^2 = a^2$.
(ii) $|a \cdot b| = |a| \cdot |b|$ and, if $b \neq 0$, then $|a/b| = |a|/|b|$.
(iii) $|a| = |b| \Leftrightarrow a = \pm b$.
(iv) *If b is a positive number, then*

$$|a| < b \Leftrightarrow -b < a < b.$$

PROOF. Parts (i) and (ii) are simple consequences of the definition of absolute value. To prove (iii) observe that if $a = \pm b$, then it follows from (i) that $|a| = |b|$. Also, if $|a| = |b|$, then $|a|^2 = |b|^2$ so that from (i) again $a^2 = b^2$ and $a = \pm b$. To establish (iv), note that the solution of the inequality $|x| < b$ is the union of the sets S_1 and S_2 where

$$S_1 = \{x: |x| < b \text{ and } x \geqslant 0\}, \qquad S_2 = \{x: |x| < b \text{ and } x < 0\}.$$

Then
$$x \in S_1 \Leftrightarrow |x| < b \text{ and } x \geqslant 0$$
$$\Leftrightarrow x < b \text{ and } x \geqslant 0 \quad (\text{since } |x| = x \text{ if } x \geqslant 0)$$
$$\Leftrightarrow x \in [0, b).$$

Similarly,
$$x \in S_2 \Leftrightarrow |x| < b \text{ and } x < 0$$
$$\Leftrightarrow -x < b \text{ and } x < 0 \quad (\text{since } |x| = -x \text{ if } x < 0)$$
$$\Leftrightarrow x \in (-b, 0).$$

The result of (iv) follows since $S_1 \cup S_2 = (-b, b)$. □

EXAMPLE 1. Find the solution of the equation
$$\left| \frac{x+2}{2x-5} \right| = 3.$$

Solution. Using the result of (ii) in Theorem A.1, we have
$$\left| \frac{x+2}{2x-5} \right| = 3 \Leftrightarrow \frac{|x+2|}{|2x-5|} = 3 \quad (x \neq 5/2)$$
$$\Leftrightarrow |x+2| = 3|2x-5|.$$

Now Part (iii) of Theorem A.1 may be used to yield
$$\left| \frac{x+2}{2x-5} \right| = 3 \Leftrightarrow x + 2 = \pm 3(2x - 5)$$
$$\Leftrightarrow x + 2 = 3(2x - 5) \text{ or } x + 2 = -3(2x - 5)$$
$$\Leftrightarrow -5x = -17 \text{ or } 7x = 13$$
$$\Leftrightarrow x = \frac{17}{5} \text{ or } \frac{13}{7}.$$

The solution consists of the two numbers $\frac{17}{5}$, $\frac{13}{7}$ which, written in set notation, is $\{\frac{17}{5}, \frac{13}{7}\}$. □

EXAMPLE 2. Find the solution of the inequality $|3x - 4| \leqslant 7$.

Solution. Using Parts (iii) and (iv) of Theorem A.1, we have
$$|3x - 4| \leqslant 7 \Leftrightarrow -7 \leqslant 3x - 4 \leqslant 7.$$

Adding 4 to each portion of this double inequality, we find
$$|3x - 4| \leqslant 7 \Leftrightarrow -3 \leqslant 3x \leqslant 11.$$

Dividing by 3, we obtain

$$|3x - 4| \leqslant 7 \Leftrightarrow -1 \leqslant x \leqslant 11/3.$$

The solution is the interval $[-1, 11/3]$.   □

EXAMPLE 3. Solve for x:

$$\left|\frac{2x - 5}{x - 6}\right| < 3.$$

Solution. Proceeding as in Example 2, we see that

$$\left|\frac{2x - 5}{x - 6}\right| < 3 \Leftrightarrow -3 < \frac{2x - 5}{x - 6} < 3 \qquad (x \neq 6).$$

The solution consists of the union of S_1 and S_2 where

$$S_1 = \left\{x: -3 < \frac{2x - 5}{x - 6} < 3 \text{ and } x - 6 > 0\right\},$$

$$S_2 = \left\{x: -3 < \frac{2x - 5}{x - 6} < 3 \text{ and } x - 6 < 0\right\}.$$

For numbers in S_1, we find

$$x \in S_1 \Leftrightarrow -3(x - 6) < 2x - 5 < 3(x - 6) \text{ and } x - 6 > 0.$$

Considering the three inequalities separately, we may write,

$$x \in S_1 \Leftrightarrow -3x + 18 < 2x - 5 \text{ and } 2x - 5 < 3x - 18 \text{ and } x - 6 > 0.$$

$$\Leftrightarrow 23 < 5x \text{ and } 13 < x \text{ and } x > 6$$

$$\Leftrightarrow 13 < x.$$

Thus $x \in S_1 \Leftrightarrow x \in (13, \infty)$.

Similarly, for $x \in S_2$ multiplication of an inequality by the negative quantity $x - 6$ reverses the direction. Therefore,

$$x \in S_2 \Leftrightarrow -3(x - 6) > 2x - 5 > 3(x - 6) \text{ and } x - 6 < 0$$

$$\Leftrightarrow 23 > 5x \text{ and } 13 > x \text{ and } x - 6 < 0$$

$$\Leftrightarrow x < 23/5.$$

Hence $x \in S_2 \Leftrightarrow x \in (-\infty, 23/5)$. The solution consists of $S_1 \cup S_2$ (see Figure A.1).   □

Figure A.1

We now prove an important theorem and corollary.

Theorem A.2. *If a and b are any numbers, then*

$$|a + b| \le |a| + |b|.$$

PROOF. Since $|a| = a$ or $-a$, we may write $-|a| \le a \le |a|$. Similarly, we have $-|b| \le b \le |b|$. Adding these inequalities (see Problem 27 at the end of Section 1.3), we get

$$-(|a| + |b|) \le a + b \le |a| + |b|.$$

The conclusion of the theorem is equivalent to this double inequality.   □

Corollary. *If a and b are any numbers, then*

$$|a - b| \le |a| + |b|.$$

PROOF. We write $a - b$ as $a + (-b)$ and apply Theorem A.2 to obtain

$$|a - b| = |a + (-b)| \le |a| + |-b| = |a| + |b|.$$

The final inequality holds since, from (i) of Theorem A.1, it is always true that $|-b| = |b|$.   □

PROBLEMS

In each of Problems 1 through 10, find the solution.

1. $|2x + 1| = 3$
2. $|4x - 5| = 3$
3. $|7 - 5x| = 4$
4. $|5 + 3x| = 2$
5. $|x - 2| = |2x + 4|$
6. $|2x - 1| = |3x + 5|$
7. $|3x - 2| = |2x + 1|$
8. $|x - 2| = |x + 4|$
9. $\left|\dfrac{2x - 3}{3x - 2}\right| = 2$
10. $\left|\dfrac{x + 2}{3x - 1}\right| = 3$

In each of Problems 11 through 22, find the solution.

11. $|x - 2| < 1$
12. $|x + 2| < 1/2$
13. $|x + 1| < 1/3$
14. $|2x + 3| < 4$
15. $|4 - 3x| < 6$
16. $|11 + 5x| \le 3$
17. $\left|\dfrac{3 - 2x}{2 + x}\right| < 2$
18. $\left|\dfrac{2x - 5}{x - 6}\right| < 3$
19. $\left|\dfrac{x + 3}{6 - 5x}\right| \le 2$
20. $|x - 1| \le |3 + x|$
21. $|x + 3| \le |2x - 6|$
22. $|3 - 2x| < |x + 4|$

23. Prove for any numbers a and b that $|a| - |b| \leqslant |a - b|$.

24. Given that a and b are positive, c and d are negative, and $a > b, c > d$. Show that

$$\frac{a}{c} < \frac{b}{d}.$$

Appendix 2. Solution of Algebraic Inequalities

In this appendix we give a method for determining the solution of inequalities which involve polynomial expressions. For example, consider the inequality

$$(x - 2)(x - 1)(x + \tfrac{1}{2}) > 0;$$

we want to find those values of x for which the inequality is valid. First, observe that the values $x = 2, 1, -\tfrac{1}{2}$ are *not* in the solution set since they make the left side of the inequality zero. For all other values of x the left side of the above inequality is either positive or negative. It is convenient to proceed geometrically (see Figure A.2) and examine the behavior of $(x - 2)(x - 1)(x + \tfrac{1}{2})$ in each of the intervals that separate the zeros of this expression. If $x > 2$, then $x - 2$ is positive, as are $x - 1$ and $x + \tfrac{1}{2}$. Therefore the polynomial expression is the product of three positive quantities, which is positive. This fact is shown in Figure A.2 by the three plus signs above the interval $(2, \infty)$. We conclude that the inequality $(x - 2)(x - 1)(x + \tfrac{1}{2}) > 0$ holds for $(2, \infty)$. In the interval $1 < x < 2$, we note that $(x - 2)$ is negative, $(x - 1)$ is positive, and $(x + \tfrac{1}{2})$ is positive. We indicate this fact by placing two plus signs and one minus sign above the interval $(1, 2)$, as shown in Figure A.2. The law of signs states that $(x - 2)(x - 1)(x + \tfrac{1}{2})$ is negative in this interval. Proceeding to the intervals $(-\tfrac{1}{2}, 1)$ and $(-\infty, -\tfrac{1}{2})$, we get the signs $- - +$ and $- - -$, respectively. The solution of the inequality

$$(x - 2)(x - 1)(x + \tfrac{1}{2}) > 0$$

is the set $S = (-\tfrac{1}{2}, 1) \cup (2, \infty)$.

More generally, suppose we have an inequality of the form

$$A(x - a_1)(x - a_2)(x - a_3)\ldots(x - a_n) > 0,$$

where A and $a_1, a_2, \ldots, a_n$ are numbers. For convenience we place the a_i, $i = 1, 2, \ldots, n$, in decreasing order (see Figure A.3), and we allow two or more of the a_i to coincide. The number A may be positive or negative. It

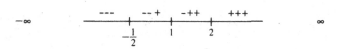

Figure A.2

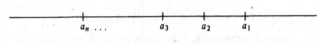

Figure A.3

is clear that $x = a_1$, $x = a_2$, ..., $x = a_n$ are values for which the inequality does *not* hold since each of these numbers makes the polynomial expression zero. For each of the intervals in between the numbers $a_1, a_2, ..., a_n$, examine the sign of each factor $(x - a_i)$, use the law of signs, and determine whether or not the inequality is valid. The solution is the union of the intervals in which the inequality holds. We illustrate with another example.

EXAMPLE 1. Solve the inequality

$$2x^2 - x > 6.$$

Solution. We rearrange the inequality so that a polynomial is on the left and zero is on the right. That is,

$$2x^2 - x > 6 \Leftrightarrow 2x^2 - x - 6 > 0.$$

We factor the polynomial, obtaining

$$2x^2 - x > 6 \Leftrightarrow (2x + 3)(x - 2) > 0.$$

Next we write the factor $2x + 3$ in the form $2(x + \frac{3}{2})$ and rearrange the terms so that the a_i are in decreasing order. Therefore

$$2x^2 - x > 6 \Leftrightarrow 2(x - 2)(x - (-\frac{3}{2})) > 0.$$

A line with the values 2, $-\frac{3}{2}$ indicated is shown in Figure A.4 and the law of signs is used to determine the validity of the inequality in each interval. We conclude that the solution is the set

$$\left(-\infty, \frac{-3}{2}\right) \cup (2, \infty). \qquad \square$$

It is a fact, although we have not proved it, that in each of the intervals separating the zeros of a polynomial the polynomial maintains one sign. Also, for values of x above a_1, the polynomial cannot change sign. The same is true for values of x below a_n. We now work another example.

EXAMPLE 2. Determine the solution of the inequality

$$2x^5 - 3x^4 > -x^3.$$

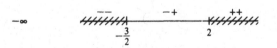

Figure A.4

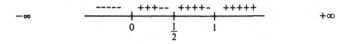

Figure A.5

Solution. We have

$$2x^5 - 3x^4 > -x^3 \Leftrightarrow 2x^5 - 3x^4 + x^3 > 0$$

$$\Leftrightarrow x^3(2x - 1)(x - 1) > 0$$

$$\Leftrightarrow 2(x - 1)(x - \tfrac{1}{2})x^3 > 0.$$

Here $a_1 = 1, a_2 = \tfrac{1}{2}, a_3 = a_4 = a_5 = 0$. The numbers $1, \tfrac{1}{2}, 0$ are marked off as shown in Figure A.5, and the signs of the factors in each of the intervals are shown. The solution set is $(0, \tfrac{1}{2}) \cup (1, \infty)$.   □

If $P(x)$ and $Q(x)$ are polynomials, then the quotient $P(x)/Q(x)$ and the product $P(x)Q(x)$ are always both positive or both negative. This fact is a restatement of Theorem 1.18, Part (ii), which asserts that if $a \neq 0, b \neq 0$ then ab and a/b always have the same sign. We conclude that

$$\left\{ \text{The solution set of } \frac{P(x)}{Q(x)} > 0 \right\} = \{ \text{The solution set of } P(x)Q(x) > 0 \}.$$

In this way the solution of inequalities involving the division of polynomials can always be reduced to a problem in polynomial inequalities.

EXAMPLE 3. Solve for x:

$$\frac{210}{3x - 2} < \frac{50}{x}.$$

Solution. We have

$$\frac{210}{3x - 2} < \frac{50}{x} \Leftrightarrow \frac{210}{3x - 2} - \frac{50}{x} < 0$$

$$\Leftrightarrow \frac{60x + 100}{x(3x - 2)} < 0$$

$$\Leftrightarrow \frac{20(3x + 5)}{x(3x - 2)} < 0$$

$$\Leftrightarrow 20(3x + 5)x(3x - 2) < 0$$

(changing the quotient of polynomials to a product)

$$\Leftrightarrow 180(x - \tfrac{2}{3})x(x - (-\tfrac{5}{3})) < 0.$$

The zeros are $a_1 = 2/3, a_2 = 0, a_3 = -5/3$. The signs of the factors are indicated in Figure A.6. The solution set is $(-\infty, -5/3) \cup (0, 2/3)$.

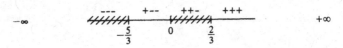

Figure A.6

PROBLEMS

In Problems 1 through 10 determine the solution set in each case.

1. $x^2 - x - 12 < 0$

2. $x^2 + 2x - 15 > 0$

3. $2x^2 + 5x + 2 > 0$

4. $60x^2 - 42x - 36 > 0$

5. $6x^5 - x^4 < x^3$

6. $x^2 > 4$

7. $x^2 < 4$

8. $x^2 + 2x - 4 < 0$ (complete the square)

9. $x^2 + 2x + 2 > 0$

10. $4x^4 < x^2$

11. Find a polynomial inequality which has the interval $(1, 2)$ as its solution set.

12. Find a polynomail inequality which has the set $S = (1, 3) \cup (4, 7)$ as its solution.

13. Find a polynomial inequality which has the set $S = (-\infty, 2) \cup (2, 3) \cup (3, \infty)$ as its solution.

14. Find a polynomial $P(x)$ of the second degree such that $(-\infty, \infty)$ is the solution of the inequality $P(x) > 0$. Is it possible to find a polynomial inequality of the third degree with the same solution set? Justify your answer.

In Problems 15 through 25 determine the solution set in each case.

15. $x^4 - 5x^2 + 4 < 0$

16. $x^3 + x^2 < x + 1$

17. $\dfrac{2x}{3} - \dfrac{x^2 - 3}{2x} + \dfrac{1}{2} < \dfrac{x}{6}$

18. $\dfrac{4}{x} + \dfrac{x - 1}{5} < \dfrac{3}{x} + 1$

19. $\dfrac{10 - 7x}{6 - 7x} < \dfrac{5x - 4}{5x}$

20. $\dfrac{x + 2}{x^2 - 3x} < 0$

21. $\dfrac{x - 1}{x^2 + x - 6} < 0$

22. $\dfrac{22}{2x - 3} + \dfrac{23x + 26}{4x^2 - 9} > \dfrac{51}{2x + 3}$

23. $\dfrac{4x}{2x - 3} - \dfrac{1}{2} > \dfrac{3x}{2x + 3}$

24. $1 + \dfrac{4}{x^2 - x - 6} < 0$

25. $\dfrac{x}{1 + x^2} < \dfrac{2}{5}$

Appendix 3. Expansions of Real Numbers in Any Base

In this appendix we develop the theory of decimal expansions of real numbers. We also describe the expansions of real numbers with an arbitrary base b, when b is any positive integer greater than 1.

Theorem A.3. *If x is any real nonnegative number, then there is a unique positive integer n such that $n - 1 \leqslant x < n$.*

PROOF. From Part (a) of Theorem 2.21, there is a positive integer k larger than x. Thus the set S of all positive integers larger than x is not empty. Hence S contains a smallest element (Theorem 1.30); call it n. If $n > 1$, the positive integer $n - 1$ is not in S and so $n - 1 \leqslant x$. Then $n - 1 \leqslant x < n$. If $n = 1$, then $0 \leqslant x < 1$. $\qquad\square$

Definition. Let b be a positive integer greater than 1. A sequence of the form

$$\frac{d_1}{b}, \frac{d_2}{b^2}, \ldots, \frac{d_n}{b^n}, \ldots$$

in which each d_i is an integer with $0 \leqslant d_i \leqslant b - 1$ is called a **development with the base b**.

Remarks. When $b = 10$ we have the familiar **decimal development**. In this case the d_i are the usual digits from 0 through 9. When $b = 2$ we have the **binary development**. In this case the d_i are always either 0 or 1.

Definitions. If, in a development with the base b, only a finite number of the d_i are different from zero, we say the development is **terminating**; otherwise, it is **nonterminating**. If all the d_i beyond some specific d_n have the value $b - 1$ we say the development is **improper**; otherwise it is called **proper**. For decimal developments, the customary notation $0.d_1 d_2 d_3 \ldots$ is used.

The decimal development $0.32799999\ldots$ is improper, while the development $0.218218218\ldots$ is proper. An improper binary development has $d_i = 1$ for all i beyond some given integer n.

Theorem A.4. *Consider an arbitrary development with the base b and define*

$$s_n = \sum_{i=1}^{n} \frac{d_i}{b^i}, \qquad n = 1, 2, \ldots.$$

(a) *Then $s_n \to a$ as $n \to \infty$ where $0 \leqslant a \leqslant 1$.*
(b) *Also, $a = 1 \Leftrightarrow d_i = b - 1$ for every positive integer i.*
(c) *If the development is improper, then there are positive integers n and p such*

that

$$a = \frac{p}{b^n}$$

with $p \leqslant b^n$.

PROOF

(a) Since $d_i \geqslant 0$ and $b > 1$, we have $s_{n+1} \geqslant s_n$ for every n. Also, since $d_i \leqslant b - 1$ for every i, we see that

$$s_n \leqslant \sum_{i=1}^{n} \frac{b-1}{b^i} = \frac{1}{b}(b-1)\left[1 + \left(\frac{1}{b}\right) + \left(\frac{1}{b}\right)^2 + \cdots + \left(\frac{1}{b}\right)^{n-1}\right].$$

Setting $r = 1/b$, we find

$$s_n \leqslant r\left(\frac{1}{r} - 1\right)\frac{1 - r^n}{1 - r} = 1 - r^n \leqslant 1 \quad \text{for every } n.$$

Hence the s_n satisfy the hypotheses of Axiom C with $M = 1$, and so $s_n \to a$ as $n \to \infty$ with $a \leqslant 1$.

(b) Note that if all the $d_i = b - 1$ then

$$s_n = \sum_{i=1}^{n} \frac{b-1}{b^i} = 1 - r^n \to 1 \quad \text{as } n \to \infty.$$

(c) Suppose that $d_i = b - 1$ for all $i \geqslant n$. Then for $k > n$,

$$s_k = \sum_{i=1}^{n-1} \frac{d_i}{b^i} + \sum_{i=n}^{k} \frac{b-1}{b^i},$$

where if $n = 1$ the first term is omitted and $a = 1$. We have

$$\lim_{k\to\infty} s_k = \sum_{i=1}^{n-1} \frac{d_i}{b^i} + (b-1)\sum_{i=n}^{\infty} \frac{1}{b^i} = \sum_{i=1}^{n-1} \frac{d_i}{b^i} + \frac{1}{b^{n-1}}.$$

The common denominator in the last expression on the right is b^{n-1}, and therefore

$$a = \frac{p}{b^{n-1}} \quad \text{with } p = \sum_{i=1}^{n-2} d_i b^{n-1-i} + (d_{n-1} + 1). \qquad \square$$

The next theorem shows that for any base $b > 1$, every real number a with $0 \leqslant a < 1$ has a unique proper development with that base.

Theorem A.5. *Let b be a positive integer greater than 1, and let a be any real number such that $0 \leqslant a < 1$. Then there is a unique proper development with base b such that $s_n \to a$ as $n \to \infty$, where $s_n = \sum_{i=1}^{n} d_i/b^i$.*

PROOF. Since ba is a real number, we may apply Theorem A.3 to assert that there is a nonnegative integer p_1 such that $p_1 \leqslant ba < p_1 + 1$. In fact, for each

i, there is a nonnegative integer p_i such that $p_i \leqslant b^i a < p_i + 1$. Then we define the d_i as follows:

$$d_1 = p_1 \quad \text{and} \quad d_i = p_i - bp_{i-1} \quad \text{for } i > 1. \tag{A.1}$$

We must show that $0 \leqslant d_i \leqslant b - 1$ for all i. Clearly, $d_1 = p_1 \leqslant ba < b$, and $d_1 \geqslant 0$. Also, for $i > 1$, it follows from the definition of p_i that

$$p_i \leqslant b^i a < p_i + 1; \qquad p_{i-1} \leqslant b^{i-1} a < p_{i-1} + 1.$$

Hence $b^{i-1} a - 1 < p_{i-1}$ or $b^i a - b < bp_{i-1}$. Therefore

$$d_i = p_i - bp_{i-1} < b^i a - (b^i a - b) = b$$

$$d_i = p_i - bp_{i-1} > (b^i a - 1) - b(b^{i-1} a) = -1.$$

Since d_i is an integer and $-1 < d_i < b$, we conclude that $0 \leqslant d_i \leqslant b - 1$.

Using the relation $p_i = d_i + bp_{i-1}$ and proceeding by induction, we find that

$$p_i = d_i + bd_{i-1} + b^2 d_{i-2} + \cdots + b^{i-1} d_1.$$

Hence,

$$p_i = \sum_{k=1}^{i} b^{i-k} d_k \leqslant b^i a < p_i + 1.$$

But $\sum_{k=1}^{i} b^{i-k} d_k = b^i \sum_{k=1}^{i} b^{-k} d_k$. Setting $s_i = \sum_{k=1}^{i} b^{-k} d_k$, we have $s_i = b^{-i} p_i$, and so $s_i \leqslant a < s_i + b^{-i}$. Therefore, $s_i \to a$ as $i \to \infty$.

Thus far we have shown that each number a has a proper development with the base b. To show that there is a *unique proper* development, let

$$\frac{d_1'}{b}, \frac{d_2'}{b^2}, \ldots, \frac{d_n'}{b^n}, \ldots$$

be a second proper development for the number a. That is,

$$a = \sum_{i=1}^{\infty} \frac{d_i'}{b^i}.$$

We may write

$$b^n a = \sum_{i=1}^{n} b^{n-1} d_i' + \sum_{i=n+1}^{\infty} b^{n-i} d_i'$$

$$= \sum_{i=1}^{n} b^{n-i} d_i' + \sum_{k=1}^{\infty} b^{-k} d_{n+k}'.$$

Since the last sum on the right is a proper development of some number, it must be a number less than 1. Hence

$$\sum_{i=1}^{n} b^{n-i} d_i' \leqslant b^n a < \sum_{i=1}^{n} b^{n-i} d_i' + 1.$$

We define $p_n = \sum_{k=1}^{n} b^{n-i} d_i'$ and obtain

$$p_n \leqslant b^n a < p_n + 1.$$

Therefore these p_n are the same as the p_n used for the development d_1/b^1, d_2/b^2, ..., d_n/b^n, ... Hence $d_i = d_i'$ for every i and the development is unique. □

The next result is frequently useful in studying properties of the real number system.

Theorem A.6. *There is a rational number between any two real numbers.*

PROOF. Let the given numbers be a and b with $a < b$. Without loss of generality assume $a \geqslant 0$, as otherwise adding a sufficiently large positive integer to a and b will make them nonnegative. Let q be the smallest positive integer larger than $1/(b - a)$. Let p be the smallest positive integer larger than qa. Then we have $p - 1 \leqslant qa$ so that

$$qa < p \leqslant qa + 1 = a < \frac{p}{q} \leqslant a + \frac{1}{q}.$$

Also, $q(b - a) > 1 \Leftrightarrow a + (1/q) < b$, and we conclude that

$$a < \frac{p}{q} < b.$$ □

Corollary. *There is an irrational number between any two real numbers.*

PROOF. Let the given real numbers be a and b. From Theorem A6 there is a rational number r between $a/\sqrt{2}$ and $b/\sqrt{2}$. Then $r\sqrt{2}$ is between a and b and $r\sqrt{2}$ is irrational (since $\sqrt{2}$ is—see Theorem A.7). □

We show now that $\sqrt{2}$ is irrational.

Theorem A.7. *There is no rational number whose square is 2.*

PROOF. Suppose there is a rational number r such that $r^2 = 2$. We shall reach a contradiction. If r is rational there are integers p and q such that $r = p/q$. Assume that p and q have no common factor, a fact which follows from the axioms of Chapter 1 (although not proved there). Since $r^2 = 2$, we have $p^2 = 2q^2$ and p^2 is even. The fact that p^2 is even implies that p is even. For if p were odd, then $p = 2l + 1$ for some integer l and then $p^2 = 4l^2 + 4l + 1$. Thus p^2 would be odd. Hence, $p = 2k$ for some integer k and $q^2 = 2k^2$. We see that q^2 is even and by the same argument as that used for p^2, it follows that q is even. However, we assumed that p and q have no common factor and then deduced that they have 2 as a common factor, a contradiction. □

PROBLEMS

*1. Given the integers $d_1, d_2, \ldots, d_n$ with $0 \leqslant d_i \leqslant b - 1$ where b is an integer > 1. We say the expansion

$$\frac{d_1}{b}, \frac{d_2}{b^2}, \ldots, \frac{d_k}{b^k}, \ldots$$

is **repeating** if and only if there is a positive integer n such that $d_{i+n} = d_i$, $i = 1, 2, \ldots$ Show that every repeating development represents a rational number.

2. Prove that $\sqrt{3}$ is irrational.

*3. Consider the numbers on $[0, 1]$ represented with the base $b = 3$. We shall write these numbers using "decimal" notation. Describe geometrically on the interval $[0, 1]$ all points corresponding to numbers of the form

$$0.1d_2 d_3 d_4 \ldots$$

where the d_i may be 0, 1, or 2. Similarly, describe the sets

$$0.01a_3 a_4 \ldots, \qquad 0.11b_3 b_4 \ldots$$

where the a_i and b_i may be 0, 1, or 2. Finally, describe the set corresponding to

$$0.c_1 c_2 c_3 \ldots$$

where all the c_i are either 0 or 2.

Appendix 4. Vectors in E_N

1. The space E_N. We recall the definition of a vector in a plane or three-space as an equivalence class of directed line segments, all having the same length and direction. We shall extend this definition to N-dimensional space and establish several of the basic properties of such vectors.

By a Euclidean N-space, denoted E_N, we mean a metric space which can be mapped onto $\mathbb{R}^N$ by an isometry. Such a mapping is called a **coordinate system** and will be denoted by a symbol such as (x). If A is a point in E_N and (x) is a coordinate system, then the point in $\mathbb{R}^N$ which corresponds to A is denoted x^A or $(x_1^A, x_2^A, \ldots, x_N^A)$. Of course, $\mathbb{R}^N$ itself is a Euclidean N-space; the plane and three-space studied in elementary geometry courses are also examples of Euclidean spaces. In this section we establish several elementary properties of E_N. The reader is undoubtedly familiar with the two- and three-dimensional versions of the results given here.

We define a **line** to be a Euclidean 1-space. If l is a line and (t) is a coordinate system on l, then the **origin** and **unit point** of (t) are the points having coordinates 0 and 1, respectively.

Theorem A.8. *Suppose that l is a line and (t) and (u) are coordinate systems on l. Let the origin of the (u) system have t-coordinate t^0 and the unit of the (u) system have t-coordinate $t^0 + \lambda$. Then for every P on l, it follows that*

$$u^P = \lambda(t^P - t^0), \qquad \lambda = \pm 1.$$

PROOF. We have $|u^P - 0| = |t^P - t^0|$ and $|u^P - 1| = |t^P - t^0 - \lambda|$ for all p. Squaring both sides of these equations and subtracting, we get

$$u^P = \lambda(t^P - t^0).$$

Choosing p and 0 as the unit point and the origin, we find

$$|u^1 - u^0| = 1 = |\lambda(t^1 - t^0)| = |\lambda| \cdot |t^1 - t^0| = |\lambda|;$$

hence $\lambda = \pm 1$. $\qquad\qquad\qquad\qquad\qquad\qquad\qquad\qquad\qquad\qquad\qquad\qquad$ $\square$

We observe that a coordinate system (t) on a line l sets up an ordering of the points in which P_1 precedes P_2 if and only if $t^1 < t^2$. We use the notation $P_1 \prec P_2$ to indicate that P_1 precedes P_2. Theorem A.8 shows that there are exactly two possible orderings on every line according as $\lambda = 1$ or -1. A line l and a specific ordering, denoted $\mathcal{O}$, determine a **directed line** $\vec{l}$. The same line with the opposite ordering is frequently denoted $-\vec{l}$. A pair of points (A, B) on a line $\vec{l}$ will determine a **directed line segment** when they are ordered with respect to $\mathcal{O}$. We use the symbol $\overrightarrow{AB}$ for such a directed line segment.

If A and B are any points on a line l, we define the **directed distance from A to B along** $\vec{l}$, denoted $\overline{AB}$, by the formula

$$\overline{AB} = |AB| \qquad \text{if } A \prec B,$$

$$\overline{AB} = -|AB| \quad \text{if } B \prec A.$$

Theorem A.9. *Let $\vec{l}$ be a directed line and (t) a coordinate system which agrees with the order on $\vec{l}$. Then*

$$\overline{AB} = t^B - t^A$$

where t^A and t^B are the (t) coordinates of A and B, respectively.

PROOF. Since (t) is a coordinate system, we have $|AB| = |t^B - t^A|$. From the definition of directed distance, it follows that $\overline{AB} = |t^B - t^A|$ if $t^B > t^A$ and $\overline{AB} = -|t^B - t^A|$ if $t^B < t^A$. $\qquad\qquad\qquad\qquad\qquad\qquad\qquad$ $\square$

Theorem A.10. *Suppose that (x) is a coordinate system on a Euclidean space E_N, l is a line in E_N, and (t) is a coordinate system on l with origin at x^0 and unit point at $x^0 + \lambda$. Then for each p on l, it follows that*

$$|\lambda| = 1, \qquad x_i^p = x_i^0 + \lambda_i t^p, \qquad i = 1, 2, \ldots, N. \qquad\qquad \text{(A.2)}$$

PROOF. Since (t) and (x) are coordinate systems,

$$1 = |1 - 0| = |x^0 + \lambda - x^0| = |\lambda|.$$

If p is any point on l there are numbers $\xi_1^p, \ldots, \xi_N^p$ such that

$$x_i^p = x_i^0 + \lambda_i t^p + \xi_i^p, \qquad i = 1, 2, \ldots, N. \qquad\qquad \text{(A.3)}$$

Then (since $|\lambda| = 1$),

$$|x^p - x^0|^2 = |\lambda|^2 (t^p)^2 + |\xi^p|^2 + 2 \sum_{i=1}^{N} \lambda_i \xi_i^p t^p = (t^p)^2$$

and

$$|x^p - x^0 - \lambda|^2 = |\lambda|^2(t^p - 1)^2 + |\xi^p|^2 + 2 \sum_{i=1}^{N} \lambda_i \xi_i^p(t^p - 1) = (t^p - 1)^2.$$

Subtracting these two equations and simplifying, we get

$$|\xi^p|^2 + 2 \sum_{i=1}^{N} \lambda_i \xi_i^p t^p = 0, \qquad |\xi^p|^2 + 2 \sum_{i=1}^{N} \lambda_i \xi_i^p(t^p - 1) = 0.$$

We conclude that $\xi_i^p = 0, i = 1, 2, \ldots, N$, which shows that (A.3) and (A.2) are the same. $\square$

Theorem A.11. *Suppose that (x) is a coordinate system on a Euclidean space E_N and that l is the range in E_N of the mapping*

$$x_i^p = x_i^0 + \lambda_i t^p, \qquad |\lambda| = 1, i = 1, 2, \ldots, N. \tag{A.4}$$

Then l is a line and the Equations (A.4) set up a coordinate system on l with origin at x^0 and unit point $x^0 + \lambda$.

PROOF. Let p and q be in the range of (A.4). Then

$$x_i^p - x_i^q = \lambda_i(t^p - t^q), \qquad i = 1, 2, \ldots, N,$$

and $|x^p - x^q| = |t^p - t^q|$. This shows that l is a line and that the equations (3) set up a coordinate system with origin $(t = 0)$ at x^0 and unit point at $x^0 + \lambda$. $\square$

It is easy to see from the above theorems that if l is a line in E_N, if (x) is a coordinate system on E_N, and if (t) and (u) are coordinate systems on l which set up the same ordering on l, then the corresponding sets λ of (t) and λ' of (u) in Equations (A.4) must be the same. Hence *the set λ depends only on the ordering of $\vec{l}$.* If $\vec{l}$ is any directed line, the set λ in Equations (A.4) is called the **direction cosines** of $\vec{l}$ corresponding to the coordinate system (x) on E_N. Two lines l_1 and l_2 are **perpendicular** if and only if l_1 and l_2 intersect at some point x^0 and

$$|x^1 - x^2|^2 = |x^1 - x^0|^2 + |x^2 - x^0|^2$$

for all points x^1 on l_1 and x^2 on l_2.

Theorem A.12. *Let (x) be a coordinate system on E_N and let l_1 and l_2 be lines in E_N with equations*

$$l_1: x_i^p = x_i^0 + \lambda_i^1 t^p, \qquad i = 1, 2, \ldots, N,$$

$$l_2: x_i^p = x_i^0 + \lambda_i^2 t^p, \qquad i = 1, 2, \ldots, N.$$

Then l_1 is perpendicular to l_2 if and only if

$$\sum_{i=1}^{N} \lambda_i^1 \lambda_i^2 = 0. \tag{A.5}$$

PROOF. Let $x^1 \in l_1$ and $x^2 \in l_2$. Then there are t^1 and t^2 such that $x_i^1 = x_i^0 + \lambda_i^1 t^1$, $x_i^2 = x_i^0 + \lambda_i^2 t^2$, $i = 1, 2, \ldots, N$. Hence

$$|x^1 - x^2|^2 = \sum_{i=1}^{N} (\lambda_i^1 t^1 - \lambda_i^2 t^2)^2 = (t^1)^2 + (t^2)^2 - 2t^1 t^2 \sum_{i=1}^{N} \lambda_i^1 \lambda_i^2$$

and

$$|x^1 - x^0|^2 + |x^2 - x^0|^2 = \sum_{i=1}^{N} [(t^1 \lambda_i^1)^2 + (t^2 \lambda_i^2)^2] = (t^1)^2 + (t^2)^2,$$

from which (A.5) follows. $\qquad\square$

Definition. Let A and B be points in E_N. Then a point P is h **of the way from** A **to** B if and only if

$$|AP| = |h| \cdot |AB| \quad \text{and} \quad |BP| = |1 - h| \cdot |AB|. \tag{A.6}$$

The number h is any real number.

Theorem A.13

(a) *Let* (x) *be a coordinate system on* E_N. *A point* P *is* h *of the way from* A *to* B *if and only if*

$$x_i^P - x_i^A = h(x_i^B - x_i^A) \quad \text{or} \quad x_i^P = (1 - h)x_i^A + hx_i^B, \quad i = 1, 2, \ldots, N. \tag{A.7}$$

(b) *A point* h *of the way from* A *to* B *coincides with the point* $(1 - h)$ *of the way from* B *to* A.

PROOF. Let P be any point in E_N. Then there are numbers $\xi_1^P, \ldots, \xi_N^P$ such that

$$x_i^P - x_i^A = h(x_i^B - x_i^A) + \xi_i^P$$

or

$$x_i^P - x_i^B = (h - 1)(x_i^B - x_i^A) + \xi_i^P, \quad i = 1, \ldots, N. \tag{A.8}$$

Part (a) results from substituting (A.8) into (A.6), squaring, subtracting, and simplifying as in the proof of Theorem A.10. Then if P is h of the way from A to B, all the ξ_i^P are zero. Part (b) is a direct consequence of (a). $\qquad\square$

Theorem A.14

(a) *Let* A *and* B *be distinct points in* E_N. *Then the set of all points* P *in* E_N *each of which is* h *of the way from* A *to* B, *where* h *is any real number, constitute a line. The set of all points* Q *which are* $(1 - h)$ *of the way from* B *to* A *as* h *takes on all real values constitutes the same line.*

(b) *There exists a unique line in* E_N *which passes through two distinct points.*

PROOF.

Let (x) be a coordinate system on E_N. If P is h of the way from A to B,

then

$$x_i^P = x_i^A + \lambda_i^{AB} t^P, \qquad i = 1, 2, \ldots, N, \quad \text{where } \lambda_i^{AB} = \frac{x_i^B - x_i^A}{|AB|}. \quad \text{(A.9)}$$

Here we set $t^P = h$ and note that $|\lambda^{AB}| = 1$. Therefore P lies on the line with Equations (A.9). On the other hand, if P is on the line given by (A.9), then P satisfies (A.7) with $t^P = h$ and so is h of the way from A to B. Now if Q is k of the way from B to A, then

$$x_i^Q - x_i^B = k(x_i^A - x_i^B)$$

or

$$x_i^Q = (1 - k)x_i^B + kx_i^A = (1 - h)x_i^A + hx_i^B = x_i^P.$$

The preceding argument shows that there is only one line through A and B. $\qquad \square$

Theorem A.15. *Let l be a line in E_N and P_1 a point in E_N not on l. Then there exists a unique line l_1 in E_N which contains P_1 and is perpendicular to l. If (x) is a coordinate system on E_N, if P_1 has coordinates $(\bar{x}_1, \ldots, \bar{x}_N)$, and if l has equations*

$$x_i = x_i^0 + \lambda_i t, \qquad i = 1, 2, \ldots, N, \quad \text{(A.10)}$$

then the (t) coordinates of the point P_2 of intersection of l and l_1 are given by

$$t^* = \sum_{i=1}^{N} \lambda_i (\bar{x}_i - x_i^0). \quad \text{(A.11)}$$

PROOF. The point of intersection P_2 will satisfy Equations (A.10) and, denoting its (x) coordinates by $(x_1^*, \ldots, x_N^*)$, we have

$$x_i^* = x_i^0 + \lambda_i t^* \quad \text{for some } t^*.$$

The differences $x_i^* - \bar{x}_i$ are proportional to the direction cosines of l_1 and so the perpendicularity condition, (A.5), implies that

$$\sum_{i=1}^{N} \lambda_i (x_i^* - \bar{x}_i) = 0.$$

Therefore $\sum_{i=1}^{N} \lambda_i (x_i^0 + \lambda_i t^* - \bar{x}_i) = 0$, from which (A.11) follows. $\qquad \square$

Definition. Let $\vec{l}$ be a directed line and $\overrightarrow{AB}$ a directed line segment in E_N. Let A' be the point of intersection with $\vec{l}$ of the line through A which is perpendicular to $\vec{l}$ (if A is on $\vec{l}$, then $A' = A$). Similarly, let B' be the point of intersection of the perpendicular to $\vec{l}$ through B. We define the **projection of $\overrightarrow{AB}$ on $\vec{l}$** as the directed distance $\overrightarrow{A'B'}$. We write

$$\text{Proj}_{\vec{l}} \, \overrightarrow{AB} = \overrightarrow{A'B'}.$$

Theorem A.16. *Let $\vec{l}$ be a directed line and $\overrightarrow{AB}$ a directed line segment in E_N. Let (x) be a coordinate system on E_N and suppose that the direction cosines of*

$\vec{l}$ are $\lambda_1, \ldots, \lambda_N$. Then

$$\text{Proj}_{\vec{l}} \ \overrightarrow{AB} = \sum_{i=1}^{N} \lambda_i (x_i^B - x_i^A).$$

PROOF. According to Theorem A.15, the (t) coordinates of A' and B', as given in the definition of projection, are

$$t^{A'} = \sum_{i=1}^{N} \lambda_i (x_i^A - x_i^0), \qquad t^{B'} = \sum_{i=1}^{N} \lambda_i (x_i^B - x_i^0).$$

Using Theorem A.9, we find

$$\text{Proj}_{\vec{l}} \ \overrightarrow{AB} = \overrightarrow{A'B'} = t^{B'} - t^{A'} = \sum_{i=1}^{N} \lambda_i (x_i^B - x_i^A). \qquad \square$$

2. *Vectors in E_N.* As in the case of two and three dimensions, we shall define a vector as an equivalence class of directed line segments. Two directed line segments $\overrightarrow{AB}$ and $\overrightarrow{CD}$ are **equivalent** if and only if

$$\text{Proj}_{\vec{l}} \ \overrightarrow{AB} = \text{Proj}_{\vec{l}} \overrightarrow{CD}$$

for every directed line $\vec{l}$ in E_N. We write $\overrightarrow{AB} \approx \overrightarrow{CD}$ for such equivalent directed line segments.

The proofs of the following elementary properties of equivalent directed line segments are left to the reader.

Theorem A.17

(a) *If (x) is a coordinate system in E_N, then $\overrightarrow{AB} \approx \overrightarrow{CD}$ if and only if $x_i^D - x_i^C = x_i^B - x_i^A, i = 1, 2, \ldots, N$.*

(b) *If $\overrightarrow{AB} \approx \overrightarrow{CD}$ then $\overrightarrow{CD} \approx \overrightarrow{AB}$.*

(c) *If $\overrightarrow{AB} \approx \overrightarrow{CD}$ and $\overrightarrow{CD} \approx \overrightarrow{EF}$, then $\overrightarrow{AB} \approx \overrightarrow{EF}$.*

(d) *If $\overrightarrow{AB}$ and P are given, then there is a unique point Q such that $\overrightarrow{PQ} \approx \overrightarrow{AB}$.*

(e) *If $\overrightarrow{AB} \approx \overrightarrow{DE}$ and $\overrightarrow{BC} \approx \overrightarrow{EF}$, then $\overrightarrow{AC} \approx \overrightarrow{DF}$.*

(f) *If $\overrightarrow{AB} \approx \overrightarrow{DE}$, if C is h of the way from A to B, and if F is h of the way from D to E, then $\overrightarrow{AC} \approx \overrightarrow{DF}$.*

Definitions. A **vector in E_N** is the collection of all ordered pairs of points (A, B), i.e., all directed line segments $\overrightarrow{AB}$ in E_N having the same magnitude and direction. The individual ordered pairs are called **representatives** of the vector containing them. We denote by $v(AB)$ the vector containing the *representative* $\overrightarrow{AB}$.

Let v and w be vectors and $\overrightarrow{AB}$ a representative of v. Then there is a representative $\overrightarrow{BC}$ of w. We define the **sum** $v + w$ as the vector u which as $\overrightarrow{AC}$

as its representative. That is,

$$u(AC) = v(AB) + w(BC).$$

If h is a real number and D is h of the way from A to B, then

$$hv = v(AD).$$

We define $|v|$ as the common length of all the representatives of v.

Let (x) be a coordinate system on E_N. We define the ith **unit point** I_i by the coordinates $(0, 0, \ldots, 0, 1, 0, \ldots, 0)$ where all the coordinates are zero except the ith which is 1. The **vector** e_i is the vector of length 1 which has as representative the directed line segment $\overrightarrow{0I_i}$, where 0 is the origin of (x).

Theorem A.18. Let (x) be a coordinate system on E_N. If A and B are any points in E_N, then

$$v(AB) = \sum_{i=1}^{N} (x_i^B - x_i^A)e_i.$$

PROOF. We prove the result for $N = 3$, the proof in the general case being similar. The vector $v(AB)$ has a representative $\overrightarrow{0P}$ where P has coordinates

$$(x_1^B - x_1^A, x_2^B - x_2^A, x_3^B - x_3^A).$$

Let $R_1, R_2, R_3, Q_1, Q_2, Q_3$ have coordinates

$$R_1 = (x_1^B - x_1^A, 0, 0), \qquad R_2 = (0, x_2^B - x_2^A, 0), \qquad R_3 = (0, 0, x_3^B - x_3^A),$$

$$Q_1 = R_1, \qquad Q_2 = (x_1^B - x_1^A, x_2^B - x_2^A, 0), \qquad Q_3 = P.$$

We observe that R_1 is $(x_1^B - x_1^A)$ of the way from 0 to I_1; R_2 is $(x_2^B - x_2^A)$ of the way from 0 to I_2 and R_3 is $(x_3^B - x_3^A)$ of the way from 0 to I_3. Also,

$$v(Q_1Q_2) = v(0R_2), \qquad v(Q_2Q_3) = v(0R_3),$$

$$v(0R_1) = (x_1^B - x_1^A)e_1, \qquad v(0R_2) = (x_2^B - x_2^A)e_2,$$

$$v(0R_3) = (x_3^B - x_3^A)e_3.$$

Hence we see that

$$v(0P) = v(0Q_1) + v(Q_1Q_2) + v(Q_2Q_3),$$

which is the desired result. $\qquad\qquad\square$

Theorem A.19. Suppose that $e_1, e_2, \ldots, e_N$ are mutually perpendicular unit vectors in E_N. Suppose that

$$v = a_1e_1 + \cdots + a_Ne_N, \qquad w = b_1e_1 + \cdots + b_Ne_N.$$

Then

$$v + w = \sum_{i=1}^{N} (a_i + b_i)e_i, \qquad hv = \sum_{i=1}^{N} ha_ie_i,$$

$$|v| = \sqrt{a_1^2 + \cdots + a_N^2}.$$

PROOF. Let (x) be a coordinate system on E_N with origin 0 and unit points I_i where $v(0I_i) = e_i$. The hypotheses of Theorem A.18 hold. Let A, B, C be the points with coordinates $(a_1, \ldots, a_N), (b_1, \ldots, b_N), (ha_1, \ldots, ha_N)$. Then C is h of the way from 0 to A and

$$v = v(0A), \qquad w = v(AB), \qquad hv = v(0C), \qquad v + w = v(0B). \qquad \square$$

The next theorem is a direct consequence of the above results on vectors.

Theorem A.20

(a) *The operation of addition for vectors satisfies Axioms A-1 through A-5 for addition of real numbers (Chapter 1).*
(b) *Let v and w be two vectors and c, d real numbers. Then*

$$(c + d)v = cv + dv, \qquad c(v + w) = cv + cw,$$

$$c(dv) = (cd)v, \qquad 1 \cdot v = v, \qquad 0 \cdot v = 0, \qquad (-1)v = -v.$$

Definition. Let u and v be vectors in E_N. The **inner** (or **scalar**) **product** of u and v, denoted $u \cdot v$, is defined by

$$u \cdot v = \tfrac{1}{2}[|u + v|^2 - |u|^2 - |v|^2].$$

The value of the inner product is independent of the coordinate system.

Theorem A.21. *Let u and v be vectors in E_N and (x) a coordinate system on E_N. Suppose $e_1, \ldots, e_N$ is a set of mutually orthogonal unit vectors and*

$$u = \sum_{i=1}^{N} a_i e_i, \qquad v = \sum_{i=1}^{N} b_i e_i.$$

Then

$$u \cdot v = \sum_{i=1}^{N} a_i b_i;$$

also,

$$u \cdot v = v \cdot u, \qquad u \cdot u = |u|^2, \qquad u \cdot (cv + dw) = c(u \cdot v) + d(u \cdot w),$$

where w is any vector in E_N.

Answers to Odd-Numbered Problems

Section 1.2

1. The inverse of T is a function if the a_i are all distinct.

7. (a) Proposition 1.3.
   (b) Propositions 1.4, 1.6, 1.7.
   (c) $\displaystyle\sum_{i=0}^{n}\sum_{j=0}^{n-i}\frac{n!}{i!\,j!\,(n-i-j)!}a^{n-i-j}b^{i}c^{j}$

11. Yes

+	0	1	2	3
0	0	1	2	3
1	1	0	3	2
2	2	3	0	1
3	3	2	1	0

×	0	1	2	3
0	0	0	0	0
1	0	1	2	3
2	0	2	3	1
3	0	3	1	2

Section 1.3

1. Yes. Each number $a + b\sqrt{7}$, a, b rational corresponds to a unique point on the line.

5. $(0, \infty)$ **7.** $(-4, \infty)$ **9.** $(-2, 3)$ **11.** $[-1, 2)$

13. $(-1, 5)$ **15.** $(-\infty, 0) \cup (\frac{15}{2}, \infty)$ **17.** $(-\infty, 1) \cup (2, \infty)$

19. $(\frac{4}{3}, 3)$

Section 1.4

9. (b) Yes. (c) Yes.

20. If S is a subset of $\mathbb{N} \times \mathbb{N}$ containing $(1, 1)$ such that for all (m, n) in S both $(m + 1, n)$ and $(m, n + 1)$ are in S, then $S = \mathbb{N} \times \mathbb{N}$.

Section 2.1

1. $\delta = 0.005$ **3.** $\delta = 0.02$ **5.** $\delta = 0.002$

7. $\delta = 0.06$ **9.** $\delta = 0.005$ **11.** $\delta = 0.01$

13. $\delta = 0.07$ **15.** $\delta = 0.09$ **17.** $\delta = 0.7$

Section 2.2

5. No.

Section 2.3

1. Yes.   **3.** No. Neither.   **5.** No. Neither.

7. No. On the right.   **9.** No. Neither.   **11.** No. Neither.

13. Limit is 1; limit is 0; limit doesn't exist.

15. If $\lim_{x \to a^+} f_1(x) = L_1$, $\lim_{x \to a^+} f_2(x) = L_2$ and $g(x) \equiv f_1(x) + f_2(x)$, then $\lim_{x \to a^+} g(x) = L_1 + L_2$.

17. Suppose $\lim_{x \to a^-} f(x) = L$, $\lim_{x \to a^-} g(x) = M$ and $f(x) \leqslant g(x)$ for all x in an interval with a as right endpoint. Then $L \leqslant M$.

Section 2.4

1. 0   **3.** $+\infty$ **5.** 0

7. $+\infty$ **9.** 2

13. Suppose that f and g are functions on $\mathbb{R}^1$ to $\mathbb{R}^1$. If f is continuous at L and $g(x) \to L$ as $x \to -\infty$, then $\lim_{x \to -\infty} f[g(x)] = f(L)$.

17. Examples are (a) $f(x) = 2x$, $g(x) = -x$;
                 (b) $f(x) = x$, $g(x) = -2x$;
                 (c) $f(x) = x + A$, $g(x) = -x$.

19. Suppose that $\lim_{x \to +\infty} f(x) = L$, $\lim_{x \to +\infty} g(x) = M$. If $f(x) \leqslant g(x)$ for all $x > A$ for some constant A, then $L \leqslant M$.

Section 2.5

1. (a) $x_n = (n + \frac{1}{2})\pi$;   (b) $y_n = 2n\pi$;   (c) $z_n = (2n + 1)\pi$.

3. *Hint*: Write $a^n = (1 + (a - 1))^n$ and use the binomial theorem.

11. 1   **13.** 0

Section 3.1

3. If $a_0 < 0$, then $f(x) \to +\infty$ as $x \to \pm\infty$; if $a_0 > 0$, then $f(x) \to -\infty$ as $x \to \pm\infty$.

7. $f(x) = x \sin(1/x)$, $c = 0$.

Section 3.2

1. 3, yes;   0, no.   **3.** 3, no; -1, no.

5. 1, no; $\frac{1}{2}$, yes.   **7.** $3\pi/2$, yes; $\pi/2$, yes.

11. (b) inf $S = \inf\{b_i\}$ provided the right-hand expression exists. Similarly, sup $S = \sup_i\{B_i\}$.

19. Example 1: Take $I_i = [(i-1)\pi, i\pi]$, $f(x) = \cos x$;
Example 2: $I_1 = [-1, 0], I_i = [1/i\pi, 1/(i-1)\pi]$, $f(x) = 0, x \leqslant 0, f(x) = \cos(1/x)$,
$x > 0$.

Section 3.3

1. No; $\{x_{2n}\}$.   **3.** No; $\{x_{2n}\}$.

5. Yes.   **7.** No; $\{x_{3n}\}$.

9. (a) $x_{k,n} = \{k + 1/n\}$, $k = 1, 2, \ldots, N; n = 1, 2, \ldots$.
(b) $x_{k,n} = \{k + 1/n\}$, $k = 1, 2, 3, \ldots, n = 1, 2, 3, \ldots$.

Section 3.5

3. $f(x) = x/(x + 1)$

Section 3.7

1. Yes.   **3.** Yes.   **5.** No; $\{x_{3n}\}$.

13. (b) A finite subfamily has a smallest interval I_N. The point $x = 1/(N + 3)$ is not covered.

15. No.

17. Finite subfamilies which cover E are $I_{4/3}$ and $I_{3/2}$.

Section 4.1

5. Let f, u, v be functions on $\mathbb{R}^1$ such that v has a derivative at x_0, u has a derivative at $v(x_0)$, and f has a derivative at $u[v(x_0)]$.

9. (b) $\eta(h) = 3 - \dfrac{3 + 3h + h^2}{(1 + h)^3}$

15. For $k \leqslant (n-1)/2$ **21.** $\frac{1}{3}$ **23.** e

Section 4.2

1. $I_1 = (-\infty, -1], I_2 = [-1, +\infty), J_1 = [1, +\infty), g_1(x) = -\sqrt{x-1} - 1, g_2(x) = \sqrt{x-1} - 1$

3. $I_1 = (-\infty, 2], I_2 = [2, +\infty), J_1 = (-\infty, 4], J_2 = (-\infty, 4], g_1(x) = 2 - \sqrt{4-x}, g_2(x) = 2 + \sqrt{4-x}$

5. $I_1 = (-\infty, -2), I_2 = (-2, +\infty), J_1 = (2, +\infty), J_2 = (-\infty, 2), g_1(x) = 2x/(2-x), g_2(x) = 2x/(2-x)$

7. $I_1 = (-\infty, -1], I_2 = [-1, +1], I_3 = [1, +\infty), J_1 = [-2, 0), J_2 = [-2, 2], J_3 = (0, 2], g_1(x) = (2 + \sqrt{4-x^2})/x, g_2(x) = (2 - \sqrt{4-x^2})/x, g_3(x) = (2 + \sqrt{4-x^2})/x$

9. $I_1 = (-\infty, +\infty), J_1 = (-\infty, +\infty)$

11. $I_1 = (-\infty, -3), I_2 = [-3, 1], I_3 = [1, +\infty)$
$J_1 = (-\infty, 31], J_2 = [-1, 31], J_3 = [31, +\infty)$

15. $f'(x) = 3(x-1)^2, g'(x) = \frac{1}{3}x^{-2/3}$

17. $f'(x) = \cos x, g'(x) = 1/\sqrt{1-x^2}$

Section 5.1

1. $S^+(f, \Delta) = \frac{11}{25}; S_-(f, \Delta) = \frac{6}{25}$

9. Note that $\sum_{i=1}^{n} [f(x_i) - f(x_{i-1})] - f(x_n) - f(x_0) = f(b) - f(a)$.

15. Yes.   **19.** *Hint*: Use the result of Problem 18.

Section 5.2

1. (c) Choose $f(x) = x, a \leqslant x < b, f(b) = b - 1$.

3. Note that $d(uv) = u\, dv + v\, du$.

5. For $0 \leqslant x \leqslant 1$ let $f(x) = 1$ for x rational, $f(x) = -1$ for x irrational.

Section 6.1

3. Not equivalent.   **5.** Yes.   **7.** No.

11. Either all x_i are zero or there is a number λ such that $y_i = \lambda x_i$ for $i = 1, 2, \ldots, n, \ldots$.

Section 6.2

3. The square with vertices at $(1, 0), (0, 1), (-1, 0), (0, -1)$.

5. No. The set $\{1/n\}, n = 1, 2, \ldots,$ is an infinite set of isolated points.

7. $\bar{A} = \{x: 0 \leqslant x \leqslant 1\}$.

11. The sets $A_n = \{(x, y): 0 \leqslant x^2 + y^2 < 1/n\}, n = 1, 2, \ldots$.

15. Define (in $\mathbb{R}^1$), $A_i = \{x: 1/i < x \leqslant 1\}$. Then $\bigcup A_i = \{x: 0 < x \leqslant 1\}; \bar{B} = \{x: 0 \leqslant x \leqslant 1\} \neq \bigcup \bar{A_i} = \{x: 0 < x \leqslant 1\}$.

Section 6.3

1. Arrange the rational points as shown in Figure 6.5.

Section 6.4

5. Choose $\rho = \frac{1}{10}$.

7. Choose $x^n = (x_1^n, x_2^n, \ldots, x_k^n, \ldots)$ so that $x_k^n = 1$ if $k = n, x_k^n = 0$ otherwise.

Section 6.5

7. Statement: Let f, g, h be functions defined on a set A in a metric space, and suppose $p_0 \in A$. If $f(p) \leqslant g(p) \leqslant h(p)$ for all $p \in A$ and if $\lim_{p \to p_0} f(p) = \lim_{p \to p_0} h(p) = L$ for $p \in A$, then $\lim_{p \to p_0} g(p) = L$ for $p \in A$.

9. (a) For every $\varepsilon > 0$ there is a $\delta > 0$ such that $d_2(f(p), f(p_0)) < \varepsilon$ whenever $d_1(p, p_0) < \delta$ where d_1, d_2 are the metrics in S_1, S_2, respectively.

(b) Same as 9(a) except that the points p must belong to A.

(c) For every $\varepsilon > 0$ there is a $\delta > 0$ such that $d_2(f(p), q_0) < \varepsilon$ whenever $d_1(p, p_0) < \delta$ and $p \in A$, where d_1, d_2 are the metrics of S_1, S_2 respectively.

11. Let $A = I_1 \cup I_2$, $I_1 = \{x: 0 \leqslant x \leqslant 1\}$, $I_2 = \{x: 2 \leqslant x \leqslant 3\}$. Define $f = 0$ on I_1, $f = 1$ on I_2.

Section 7.1

5. $H_{,3}(x) = -4x_3(x_1^2 + x_4^2) + 2x_1^2(x_1^2 x_3 + x_4) + 2[\cos(x_1 + x_3) - 2x_4]\sin(x_1 + x_3)$

Section 7.2

1. $f_{,1,1} = \dfrac{2x_1^2 - x_2^2 - x_3^2}{(x_1^2 + x_2^2 + x_3^2)^{5/2}}$; f is symmetric in x_1, x_2, x_3.

7. $f(x) = f(a) + D_1 f(a)(x_1 - a_1) + D_2 f(a)(x_2 - a_2) + D_3 f(a)(x_3 - a_3)$

$\qquad + \frac{1}{2}D_1^2 f(a)(x_1 - a_1)^2 + D_1 D_2 f(a)(x_1 - a_1)(x_2 - a_2)$

$\qquad + D_1 D_3 f(a)(x_1 - a_1)(x_3 - a_3) + \frac{1}{2}D_2^2 f(a)(x_2 - a_2)^2$

$\qquad + D_2 D_3 f(a)(x_2 - a_2)(x_3 - a_3) + \frac{1}{2}D_3^2 f(a)(x_3 - a_3)^2$

$\qquad + \frac{1}{6}D_1^3 f(\xi)(x_1 - a_1)^3 + \frac{1}{2}D_1^2 D_2 f(\xi)(x_1 - a_1)^2(x_2 - a_2)$

$\qquad + \frac{1}{2}D_1^2 D_3 f(\xi)(x_1 - a_1)^2(x_3 - a_3) + \frac{1}{2}D_1 D_2^2 f(\xi)(x_1 - a_1)(x_2 - a_2)^2$

$\qquad + D_1 D_2 D_3 f(\xi)(x_1 - a_1)(x_2 - a_2)(x_3 - a_3)$

$\qquad + \frac{1}{2}D_1 D_3^2 f(\xi)(x_1 - a_1)(x_3 - a_3)^2 + \frac{1}{6}D_2^3 f(\xi)(x_2 - a_2)^3$

$\qquad + \frac{1}{2}D_2^2 D_3 f(\xi)(x_2 - a_2)^2(x_3 - a_3) + \frac{1}{2}D_2 D_3^2 f(\xi)(x_2 - a_2)(x_3 - a_3)^2$

$\qquad + \frac{1}{6}D_3^3 f(\xi)(x_3 - a_3)^3$

9. $\left(-\frac{9}{31}, \frac{55}{62}, \frac{18}{31}, -\frac{41}{31}\right)$

11. Positive definite.

13. Negative definite.

Section 7.3

11. (a) Definition: The derivative of f at a is the linear function $L: \mathbb{R}^N \to \mathbb{R}^M$ such that

$$\lim_{x \to a} \frac{d_M(f(x), L(x))}{d_N(x, a)} = 0.$$

Section 8.1

1. $S = \{x: 0 \leqslant x_i \leqslant 1, x_i \text{ is rational}, i = 1, 2, \ldots, N\}$.

3. The result does not hold even in $\mathbb{R}^2$ with $S = \{(x_1, x_2): \frac{1}{3} \leqslant x_i \leqslant \frac{2}{3}, i = 1, 2\}$.

7. $S_1 = \{x: 0 \leqslant x_i \leqslant 1, x_i \text{ rational}\}$, $S_2 = \{x: 0 \leqslant x_i \leqslant 1, x_i \text{ irrational}\}$. $S_1 \cup S_2$ is a figure, $S_1 \cap S_2 = \varnothing$. Also $S_1 - S_2 = \varnothing$; hence all are figures.

Section 8.2

1. Let $F = \{x: 0 \leqslant x_i \leqslant 1, i = 1, 2, \ldots, N\}$. Define $f: F \to \mathbb{R}^1$ such that $f(x) = 0$ if x_1 is rational and $f(x) = 1$ if x_1 is irrational.

Section 8.3

3. Let F_1 be a regular figure in $\mathbb{R}^N$ and G a figure with zero N-dimensional volume. Define $F = F_1 \cup G$ and suppose $f: F \to \mathbb{R}^1$ is Riemann integrable on F_1 and unbounded on G.

7. (b) Observe that for every subdivision into subsquares of S, we have $S^+(f, \Delta) = 1$ and $S^-(f, \Delta) = 0$.

Section 9.1

1. Convergent.	3. Convergent.	5. Convergent.
7. Convergent.	9. Convergent.	13. $p > 1$.
17. Divergent.		

Section 9.2

1. Absolutely convergent.	3. Conditionally convergent.
5. Divergent.	7. Divergent.
9. Conditionally convergent.	11. Conditionally convergent.
13. Divergent.	15. Convergent.
17. $-1 < x < 1$	19. $-\frac{2}{3} \leqslant x < \frac{2}{3}$
21. $1 < x < 5$	23. $-\frac{3}{2} < x < -\frac{1}{2}$

29. (i) $\displaystyle\sum_{n=0}^{\infty} n!\, x^n$; (ii) $\displaystyle\sum_{n=0}^{\infty} \frac{x^n}{n!}$.

Section 9.3

1. Uniform.	3. Uniform.	5. Uniform.
7. Not uniform.	9. Uniform.	

Section 9.4

1. $h < 1$	3. $h < 1$	5. $h < 3/2$
7. $h < 1$	9. $h < 1$	

23. $1 + \displaystyle\sum_{n=1}^{\infty} \frac{(-1/2)(-3/2)\ldots(-1/2 - n + 1)}{n!} x^{2n}, \quad |x| < 1$

25. $1 + \displaystyle\sum_{n=1}^{\infty} \frac{(-1)^n(-3)(-4)\ldots(-3 - n + 1)}{n!} x^n, \quad |x| < 1$

27. $1 + \sum\limits_{n=1}^{\infty} \dfrac{(-1)^n(-3)(-4)\ldots(-3-n+1)x^{2n}}{n!}$, $\quad |x| < 1$

29. $x^3 + \sum\limits_{n=1}^{\infty} \dfrac{(-1)^n(-1/2)(-3/2)\ldots(-1/2-n+1)}{n!(2n+1)} x^{6n+3}$, $\quad |x| < 1$

31. 0.47943   **33.** 0.89837

35. 1.99476   **37.** 0.23981

Section 9.5

3. No.   **5.** No.

Section 9.6

3. No.

9. No. In fact, the sum does not converge.

13. No. In fact, the sum does not converge.

Section 9.7

1. $1 + x + \dfrac{x^2}{2} - \dfrac{y^2}{2} + \dfrac{x^3}{6} - \dfrac{xy^2}{2}$

3. $1 - x + \dfrac{x^2}{2} + \dfrac{y^2}{2} - \dfrac{x^3}{6} - \dfrac{xy^2}{2}$

5. 1

Section 10.1

1. $\dfrac{1}{2} + \dfrac{2}{\pi}\left[\dfrac{\sin x}{1} + \dfrac{\sin 3x}{3} + \cdots + \dfrac{\sin(2k+1)x}{2k+1} + \cdots \right]$, $\quad -\pi \leqslant x \leqslant \pi$.

3. $\dfrac{\pi^2}{3} + 4\left[-\dfrac{\cos x}{1^2} + \dfrac{\cos 2x}{2^2} - \dfrac{\cos 3x}{3^2} + \cdots + \dfrac{(-1)^k \cos kx}{k^2} + \cdots \right]$, $\quad -\pi \leqslant x \leqslant \pi$.

5. $\dfrac{2}{\pi} + \dfrac{4}{\pi}\left[\dfrac{\cos 2x}{3} - \dfrac{\cos 4x}{15} + \cdots + \dfrac{(-1)^{k+1} \cos 2kx}{4k^2 - 1} \right]$, $\quad -\pi \leqslant x \leqslant \pi$.

7. $\dfrac{e^{2\pi} - e^{-2\pi}}{4\pi} + \dfrac{e^{2\pi} - e^{-2\pi}}{\pi}\left[\dfrac{-2\cos x}{5} + \dfrac{\sin x}{5} + \dfrac{2\cos 2x}{8} - \dfrac{2\sin 2x}{8} + \dfrac{(-1)^k 2\cos kx}{k^2 + 4} \right.$

$\left. - \dfrac{(-1)^k k \sin kx}{k^2 + 4} + \cdots \right]$, $\quad -\pi \leqslant x \leqslant \pi$.

9. $\dfrac{1}{2} - \dfrac{\cos 2x}{2}$

11. $\dfrac{4}{\pi}\sum\limits_{n=1}^{\infty} \dfrac{1}{2n-1} \sin(2n-1)x$, $\quad -\pi \leqslant x \leqslant \pi$.

13. $\frac{3}{4}\cos x + \frac{1}{4}\cos 3x, \quad -\pi \leqslant x \leqslant \pi.$

19. (a) $\dfrac{\pi^3}{3} - \dfrac{4\cos x}{1^2} + \dfrac{2\sin x}{1} + \dfrac{4\cos 2x}{2^2} - \dfrac{2\sin 2x}{2} + \cdots + \dfrac{4(-1)^k\cos kx}{k^2}$

$$- \dfrac{2(-1)^k\sin kx}{k} + \cdots, \quad -\pi \leqslant x \leqslant \pi.$$

Section 10.2

1. $\dfrac{1}{2} + \dfrac{2}{\pi}\left[\dfrac{\cos x}{1} - \dfrac{\cos 3x}{3} + \dfrac{\cos 5x}{5} + \cdots\right], \quad 0 \leqslant x \leqslant \pi.$

3. $\dfrac{\pi}{2} - \dfrac{4}{\pi}\left[\dfrac{\cos x}{1^2} + \dfrac{\cos 3x}{3^2} + \dfrac{\cos 5x}{5^2} + \cdots\right], \quad 0 \leqslant x \leqslant \pi.$

5. $\dfrac{8}{\pi} + \left[\dfrac{\sin 2x}{2} + \dfrac{\sin 6x}{2} + \dfrac{\sin 10x}{2} + \cdots\right], \quad 0 \leqslant x \leqslant \pi.$

7. $2\left[\dfrac{\sin x}{1} - \dfrac{\sin 2x}{2} + \dfrac{\sin 3x}{3} - \cdots\right], \quad 0 \leqslant x \leqslant \pi.$

9. $-\dfrac{4}{\pi}\left[\sin\dfrac{1}{2}\pi x + \dfrac{\sin\frac{3}{2}\pi x}{3} + \dfrac{\sin\frac{5}{2}\pi x}{5} + \cdots\right], \quad -2 \leqslant x \leqslant 2.$

11. $\dfrac{1}{3} + \dfrac{4}{\pi^2}\left[\dfrac{\cos x\pi}{1^2} + \dfrac{\cos 2\pi x}{2^2} - \dfrac{\cos 3\pi x}{3^2} + \cdots\right], \quad 0 \leqslant x \leqslant \pi.$

Section 10.3

1. $4\left[\dfrac{\sin x}{1^3} - \dfrac{\sin 2x}{2^3} + \dfrac{\sin 3x}{3^3} - \cdots\right], \quad -\pi \leqslant x \leqslant \pi.$

3. $-\dfrac{4}{\pi}\left[\dfrac{\sin x}{1^3} + \dfrac{\sin 3x}{3^3} + \dfrac{\sin 5x}{5^3} + \cdots\right], \quad -\pi \leqslant x \leqslant \pi.$

5. $f(x) \sim \dfrac{2}{\pi} - \dfrac{4}{\pi}\left[\dfrac{\cos 2x}{2^2 - 1} - \dfrac{\cos 4x}{4^2 - 1} + \dfrac{\cos 6x}{6^2 - 1} + \cdots\right], \quad -\pi \leqslant x \leqslant \pi.$

$$F(x) \sim -\dfrac{4}{\pi}\left[\dfrac{\sin 2x}{2(2^2 - 1)} + \dfrac{\sin 4x}{4(4^2 - 1)} + \dfrac{\sin 6x}{6(6^2 - 1)} + \cdots\right], \quad -\pi \leqslant x \leqslant \pi.$$

7. $\dfrac{a_n}{n^k} \to 0, \dfrac{b_n}{n^k} \to 0$ as $n \to \infty$, for all positive k.

11. $\dfrac{n + \frac{1}{2}}{\pi}$

Section 11.1

1. $\sin x - \displaystyle\int_0^1 \dfrac{\cos xt}{1 + t}\, dt$

3. $1/(x + 1), x \neq -1$

5. $2\sin(x^2) - 3x\sin(x^3) + \dfrac{\cos(x^2) - \cos(x^3)}{x^2}$, $\quad x \neq 0$, $\phi'(0) = 0$.

7. $\dfrac{2xe^{-1-x^2}}{1 + x + x^3} + \dfrac{\sin xe^{-\cos x}}{1 + x\cos x} - \dfrac{e^{-1-x^2}}{x^2(1 + x + x^3)} + \dfrac{e^{-\cos x}}{x^2(1 + x\cos x)}$

$\quad - \dfrac{1 + x}{x^2}\displaystyle\int_{\cos x}^{1+x^2} \dfrac{e^{-t}}{1 + xt}\,dt$, $\quad x \neq 0$.

9. $\dfrac{2\sin(x^2) - 3\sin(x^3)}{x}$

11. $\dfrac{\sqrt{1 - x^2} - 1}{x^2} + \dfrac{1}{\sqrt{1 - x^2}}$, $\quad x \neq 0, 1$, $\phi'(0) = 0$.

13. $\dfrac{3e^{x^3} - e^{xy}}{x}$

15. $-4x^4y - 4y^5 - 8x^2y + 8y^3$

17. $F_{,1} = f[x, y, h_1(x, y)]h_{1,1}(x, y) - f[x, y, h_0(x, y)]h_{0,1}(x, y) + \displaystyle\int_{h_0(x,y)}^{h_1(x,y)} f_{,1}(x, y, t)\,dt.$

$\quad F_{,2} = f[x, y, h_1(x, y)]h_{1,2}(x, y) - f[x, y, h_0(x, y)]h_{0,2}(x, y) + \displaystyle\int_{h_0(x,y)}^{h_1(x,y)} f_{,2}(x, y, t)\,dt.$

19. $\phi_{,3} = f[x, y, z, g_1(x, y, z)]g_{1,3}(x, y, z) - f[x, y, z, g_0(x, y, z)]g_{0,3}(x, y, z)$

$\quad + \displaystyle\int_{g_0(x,y,z)}^{g_1(x,y,z)} f_{,3}(x, y, z, t)\,dt.$

Section 11.2

1. Convergent. **3.** Convergent. **5.** Convergent.

7. Divergent. **9.** Convergent. **11.** Convergent.

Section 11.3

1. $\phi'(x) = -\dfrac{1}{x} + \displaystyle\int_0^\infty \dfrac{e^{-xt}}{1 + t}\,dt$ **3.** $\phi'(x) = -\displaystyle\int_0^\infty \dfrac{t\sin xt}{1 + t^3}\,dt$

5. $\phi'(x) = -\dfrac{1}{x}\displaystyle\int_0^1 \dfrac{\sin xt}{t}\,dt$

7. $\phi'(x) = \dfrac{2}{x + 1} - \dfrac{2x + 1}{(x^2 + x)^{3/2}}\operatorname{arctanh}\dfrac{x}{\sqrt{x^2 + x}}$

15. $\phi'(x) = \dfrac{1}{x + 1}$, $\quad \phi(x) = \log(x + 1)$

Section 12.1

3. 4 **5.** 2

11. (b) Example: $f(x) = x\sin(1/x)$, $\quad 0 < x \leqslant 1$, $f(0) = 0$.

Section 12.2

1. (a) $\displaystyle\int_a^b f\,dg = 0$ (b) $\displaystyle\int_a^b f\,dg = (d_n - 1)f(b) + (1 - d_1)f(a)$

3. 2   7. 0

13. Conditions under which the "integration by parts" formula holds.

Section 13.1

1. 1.581   3. -1.24

5. $1.30, 0.67, -2.32$

7. $x_{n+1} = \dfrac{10 + 4x_n^5}{10 + 5x_n^4};\quad x_1 = 1,\ x_2 = 0.933,\ x_3 = 0.9303$

13. 0.169

Section 13.2

5. $1 + x + x^2 + \frac{4}{3}x^3 + \frac{7}{6}x^4 + \frac{6}{5}x^5$

Section 14.1

1. $f' = 2x/(3y^2 + 1)$

3. $f' = -\dfrac{xy^2 + 2y}{x^2y + 3x}$

5. Example: $F(x, y) = y^3 - x,\ (x_0, y_0) = (0, 0)$

7. $f_{,1}(1, 0) = 2,\ f_{,2}(1, 0) = 0$

9. $f_{,1}(0, \frac{1}{2}) = -\frac{3}{4},\ f_{,2}(0, \frac{1}{2}) = -1$

13. Example: $F(x_1, x_2, y) = y^3 - x_1^3,\ (x_1^0, x_2^0, y) = (0, 0, 0)$

15. $f_{,1,1} = \left(\dfrac{\partial F}{\partial x_1}\dfrac{\partial^2 F}{\partial x_1 \partial y} - \dfrac{\partial F}{\partial y}\dfrac{\partial^2 F}{\partial x_1^2} - \dfrac{\partial F}{\partial x_1}\dfrac{\partial^2 F}{\partial y^2} + \dfrac{\partial F}{\partial y}\dfrac{\partial^2 F}{\partial x_1 \partial y}\right)\bigg/\left(\dfrac{\partial F}{\partial y}\right)^3$

$f_{,1,2} = \left(\dfrac{\partial F}{\partial x_2}\dfrac{\partial^2 F}{\partial x_2 \partial y} - \dfrac{\partial F}{\partial y}\dfrac{\partial^2 F}{\partial x_1 \partial x_2} - \dfrac{\partial F}{\partial x_2}\dfrac{\partial^2 F}{\partial y^2} + \dfrac{\partial F}{\partial y}\dfrac{\partial^2 F}{\partial x_2 \partial y}\right)\bigg/\left(\dfrac{\partial F}{\partial y}\right)^3$

$f_{,2,2} = \left(\dfrac{\partial F}{\partial x_2}\dfrac{\partial^2 F}{\partial x_2 \partial y} - \dfrac{\partial F}{\partial y}\dfrac{\partial^2 F}{\partial x_2^2} - \dfrac{\partial F}{\partial x_2}\dfrac{\partial^2 F}{\partial y^2} + \dfrac{\partial F}{\partial y}\dfrac{\partial^2 F}{\partial x_2 \partial y}\right)\bigg/.\left(\dfrac{\partial F}{\partial y}\right)^3$

Section 14.2

1. $f = (f_1, f_2);\quad f_{1,1} = -\frac{5}{3},\ f_{1,2} = \frac{4}{3},\ f_{2,1} = \frac{1}{3},\ f_{2,2} = -\frac{5}{3}$

3. $f = (f_1, f_2);\quad f_{1,1} = \frac{2}{3},\ f_{1,2} = \frac{4}{3},\ f_{2,1} = -\frac{5}{3},\ f_{2,2} = \frac{2}{3}$

5. $\dfrac{\partial f^2}{\partial x_1} = \left(\dfrac{\partial F^1}{\partial x_1}\dfrac{\partial F^2}{\partial y_1} - \dfrac{\partial F^1}{\partial y_1}\dfrac{\partial F^2}{\partial x_1}\right)\bigg/\left(\dfrac{\partial F^1}{\partial y_1}\dfrac{\partial F^2}{\partial y_2} - \dfrac{\partial F^1}{\partial y_2}\dfrac{\partial F^2}{\partial y_1}\right)$

$$\frac{\partial f^1}{\partial x_1} = \left(\frac{\partial F^1}{\partial y_2}\frac{\partial F^2}{\partial x_1} - \frac{\partial F^1}{\partial x_1}\frac{\partial F^2}{\partial y_2}\right)\Bigg/\left(\frac{\partial F^1}{\partial y_1}\frac{\partial F^2}{\partial y_2} - \frac{\partial F^1}{\partial y_2}\frac{\partial F^2}{\partial y_1}\right)$$

$$\frac{\partial f^1}{\partial x_2} = \left(\frac{\partial F^1}{\partial y_2}\frac{\partial F^2}{\partial x_2} - \frac{\partial F^1}{\partial x_2}\frac{\partial F^2}{\partial y_2}\right)\Bigg/\left(\frac{\partial F^1}{\partial y_1}\frac{\partial F^2}{\partial y_2} - \frac{\partial F^1}{\partial y_2}\frac{\partial F^2}{\partial y_1}\right)$$

$$\frac{\partial f^2}{\partial x_2} = \left(\frac{\partial F^1}{\partial x_2}\frac{\partial F^2}{\partial y_1} - \frac{\partial F^1}{\partial y_1}\frac{\partial F^2}{\partial x_2}\right)\Bigg/\left(\frac{\partial F^1}{\partial y_1}\frac{\partial F^2}{\partial y_2} - \frac{\partial F^1}{\partial y_2}\frac{\partial F^2}{\partial y_1}\right)$$

9. $x_1 = y_1, x_2 = y_2 - y_1^2$

11. $x_1 = \dfrac{y_1}{1 - y_1 - y_2}, x_2 = \dfrac{y_2}{1 - y_1 - y_2}$

13. All $P(x_1, x_2, x_3)$ such that $x_3 \neq n\pi$ for integer values of n.

Section 14.3

1. $\int_F K(x_1, x_2)\, dV_2 = \frac{22}{3}$, $u_1 = (3x_1 - x_2)/8, u_2 = (3x_2 - x_1)/8$

3. $\int_F K(x_1, x_2)\, dV_2 = 0$, $u_1 = 2x_1 + x_2 - 2x_2^2, u_2 = x_1 + x_2 - x_1^2$

5. $\int_F K(x_1, x_2)\, dV_2 = -\frac{1}{105}$, $u_1 = x_1 - x_2, u_2 = x_1 + x_2$

9. $\int_F x_3\, dV_3 = \pi/8$

Section 14.4

1. 15   **3.** 45   **5.** 5/2

7. $(2, 1), (-2, -1)$ **9.** $(\sqrt{2}, -\sqrt{2}), (-\sqrt{2}), \sqrt{2})$

11. (a) $n^{1/n}$

Section 15.1

1. $M(S)$ is a metric space.

7. Example: Let $A = [0, \frac{1}{2}) \cup (\frac{1}{2}, 1]$ and $f(x) = \begin{cases} 0 & \text{for } 0 \leqslant x < \frac{1}{2} \\ 1 & \text{for } \frac{1}{2} < x \leqslant 1 \end{cases}$

Section 15.3

13. $f(x) = 0$ for $-\infty < x_1 \leqslant 0$; $f(x) = 0$ for $-\infty < x_2 \leqslant 0$;
$f(x) = 1$ for $1 \leqslant x_1 < +\infty$ and $1 \leqslant x_2 < +\infty$; $f(x) = x_1$ for $0 \leqslant x_1 \leqslant 1$,
$1 \leqslant x_2 < +\infty$;
$f(x) = x_2$ for $0 \leqslant x_2 \leqslant 1, 1 \leqslant x_1 < +\infty$.

Section 15.4

9. Example: The function $f(x) = \sin(1/x)$ is continuous on I.

13. It is sufficient that f is integrable on $0 \leqslant y \leqslant 1$.

Section 16.1

1. $x_i = x_2^{1/2}; \quad [0, \infty)$

7. Not rectifiable.

9. $T = (\sqrt{3}/3)(e_1 + e_2 + e_3);$ $\quad N = (\sqrt{2}/2)(-e_1 + e_2)$
$B = (\sqrt{6}/6)(-e_1 - e_2 + 2e_3);$ $\quad \kappa = \sqrt{2/3};$ $\quad \tau = 1/3$

11. $T = \frac{1}{19}(e_1 + 6e_2 + 18e_3);$ $\quad N = \frac{1}{19}(-6e_1 - 17e_2 + 6e_3)$
$B = \frac{1}{19}(18e_1 - 6e_2 + e_3);$ $\quad \kappa = \frac{3}{361};$ $\quad \tau = \frac{3}{361}$

15. 14

17. $\dfrac{\pi}{32}\sqrt{\pi^2 + 32} + \ln\left(\dfrac{\pi}{4} + \dfrac{1}{4}\sqrt{\pi^2 + 32}\right) - \dfrac{1}{2}\ln 2$

Section 16.2

3. $\int_F u\, dV = 0$

5. $(2x_1 + x_2)e_1 + (x_1 - 2x_2)e_2 + 2x_3 e_3;$ $\quad D_a f(\bar{x}) = 27/7$

7. $2x_1 \log(1 + x_2^2)e_1 + \dfrac{2x_1^2 x_2}{1 + x_2^2}e_2 - 2x_3 e_3;$ $\quad D_a f(\bar{x}) = 4/\sqrt{10}$

9. $D_a w(x) = -\dfrac{1}{\sqrt{14}}e_1 + \dfrac{-x_2 + 2x_3}{\sqrt{14}}e_2 + \dfrac{4x_2 - 2x_3}{\sqrt{14}}e_3$

$D_a w(\bar{x}) = \dfrac{1}{\sqrt{14}}(-e_1 + 7e_2 - 2e_3)$

13. $\operatorname{div} v(\bar{x}) = 0$

15. $\operatorname{div} v(\bar{x}) = -12$

17. $\operatorname{div} v(\bar{x}) = -a \operatorname{curl} r(\bar{x})$

19. $\operatorname{curl} v = 0;$ $\quad f(x) = x_1 x_2 x_3 + \frac{1}{3}(x_1^3 + x_2^3 + x_3^3)$

21. $\operatorname{curl} v = 0$

23. $\left(\dfrac{\partial^2 u_2}{\partial x_1 \partial x_2} + \dfrac{\partial^2 u_3}{\partial x_1 \partial x_3} - \dfrac{\partial^2 u_1}{\partial x_2^2} - \dfrac{\partial^2 u_1}{\partial x_3^2}\right)e_1 + \left(\dfrac{\partial^2 u_1}{\partial x_1 \partial x_2} + \dfrac{\partial^2 u_3}{\partial x_2 \partial x_3} - \dfrac{\partial^2 u_2}{\partial x_1^2} - \dfrac{\partial^2 u_2}{\partial x_2^2}\right)e_2$

$+ \left(\dfrac{\partial^2 u_1}{\partial x_1 \partial x_3} + \dfrac{\partial^2 u_2}{\partial x \partial x_3} - \dfrac{\partial^2 u_3}{\partial x_1^2} - \dfrac{\partial^2 u_3}{\partial x_2^2}\right)e_3$

Section 16.3

1. $\displaystyle\int_{\bar{C}} g \cdot dr = \frac{1}{3}$

3. $\displaystyle\int_{\bar{C}} g \cdot dr = -2$

5. $\displaystyle\int_{\bar{C}} g \cdot dr = \frac{11}{15}$

7. $\displaystyle\int_{\bar{C}} g \cdot dr = \dfrac{a_1^2 a_4}{3} + \dfrac{a_2^2}{2} - \dfrac{a_2 a_3 a_4}{3} + \dfrac{a_3 a_4}{2}$

9. $\displaystyle\int_{\bar{C}} \nabla u \cdot dr = 3$

Section 16.4

11. -24π

13. 2π

15. 2π

Section 16.5

5. No.

Section 16.6

1. $\sqrt{3}/6$

3. $\dfrac{15\sqrt{2}\pi}{4}$

5. 2π

7. $\pi^4/192$

9. $(17 + 15\sqrt{2})\pi/4$

11. $1024\delta/45$

13. $(2 + \sqrt{3})\delta/6$

15. $(0, 0, 0)$

19. Case 1: $4\pi a^2 \rho/c$;   Case 2: $4\pi a\rho$

21. $2\pi\rho$

Section 16.7

1. If the edges of the cube are parallel to the coordinate axes, then $n = \pm e_1$ or $\pm e_2$ or $\pm e_3$ for each of the appropriate faces.

3. $n = \dfrac{1}{\sin^2\phi}[\cos\theta\sin^2\phi e_1 + \sin\theta\sin^2\phi e_2 + \sin\theta\cos\phi e_3]$

Section 16.8

1. 0

3. $15\pi/2$

5. $\frac{4}{15}$

13. 0

15. $\pi/2$

Section 16.9

1. $\frac{1}{8}$

3. $32\pi/3$

5. $12\sqrt{3}\pi$

7. 0

9. 2π

11. 0

13. 5π

Appendix 1

1. $\{-2, 1\}$

3. $\{\frac{3}{5}, \frac{11}{5}\}$

5. $\{-6, -\frac{2}{3}\}$

7. $\{\frac{1}{5}, 3\}$

9. $\{\frac{1}{4}, \frac{7}{8}\}$

11. $(1, 3)$

13. $(-\frac{4}{3}, -\frac{2}{3})$

15. $(-\frac{2}{3}, \frac{10}{3})$

17. $(-\frac{1}{4}, \infty)$

19. $(-\infty, \frac{9}{11}) \cup [\frac{5}{3}, \infty)$

21. $(-\infty, 1] \cup [9, \infty)$

Appendix 2

1. $(-3, 4)$

3. $(-\infty, 2) \cup (-\frac{1}{2}, \infty)$

5. $(-\infty, -\frac{1}{3}) \cup (0, \frac{1}{2})$

7. $(-2, 2)$

9. $(-\infty, \infty)$

11. Example: $x^2 - 3x + 2 \leqslant 0$

13. Example: $x^4 - 10x^3 + 37x^2 - 60x + 36 > 0$

15. $(-2, 1) \cup (1, 2)$ 17. $(-3, 0)$

19. $(-\infty, 0) \cup (\frac{6}{7}, 3)$ 21. $(-\infty, -3) \cup (1, 2)$

23. $(-\frac{3}{2}, -\frac{3}{14}) \cup (\frac{3}{2}, \infty)$ 25. $(-\infty, \frac{1}{2}) \cup (2, \infty)$

Index

Undergraduate Texts in Mathematics

(continued from page ii)

Undergraduate Texts in Mathematics